Analysis and Design of Electrical Power Systems

Analysis and Design of Electrical Power Systems

A Practical Guide and Commentary on NEC and IEC 60364

Ismail Kasikci

Author

Prof. Ismail Kasikci
Biberach University of Applied Sciences
Institut für Gebäude- und Energiesysteme
Karlstraße 11
88400 Biberach
Germany

Cover Images: Photograph by Ismail Kasikci, High Voltage icon © OGdesign/Shutterstock

Library of Congress Card No.: applied for

British Library Cataloguing-in-Publication Data
A catalogue record for this book is available from the British Library.

Bibliographic information published by the Deutsche Nationalbibliothek
The Deutsche Nationalbibliothek lists this publication in the Deutsche Nationalbibliografie; detailed bibliographic data are available on the Internet at <http://dnb.d-nb.de>.

Print ISBN: 978-3-527-34137-5
ePDF ISBN: 978-3-527-80340-8
ePub ISBN: 978-3-527-80343-9
oBook ISBN: 978-3-527-80342-2

Typesetting Straive, Chennai, India

Printing and Binding
CPI Group (UK) Ltd, Croydon CR0 4YY

Printed on acid-free paper

C064315_040122

Contents

Preface

For the design, calculation, dimensioning, and evaluation of an electrical power system, the electrical engineer and technician need not only comprehensive theoretical knowledge but also a reference book to make his work easier. This book is intended to fulfill this task.

This book is a follow-up to "Analysis and Design of Low-Voltage Power Systems", published in 2004, and contains didactical improvements and new topics such as power flow, generators, earthing in electrical networks, and relays. Each topic has been written in such a way that the readers can accomplish their tasks with the help of this book without much effort.

Many practical examples, tables, diagrams, and a comprehensive collection of literature make the compendium a complete tool. Planning values and equations required for the calculation process can be extracted from the numerous tables and diagrams. The book is well suited for teaching as well as for practical use. Special emphasis has been placed on deepening the theory, practice, and standards.

For this reason, the present book, intended as a help for the planning engineer in the solution of problems in electrical networks, also presents a detailed discussion of the current situation in regard to standards.

Following the theoretical part and the discussion of regulations and standards, a wide range of examples taken largely from practice is worked out fully.

The readers will be systematically familiarized with the structure, design, behavior, protection, calculation, planning, and design of electrical networks and switchgear. The following questions are covered in this book:

- How can I design the electrical system?
- Which regulations do I have to observe?
- Which calculations do I have to perform?
- Which methods/CAD can I use?
- Which protective measures do I have to consider?
- Which requirements and conditions apply for project planning?
- How can persons and animals be protected against electrical shock?
- Which operational components shall I select?
- Are there special problems with regard to planning?

For calculation, dimensioning, and evaluation of a system, in addition to extensive professional knowledge the planning engineer requires above all CAD experience and a knowledge of all relevant standards and regulations. Due to the great number of standards and their revision in regular intervals and also due to their increasing international harmonization, maintaining this knowledge is becoming more and more difficult.

This book will give engineers, technicians, master electricians, industry professionals, students convincing insights into the immense complexity of electrical power systems and networks and the breadth and depth of power engineering.

It is not possible to present all topics in one book with theory, practice, and standards. Individual topics can still be deepened with the literature given at the end of the book.

I wish you much success and enjoy reading this book.

Weinheim
1 July 2021

Ismail Kasikci

Acknowledgments

I wish to extend my heartfelt thanks to all my professional colleagues and acquaintances who supported me through their ideas, criticism, and suggestions.

I am especially indebted to Dr. Martin Preuss and Mrs Stefanie Volk for critically reviewing the manuscript and for valuable suggestions.

At this point, I would also like to express my gratitude to all those colleagues who supported me with their ideas, criticism, suggestions, and corrections. My heartiest appreciation is due to Wiley-VCH for the excellent cooperation and their support in the publication of this book.

I would like to thank the companies Siemens and ABB for their figures, pictures, and technical documentation. In particular, as a member, I am also indebted to the VDE (Association for Electrical, Electronic and Information Technologies) for their support and release of different kind of tables and data.

Furthermore, I welcome every suggestion, criticism, and idea regarding the use of this book from those who read the book.

Finally, I appreciate the feedbacks from designers, planners, and readers for their useful recommendations and critics.

This book describes the most important theory, practical terms, and definitions with respect to IEC or EN standards and useful examples. This book is structured as follows:

Chapter 1 is an introduction into the power system today.
Chapter 2 gives an overview of electrical power systems.
Chapter 3 describes the scope of DC current installations.
Chapter 4 gives a small introduction into smart grids.
Chapter 5 explains planning and project management briefly.
Chapter 5 deals with three phase alternating current.
Chapter 6 gives an overview of the network forms for low and medium voltage.
Chapter 7 describes the method of symmetrical components.
Chapter 8 explains the type of short-circuit currents in three-phase networks and the meaning, tasks, and origin of DIN EN 60909-0.
Chapter 9 describes the relays in electrical power systems generally.

Chapter 10 explains the load flow calculation.
Chapter 11 explains the type of neutral point treatment and substation earthing in high-voltage power installations.
Chapter 12 presents the protection against electric shock.
Chapter 13 deals with the most important overcurrent protection devices with the time–current characteristics.
Chapter 14 discusses the current carrying capacity of conductors and cables.
Chapter 15 gives an overview of the selectivity and back-up protection.
Chapter 16 presents the voltage drop calculations.
Chapter 17 gives a brief overview of the switchgear combinations.
Chapter 18 explains the compensation for reactive power.
Chapter 19 describes lightning protection systems.
Chapter 20 gives a brief overview of lighting systems.
Chapter 21 explains generators briefly.
Chapter 22 describes transformers.
Chapter 23 presents low voltage motors.
Chapter 24 asks some questions of each topic.

Symbols

a	center-to-center distance between bus bars, costs of electrical energy, room length, center-to-center distance between conductors, near-to-generator short-circuit
a_i	utilization factor for motors
$A_{W,D}$	surface area of walls and ceiling
A	acquisition price, floor area of room, air intake, and exhaust opening, initial value of DC component
A_2	area of circle
A_e	effective cooling area of housing, equivalent collecting area of stand-alone structure
A_0	individual surface areas of external side of housing
c_{max}	voltage factor
c	voltage factor, temperature distribution factor, specific heat capacity of conducting material, smallest power step
C_{str}	phase capacitance
C_e	Environmental coefficient
C'_E	ground capacitance
C	capacitor power
C_p	rated power
D	separation distance
E	light intensity
E_m	average light intensity
E_n	rated light intensity
f_1	stator frequency
f_2	rotor frequency
f_n	network frequency
F	electrodynamic force between conductors
g_i	coincidence factor
GD	moment of inertia

h	height difference, distance between lighting elements and evaluation level
H	magnetic field strength, height of housing
i_{DC}	decaying DC current component
i_p	peak short-circuit current
I	current, light intensity
I_0	no-load current
I_{thr}	rated short time current
I_{an}	starting current of motor
I_{rM}	rated current for motors in a group
I''_{kQ}	initial symmetrical short-circuit current
I_a	cutoff current of overcurrent protective equipment
I_A	starting current
I_d	leakage current
$I_{\Delta n}$	rated differential current of RCD
I_f	fault current (smallest short-circuit current)
I_{an}/I_{rM}	ratio of starting current to rated current for motor
I_n	nominal current
I_e	current setting
I_{rm}	magnetic current setting
I''_{k1}	single-pole short-circuit current
I''_{k2}	two-pole short-circuit current
I''_{k3}	three-pole short-circuit current
I''_{k2E}	two-pole short-circuit with contact to ground
I''_{kEE}	double ground fault short-circuit current
I_k	steady state short-circuit current
I_{th}	thermal short-circuit current
I_B	load current
I_n	rated current of protective equipment
I_z	permissible current loading of cable or conductor
I_r	rated current
I_Δ	current for delta connection
I_Y	current for star connection
I_2	large test current
I''_k	initial symmetrical short-circuit current
J_L	mass moment of inertia of load
J_{thr}	rated short-time current density
k	housing constant, material factor, or specific conductivity factor, transformation ratio of transformer, material coefficient, correction factor for operating conditions

k	1.06 for oil transformers, 1.2 for resin-encapsulated transformer
k_a	costs of work
k_c	current distribution coefficient, dependent on geometrical arrangement
k_L	power costs
k_m	depends on material of isolation path
k_i	depends on lightning protection class
K_1	costs of a light fixture, capacity costs for amortization and interest
K_2	costs of installation and installation material
K_3	price of a lamp
K_4	costs of replacing a lamp
K_J	annual costs
K_A	proportionate acquisition costs
$K(P_0/P_k)$	operating costs resulting from no-load and short-circuit losses
K_u	maintenance costs
$K_{W,D}$	heat transfer coefficient
l	length
l_h	length of horizontal grounding electrode
l_v	length of vertical grounding electrode
l_1	minimum length of grounding electrode
m	decaying DC component, thermal effect of DC component with three-phase AC current and single-phase AC current
M_M	motor torque
$M_{M\Delta}$	motor torque for direct startup
M_{MY}	motor torque for star-delta startup
M_L	load torque (counter-torque)
M_{L0}	load breakaway torque
M_N	rated torque
M_A	pull-up torque
M_acc	accelerating torque
M_S	pull-up torque
M_K	breakdown torque
$M_L(M)$	load moment relative to motor shaft
n	speed of rotation, calculated number of lighting elements, thermal effect of AC current component with three-pole short-circuit, number of internal horizontal partitions, number of transformers in parallel, decaying AC current component, amortization time in years, number of loads

n_1	total number of lamps
n_2	number of lamps per lighting element
n_S	synchronous speed of rotation
n_M	speed of rotation of motor
n_L	speed of rotation of load
N	number of windings
N_c	permissible number of critical lightning strikes
N_d	strike frequency of the structural installation
N_g	lightning density
p	rate of interest
P_n	rated power
P_v	transformer power loss, control gear power loss
P	effective power, effective power loss of operational equipment installed in housing, power consumption of one lamp + control gear
P_L	lamp power
P_k	Short-circuit losses
$P_{\max}$	power requirement
P_i	installed power
P_{input}	power input
P_{output}	power output
P_d	output
P_0	no-load losses
P_{Vr}	equipment power losses
P_{Fe}	core losses
P_{Cu}	load losses
P_{rM}	rated power of motor
q	factor for the calculation of breaking currents of asynchronous motors
Q	reactive power
Q_v	dissipated losses
$Q_{W,D}$	losses dissipated through walls and ceiling
Q_{v1}	proportion in natural air stream
Q_{v2}	proportion through walls and ceiling
Q_{v3}	proportion in forced air stream
Q_T	no-load reactive power of transformer
r	average radius, percent capital costs from interest and amortization
R_A	sum of resistances of grounding electrode and protective conductor

R'_L	relative effective resistance of a conductor
R_E	grounding resistance
R_l	conductor resistance
R	pure resistance, equivalent resistance, costs of cleaning per light and per year
R_Q, X_Q	ohmic, inductive resistance of control gear network
R_T, X_T	ohmic, inductive resistance of transformer
R_L, X_L	ohmic, inductive resistance of network
R_{0T}, X_{0T}	ohmic, inductive no-load resistance of transformer
R_{0L}, X_{0L}	ohmic, inductive no-load resistance of network
R_G	resistance of generator
s	slip, protection ratio
S	apparent power, cross section of conductor
S''_k	short-circuit power
S_{rT}	rated power of individual transformer
S_{st}	load starting capability
$\sum P_{rM}$	sum of rated effective powers
$\sum S_{rT}$	sum of rated apparent powers
S''_{kQ}	initial symmetrical short-circuit apparent power
S_0	no-load apparent power of transformer
t	time
t_{ab}	cutoff time of overcurrent protection equipment
t_{zu}	permissible cutoff time
t_L	economic life of lamp
t_B	yearly time in use
t_a	cutoff time
T_B	operating time in years
T_a	starting temperature
T_e	end temperature
T_B	operating time
Δt	overtemperature of air in housing, general
$\Delta t_{0,5}$	overtemperature of air, internal, at half height of housing
$\Delta t_{0,75}$	overtemperature of air, internal, at three-quarters height of housing
$\Delta t_{1,0}$	overtemperature of air, internal, at upper edge of housing
Δu	percent voltage drop
ΔP	power loss
ΔU	voltage drop
U_E	ground potential rise

U_{T1}	touch voltage without potential grading (on concrete-footing grounding electrode)
U_{T2}	touch voltage without potential control (on concrete-footing grounding electrode + potential grading grounding electrode)
U_0	line-to-ground voltage
U_{T}	touch voltage
U_{S}	step voltage
U_{nQ}	rated voltage of network at connecting point Q
U	rated AC voltage between external lines, charging voltage
U_n	rated voltage of network
U_{rG}	rated voltage of generator
U_{rM}	rated voltage of motor
$''u$	transformation ratio
$''u_f$	fictitious transformation ratio
$''u_r$	rated value of transformation ratio for transformer with step switch at principal tapping
v	depreciation factor
V_L	amount of air
x	exponent
X	reactance, distance from concrete-footing grounding electrode
X''_d	subtransient reactance
X'_L	relative reactance of a conductor
Z	impedance
Z_1	positive-sequence impedance
Z_2	negative-sequence impedance
Z_0	zero-sequence impedance
Z_E	impedance of grounding electrode system
Z_Q	impedance of control network
Z_{PE}	impedance of protective conductor
Z_T	impedance of transformer
Z_v	source impedance
Z'	relative impedance
Z_{F}	fault impedance
Z_k	body impedance
Z_{st}	site impedance
Z_S	ground fault loop impedance
Z'_S	ground fault loop impedance, consisting of neutral conductor and protective conductor of circuit

Z_{TLV}	impedance of transformer (low voltage side)
Z_{THV}	impedance of transformer (high voltage side)
Z_{KW}	Corrected impedance of power plant block, relative to high voltage side
Z_G	impedance of generator
Z_M	short-circuit impedance of a motor
Z_{GK}	corrected impedance of generator
α	temperature coefficient
δ	loss factor
η	efficiency of gear system
η_b	lighting utilization factor
η_i	utilization factor
η_B	lighting utilization factor according to data sheet
ϑ	temperature
$\Delta\vartheta$	temperature rise
θ	conductor temperature
Θ	current linkage
Θ_{max}	highest temperature attained
κ	conductivity
ρ_m	density of conductor material
φ_{rG}	Phase angle between $U_{\mathrm{rG}}/\sqrt{3}$ and I_{rG}
$\cos\varphi$	power factor
$\sin\varphi$	reactive factor
Φ_n	lumens per lamp per lighting element × 0.95 correction factor
μ	factor for the calculation of the symmetrical short-circuit current

Abbreviations

A	aluminum conductor
AC1	non-inductive or weakly inductive load, resistance furnace
AC2	slipring motors: starting, switching off
AC3	squirrel cage motors: starting, switching off while running
AC4	squirrel cage motors: switching on, breaking by plugging, jogging
ASM	asynchronous motor
B	mine-type installations
BHKW	block heating power plant
CENELEC	European Committee for Electrotechnical Standards
CW	wave-shaped concentric conductor
DIN	German Standards Institute
DKE	German Electrotechnical Commission
ED	ON period
EN	European Norm
EPR	ethylene–propylene–rubber insulation
FE	concrete-footing grounding electrode
G	rubber insulation or generator
HKS	heating, climate, sanitary
HV	high voltage
IEC	International Electrotechnical Commission
L_1, L_2, L_3	external conductor
LEMP	lightning electromagnetic pulse
LV	low voltage
LVMD	main low voltage distribution panel
M	motor, switchgear
MDP	main distribution panel
MGT	main grounding terminal
MV	medium voltage
N	neutral conductor
OPE	overcurrent protection equipment
PE	protective conductor
PV	primary voltage (transformer) or harmonics
PVC	polyvinyl chloride insulation

R	semiconductor
RCD	residual current protective device
SEMP	switching electromagnetic pulse
SE	grading grounding electrode
SV	secondary voltage (transformer)
T	transformer
TAB	technical conditions for connection
UMZ	independent maximum time relays or independent maximum current protection (UMZ relays)
UVV	accident prevention regulations
VBG	accident prevention regulations of the BG
VDE	Union of Electrotechnical, Electronics and Information Technology
VdS	union of property insurers
VPE	cross-linked polyethylene insulation
Y	PVC insulation

1
Introduction

Energy turnaround, smart grid, smart meters, smart buildings, smart homes, smart cities, renewable energies (RE), digitization, data protection, data security, efficient use of energy, control and regulation with digital technologies, industry 4.0, decentralized energy supply, demand side management (DSM), electric filling stations, and new business models for the electricity market are the new topics, tasks, and challenges we will deal with in the coming years. Figure 1.1 shows the path of electrical energy from the power plant to the consumers.

Electrical networks and switchgear and their planning and project planning are very much affected by this. They must perform various tasks relating to the transmission and distribution of electrical energy safely and economically. Transmission networks are highly meshed. They are used for large-scale energy transmission, and ensure mutual grid support. Thermal power plants and onshore and offshore wind farms feed into the high-voltage grids. Only a few major customers are connected to the grid. The distribution networks are meshed. These networks are fed by smaller thermal and industrial power plants and wind farms. Typical followers are customers from the big industry.

The high-voltage networks are subordinated to the medium-voltage and low-voltage networks. Smaller decentralized systems feed into the medium-voltage and low-voltage networks. Energy generation plants are based on fossil or renewable fuels such as fuel oil, natural gas, vegetable oil, biodiesel or biogas as well as wind energy plants, photovoltaic (PV) systems, or combined heat and power (CHP) plants. Medium-voltage networks supply industrial, commercial, offices, and department stores, while low-voltage networks supply households, agriculture, and small businesses. The legal, political, and social requirements for electrical energy supply are laid down in the Energy Industry Act (EIA). The EIA requires the planning of safe, reliable, inexpensive, and reliable environmentally compatible networks. In addition, the Renewable Energies Act and the Combined Heat and Power Act promote the expansion of renewable energies and combined heat and power generation. Further advances in heat coupling.

In the distribution grids, the increase in regenerative feeds from wind and sun also lead to variable load flows. The wind and solar energy completely covers local consumption at times, so that the grids can be fed back into the grid by means of regenerative braking, thus endangering network security. Today's networks are not

Analysis and Design of Electrical Power Systems: A Practical Guide and Commentary on NEC and IEC 60364, First Edition. Ismail Kasikci.

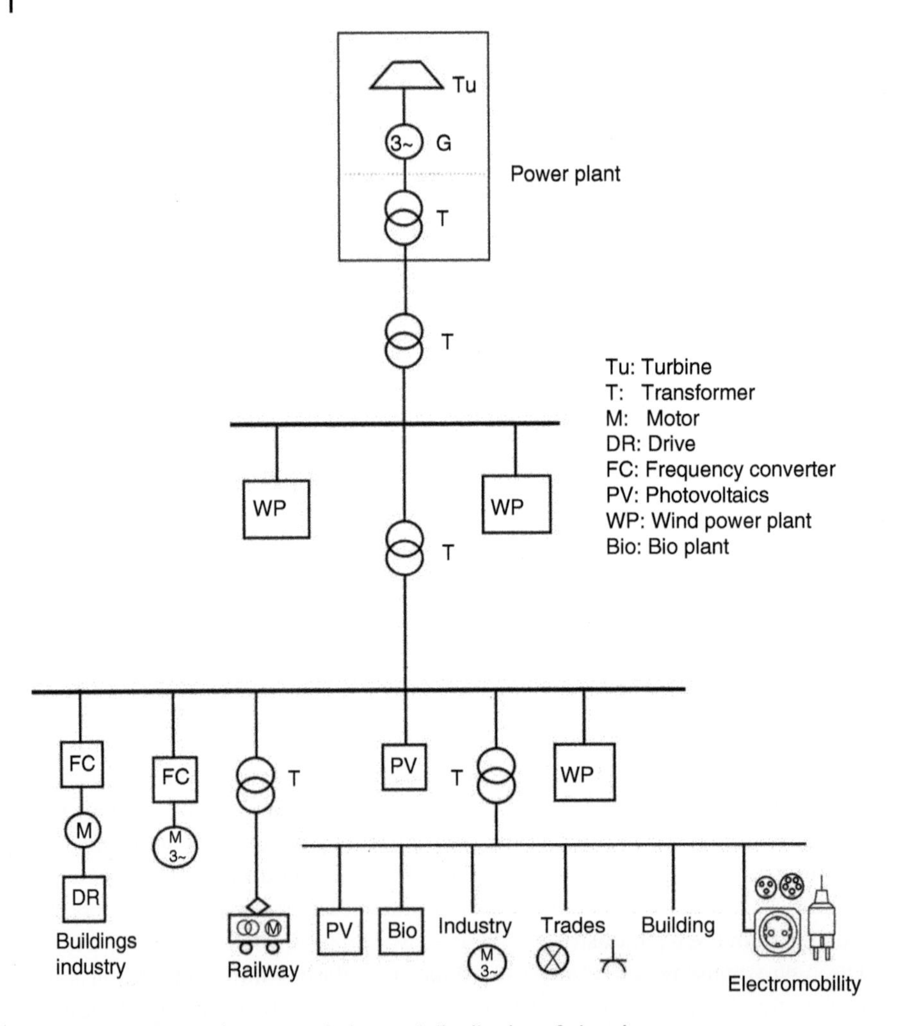

Figure 1.1 Generation, transmission, and distribution of electric power.

designed for feed-ins from regenerative feed-ins. The power quality is impaired at the feed-in node. In addition, the changing load flow directions cause voltage fluctuations that cannot be regulated by the transformers. The use of decentralized energy management systems (DEMS) in the distribution networks will become very important in the future. They coordinate the energy use of the decentralized generation plants with the energy consumption of the consumers and control their energy feed or acceptance. In addition to electricity, a communication network is required that enables the exchange of information between producers, consumers, and storage facilities.

In combination with smart metering in buildings (use of intelligent, electric electricity meters), the so-called Smart Grids (intelligent power grids), which provide load management via bidirectional data communication, are already installed in various plants. Every electrical system must not only satisfy the normal operating condition, but also be designed for faults and must be able to handle both faultless and faulty operating conditions. Therefore, electrical systems must be dimensioned in such a way that neither persons nor material assets are endangered.

The dimensioning, efficiency, and safety of the systems are strongly dependent on the control of short-circuit currents. With increasing installed capacity, the calculation of short-circuit currents also gained in importance. A three-phase system can be temporarily or permanently disturbed by faults, especially short-circuits, circuit measures, or by consumers. Calculation models and solution algorithms for power generation, transmission, and distribution systems provide a comprehensive tool as a calculation and dimensioning program for the planning, design, analysis, optimization, and management of any power supply network. Owing to the liberalization of the energy markets and in particular the rapid expansion of renewable energies, the requirements for network planning and operational management processes are becoming increasingly complex.

Transformers (with or without medium voltage), generators, and neutral grid feeds are available. A neutral mains supply can be mapped by specifying the impedances, the loop impedance, or the short-circuit currents. In supply circuits, an individual fuse of parallel cables with several protective devices can optionally be calculated and dimensioned in addition to the fuse protection of parallel cables by a protective device. The selected feeds can be connected to each other via directional or nondirectional couplings. By defining the different operating modes required (e.g. normal operation, emergency operation ...), the network supply can be represented in a practical way and included in the calculation.

Sub-distributors, group switches, busbar trunking systems, busbar trunking systems with central supply, or distributors with replacement impedances are available as distributors. When selecting these elements, certain specifications must also be made with regard to the design, e.g. whether the connecting cable is to be designed as a busbar or cable and which and how many switching devices are to be used. If a cable section is selected, the intended type of installation must also be specified so that the values of the current-carrying capacity influenced by this are taken into account in the dimensioning. The distributors are always inserted into the graphic on a busbar. This can be the busbar that symbolizes the feed-in point or the busbar of an already connected distributor or the representation of a busbar string, so that the network can be branched further as a radiation network.

For final circuits, consumers with fixed connections, socket outlet circuits, motors, charging units, capacitors, and equivalent loads are available as elements. These are in turn connected to the busbar of existing sub-distribution boards or the representation of a busbar line or directly to the busbar symbolizing the feed-in point. There are also various options for placing these elements in the mesh graphic. These are offered in the selection window specific to the element. Simultaneity factors or utilization factors can also be specified for the different circuits, which are also then taken into account during dimensioning. Once the network structure is completely constructed in this way, the actual calculation and thus the dimensioning and selection of the elements can be initiated.

The results of this dimensioning can be viewed and documented in the various available view variants of the network graphic. In addition to the possibility of individually configuring the labeling of the network graphic, standardized labeling variants (device parameters, load flow/load distribution, short-circuit load, energy balance) are available, so that all parameters relevant for the network calculation are clearly displayed.

In practice, selectivity verification is often mandatory, e.g. for safety power supply systems. Back-up protection can also be taken into account when selecting switchgear, i.e. the switching capacity of a downstream switch can be increased by the fact that the upstream switch trips simultaneously and thereby limits the current.

Suitable programs can be used for the design and selection of electrical equipment, calculation of mechanical and thermal short-circuit resistance, calculation of short-circuit currents, selectivity and back-up protection for the selection of overcurrent protection devices, and calculation of the temperature increase in control cabinets.

2

Electrical Systems

2.1 High-Voltage Power Systems

For the distribution and transmission of electric power, standardized voltages according to IEC 60038 are used worldwide. For three-phase AC applications, two voltage levels are given as follows:

- *Low voltage*: Up to and including 1 kV AC (or 1500 V DC)
- *High voltage*: Above 1 kV AC (or 1500 V DC)

In Europe, 245 kV lines were used for the interconnection of power supply systems. Long-distance transmission, for example, between the hydro-power plants and consumers, was done by 245 kV lines. Nowadays, the importance of 245 kV lines is decreasing due to the existence of the 420 kV transmission system. The 420 kV level represents the highest operation voltage used for AC transmission. It typically interconnects the power supply systems and transmits energy over long distances. Some 420 kV lines connect the national grids of individual European countries, enabling interconnected network operation (UCTE = Union for the Coordination of Transmission of Electricity) throughout Europe.

When considering power transmission over long distances, a more economical solution is the high-voltage direct current (HVDC) technology. DC voltages vary from the voltage levels recommended in the abovementioned standardized voltages used for AC.

Figure 2.1 shows an overview of a high- and a low-voltage power system.

Medium voltage systems are defined up to 52 kV. The values vary greatly from country to country, depending on the historical development of technology and the local conditions. Transmission is normally with the use of buried cables, which connect the individual stations to each other in a ring structure with the help of isolating points.

The lines between the stations can be isolated in the event of a disturbance (Figure 2.2).

Analysis and Design of Electrical Power Systems: A Practical Guide and Commentary on NEC and IEC 60364, First Edition. Ismail Kasikci.

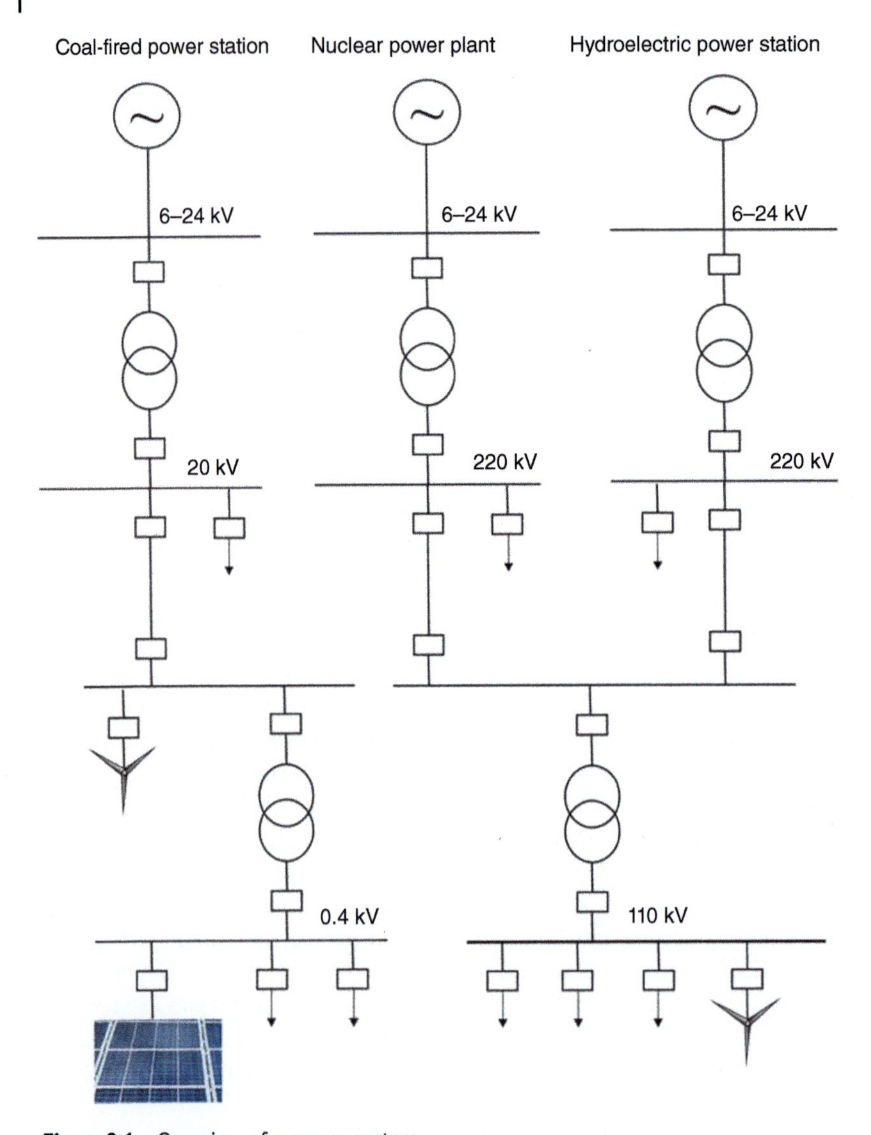

Figure 2.1 Overview of a power system.

In power supply and distribution systems, medium-voltage equipment is available in the following:

- Power plants, for generators and station supply systems.
- Transformer substations of the primary distribution level (public supply systems or systems of large industrial companies), in which power supplied from the high-voltage system is transformed to medium voltage.
- Local supply, transformer, or customer transfer substations for large consumers (secondary distribution level), in which the power is transformed from medium to low voltage and distributed to the consumer.

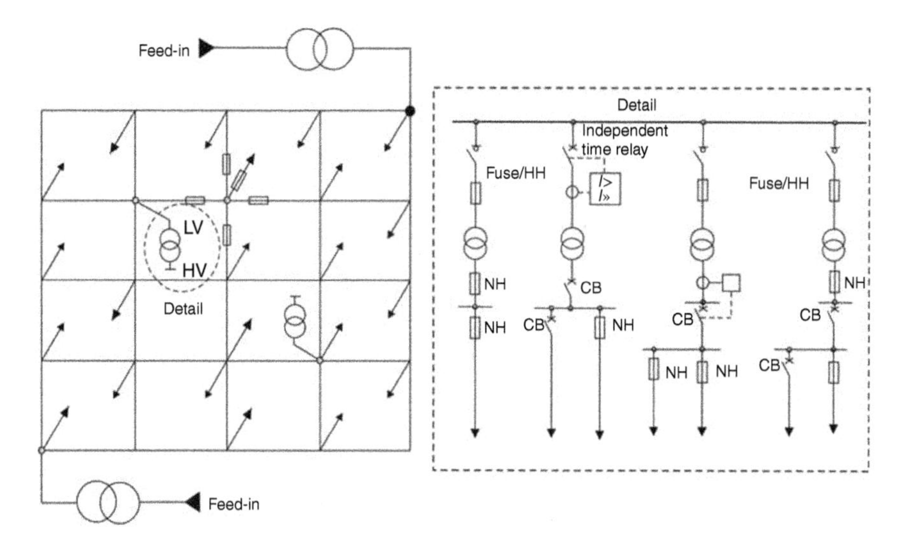

Figure 2.2 Feed-in and load field.

- *Circuit-breakers*: While the number of mechanical operating cycles is specifically stated in the M classes, the circuit-breaker standard IEC 62271-100/VDE 0671-100 does not define the electrical endurance of the E classes by specific numbers of operating cycles; the standard remains vague on this. The short-circuit type tests provide an orientation as to what is meant by normal electrical endurance and extended electrical endurance.
  Modern vacuum circuit-breakers can generally make and break the rated normal current up to the number of mechanical operating cycles. The switching rate is not a determining selection criterion, because circuit-breakers are always used where short-circuit breaking capacity is required to protect the equipment.
- *Disconnectors*: Disconnectors do not have any switching capacity (switches for limited applications must only control some of the switching duties of a general-purpose switch). Switches for special applications are provided for switching duties such as switching of single capacitor banks, paralleling of capacitor banks, switching of ring circuits formed by transformers connected in parallel, or switching of motors in normal and locked condition. Therefore, classes are only specified for the number of mechanical operating cycles.
- *Earthing switches*: With earthing switches, the E classes designate the short-circuit making capacity (earthing on applied voltage). E0 corresponds to a normal earthing switch; switches of the E1 and E2 classes are also called make-proof or high-speed earthing switches. The standard does not specify how often an earthing switch can be actuated purely mechanically; there are no M classes for these switches.
- *Contactors*: The standard has not specified any endurance classes for contactors yet. Commonly used contactors today have mechanical and electrical endurance in the range of 250 000 to 1 000 000 operating cycles. They are used wherever switching operations are performed very frequently, e.g. more than once per hour.

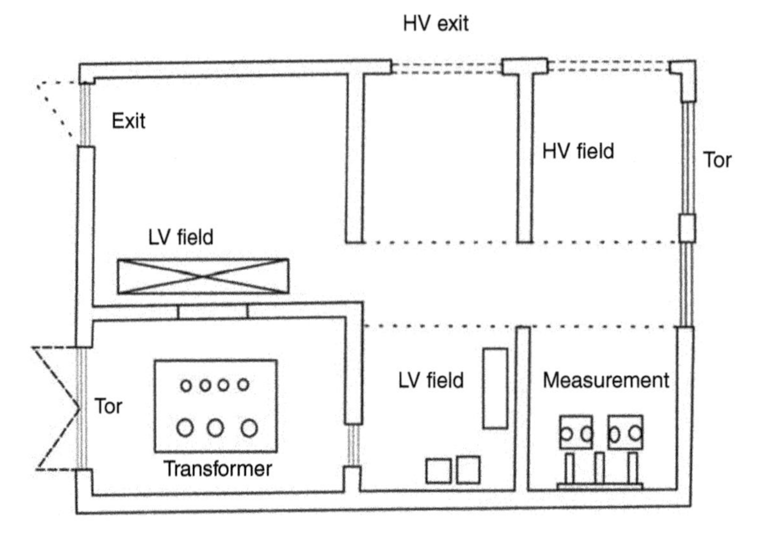

Figure 2.3 Arrangement of a transformer room.

The distribution stations can be housed in precast concrete cells, containers, or special rooms (Figure 2.3).

Low-voltage lines (up to 1 kV AC) serve households and small business consumers. Lines on the medium-voltage level supply small settlements, individual industrial plants, and large consumers; the transmission capacity is typically less than 10 MVA per circuit.

System configurations of low-voltage power systems are as follows:

- *Simple radial system*: All consumers are centrally supplied from one power source. Each connecting line has an unambiguous direction of energy flow.
- *Radial system with changeover connection*: All consumers are centrally supplied from 2 to n power sources (stand-alone operation with open couplings) (Figure 2.4).

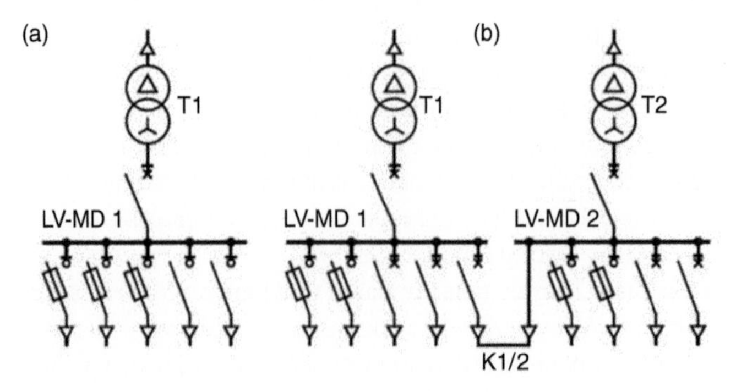

Figure 2.4 Simple radial system, radial system with changeover connection. (a) Simple radial network without power outage reserve. (b) Radial network with changeover reserve.

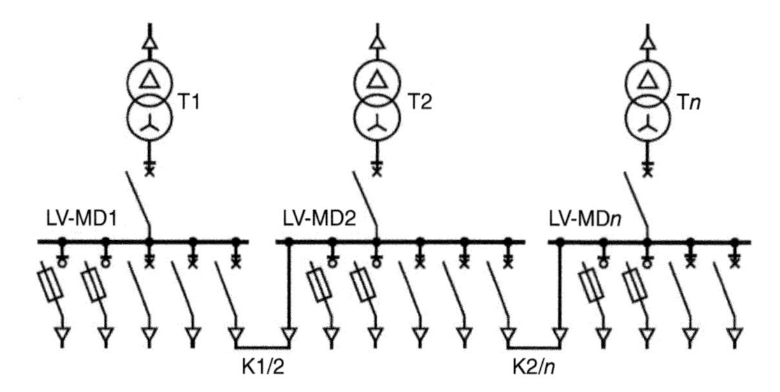

Figure 2.5 Radial system in an interconnected network.

- *Radial system in an interconnected network*: Individual radial systems, in which the connected consumers are centrally supplied by one power source, are additionally coupled electrically with other radial systems by means of coupling connections. All couplings are normally closed (Figure 2.5).

Depending on the rating of the power sources in relation to the total load connected, the application of the $(n - 1)$ principle can ensure continuous and faultless power supply of all consumers by means of additional connecting lines. Radial system with power distribution via busbars that can be operated in an interconnected network and busbar trunking systems are used instead of cables. The network stations often incorporate two network transformers in a room (for greater availability) and the main low-voltage main distribution (LVMD), which supplies the loads in the form of a radial network. The rated transformer powers cover the range from 100 to 2500 kVA.

2.2 Transformer Selection Depending on Load Profiles

In the EU Commission Regulation No. 548/2014 the eco-design requirements for power transformers were requested. It applies to transformers with a minimum rated power of 1 kVA. The regulation is already the second stage of specifying the efficiency improvements, or loss reductions, for placing the transformers on the market. The requirements described therein include the determination of the maximum permissible limits for short-circuit losses and no-load losses or the minimum value for the maximum efficiency of the transformer. The reduction of transformer losses should reduce the annual CO_2 emissions during operation [1].

A distinction is made between simple efficiency class 1 and high efficiency class 2. The selection of an efficient transformer therefore takes into account both its specific characteristics (load-dependent efficiency data) and its operating load (load profile).

In the standard IEC 60364-8-1: Erection of Low-voltage systems – Part 8-1: Energy efficiency, the load dependence of the efficiency of transformers is explicitly indicated. The environmental impact of the project also includes transformers from the operating point of view, and depends on the load dependence of the losses.

The total power loss (PV) during operation is calculated from the sum of idling and load losses at a specific Load (S_{Load}). The no-load losses are Load independent and occur as soon as voltage is applied to one of the attached transformer windings (primary or secondary side). The load losses, however, are dependent on the square load ratio ($S_{\text{Load}}/S_{\text{rT}}$) and the load loss values Pk. The load ratio is the ratio of load (S_{Load}) to the rated apparent power (S_{rT}) of the transformer.

$$P_{\text{Loss}} = P_0 + \left(\frac{S_{\text{Load}}}{S_{\text{rT}}}\right)^2 \cdot P_k \tag{2.1}$$

The no-load and short-circuit losses of the transformers are of great importance for the loss and efficiency estimation. In this section, these losses are briefly explained. No-load losses consist of the losses in the iron core and dielectric as well as on the losses caused by the no-load current in the windings. The losses in the dielectric and windings are generally insignificant. The iron losses are composed of the hysteresis losses and the eddy current losses. The hysteresis losses are created by turning over the microcrystals, which are elementary magnets. They offer resistance with each rotation and alignment. The related work is generally not recoverable, and occurs as heat loss. The heat losses occur in iron in addition to the hysteresis losses. Eddy current losses are caused by the voltages induced by the time-varying magnetic field in iron. As a result of these voltages, currents run in vortex-shaped paths. Ohmic resistance of iron and the eddy currents result in eddy current losses due to the relation $I^2 R$. By using particularly thin iron core and dielectric, insulated from each other, the eddy current losses can be kept low.

The short-circuit losses consist of the current heat losses in the ohmic resistors ($I^2 R$) and the additional losses caused by eddy currents in the windings and in the structural parts [2].

A minimum of the losses is achieved when no-load and load losses are equal. This results in the so-called load factor k

$$k = \sqrt{\frac{P_0}{P_k}} \tag{2.2}$$

2.3 Low-Voltage Power Systems

Electrical switchgear comprises a multitude of various operational equipment and components. The generation, transmission, and distribution of electrical energy takes place with these elements as shown in Figure 2.6. When all these components are adequately dimensioned for their intended purposes under all operating conditions, economical and stable operation can be expected.

For planning the supply network, it is necessary to thoroughly consider the network design in order that the regulations and requirements are satisfied.

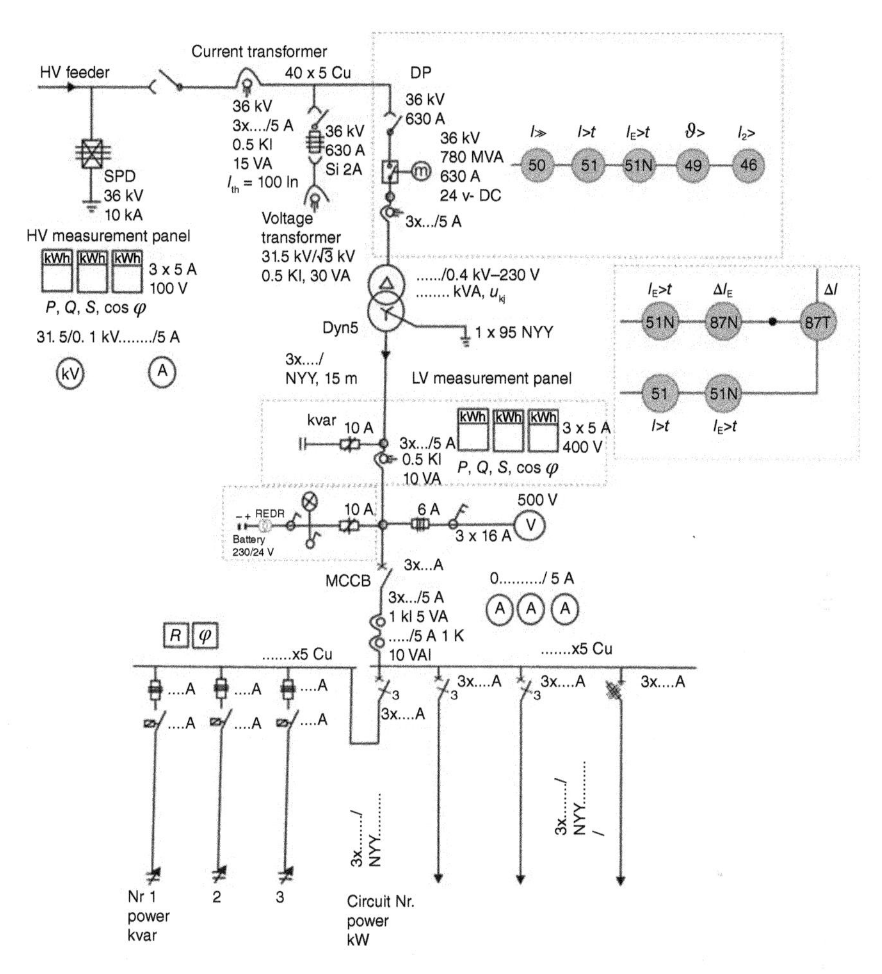

Figure 2.6 Principle of low-voltage (LV) distribution.

The following requirements can be summarized here:

- The network should be as simple as possible and its design should be easy to oversee.
- There must be optimum protection for the equipment installed.
- Good security of supply and low network losses must be guaranteed.
- Operation and maintenance of the network must be convenient.
- The network must have good supply quality and minimum harmonics.
- The network should be designed as well as possible as a radial system.
- Project-related regulations must be observed.

For the planning of networks and electrical systems, the total power of the system, the size and number of transformers, the dimensioning of cables and lines, the size and distribution of operating and short-circuit currents, operational failures, and changes in loads must all be clarified as soon as possible. The thermal and dynamic short-circuit currents and the making and breaking capacities of the protective equipment, the contributions of low-voltage motors to the initial symmetrical short-circuit current, to the peak short-circuit current, and to the breaking current, the voltage drop, and the system perturbations determine the stability of the system. Electrical equipment must be designed so that its operation is not impaired by external electromagnetic influences and so that the equipment itself does not become a source of disturbance for other equipment [5]. For this reason, during the planning of systems it is necessary to clarify possible sources of disturbance as early as possible and initiate appropriate countermeasures. The following characteristic specifications are part of planning a low-voltage system:

- Type of supply
- Power capacity
- Coincidence factor
- System forms
- External influences on operational equipment
- Compatibility and serviceability
- Maximum transmission length
- Number of power supply units connected
- Cross sections for the loads
- Fusing of circuits

The planning and installation of an electrical system requires meticulous calculation of the power demands for which the system is to be designed. This is determined by the equation

$$P_{\max} = \sum (P_i\, g_i) \tag{2.3}$$

The coincidence factor or demand factor g_i indicates how many consumers are in operation at the same time (Tables 2.1 and 2.2). It is an important factor for determining the feed-ins. When more motor drives are connected in the system, it is also necessary to consider the utilization factor a_i and the efficiency η_i in this calculation. The maximum power demand is then [3].

$$P_{\max} = \sum \frac{P_{\mathrm{rM}}\, g_i\, a_i}{\eta_i} \tag{2.4}$$

The meanings of the symbols are as follows:

$P_{\max}$	Power demand
P_i	Installed power
g_i	Coincidence factor
a_i	Utilization factor of motor
η_i	Efficiency of motor
P_{rM}	Rated power of motor

The apparent power of the network input can be calculated from the calculated $P_{\max}$ and the average power factor $\cos\varphi$. Once the total power, with a reserve factor, has

Table 2.1 Coincidence factors for the main feed-in.

Building type	Factor
Residential	0.4
Apartment blocks	
With electrical heating	0.8–1
Without electrical heating	0.6
High-rise office building	
Ventilation, heating	1
Data processing	1
Lighting	1
Sprinkler system	1
Sanitation facilities	0.8
Elevators	0.7
Cooling system	1
Schools	0.6–0.7
Assembly rooms, theater, restaurants, etc.	0.6–0.8
Stores	0.6–0.7
Traffic systems	1
Administrative offices, banks	0.7–0.9
Kindergartens	0.6–0.9
Carpenters' shops	0.2–0.6
Butchers	0.5–0.8
Bakeries	0.4–0.8
Construction sites	0.2–0.4
Cranes	0.7 per crane

been defined the size and type of the transformer can be established. Supply takes place in either a ring network or a radial network (Figure 2.7).

Here, it is necessary to give some thought to the characteristics of the operational electrical equipment that can affect other operational equipment, such as harmonics, reactive power compensation, overvoltages, electromagnetic fields, voltage quality, and power system protection.

The operational electrical equipment, such as cables and lines, fuses, Circuit-breakers, and transformers must be optimally chosen and dimensioned both economically and technically. If no information is available, Table 2.3 can be used for this purpose. The minimum fuse protection for a residential installation is 63 A.

Table 2.2 Coincidence factors for important consumers.

Consumer groups	Office buildings	Hospitals	Department stores
Lighting	0.85–0.95	0.7–0.9	0.85–0.95
Electrical outlets	0.1–0.15	0.1–0.2	0.2
Kitchens	0.5–0.85	0.6–0.8	0.6–0.8
Air conditioners	1	1	1
Elevators and escalators	0.7–1	0.5–1	0.7–1

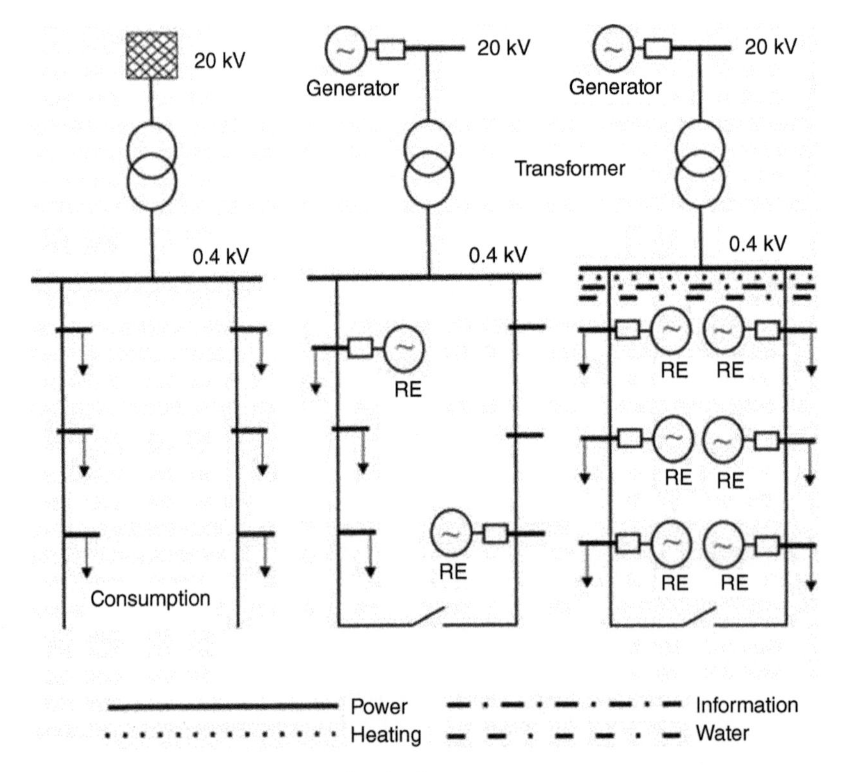

Figure 2.7 Low-voltage concept for the future.

Table 2.3 Planning values for networks.

Category	Power demand (W/m^2)	Reference area
Industry	20–150	Site area
Supermarkets	15–80	Shopping area
Department stores	30–100	Shopping area
Offices	30–60	Total business area
Housing		Figure 2.8

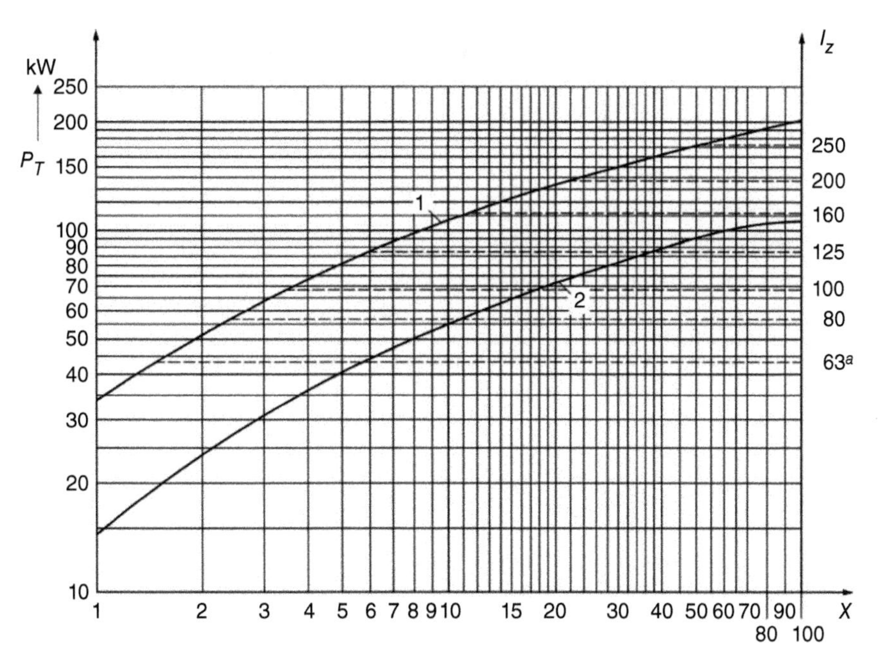

Figure 2.8 Design basis for main lines in residential buildings without electric heating; nominal voltage 230/400 V. Source: Ref. [4]. 1 with electric hot water preparation for bathing or showering purposes; 2 without electric water heating for bathing or showering purposes; I_z minimum required current-carrying capacity, in A; P_T power resulting from the required current-carrying capacity and the nominal voltage, an assumed cos phi of 1; X number of apartments; a minimum fuse to ensure the selectivity of fuses.

For larger buildings, with or without electrical water heating for bathing/showering, the effective powers for the dimensioning of the main lines can be taken from the diagram in Figure 2.8.

Each project planning task begins with the collection of all data required for the calculations and design. The coordination between all those concerned (owner, architect, and electrical specialists) plays an important part. The applicable standards and regulations, such as IEC and EN, as well as any country-specific regulations, must always be taken into account. A few of the most important of these are given here:

- IEC 60 364 Installation of power current systems up to 1000 V
- IEC 60 909 Calculation of short-circuit currents in three-phase networks
- IEC 523 Current-carrying capacity of lines and cables
- Conversion factors for the current-carrying capacity of lines and cables
- Accident prevention regulations of the trade union for electrotechnology
- Planning of electrical systems in residential buildings
- Regulations of fire alarm installations
- IEC 1024 Lightning protection systems
- EN 60204-1 Equipping of industrial machines
- Electrotechnical operating, service panel

- Lighting engineering
- Regulations for construction of the German federal states
- Accident prevention regulations
- Circuit documentation
- Power economics law
- Supply service
- Concrete-footing grounding electrodes

In practice, we can speak of the following cases:

Case I:

1. The short-circuit power is known or can be obtained from the power supplier.
2. The power and the short-circuit voltage of the transformer are given.
3. The consumers and the connection locations are known.
4. The coincidence factor is given or can be taken from the tables.
5. It is therefore possible to calculate the cross section of the main line, the short-circuit currents, and the voltage drops.

Case II:

1. The source impedance of the source network and the length up to the feed-in of the main distribution are known.
2. The consumer data are known and are drawn in on the planimetric map.
3. The coincidence factor is given.
4. The protective measures (TN or TT system) are known.
5. The installed total effective and reactive powers must be calculated.
6. The operating current is calculated from the total effective power.
7. The cross section of the feeder line is calculated, considering the type of cable installation and the voltage drop.
8. The main fuse or circuit-breaker is determined from the cabling selected.
9. For the calculation of short-circuit currents, it is necessary to determine the impedances of the individual lines and cables.
10. The short-circuit currents for three-pole and single-pole short -circuits are calculated.
11. The cross sections of the lines and cables are determined in accordance with existing regulations (IEC 60 364 Parts 41, 43, 52 and 54).
12. The overcurrent protection equipment (OPE) must be chosen for these cross sections.
13. The break times of the individual fuses are read from either the characteristic curves or from IEC 60 364 Part 61.
14. The selectivity of the OPE must be determined.
15. All data compiled can then be transferred to the overview, circuit diagrams, and the planimetric map.

2.4 Examples of Power Systems

2.4.1 Example 1: Calculation of the Power

In a factory a total of 250 kW capacity is installed. The simultaneity factor is 0.65. What is the expected power requirement?

$$P = g \cdot P_{\text{inst.}} = 0.65 \cdot 250\ \text{kW} = 162.5\ \text{kW}$$

2.4.2 Example 2: Calculation of the Main Power Line

Given is a building with 22 floors. Determine the required electrical power.

We can read the required power without electric water heating for bathing or showering purposes in Figure 2.8, which gives 75 kW. The current-carrying capacity must be at least $I_z = 139$ A and $I_n = 125$ A (Table 14.18).

In this case, the demand factor for 22 floors is $\frac{75\ \text{kW}}{22 \cdot 15\ \text{kW}} = 0.227$.

2.4.3 Example: Power Supply of a Factory

A factory is supplied by a transformer and an external power supply. For a new production plant a sub-distribution system is to be planned and extended (Figure 2.9).

The following data is given by the customer. High voltage 20 kV, low voltage 400/230 V, 50 Hz. The data of the consumers include the following:

1. *Lighting system*: 25 kW. The simultaneity factor is 1.
2. *12 pieces motor*: 12 motors per 15 kW, $\cos\varphi = 0.82$, $\eta = 0.9$.
3. *Thermal output*: Total 250 kW.
4. Six socket circuit, each 15 kW, $\cos\varphi = 0.8$, simultaneity factor is 0.4.

Calculate all the operating currents:

Solutions

1. The total power of the lighting system is 25 kW. The simultaneity factor is 1.

$$I_B = \frac{P}{\sqrt{3} \cdot U_n \cos\varphi} = \frac{25\ \text{kW}}{\sqrt{3} \cdot 400\ \text{V} \cdot 0.85} = 42.45\ \text{A}$$

2. The total power of the motors is 12×15 kW = 180 kW.

$$I_{rM} = \frac{P_{rM}}{\sqrt{3} \cdot U_n \cos\varphi \cdot \eta} = \frac{15\ \text{kW}}{\sqrt{3} \cdot 400\ \text{V} \cdot 0.82 \cdot 0.9} = 29.3\ \text{A}$$

$12 \times 29.3\ \text{A} = 351.6\ \text{A}$

3. Thermal heating power

$$I_T = \frac{P}{\sqrt{3} \cdot U_n \cos\varphi} = \frac{250\ \text{kW}}{\sqrt{3} \cdot 400\ \text{V} \cdot 1} = 360.84\ \text{A}$$

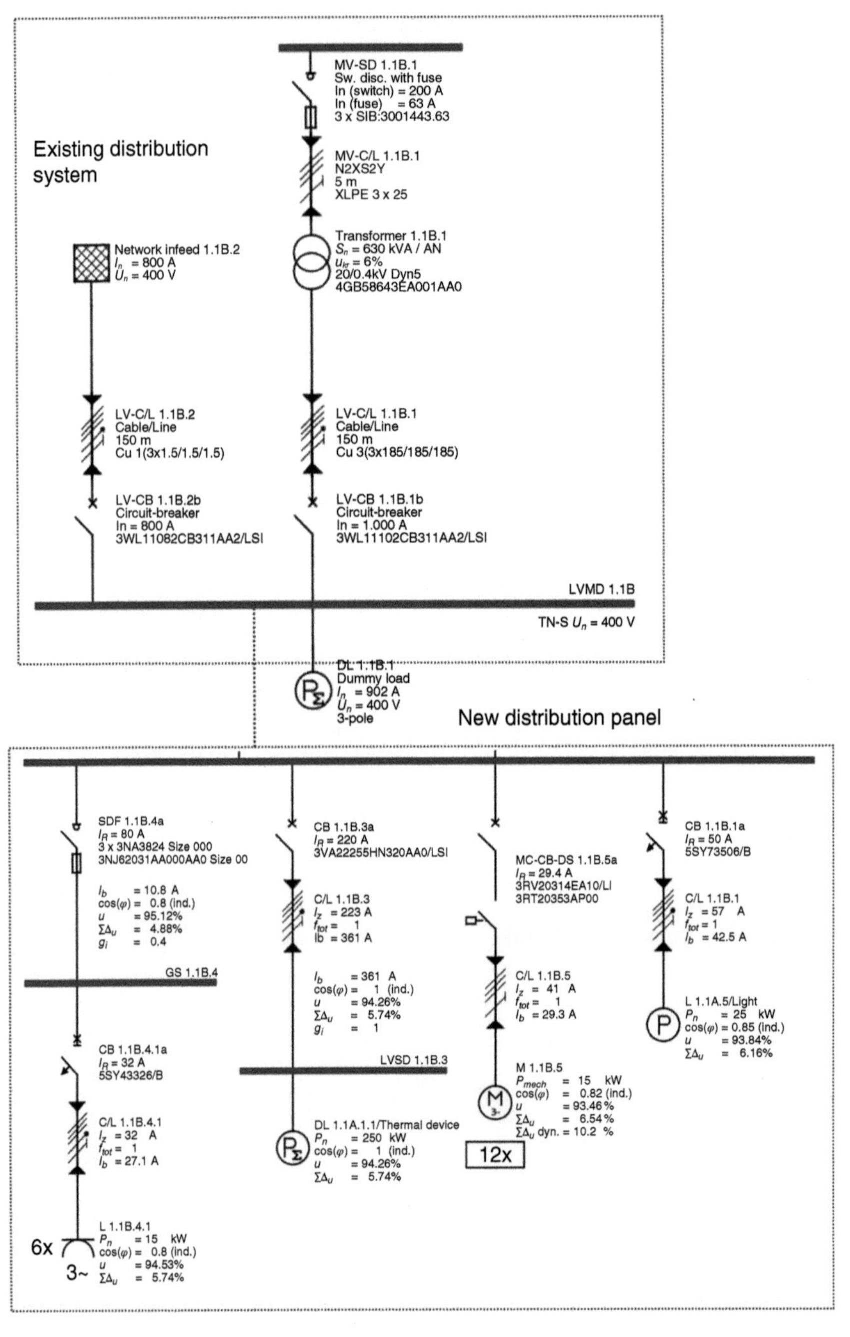

Figure 2.9 Design of a new distribution panel.

4. Sockets

$$I_{\mathrm{B}} = \frac{P}{\sqrt{3} \cdot U_n \cdot \cos\varphi} = \frac{15\ \mathrm{kW}}{\sqrt{3} \cdot 400\ \mathrm{V} \cdot 0.8} = 27.06\ \mathrm{A}$$

Simultaneity factor is $0.4 \times 6 \times 27.06\ \mathrm{A} = 65\ \mathrm{A}$.
Total operating current

$$I_T = 42.45\ \mathrm{A} + 351.6\ \mathrm{A} + 360.84\ \mathrm{A} + 65\ \mathrm{A} = 819.89\ \mathrm{A}$$

3

Design of DC Current Installations

Design and application of photovoltaic systems in buildings under 10 kWp can become a reality in the future due to distributed electrical sources using renewable energy. The power supply on direct current electrical installations, which is not intended to be connected to Public Distribution Networks, would be a great benefit because of the access to electrical power for people living in poor areas (Figure 3.1).

But, supply from these renewable energies is not constant; photovoltaic panels do not operate at night, and wind turbines require wind for generating electrical energy. Therefore, the use of storage units becomes a necessity. It is noted that manufacturers of stationary secondary batteries are investing a lot in these technologies and prices will soon become affordable to those people having no access to electricity.

In addition to that, new technologies such as light-emitting diodes (LEDs) or electronic devices generally operate on direct current. Connecting these types of current-using equipment to electricity sources using renewable energy through dc electrical installation is more and more realistic. For changing the DC voltage, DC/DC converters are available now.

The concept of installing any low-voltage electrical installation is to be considered as a set of electrical equipment having the following functions:

- Supply (e.g. local generator, photovoltaic systems, wind turbine, batteries)
- Distribution (e.g. distribution board, wiring systems, socket-outlets)
- Consumption (e.g. fans, lighting, appliances, pumps, batteries)

3.1 Earthing Arrangement

Earthing arrangement in a TN system: The midpoint or one polarity (positive or negative) of DC power supplies shall be connected to earth at one point only, and the exposed conductive parts shall be connected to this earthing arrangement. In an individual installation, the earth connection shall preferably be made on the main busbar/terminal, thus ensuring the connection of all possible power sources to earth.

Any protective measure against electric shock shall be an adequate combination of two separate types of protection (basic and fault), or an enhanced protection, combining both types of protection in one single measure (IEC 60364-4-41).

Figure 3.2 shows the electrical installation of TN-S system in Buildings.

Analysis and Design of Electrical Power Systems: A Practical Guide and Commentary on NEC and IEC 60364, First Edition. Ismail Kasikci.

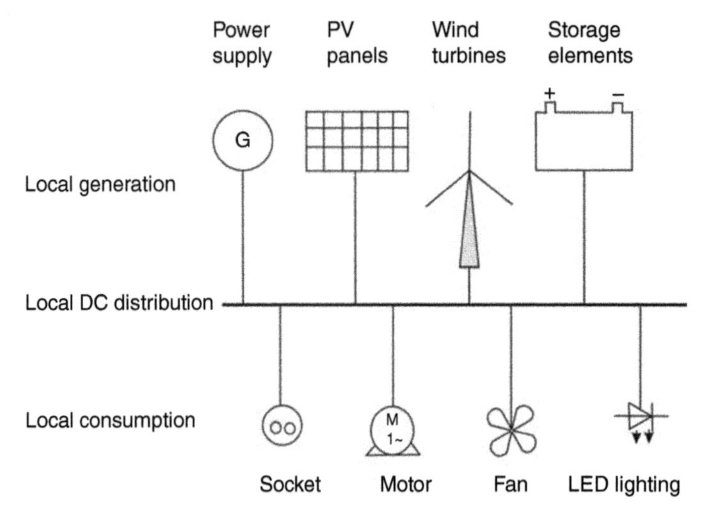

Figure 3.1 Concept of DC low-voltage electrical installation.

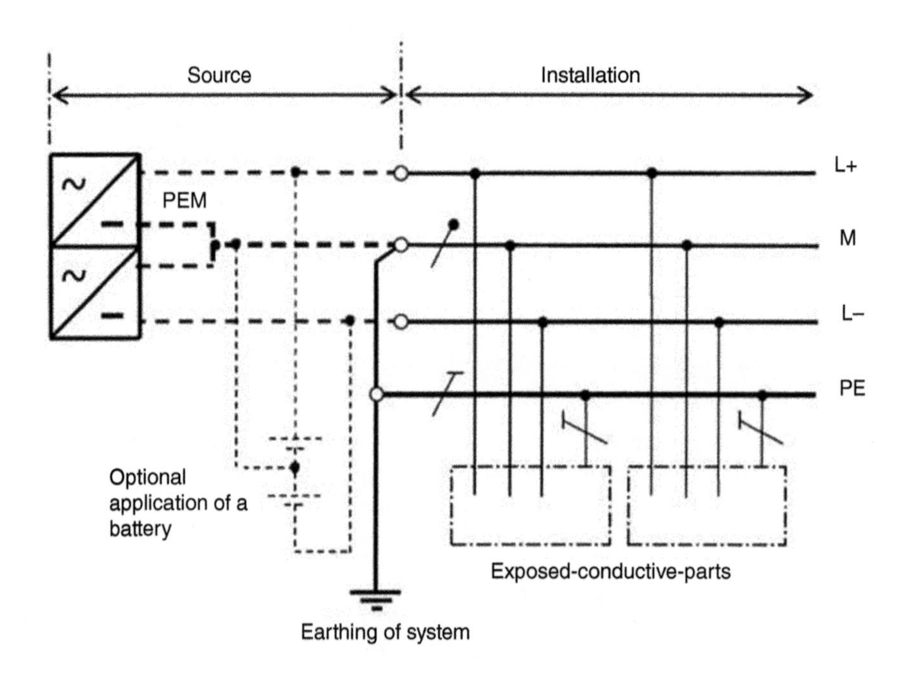

Figure 3.2 Example of electrical installation in TN-S system.

3.2 Protection Against Overcurrent

Protection against thermal effects and electric arc: Direct current once established does not cross the zero value as is the case for alternating current. In the case of a fault, an arc is quickly created between a faulty live part and another part, and due to the high impedance of the arc and low level of current drawn, arc faults could not be cleared by overcurrent protective devices. Additional care is necessary to reduce the risk of arcing, e.g. selection of electrical equipment.

Risk of explosion with batteries: Hydrogen and oxygen gases can be released from a lead acid battery during normal operation and also in the case of excess of charging current (overcharge). Hydrogen concentration between 4% and 72% is considered flammable and may provoke explosion in case of ignition. The ventilation system in a stationary battery location should be designed to keep hydrogen concentration below 1%. Stationary secondary batteries shall be installed in a room with sufficient space and well ventilated to avoid risk of explosion.

Each source and final circuit shall be individually protected against overcurrent by an overcurrent protective device (OCPD) suitable for DC (circuit-breaker or fuse). The minimum breaking capacity of the overcurrent protective device shall be higher than the maximum short-circuit current at the point where the device for short-circuit protection is installed.

3.3 Architecture of Installations

Individual installation corresponds to a single consuming and/or producing electrical installation (Figure 3.3). Such an installation always includes current-using equipment (or commonly named loads), and may also include local power supply units (photovoltaic panels, wind turbine) and local storage units (e.g. stationary

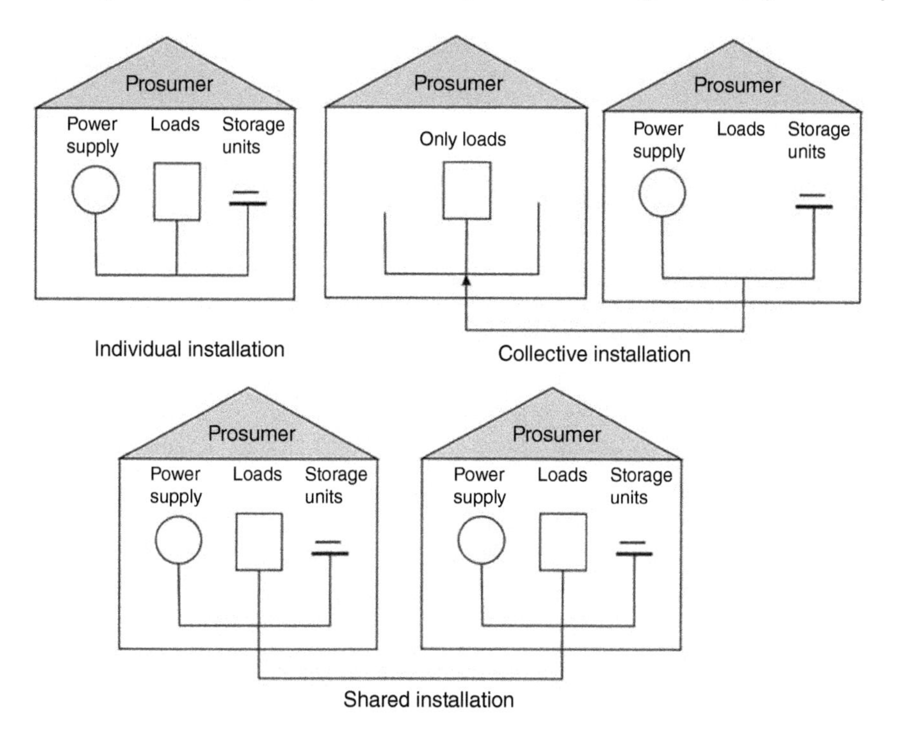

Figure 3.3 Example of architecture and operating modes of installations.

secondary batteries). Usually such installation may include several single electrical energy meters or measuring equipment. Operating modes of individual installation can be a direct feeding mode, in which the electrical DC installation is connected to the other electrical installations, and partly or completely fed from them; a reverse feeding mode, in which the electrical DC installation is connected to the other electrical installations, and partly or completely feed them; and an autonomous mode, in which the electrical DC installation is not connected to the other electrical installations, and is fed directly from local power sources.

Figure 3.3 shows an example of architecture and operating modes of installations.

4

Smart Grid

A smart grid is an electrical network that coordinates the actions of all users connected – intelligently and to ensure that efficiency in sustainable, ecological, economical and reliable power supply – the producers and the consumers (VDE).

According to the National Electrical Manufacturers Association (NEMA), the basic concept of Smart Grid is to add monitoring, analysis, control, and communication capabilities to the national electric grid in order to improve reliability, maximize throughput, increase energy efficiency, provide consumer participation, and allow diverse generation and storage options.

The main problems to be solved are seen in the following:

1. The increasing impact of emissions on the environment
2. The dependency of imported energy sources
3. The decreasing availability of fossil and nuclear energy and the related, extremely increasing processes involved
4. Maximum voltage drop at the feeder
5. Control of transformer power station (research)
6. Cable type and length

For a safe energy supply:

1. The establishment of Smart Grids (SG) is a mandatory task to ensure the sustainability of electric power supply in the future.
2. The way to SG requires high investments and the consistent development of new market design.
3. New technologies and efficiency have to be applied at all systems levels.
4. Communication systems will penetrate the distribution systems.

In a distributed and differentiated energy system, power generation and consumption must be balanced to the extent that today's quality standards (EN 60150) retain their validity.

The flow of energy from a large power station to the consumer by way of the transmission and distribution grids is increasingly being replaced by distributed power generation in small units within the distribution grid. The flow of energy may even be reversed with it being fed from the distribution grid into the transmission grid [6].

Analysis and Design of Electrical Power Systems: A Practical Guide and Commentary on NEC and IEC 60364, First Edition. Ismail Kasikci.

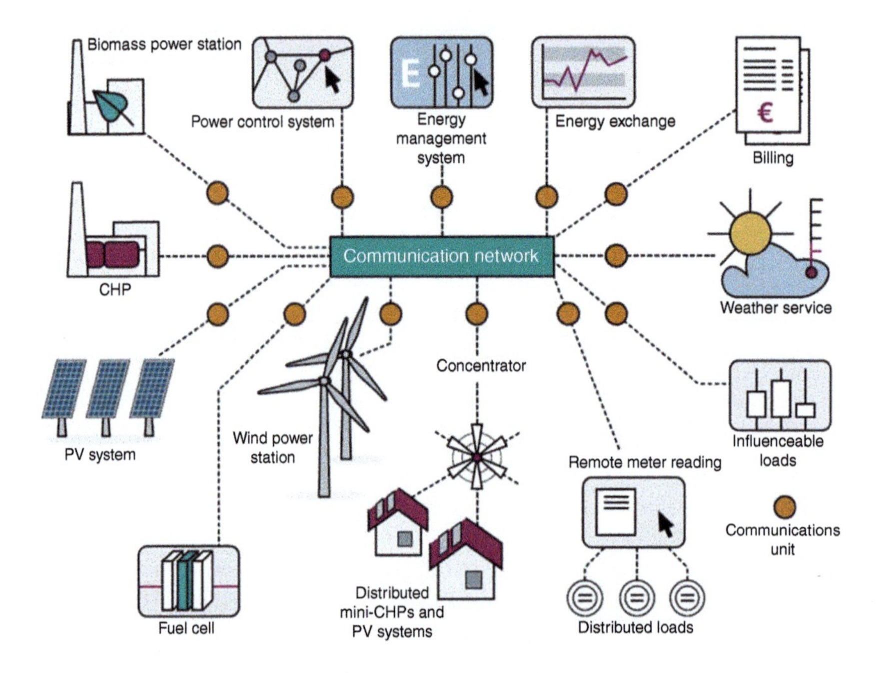

Figure 4.1 Energy management in the smart grid networks [6].

What is absolutely needed for the smart grid is effective communication between the energy suppliers involved. Figure 4.1 shows energy management in the smart grid through communication effected across all energy networks.

5

Project Management

5.1 Guidelines for Contracting

For the planning and project management of public contracts, the Regulations for Contracting System Installations and the Guidelines for the Remuneration of Architects and Engineers are used exclusively. Invitations for bids, awarding of contracts, performing the work contracted, and invoicing of constructional services are summarized in the Construction Services and serve as the legal basis for private construction contracts. Of particular interest here are the Agreement on Construction Services (electrical cable and line systems in buildings), General Regulations for Constructions of all Types, and Lightning Protection Systems [1].

The Regulations for Contracting System Installations regulates the payments and describes the services in detail, apart from specifying acceptable ranges of remuneration (Table 5.1).

The Engineering Services Manual deals with all contractors' services. The electrical system planners determine their costs, following the prescribed cost groups. Engineering services can be described either in the form of a directory of services (scope book) or in the form of a system description. The individual items include details by the number of times applicable, meter, or lump sums.

The system description contains only the equipping and function of the electrical systems. The contracting company is obligated to submit any criticism of the proposed work in writing prior to beginning the work. The contracting company must take into proper account the validity of regulations and the transitional periods defined for new regulations.

After completing the contracted work, the contractor must also supply overview circuit diagrams and installation plans. Prior to commissioning, IEC 60 364 Part 61 requires that the system be tested with respect to operational capability. Detailed descriptions can be found in the Expert's Planning Guide [6].

The project is carried out, e.g. according to the following sequential steps:

- Customer consultation
- Preliminary design and submission of bid
- Cost estimate
- Negotiations

Analysis and Design of Electrical Power Systems: A Practical Guide and Commentary on NEC and IEC 60364, First Edition. Ismail Kasikci.

Table 5.1 Acceptable ranges of remuneration.

Remuneration range I	Simple low voltage and telephone installations
Remuneration range II	Compact stations, low-voltage installations, and distribution systems telecommunication installation not belonging to remuneration ranges I and III, lightning protection systems, lighting systems
Remuneration range III	High-voltage and medium-voltage systems, low-voltage switchgear, current generating and converting systems, low-voltage line systems, and lighting systems requiring extensive planning costs, large telephone systems and networks

- Beginning of project, processing
- Creation of project plan, invitation for bids
- Writing contracts for the system
- Carrying out, monitoring
- Installation of the system
- Commissioning, measurements
- Turnover, documentation, warranty

5.2 Guidelines for Project Planning of Electrical Systems

The following circuit diagrams, drawings, and technical documents are required for the installation of electrical systems:

1. *Power balance of entire system*:
   - low voltage (230 V, 400 V, 690 V)
   - medium voltage (6 kV, 10 kV, 20 kV)
   - high voltage, as required (110 kV, 220 kV, 380 kV)
2. *Motor lists in accordance with IEC 60034*:
   - rated power in kW
   - rated voltage in V

- rated motor speed in rpm
- rated torque in Nm
- moment of inertia in kg/m^2
- ambient temperature in °C
- motor type and class
- design, size, and type of protection
- manufacturer, accessories
- terminal designations
- load data, mode of operation
- breakaway torque, starting current
- efficiency
- operating cycle

3. *Overview circuit diagrams*: An overview diagram is the simplified representation of a circuit. It shows the functional principle and the structure of an electrical system and includes, e.g. voltages, frequency, power, terminal designations, and transformer data.
4. *Network diagrams*: The network diagram shows all connections and parts of a network, not drawn to scale.
5. *Single-line or three-line diagrams*: These diagrams include, e.g.
   - high-voltage switchgear
   - medium-voltage switchgear
   - low-voltage switchgear
   - light distributions
   - power distributions
   - weak current distributions
   - DC current distributions
   - Standby power supply.
6. *Location diagram*: The location diagram includes the installation locations of all operational equipment.
7. *Routing diagrams*: Routing diagrams show the type of cable installation within a building or system part, e.g. on cable racks.
8. *Construction details*: All electrical objects, such as transformers, switch rooms, cable ducts, cutouts, and cable and line installation, are shown in their correct locations.
9. *Functional descriptions*: The function of the system, all operating conditions, the number and type of controls in the form of a logic diagram, functional diagram, and structogram or equipotential bonding diagram are described in detail. Circuit diagrams can be created on the basis of these.
10. *Circuit diagrams*: A circuit diagram is the detailed representation of a system or circuit with its details. All overcurrent protection equipment and units, terminals and terminal strips, cable installation types, cross sections, power ratings, voltages, and frequencies must be entered. The list of components includes all units shown in the circuit diagram.
11. *Terminal and cross-connection diagrams (execution company)*: Terminal and cross-connection diagrams are connection diagrams for terminal strips in

cross-connection fields. They represent the electrical connections of the system. Terminal numbers, destination names, and cable types must be specified.

12. *Cable list*: The cable list shows the cable number, type, cross section, voltage, number of cores, and interconnections of grounding electrodes.
13. *Weak current systems*: Weak current systems, such as intercom, telephone, and fire alarm systems, must be shown on separate installation diagrams.
14. *Material lists*: Material lists include all documentation for the quality of the electrical material and for the specification of installations as described in the project.
15. *Installation schedule*: The entire installation schedule, including periods for the specified tasks and time in hours, is summarized.
16. *Documentation*: Following installation of the systems, all drawings, documents, and measurement certificates must be checked for correctness and if necessary corrected.

6

Three-Phase Alternating Current

6.1 Generation of Three-Phase Current

Three-phase current is generated almost exclusively with three-phase generators that are operated in the regenerative operating range. Synchronous generators are rotating energy converters (electrical machines) that convert mechanical energy into electrical energy. For example, energy is generated in power plants by using huge synchronous generators (with, in some cases, over one thousand megawatt output) mechanically driven by a turbine driven by water vapor; the mechanical drive energy is finally dissipated by the generator at high speed and efficiency into electrical energy. Figure 6.1 shows the basic structure of three-phase synchronous machines. The machine consists of a rotating part (called rotor) and a fixed part (called stator).

The rotor has the task of generating a magnetic DC field. This is done either by permanent magnets in the rotor (permanently excited synchronous machine) or by a winding, which is supplied with direct current via slip rings (electrically excited synchronous machine). The stator contains a three-phase winding. In Figure 6.1 the coil axes u, v, and w are the three stator coils. It can be seen that the three coils are arranged at a spatial angle of 120°. When the rotor is set in rotation, the magnetic field generated by it also rotates over the fixed stator coils at the same angular velocity. Since the stator coils are exposed to a variable magnetic field, voltage is induced in accordance with the induction law. Owing to the spatial offset of the coil position by 120°, these voltages are also electrically phase shifted by 120°. This results in three voltages in the three stator windings, which are phase-shifted by 120°, i.e. a three-phase voltage system.

6.2 Advantages of the Three-Phase Current System

Compared to direct current, AC systems have the advantage that the AC voltage can be transformed to almost any voltage level using transformers. This is especially important for economic energy transmission (low currents, i.e. low $R \cdot I^2$ losses but high voltages).

Analysis and Design of Electrical Power Systems: A Practical Guide and Commentary on NEC and IEC 60364, First Edition. Ismail Kasikci.

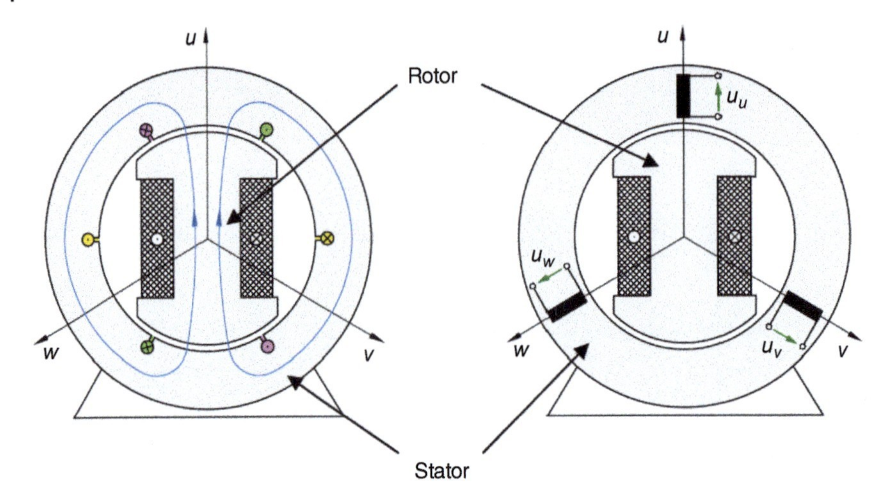

Figure 6.1 Three-phase synchronous machines.

However, compared to the single-phase AC system, the three-phase system still has the following additional benefits:

1. Simple generation using three-phase synchronous generators.
2. Two different voltages are available (star and conductor voltages).
3. Direct operation of three-phase motors is possible.
4. In symmetrical conditions the power is constant in time, i.e. $p(t) = P$.
5. For the same transmitted active power P, the cost of conductor material compared to the single-phase system is only 50%!

6.3 Conductor Systems

Three-phase current is a system of three interlinked alternating currents or alternating voltages. A distinction is made between the three-wire three-phase system consisting of the three phase conductors $L1, L2$, and $L3$ (Figure 6.2). The conductor voltages $\underline{U}_{12}$, $\underline{U}_{23}$, and $\underline{U}_{31}$ can be measured to each other. With a three-phase system, the voltage level is determined by the RMS value of the conductor voltages indicated.

In the four-wire three-phase system, the three outer conductors are supplemented by another conductor, called the neutral conductor N (Figure 6.3).

In addition to the conductor voltages, there are also voltages between the $L1$-N, $L2$-N, and $L3$-N conductors. These voltages are called star voltages $\underline{U}_1$, $\underline{U}_2$, and $\underline{U}_3$. For example, the star voltages with the effective value $\underline{U}_S$ can be displayed as time functions and can be written as follows:

$$\begin{aligned} u_1(t) &= \sqrt{2} \cdot U_S \cdot \sin(\omega \cdot t) \\ u_2(t) &= \sqrt{2} \cdot U_S \cdot \sin\left(\omega \cdot t - \frac{2 \cdot \pi}{3}\right) \\ u_3(t) &= \sqrt{2} \cdot U_S \cdot \sin\left(\omega \cdot t - \frac{4 \cdot \pi}{3}\right) \end{aligned} \tag{6.1}$$

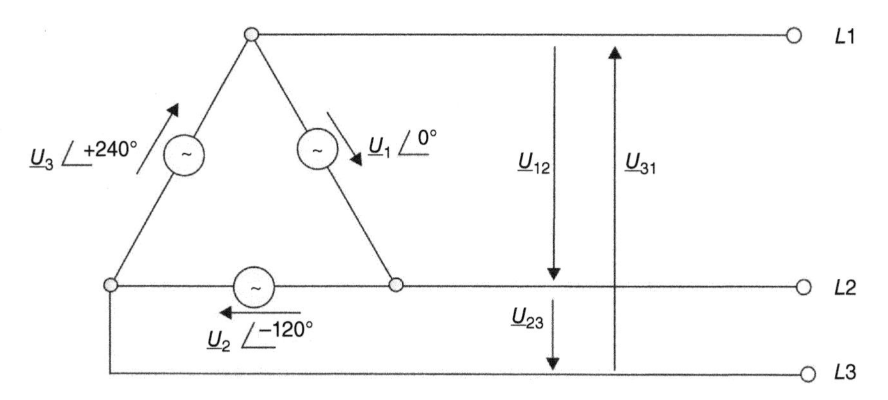

Figure 6.2 System of three conductors.

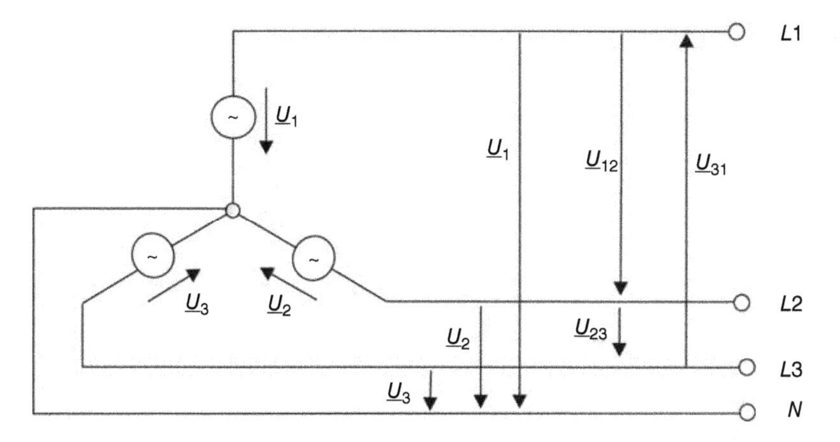

Figure 6.3 Four-wire system.

The corresponding time functions of the star voltages according to Eq. (6.1) are shown in Figure 6.4.

From these equations, the symmetry of three-phase system becomes apparent.

1. The effective values of the three voltages are equal to $U_1 = U_2 = U_3 = U_S$ or $U_{12} = U_{23} = U_{31} = U_L$.
2. All voltages have the same frequency f or angular frequency ω.
3. The phase shift between the respective voltages is 120° and $\frac{2\pi}{3}$ respectively.

Technically this means that the three-phase system is symmetrical

In complex pointer representation the star voltages can be calculated as follows:

$$\underline{U}_1 = U = U_S \cdot e^{j \cdot 0°}$$

$$\underline{U}_2 = \underline{U} \cdot e^{-j \cdot 120°} = U_S \cdot e^{-j \cdot 120°} \tag{6.2}$$

$$\underline{U}_3 = \underline{U} \cdot e^{-j240°} = U_S \cdot e^{-j240°}$$

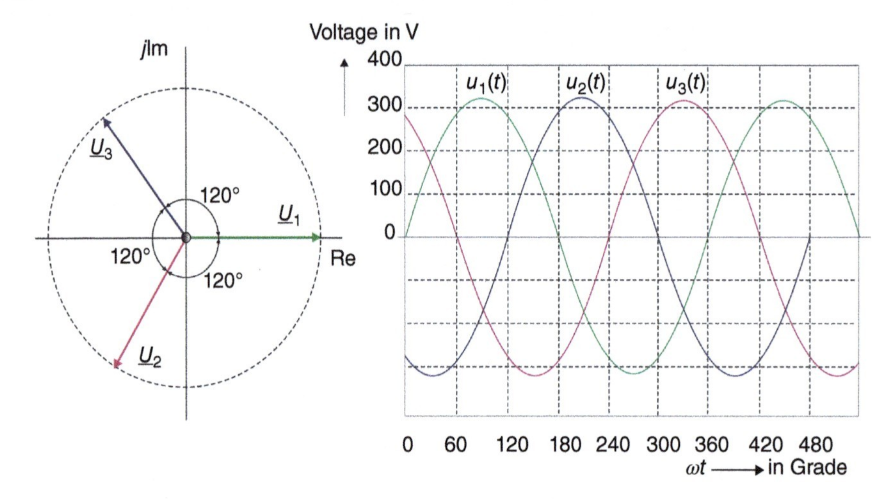

Figure 6.4 Pointer and line diagrams of star stresses.

From the figure and Eq. (6.1) the results for the conductor voltages with the effective value U_L are

$$\underline{U}_{12} = \underline{U}_1 - \underline{U}_2 = U_S - U_S e^{-j\cdot 120^\circ} = \sqrt{3} \cdot U_S \cdot e^{j\cdot 30^\circ} = U_L e^{j\cdot 30^\circ} \tag{6.3}$$

$$\underline{U}_{23} = \underline{U}_2 - \underline{U}_3 = U_S \cdot e^{-j120^\circ} - U_S \cdot e^{-j240^\circ} = \sqrt{3} \cdot U_S \cdot e^{-j90^\circ} = U_L e^{-j\cdot 90^\circ} \tag{6.4}$$

$$\underline{U}_{31} = \underline{U}_3 - \underline{U}_1 = U_S \cdot e^{-j\cdot 240^\circ} - U_S \cdot e^{-j0^\circ} = \sqrt{3} \cdot U_S \cdot e^{-j210^\circ} = U_L \cdot e^{-j210^\circ} \tag{6.5}$$

In the usual pointer display, the star and conductor voltages of the three-phase system are as follows (Figure 6.5):

From Eqs. (6.3)–(6.5) the following relationship between amounts (effective values) of the conductor voltages U_L and the amounts (effective values) of the Star voltages U_S results:

$$U_L = \sqrt{3} \cdot U_S \tag{6.6}$$

The conductor voltages precede the star voltages of the same name by 30° each.

RMS values of conductor voltages U_L are $\sqrt{3}$-times RMS values of the Star voltages U_S.

There are further representations of the star and conductor voltages in the pointer diagram possible (Figure 6.6).

Both from the temporal progressions and from the preceding pointer diagrams it can be seen that

$$u_1(t) + u_2(t) + u_3(t) = 0 \quad \text{respectively} \quad u_{12}(t) + u_{23}(t) + u_{31}(t) = 0 \tag{6.7}$$

$$\underline{U}_1 + \underline{U}_2 + \underline{U}_3 = 0 \quad \text{respectively} \quad \underline{U}_{12} + \underline{U}_{23} + \underline{U}_{31} = 0 \tag{6.8}$$

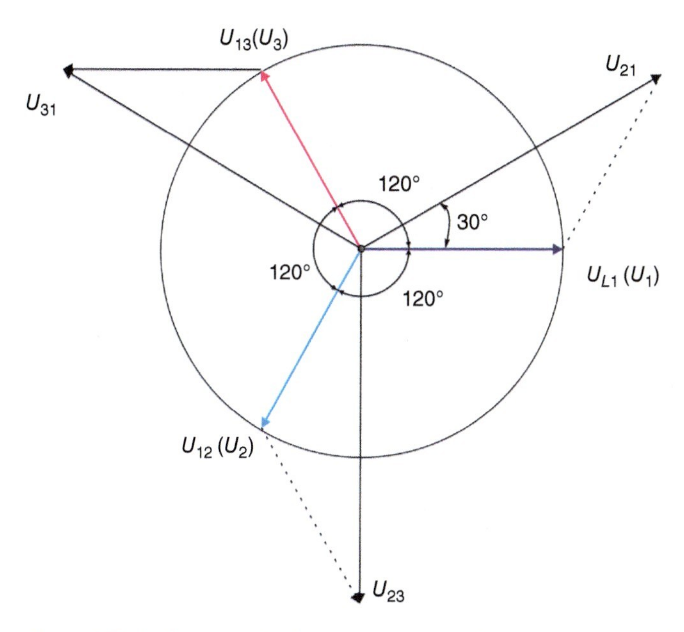

Figure 6.5 Voltages of the three-phase system in pointer representation.

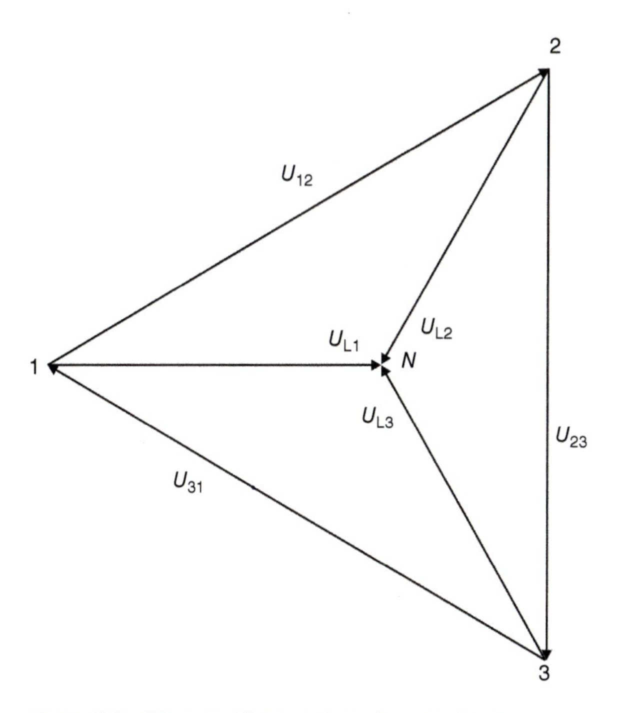

Figure 6.6 Diagram of star and conductor voltages.

Connection of a generator and consumer to the three-phase system.

Three-phase generators (e.g. synchronous generators) and three-phase consumers (e.g. furnace, three-phase motor) consist of three strings. Each of the three strands has a symmetrical three-phase consumer of the same impedance Z_S, i.e. it applies

$$\varphi_1 = \varphi_2 = \varphi_3 = \varphi \quad \text{and} \quad Z_{S1} = Z_{S2} = Z_{S3}$$

Each string is a two-pole, so that in the general case the three-phase generator or three-phase consumer has a total of five connections.

In many cases, three-phase consumers have only three connections and in some cases four connections. Both the three-phase generators and the three-phase consumers can be connected to the three-wire or four-wire system in different ways. The two most important circuit types are presented below.

6.4 Star Connection

The star connection of a consumer is shown in Figure 6.7. The star point here is connected at the neutral conductor N.

The following designations were introduced for the currents:

Currents flowing in the outer conductors (conductor currents I_L): I_1, I_2, I_3
Currents flowing in the consumer strings (string currents I_S): I_{S1}, I_{S2}, I_{S3}

From Figure 6.7 it follows for the star connection with neutral that the string currents are equal to the phase currents:

$$\underline{I}_{S1} = \underline{I}_1, \quad \underline{I}_{S2} = \underline{I}_2, \quad \underline{I}_{S3} = \underline{I}_3 \tag{6.9}$$

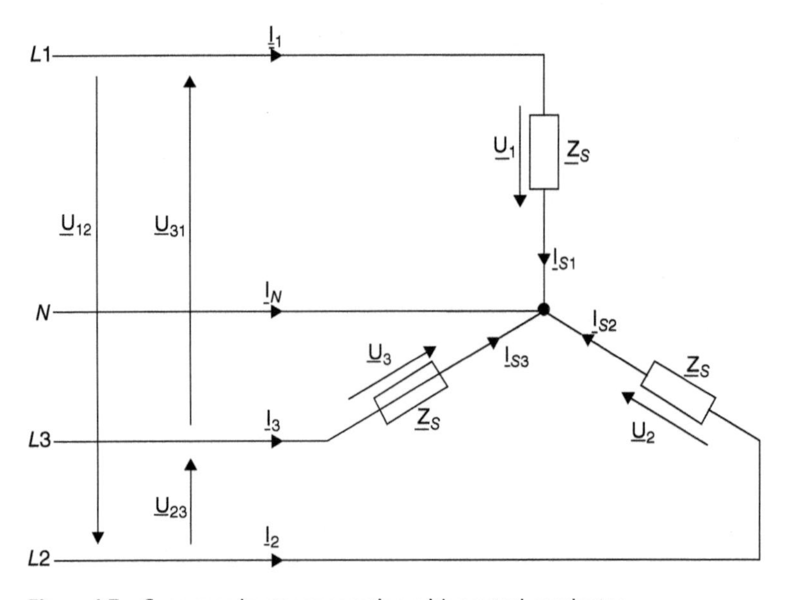

Figure 6.7 Consumer in star connection with neutral conductor.

$$\underline{I}_{S1} = \frac{\underline{U}_1}{\underline{Z}_{S1}}, \quad \underline{I}_{S2} = \frac{\underline{U}_2}{\underline{Z}_{S2}}, \quad \underline{I}_{S3} = \frac{\underline{U}_3}{\underline{Z}_{S3}} \tag{6.10}$$

$$U_1 = U_2 = U_3 = U_S \quad \text{with Eq. (6.6)} \quad U_S = \frac{U_L}{\sqrt{3}} \tag{6.11}$$

The current in the neutral is calculated with the node rule:

$$\underline{I}_0 = \underline{I}_1 + \underline{I}_2 + \underline{I}_3 \quad I_0 = I_N \tag{6.12}$$

With symmetrical conditions this results accordingly:

$$\underline{I}_0 = 0 \tag{6.13}$$

Since the current in the neutral conductor is equal to zero, the neutral conductor can be omitted in the case of symmetrical loads.

For symmetrical loads, the following applies in summary to the RMS values:

$$U_1 = U_2 = U_3 = U_S \quad \text{with Eq. (6.6)} \quad U_S = \frac{U_L}{\sqrt{3}} \tag{6.14}$$

$$I_{S1} = I_{S2} = I_{S3} = I_S = \frac{U_S}{Z_S} \tag{6.15}$$

$$I_0 = 0 \tag{6.16}$$

6.5 Triangle Circuit

The delta connection of a three-phase consumer is created when the three load strings with string impedances $\underline{Z}_{S1}$, $\underline{Z}_{S2}$, and $\underline{Z}_{S3}$ are connected as in Figure 6.8 to the three-phase system. One immediately recognizes that with this circuit variant no neutral conductor is required or is present; three connecting conductors are therefore sufficient!

From Figure 6.8 it is immediately apparent that the string voltages equal the respective conductor voltages:

$$\underline{U}_{S1} = \underline{U}_{12}, \quad \underline{U}_{S2} = \underline{U}_{23}, \quad \underline{U}_{S3} = \underline{U}_{31} \tag{6.17}$$

The following applies to the currents in the three strings of the consumer:

$$\underline{I}_{12} = \frac{\underline{U}_{12}}{\underline{Z}_{S1}} \tag{6.18}$$

It follows from the circuit that the following applies to the phase currents:

$$\underline{I}_{12} = \frac{\underline{U}_{12}}{\underline{Z}_{S1}} \tag{6.19}$$

$$\underline{I}_1 = \underline{I}_{12} - \underline{I}_{31} \quad \underline{I}_2 = \underline{I}_{23} - \underline{I}_{12} \quad \underline{I}_3 = \underline{I}_{31} - \underline{I}_{23} \tag{6.20}$$

For symmetrical loads, the following applies in summary to the RMS values:

$$U_S = U_L \tag{6.21}$$

$$I_{12} = I_{23} = I_{31} = I_S \text{ respectively } I_1 = I_2 = I_3 = I_L \text{ with } I = \sqrt{3} \cdot I_S \tag{6.22}$$

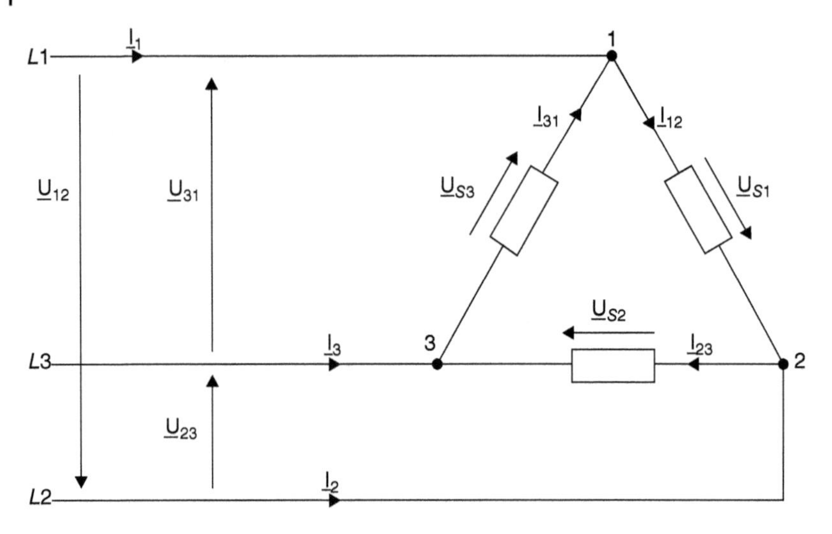

Figure 6.8 Consumer in star connection with neutral conductor.

6.6 Three-Phase Power

In the general case, the total power (apparent power) consists of the three string services together:

$$S = U_1 \cdot I_1^* + U_2 \cdot I_2^* + U_3 \cdot I_3^* = P + j \cdot Q \tag{6.23}$$

In the case of symmetric conditions the following applies:

Symmetry

$$I_1 = I_2 = I_3 = I_L$$

$$U_1 = U_2 = U_3 = U_S \tag{6.24}$$

$$\cos(\varphi_1) = \cos(\varphi_2) = \cos(\varphi_3) = \cos(\varphi)$$

The following therefore applies to the amount of apparent power S:

$$S = 3 \cdot U \cdot I_S \tag{6.25}$$

The above results are independent of the circuit with the conductor sizes:

$$S = \sqrt{3} \cdot U \cdot I_L \tag{6.26}$$

In general, the following relationship applies to apparent power, active power, and reactive power:

$$\underline{S} = P + j \cdot Q \tag{6.27}$$

The active power is already calculated as known from single-phase alternating current:

$$P = S \cdot \cos(\varphi) = \sqrt{3} \cdot U_L \cdot I_L \cdot \cos(\varphi) \tag{6.28}$$

The reactive power is calculated analogously:

$$Q = S \cdot \sin(\varphi) = \sqrt{3} \cdot U_L \cdot I_L \cdot \sin(\varphi) \tag{6.29}$$

6.7 Example: Delta Connection

A consumer is connected to a three-phase network (400 V conductor voltage), which consists of three resistors connected in a triangle (Figure 6.9). The total power drawn from the grid is $P = 3.6$ kW.

1. What is the current I_R that flows in each resistor?
2. How large is each of the phase currents I_L?
3. What is the value of each resistor R?

Solution:
The mains voltage U_N (also called chained voltage or conductor voltage) here is preset. It is $U_L = 400$ V (so the star or center point voltages $U_Y = 230$ V).

A balanced, ohmic load in delta connection is connected to the mains, whose total power consumption is $P = 3.6$ kW. Symmetrical consumer means that the resistance (by amount and phase) is the same in each of the three consumer cords. This also means that all string currents and all phase currents are equal in amount and are shifted by 120° against each other.

Symmetrical net: $U_{12} = U_{23} = U_{31} (400\ \text{V})$
Symmetrical consumer: $I_{12} = I_{23} = I_{31}$
$I_1 = I_2 = I_3$

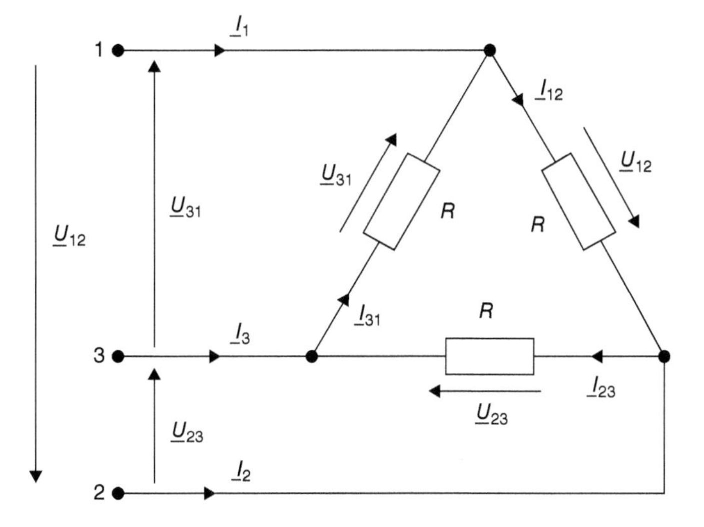

Figure 6.9 Three-phase network in delta connection.

The following generally applies to the power in a line (*L*-*N*):

$$P_Y = U_Y \cdot I_Y \cdot \cos(\varphi)$$

For an ohmic load, the following applies in particular:

$$P_Y = U_Y \cdot I_Y \quad \text{(active power only)}$$

Because of the (amount) equal voltages and currents in each string, the service in each string is the same, i.e. 1/3 of the total service.

$$P_Y = \frac{P_{\text{Total}}}{3} = \frac{3600\ \text{W}}{3} = 1200\ \text{W}$$

(a) Calculation of string currents
From $P_Y = U_Y \cdot I_Y$ results for the string currents:

$$I_Y = \frac{P_Y}{U_Y} = \frac{1200\ \text{W}}{=} 5.21\ \text{A}$$

$$I_{12} = I_{23} = I_{31} = 5.21\ \text{A}$$

(b) Calculation of phase currents
For the calculation of the phase currents the node law could be applied in principle, for example,

$$I_1 - I_{12} + I_{31} = 0, \quad \text{results:} \quad I_1 = I_{12} - I_{31}$$

For this, however, the phase position of the currents must be known.
If there is a symmetrical consumer and you are only interested in the amounts of the currents, you can use a simpler way.
The following applies to the total power in a symmetrical consumer:

$$P = \sqrt{3} \cdot U_L \cdot I_L \cdot \cos(\varphi)$$

For an ohmic load, $\cos\varphi = 0°$, i.e. $\cos\varphi = 1$, so in this case,

$$P = \sqrt{3} \cdot U_L \cdot I_L$$

The conductor current is calculated from this:

$$I_L = \frac{P_{\text{Total}}}{\sqrt{3} \cdot U_L} = \frac{3600\ \text{W}}{\sqrt{3} \cdot 400\ \text{V}}$$

For symmetry reasons $I_1 = I_2 = I_3 = 5.2\ \text{A}$
Second possibility:
For a symmetrical consumer:

$$I_L = \sqrt{3} \cdot I_Y = \sqrt{3} \cdot 3\ \text{A} = 5.2\ \text{A}$$

(c) Calculation of resistances:
- According to Ohm's law:

$$R = \frac{U_Y}{I_Y} = \frac{400\ \text{V}}{3\ \text{A}} = 133.3\ \Omega$$

- About the performance:

$$P_Y = \frac{U_Y^2}{R} \Longrightarrow R = \frac{U_Y^2}{P_Y} = \frac{(400\ \text{V})^2}{1200\ \text{W}} = 133.3\ \Omega$$

or

$$P_Y = I_Y^2 \cdot R \Longrightarrow R = \frac{P_Y}{I_Y^2} = \frac{1200\ \text{W}}{(3\ \text{A})^2} = 133.3\ \Omega$$

6.8 Example: Star Connection

The following resistors are connected to the adjacent three-phase mains (230/400 V) (Figure 6.10):

$$R_1 = 110\ \Omega, R_2 = 100\ \Omega, R_3 = 60\ \Omega, X_L = 80\ \Omega, X_C = -90\ \Omega$$

Determine the four currents

$$I_1, I_2, I_3, \text{ and } I_N.$$

Solution:
It is an unbalanced load in $Y - \Delta$ circuit. The load star point is connected to the generator star point (mains star point). This means that the individual load strings at the star voltages of the grid and it applies:

$$U_{1N} = U_{2N} = U_{3N} = 230\ \text{V}$$

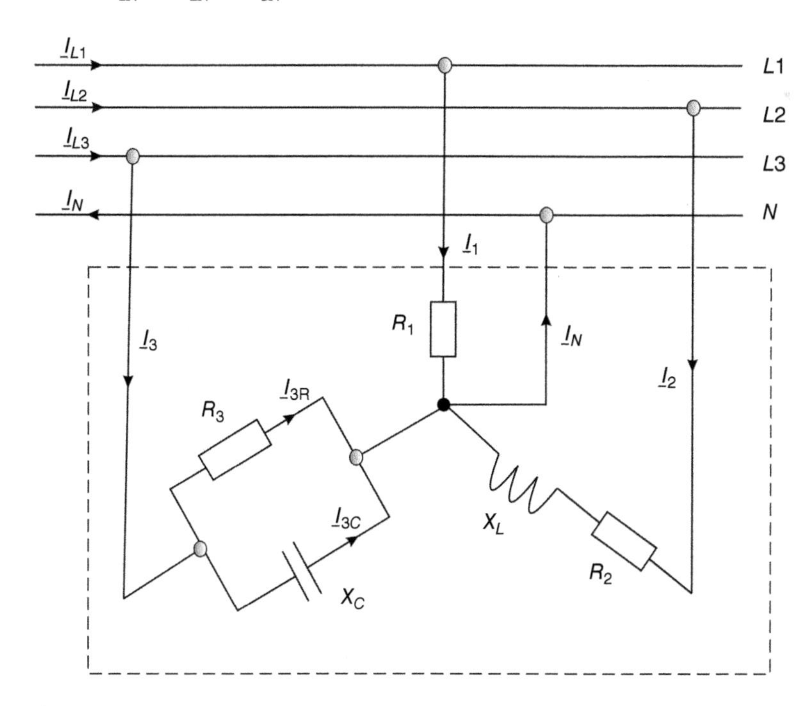

Figure 6.10 Three-phase network in star connection.

The phase positions of the voltages must be known for the calculation. If U_{1N} is selected as reference value and U_{1N} is placed in the real axis, the pointer image of the star voltages is shown opposite:

This applies to the star voltages:

$$\underline{U}_{1N} = 230\ \text{V} \cdot e^{j0^\circ}$$

$$\underline{U}_{2N} = 230\ \text{V} \cdot e^{j240^\circ} = 230\ \text{V} \cdot e^{-120^\circ}$$

$$\underline{U}_{3N} = 230\ \text{V} \cdot e^{j120^\circ} = 230\ \text{V} \cdot e^{-240^\circ}$$

The same current passes through X_L and R_2, and they are in series. X_C and R_3 are due to the same voltage and are connected in parallel.

The calculation of the currents is done with the ohmic law in complex form:

$$\underline{U} = \underline{I} \cdot \underline{Z}$$

Determination of $\underline{I}_1$:

$$\underline{I}_1 = \frac{\underline{U}_{1N}}{\underline{Z}_R} = \frac{230\ \text{V} \cdot e^{j0^\circ}}{110\ \Omega} = 2.09\ \text{A} \cdot e^{j0^\circ}$$

Determination of $\underline{I}_2$:

$$\underline{I}_2 = \frac{\underline{U}_{2N}}{\underline{Z}_2}$$

Calculation of $\underline{Z}_2$:

$$\underline{Z}_2 = R_2 + jX_L = (100 + j80)\Omega = \sqrt{(100\ \Omega)^2 + (80\ \Omega)^2} = 128.06\ \Omega$$

This makes $\underline{Z}_2$ as

$$|\underline{Z}_2| = Z_2 = \sqrt{R_2^2 + X_L^2} = \sqrt{(100\Omega)^2 + (80\Omega)^2} \cdot e^{j\arctan\frac{80}{100}} = 128.06\Omega \cdot e^{j38.66^\circ}$$

The current $\underline{I}_2$ is calculated as follows:

$$\underline{I}_2 = \frac{230\ \text{V} \cdot e^{j240^\circ}}{128.06\ \Omega \cdot e^{38.66^\circ}} = \frac{230\ \text{V}}{128.06\ \Omega} \cdot e^{j(240^\circ - 38.66^\circ)}$$

$$\underline{I}_2 = 1.80\ \text{A} \cdot e^{j201.34^\circ} = 1.80\ \text{A} \cdot e^{-j158.66^\circ} = (-1.68 - j0.66)\text{A}$$

Determination of $\underline{I}_{32}$:

$$\underline{I}_3 = \frac{\underline{U}_{3N}}{\underline{Z}_3}$$

Calculation of $\underline{Z}_3$:

$$\underline{Z}_3 = \frac{1}{\frac{1}{R_3} + \frac{1}{jX_C}} = \frac{1}{\frac{jX_C + R_3}{jR_3X_C}} = \frac{jR_3X_C}{R_3 + jX_C}$$

For the calculation of amount and angle $\underline{Z}_3$ must be split into real and imaginary parts. To do this, the denominator must be made real.

$$\underline{Z}_3 = \frac{jR_3X_C}{R_3 + jX_C} = \frac{jR_3X_C \cdot (R_3 - jX_C)}{R_3 + jX_C \cdot (R_3 - jX_C)} = \frac{R_3X_C \cdot (R_3 - jX_C)}{R_3^2 + X_C^2}$$

$$\underline{Z}_3 = \frac{R_3X_C}{R_3^2 + jX_C^2} = \frac{R_3X_C}{R_3^2 + X_C^2} + j\frac{R_3^2X_C}{R_3^2X_C} = (\text{real part}) + j \cdot (\text{imaginary part})$$

With the given numerical values the result is

$$\underline{Z}_3 = (41.5 - j27.7)\Omega = 50\ \Omega \cdot e^{j33.72°}$$

The current is then calculated to be

$$\underline{I}_3 = \frac{230\ \text{V} \cdot e^{j120°}}{50\ \Omega \cdot e^{-j33.72°}} = \frac{230\ \text{V}}{50\ \Omega} \cdot e^{j120° + 33.72°} = 4.6\ \text{A} \cdot e^{j153,72°} = (-4.12 + j2.04)\ \text{A}$$

Another way to calculate $\underline{I}_3$:

Definition of partial streams $\underline{I}_{3R}$ and $\underline{I}_{3C}$ and summary under the Node Act:

$$\underline{I}_3 = \underline{I}_{3R} + \underline{I}_{3C}$$

$$\underline{I}_{3R} = \frac{\underline{U}_{3N}}{R_3} = \frac{230\ \text{V} \cdot e^{j120°}}{60\ \Omega \cdot e^{j0°}} = 3.83\ \text{A} \cdot e^{j120°} = (-1.92 + j3.32)\ \text{A}$$

$$\underline{I}_{3C} = \frac{\underline{U}_{3N}}{-jX_C} = \frac{230\ \text{V} \cdot e^{j120°}}{90\ \Omega \cdot e^{-j90°}} = 2.56\ \text{A} \cdot e^{j210°} = (-2.22 - j1.11)\ \text{A}$$

$$\underline{I}_3 = \underline{I}_{3R} + \underline{I}_{3C} = (-1.92 - 2.22 + j3.32 - j1.28)\ \text{A} = (4.14 + j2.04)\ \text{A} \cdot e^{j153.77°}$$

Determination from $\underline{I}_N$:

$$\underline{I}_N = \underline{I}_1 + \underline{I}_2 + \underline{I}_3 = (2.09 + j0)\ \text{A} + (-1.68 - j0.66)\ \text{A} + (-4.12 + j2.04)\ \text{A}$$

Combine real and imaginary parts separately.

$$\underline{I}_N = (2.09 - 168 - 4.12)\ \text{A} + j(0 - 0.66 + 2.04)\ \text{A}$$

$$\underline{I}_N = (-3.71 + j1.38)\ \text{A} = 3.96\ \text{A} \cdot e^{-159.6°}$$

Note the position of the pointer in the complex plane when converting into Euler's form!

6.9 Example: Three-Phase Consumer

The following consumers are connected to the adjoining three-phase mains (230/400 V) (Figure 6.11).

Calculate all currents in the outer conductors.

The following applies to phase conductors $L1$ and $L2$:

$$S_{L1a} = \frac{1}{3}3 \cdot S = \frac{1}{3}3 \cdot 30\ \text{kVA} = 10\angle 36.9°\ \text{kVA}$$

400 V

30 kVA
0.8 ind.

15 kVA
0.6 kap.

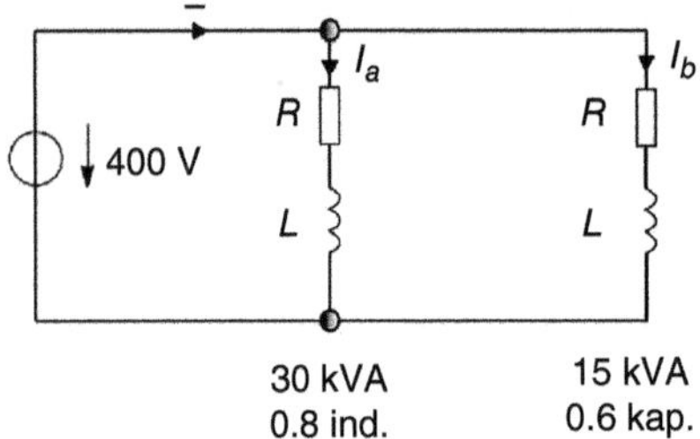

Figure 6.11 Three-phase consumer.

$$S_{L1b} = 50\angle - 53.1^\circ \text{ kVA}$$

$$I_{L1a} = \frac{S_{L1}}{U} = \frac{10\angle 36.9^\circ \text{ kVA}}{400\angle 0^\circ \text{ V}} = 25\angle - 36.9^\circ \text{ A}$$

$$I_{L1b} = \frac{S_{L1}}{U} = \frac{5\angle - 53^\circ \text{ kVA}}{400\angle 0^\circ \text{ V}} = 12.5\angle - 53^\circ \text{ A}$$

For the impedances:

$$Z_{L1a} = \frac{U}{I_{L1}} = \frac{230\angle 0^\circ \text{ V}}{25\angle - 36.9^\circ \text{ A}} = 9.2\angle + 53^\circ \ \Omega = (5.53 + j7.34)\Omega$$

$$Z_{L1b} = \frac{U}{I_{L1}} = \frac{230\angle 0^\circ \text{ V}}{12.5\angle - 53^\circ \text{ A}} = 18.4\angle + 53^\circ \ \Omega = (11 - j14.7)\Omega$$

Total current:

$$I = I_{L1a} + I_{L1b} = 25\angle - 36.9^\circ \text{ A} + 12.5\angle - 53^\circ \text{ A}$$

$$19.99 - j15 + 7.5 + j9.98\text{A} = (27.49 + j24.98)\text{A} = 37.14 \cdot \angle 24.67^\circ$$

If the system is symmetrical:

$$I_{L2} = (37.14 \cdot \angle 24.67^\circ - 120^\circ)A = 37.14 \cdot \angle - 95.33$$

$$I_{L3} = (37.14 \cdot \angle 24.67^\circ - 240^\circ)A = (37.14 \cdot 215.33^\circ)\text{A}$$

6.10 Example: Network Calculation

Given is the following network with the data: $R = 1\ \Omega$, $U = 16$ V, $L = 9$ mH, $C = 2200\ \mu$F, $f = 50$ Hz (Figure 6.12).

Determine all the occurring voltage drops and currents by amount and phase.

With the mesh equation, mesh M1:

$$U_C = U = 16 \text{ V}$$

From Ohm's law, the partial currents I_1 and I_2 are obtained:

$$\underline{I}_2 = \frac{\underline{U}_C}{\underline{Z}_C} = \underline{U}_C \cdot j\omega C$$

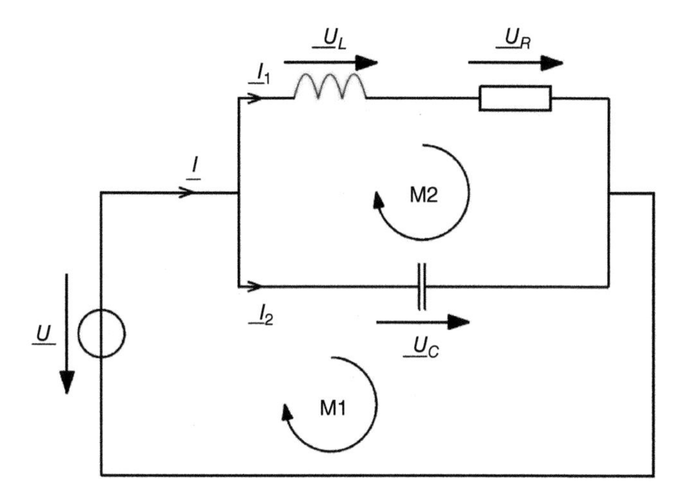

Figure 6.12 Example circuit.

$$\underline{I}_2 = j11.06\text{ A} \cdot e^{j90°}$$

$$\underline{I}_1 = \frac{\underline{U}}{\underline{Z}_R + \underline{Z}_L} = \frac{16\text{ V}}{R + j\omega L} = \frac{16\text{ V}}{1\Omega + j2.83\Omega} = (1.78 - j5.03)\text{ A}$$

The total current results from the node equation:

$$\underline{I} = \underline{I}_1 + \underline{I}_2 = j11.06\text{ A} + 1.78\text{ A} - j5.03\text{ A} = 1.78\text{ A} + j6.03\text{ A} = 6.29\text{ A} \cdot ej73.6°$$

The voltage drop across the resistance results from Ohm's law:

$$\underline{U}_R = \underline{Z}_R \cdot \underline{I}_1 = 1\ \Omega \cdot 5.33\text{ A} \cdot e^{-j70.52°} = 5.33\text{ V} \cdot e^{-j70.52°}$$

The voltage drop across the inductance can be calculated using mesh M2:

$$\underline{U}_L = \underline{U} - \underline{U}_R = 16\text{ V} - 5.33\text{ V}\ e^{-j70.52°} = 16\text{ V} - (1.78\text{ V} - j5.03\text{ V})$$
$$= 15.08V \cdot e^{j19.48°}$$

6.11 Example: Network

Given is the following network with the data: $R_1 = 10\ \Omega$, $R_2 = 5\ \Omega$, $R_3 = 9\ \Omega$, $X_{C1} = -4\ \Omega$, $X_{C2} = -6\ \Omega$, $UX_0 = 100$ V, $f = 50$ Hz (Figure 6.13).

1. Determine the complex resistances and the total resistance of the circuit. Enter the resistance values by amount and phase.
2. What is the apparent, active, and reactive power absorbed by the network?
3. Calculate the voltages by amount and phase.
4. Calculate the currents by amount and phase.
5. Between terminals 1 and 4 of the network an additional component with the reactive conductance is included so that the active factor of the new

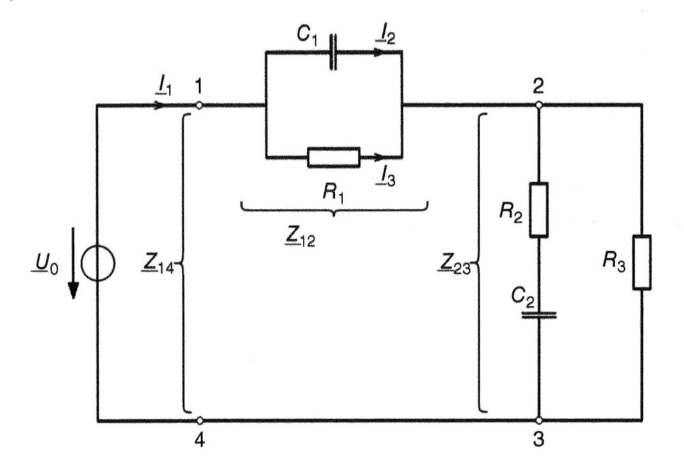

Figure 6.13 Example circuit.

network is 1 (reactive power compensation). What is the required reactive conductance?

6. What is the effective power absorbed by the new network?

Solutions:

1. Summary of load impedances, calculation of Z_{12}

$$\underline{Z}_{12} = \frac{R1 \cdot jX_{C1}}{R1 + jX_{C1}} = \frac{10\ \Omega \cdot (-j4\ \Omega)}{10\ \Omega - j4\ \Omega} = \frac{40\ \Omega^2 \cdot e^{-j90°}}{10.77\ \Omega \cdot e^{-j21.8°}} = 3.71\ \Omega \cdot e^{-j68.2°}$$
$$= 1.38\ \Omega - j3.45\ \Omega$$

2. Calculation of Z_{23}:

$$\underline{Z}_{23} = \frac{R3 \cdot (R2 + jX_{C2})}{R3 + (R2 + jX_{C2})} = \frac{9\ \Omega(5\ \Omega - j6\ \Omega)}{9\ \Omega + 5\ \Omega - j6\ \Omega} = \frac{45\ \Omega - j54\ \Omega}{14\ \Omega - j6\ \Omega}$$
$$= \frac{70.29\ \Omega \cdot e^{-j50.19°}}{15.23\ \Omega \cdot e^{-j23.2°}} = 4.61\ \Omega \cdot e^{-j27.0°} = 4.11\ \Omega - j2.09\ \Omega$$

3. Calculation of Z_{14}

$$\underline{Z}_{14} = \underline{Z}_{12} + \underline{Z}_{23} = 1.38\ \Omega - j3.44\ \Omega + 4.11\ \Omega - j2.09\ \Omega$$
$$= 5.49\ \Omega - j5.54\ \Omega = 7.80\ \Omega \cdot e^{-j45.3°}$$

To calculate the power the current I_1 is determined first:

$$\underline{I}_1 = \frac{\underline{U}_0}{\underline{Z}_{14}} = \frac{100\ \text{V}}{7.80\ \Omega \cdot e^{-j45.3°}} = 12.82\text{A} \cdot e^{j45.3°}$$

Now the services can be calculated:

$$\underline{S} = \underline{U}_0 \cdot \underline{I}_1^* = 100\ \text{V} \cdot 12.82\ \text{A} \cdot e^{-j45.3°} = 1281.62\ \text{VA} \cdot e^{-j45.3°}$$
$$= (902.0 - j910.5)\text{VA}$$

$$P = \text{Re}(\underline{S}) = 902.0\ \Omega$$

$$Q = \text{Im}(\underline{S}) = -910.5\ \text{var}$$

4. Calculation of voltages U_{12} and U_{23}:

$$\underline{U}_{12} = \underline{I}_1 \cdot \underline{Z}_{12} = 12.82\ \text{A} \cdot e^{j45.3^\circ} \cdot 3.71\ \Omega \cdot e^{-j68.2^\circ} = 43.84\ \text{V} - j18.54\ \text{V}$$
$$= 47.60\ \text{V} \cdot e^{-j22.9^\circ}$$

$$\underline{U}_{23} = \underline{I}_1 \cdot \underline{Z}_{23} = 12.82\ \text{A} \cdot e^{j45.3^\circ} \cdot 4.61\Omega \cdot e^{-j27.0^\circ} = 56.16\ \text{V} + j18.54\ \text{V}$$
$$= 59.15\ \text{V} \cdot e^{j18.3^\circ}$$

5. Calculation of currents:

$$\underline{I}_1 = 9.02\ \text{A} + j9.10\ \text{A} = 12.82\ \text{A} \cdot e^{j45.3^\circ}$$

$$\underline{I}_2 = \frac{\underline{U}_{12}}{-jX_{C1}} = \frac{47.5\ \text{V} \cdot e^{-j22.9^\circ}}{4\ \Omega \cdot e^{-j90^\circ}} = 4.64\ \text{A} + j10.96\ \text{A} = 11.90\ \text{A} \cdot e^{j67.1^\circ}$$

$$\underline{I}_3 = \frac{\underline{U}_{12}}{R_1} = \frac{47.5\ \text{V} \cdot e^{-j22.9^\circ}}{10\ \Omega} = 104.38\ \text{A} - j1.85\ \text{A} = 4.76A \cdot e^{-j22.9^\circ}$$

6. Calculation of the reactive conductance: the conductance is the inverse of the impedance and is calculated from current and voltage.

$$\underline{Y} = \frac{I_1}{U_0} = \frac{12.82\ \text{A} \cdot e^{j45.3^\circ}}{100\ \text{V}} = 0.13\frac{\text{A}}{\text{V}} \cdot e^{j45.3^\circ} = 0.13S \cdot e^{j45.3^\circ} = 0.09S + j0.09S$$

The reactive component of the conductance corresponds to the imaginary component:

$$B = Y \cdot \sin\varphi = 0.13S \cdot \sin(45.3^\circ) = 91\ \text{mS}$$

To ensure that the circuit only absorbs active power, the reactive component (imaginary component) must be the impedance by parallel connection of an impedance with an equal-amount reactance but with the opposite sign. The susceptance of the impedance to be switched in parallel is therefore:

$$B_Z = -B = -91\ mS$$

7. Calculation of the new active power:
Since the active resistance is not changed by the reactive power compensation the active power does not change either:

$$P_{\text{new}} = P_{\text{old}} = 902\ \text{W}$$

6.12 Example: Star Connection

Three impedances $\underline{Z}_1 = 3\ \Omega + j4\ \Omega = 5\ \Omega \cdot e^{j53.1^\circ}$, $\underline{Z}_2 = 4\ \Omega - j3\ \Omega = 5\ \Omega \cdot e^{-j36.9^\circ}$, $\underline{Z}_3 = 5\ \Omega$ are star-connected. The string voltage is 25 V. Determine the amount of current I_N in the neutral.

Solution:

$$\underline{I}_1 = \frac{U_{1N}}{\underline{Z}_1} = \frac{25\ \text{V} \cdot e^{j0^\circ}}{5\ \Omega \cdot j^{53.1^\circ}} = 5\ \text{A} \cdot e^{-j53.1^\circ}$$

$$\underline{I}_2 = \frac{\underline{U}_{2N}}{\underline{Z}_2} = \frac{25\ \text{V} \cdot e^{-120°}}{5\ \Omega \cdot j^{-36.9°}} = 5A \cdot e^{-j83.1°}$$

$$\underline{I}_3 = \frac{\underline{U}_{3N}}{\underline{Z}_3} = \frac{25\ \text{V} \cdot e^{120°}}{5\ \Omega \cdot j^{0°}} = 5\ \text{A} \cdot e^{120°}$$

Neutral current:

$$\underline{I}_N = \underline{I}_1 + \underline{I}_2 + \underline{I}_3$$

$$\underline{I}_N = 5\ \text{A} \cdot e^{-j53.1°}$$
$$+ 5\ \text{A} \cdot e^{-j83.1°} + 5\ \text{A} \cdot e^{j120°}$$

$$I_N = 5\ \text{A} \cdot (\cos(-53.1°) + \cos(-83.1°)$$
$$+ \cos(120°))j5\ \text{A} \cdot (\sin(-53.1°) + \sin(-83.1°) + \sin(120°))$$

$$\underline{I}_N = 1.103\ \text{A} - j4.632\ \text{A} = 4.762\ \text{A} \cdot e^{-j76.6°}$$

7

Symmetrical Components

The three-phase networks of electrical energy distribution are set in normal operation symmetrical three-phase systems in which the currents and voltages of the three conductors L1, L2, and L3 are equal in magnitude. The symmetrical three-phase system can therefore also be simulated by a single-pole replacement circuit, which makes the calculation very easy. It is sufficient to supply voltages, currents, and power in one phase. This symmetry can be disturbed by errors, e.g. a single-pole earth short-circuit; by switching actions, e.g. a single-pole short sub-calculation; or by asymmetrical load. A single-pole replacement circuit is therefore no longer possible. For this the method of "Symmetrical components" is used. The separation of a three-phase current system into symmetrical components (with, without, and without system) is done with the help of the complex calculation [28, 29].

7.1 Symmetrical Network Operation

The designations R, S, and T are used for the design of symmetrical components. According to IEC 60027 the active conductors are designated L1, L2, and L3 (Figure 7.1). Single-phase equivalent circuit diagrams are sufficient for calculating symmetrical three-phase networks. Therefore, the linear AC network theory can be applied.

The concatenation factor is taken into account when calculating sizes 3 and $\sqrt{3}$. Electrical networks consist of three- and four-wire networks. In low-voltage networks, the fourth conductor is designed as a neutral conductor (N) or neutral conductor + protective earth conductor (PEN). In high voltage networks, on the other hand, the earth conductor or soil can be used as the fourth conductor.

In symmetrical network operation the three-phase network is symmetrically loaded, i.e. it applies to impedances $\underline{Z}_R = \underline{Z}_S = \underline{Z}_T = \underline{Z}$.

The following relationships also apply:

$$\underline{I}_R + \underline{I}_S + \underline{I}_T = 0; \quad |\underline{I}_R| + |\underline{I}_S| + |\underline{I}_T| \tag{7.1}$$

$$\underline{U}_{RS} + \underline{U}_{ST} + \underline{U}_{TR} = 0; \quad \underline{U}_{RN} + \underline{U}_{SN} + \underline{U}_{TN} = 0 \tag{7.2}$$

Analysis and Design of Electrical Power Systems: A Practical Guide and Commentary on NEC and IEC 60364, First Edition. Ismail Kasikci.

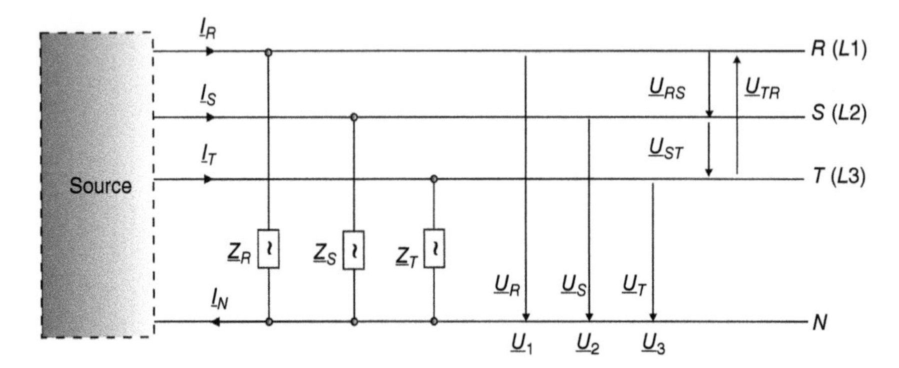

Figure 7.1 View of a three-phase network.

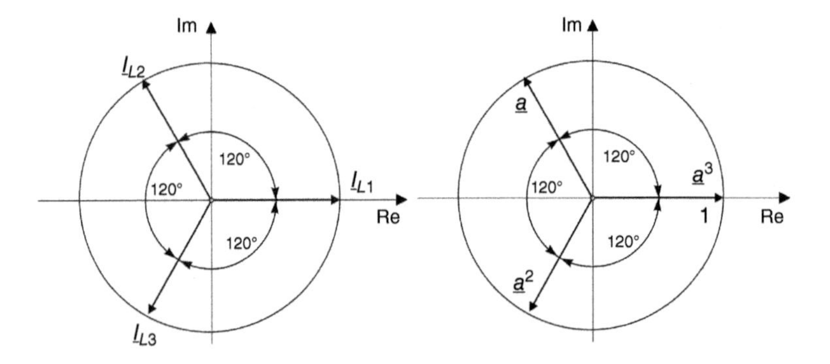

Figure 7.2 Three-phase symmetrical systems.

and

$$|\underline{U}_{RS}| = |\underline{U}_{ST}| = |\underline{U}_{TR}| = \sqrt{3}|\underline{U}_{RN}| = \sqrt{3}|\underline{U}_{SN}| = \sqrt{3}|\underline{U}_{TN}| \tag{7.3}$$

In a symmetrical three-phase network, the sum of voltages and currents is zero at all times. Figure 7.2 shows a symmetrical system consisting of three active conductors and unit hands moved 120° against each other.

The following calculation rules must be observed for the rotary generators a:

$$\underline{a} = e^{j2\pi/3} = e^{j120°} = -\frac{1}{2} + j\frac{\sqrt{3}}{2}$$

$$\underline{a}^2 = \underline{a}^* = \underline{a}^{-1} = e^{j4\pi/3} = e^{-j2\pi/3} = e^{j240°} = e^{-j120°} = -\frac{1}{2} - j\frac{\sqrt{3}}{2}$$

$$\underline{a}^3 = 1, \quad \underline{a}^4 = a$$

$$\underline{a} - \underline{a}^2 = j\sqrt{3}, \quad \underline{a}^2 - \underline{a} = -j\sqrt{3}$$

$$1 + \underline{a} + \underline{a}^2 = 0$$

The introduction of the rotary generator results in the three asymmetrical phase currents described in Chapter 8.

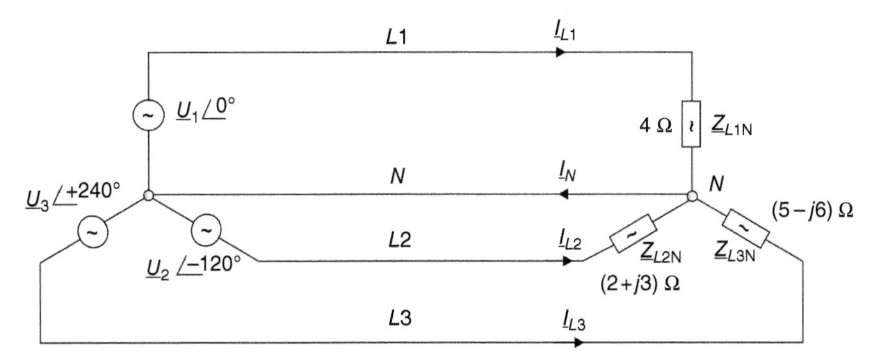

Figure 7.3 Unbalanced network operation.

7.2 Unsymmetrical Network Operation

If the three-phase network is loaded asymmetrically, i.e. $\underline{Z}_R \neq \underline{Z}_S \neq \underline{Z}_T$, then it is an unbalanced network operation (Figure 7.3). The following are the possible causes:

- Uneven conductor–earth capacities (overhead lines)
- Uneven loads (at the LV level)
- Message in the network
  - earth fault, earth short
  - double earth fault (different affected conductors and fault locations)
  - double-pole short-circuit with or without earth contact
  - unipolar conductor interruptions (defective switch poles)

7.3 Description of Symmetrical Components

The method of symmetrical components is made possible through the additive superposition (superposition or superposition principle) of any three symmetrical systems on symmetrical and unsymmetrical pointers at one network location. A system consisting of symmetrical or unsymmetrical hands is divided into the three symmetrical systems, co system, negative sequence, and zero system, which are generally phase-shifted.

Figure 7.4 shows the decomposition of a pointer system. Stationary operation is always assumed, as no transient processes can be described with the complex pointers of the AC current gauge, only the steady state. Three symmetrical systems, which can be described with single-phase equivalent circuit diagrams, are the results of decomposition into positive, negative, and zero systems. This requires the positive, negative, and zero impedances of the individual equipment. The zero system is only required if the sum of the pointers, e.g. I_R, I_S, and I_T is not zero, i.e. especially for connections with neutral and faults with earth contact.

In three-phase electrical systems, unbalanced faults, e.g. I''_{k1}, are common as symmetrical faults I''_{k3}. These errors are calculated using the symmetrical components

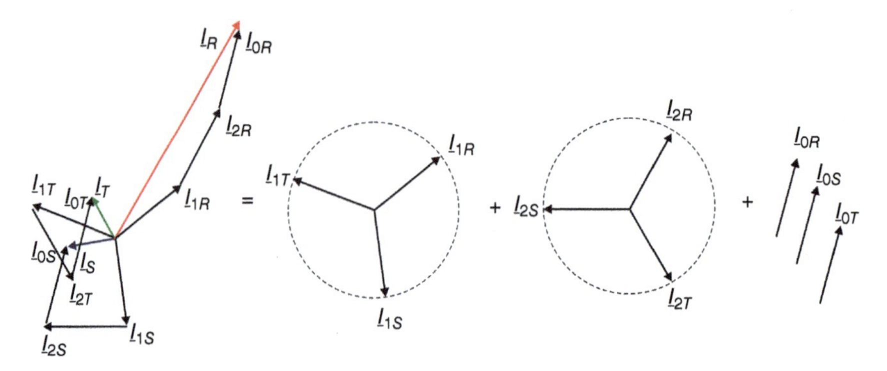

Figure 7.4 View of an asymmetrical system of positive, negative, and zero systems.

method. The unbalanced three-phase current system is broken down into components, resulting in positive, negative, and zero systems. The results are then superimposed on the overall solution.

After the transformation, the operating behavior of the three-phase current system is described by three symmetrical operating states, each of which can be described by a single-phase equivalent circuit diagram. This allows a considerable simplification of the calculation effort.

If no neutral conductor is present or if there is no ground contact, the positive and negative sequence is sufficient to describe the operating behavior, i.e. no zero system is present.

For asymmetric devices, the impedance matrix in the 120 range $[\underline{Z}_{120}]$ also contains elements outside the main diagonals. Therefore, such equipment cannot be simplified by using symmetrical components, since the impedances of the positive, negative, and zero systems are no longer sufficient to describe the operating behavior. This problem also occurs with equipment with rotating parts (e.g. synchronous generator). Suitable simplifications must then be made for mathematical treatment.

At symmetrically constructed devices the positive impedance $\underline{Z}_1$ is equal to the negative impedance $\underline{Z}_2$ as the phase sequence has no influence on current magnitude.

The decomposition into positive, negative, and zero systems shown can be described by the following equations (the example of currents also applies equivalently to voltages).

$$\underline{I}_R = \underline{I}_{1R} + \underline{I}_{2R} + \underline{I}_{0R} \tag{7.4}$$

$$\underline{I}_S = \underline{I}_{1S} + \underline{I}_{2S} + \underline{I}_{0S} \tag{7.5}$$

$$\underline{I}_T = \underline{I}_{1T} + \underline{I}_{2T} + \underline{I}_{0T} \tag{7.6}$$

The system of equations can be further simplified by taking advantage of the fact that the three hands of the co- and counter system are phase-shifted by 120° and that the three hands of the zero system are identical. These correlations can be taken into account if one pointer in each of the three systems is taken as the reference value.

Usually the component pointers $\underline{I}_{1R}, \underline{I}_{2R}, \underline{I}_{0R}$ of conductor R are chosen as reference values. The phase position can be expressed as follows:

$$1 = e^{j0°} \quad \underline{a} = e^{j2\pi/3} = e^{j120°} \quad \underline{a}^2 = e^{j4\pi/3} = e^{j240°} = e^{-j120°} \tag{7.7}$$

This results in the following relationships:

$$\underline{I}_{1R} = 1\underline{I}_{1R} \qquad \underline{I}_{2R} = 1\underline{I}_{2R} \qquad \underline{I}_{0R} = 1\underline{I}_{0R} \tag{7.8}$$

$$\underline{I}_{1S} = \underline{a}^2\underline{I}_{1R} \qquad \underline{I}_{2S} = \underline{a}\underline{I}_{2R} \qquad \underline{I}_{0S} = 1\underline{I}_{0R} \tag{7.9}$$

$$\underline{I}_{1T} = \underline{a}\underline{I}_{1R} \qquad \underline{I}_{2T} = \underline{a}^2\underline{I}_{2R} \qquad \underline{I}_{0S} = 1\underline{I}_{0R} \tag{7.10}$$

The above equation system can be written as

$$\underline{I}_R = \underline{I}_{1R} + \underline{I}_{2R} + \underline{I}_{0R} \tag{7.11}$$

$$\underline{I}_S = \underline{a}^2\underline{I}_{1R} + \underline{a}\underline{I}_{2R} + \underline{I}_{0R} \tag{7.12}$$

$$\underline{I}_T = \underline{a}\underline{I}_{1R} + \underline{a}^2\underline{I}_{2R} + \underline{I}_{0R} \tag{7.13}$$

On the basis of a choice of component pointers $\underline{I}_{1R}, \underline{I}_{2R}, \underline{I}_{0R}$ the head R as reference values can be written simply:

$$\underline{I}_1 = \underline{I}_{1R} \quad \underline{I}_2 = \underline{I}_{2R} \quad \underline{I}_0 = \underline{I}_{0R} \tag{7.14}$$

In matrix notation, this results in the following transformation equation:

$$[\underline{I}_{RST}] = [\underline{T}] \cdot [\underline{I}_{120}] \tag{7.15}$$

$$\begin{bmatrix} \underline{I}_R \\ \underline{I}_S \\ \underline{I}_T \end{bmatrix} = \begin{bmatrix} \underline{1} & 1 & 1 \\ \underline{a}^2 & \underline{a} & 1 \\ \underline{a} & \underline{a}^2 & \underline{1} \end{bmatrix} \cdot \begin{bmatrix} \underline{I}_1 \\ \underline{I}_2 \\ \underline{I}_0 \end{bmatrix} \tag{7.16}$$

Here the matrix $[\underline{T}]$ transforms the component currents or voltages $[\underline{T}_{120}]$ or $[\underline{I}_{120}]$ into the actual phase currents respectively. This transformation is therefore the back transformation from the image area to the original area (desymmetry). The transformation from the original area to the image area (symmetry) is done with the transformation matrix $[\underline{S}] = [\underline{T}]^{-1}$.

The asymmetrical phase currents are calculated from the equations.

$$\underline{I}_{0R} = \frac{1}{3} \cdot (\underline{I}_R + \underline{I}_S + \underline{I}_T) \tag{7.17}$$

$$\underline{I}_{1R} = \frac{1}{3} \cdot (\underline{I}_R + \underline{a} \cdot \underline{I}_S + \underline{a}^2 \cdot \underline{I}_T)$$

$$\underline{I}_{2R} = \frac{1}{3} \cdot (\underline{I}_R + \underline{a}^2 \cdot \underline{I}_S + \underline{a} \cdot \underline{I}_T)$$

In matrix notation, then

$$\begin{bmatrix} \underline{I}_{1R} \\ \underline{I}_{2R} \\ \underline{I}_{0R} \end{bmatrix} = \frac{1}{3} \cdot \begin{bmatrix} \underline{1} & \underline{a} & \underline{a}^2 \\ \underline{1} & \underline{a}^2 & \underline{a} \\ 1 & 1 & 1 \end{bmatrix} \cdot \begin{bmatrix} \underline{I}_R \\ \underline{I}_S \\ \underline{I}_T \end{bmatrix} \tag{7.18}$$

7.4 Examples of Unbalanced Short-Circuits

7.4.1 Example: Symmetrical Components

A three-phase current system consists of the currents:

$$\underline{I}_R = (4.7 + j2.8)\,A$$
$$\underline{I}_S = (1.6 - j2.3)\,A$$
$$\underline{I}_T = (-1.8 - j1.7)\,A$$

Calculate and draw the symmetrical components with

$$a = -0.5 + j0.866 \qquad a^2 = -0.5 + j0.866$$

We can calculate the symmetrical components:

$$\underline{I}_0 = \frac{1}{3} \cdot (\underline{I}_R + \underline{I}_S + \underline{I}_T) = \frac{1}{3} \cdot (4.7 + j2.8 + 1.6 - j2.3 - 1.8 - j2.7)\,A$$
$$= \frac{1}{3} \cdot (4.5 - j1.2)\,A = (1.5 - j0.4)\,A$$
$$\underline{I}_1 = \frac{1}{3} \cdot (\underline{I}_R + a \cdot \underline{I}_S + a^2 \cdot \underline{I}_T)$$
$$= \frac{1}{3} \cdot (4.7 + j2.8 + (-0.5 + j0.866) \cdot (1.6 - j2.3)$$
$$+ (-0.5 - j0.866) \cdot (-1.8 - j1.7))\,A$$
$$= \frac{1}{3} \cdot (4.7 + j2.8 - 0.8 + j1.15 + j1.385 + 1.99 + 0.9$$
$$+ j0.85 + j1.56 - 1.47)\,A$$
$$= \frac{1}{3} \cdot (5.32 + j6.745)\,A = (1.77 + j2.024)\,A$$
$$\underline{I}_2 = \frac{1}{3} \cdot (\underline{I}_R + a^2 \cdot \underline{I}_S + a \cdot \underline{I}_T)$$
$$= \frac{1}{3} \cdot (4.7 + 2.8 + (-0.5 - j0.866) \cdot (1.6 - j2.3)$$
$$+ (-0.5 + j0.866) \cdot (-1.8 - j1.7))\,A$$
$$= \frac{1}{3} \cdot (4.7 + j2.8 - 0.8 + j1.15 - j1.385 - 1.99 + 0.9$$
$$+ j0.85 - j1.56 + 1.47)\,A$$
$$= \frac{1}{3} \cdot (4.28 + j1.855)\,A = (1.43 + j0.62)\,A$$

The calculated currents and the symmetrical components are shown in Figure 7.5.

7.4.2 Example: Symmetrical Components

How large are the symmetrical components of an AC voltage of 225 V?
Given: $\underline{U}_R = 225\,\text{V}$, $\underline{U}_S = \underline{U}_T = 0$.

Solution:

$$\underline{U}_0 = \frac{1}{3} \cdot (\underline{U}_R + \underline{U}_S + \underline{U}_T) = \frac{1}{3} \cdot 225\,\text{V} = 75\,\text{V}$$

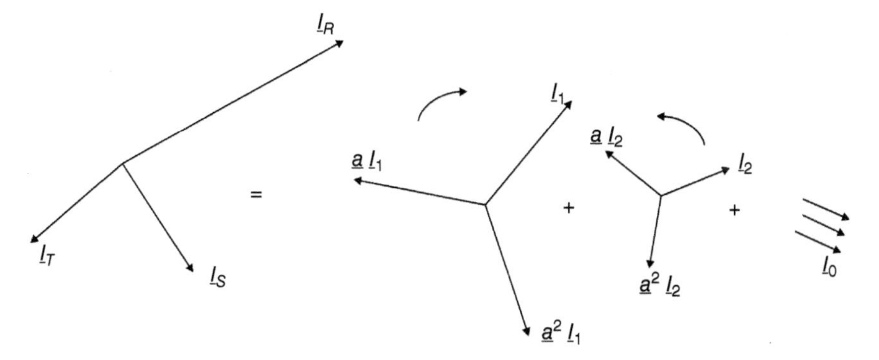

Figure 7.5 Symmetrical components.

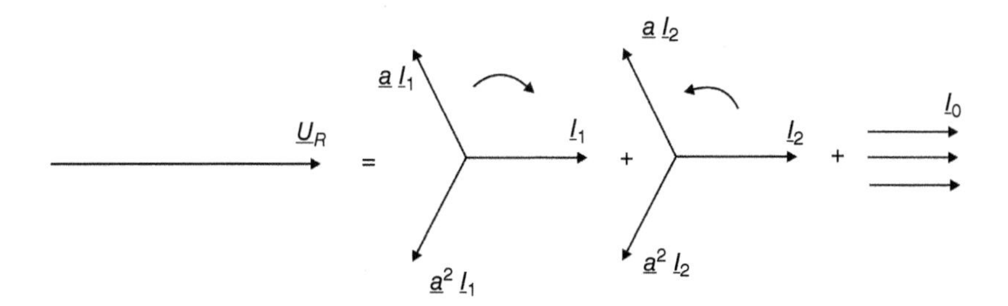

Figure 7.6 Results of symmetrical components.

$$\underline{U}_1 = \frac{1}{3} \cdot (\underline{U}_R + a \cdot \underline{U}_S + a^2 \cdot \underline{U}_T) = \frac{1}{3} \cdot 225\ \text{V} = 75\ \text{V}$$
$$\underline{U}_2 = \frac{1}{3} \cdot (\underline{U}_R + a^2 \cdot \underline{U}_S + a \cdot \underline{U}_T) = \frac{1}{3} \cdot 225\ \text{V} = 75\ \text{V}$$

The results are shown in Figure 7.6.

7.4.3 Example: Symmetrical Components

The three-phase current system is given:

$$\underline{I}_R = 200\ \text{A} \qquad \underline{I}_S = \underline{I}_T = -100\ \text{A}$$

How large are the symmetrical components?

Solution:

$$\underline{I}_0 = \frac{1}{3} \cdot (\underline{I}_R + \underline{I}_S + \underline{I}_T) = (200 - 100 - 100)\ \text{A} = 0$$
$$\underline{I}_1 = \frac{1}{3} \cdot (\underline{I}_R + a \cdot \underline{I}_S + a^2 \cdot \underline{I}_T)$$
$$= \frac{1}{3} \cdot (200 + (-0.5 + j0.866) \cdot (-100) + (-0.5 - j0.866) \cdot (-100))\ \text{A}$$
$$= \frac{1}{3} \cdot (200 + 50 - j86.6 + 50 + j86.6)\ \text{A} = \frac{1}{3} \cdot 300\ \text{A} = 100\ \text{A}$$

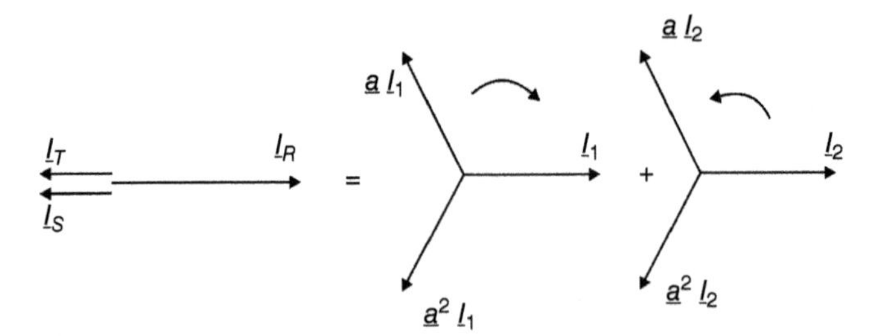

Figure 7.7 Symmetrical components.

$$
\begin{aligned}
\underline{I}_2 &= \frac{1}{3} \cdot (\underline{I}_R + a^2 \cdot \underline{I}_S + a \cdot \underline{I}_T) \\
&= \frac{1}{3} \cdot (200 + (-0.5 - j0.866) \cdot (-100) + (-0.5 + j0.866) \cdot (-100))\ \text{A} \\
&= \frac{1}{3} \cdot (200 + 50 + j86.6 + 50 - j86.6)\ \text{A} = \frac{1}{3} \cdot 300\ \text{A} = 100\ \text{A}
\end{aligned}
$$

The results are shown in Figure 7.7.

8

Short-Circuit Currents

8.1 Introduction

In addition to load flow calculations, short-circuit current calculations are among the basic investigations in the planning of a three-phase network. Short-circuit current calculation examines the following network planning parameters:

- Determination of the dynamic and thermal stress of the equipment
- Design and adjustment of the mains protection
- Calculation of the required opening and breaking capacity of the circuit-breakers
- Assessment of the type of neutral point treatment
- Check of the permissible grounding and contact voltages

Short-circuit current calculations, i.e. the calculation of short-circuit currents and their effects, are incorporated in various parts of the standards DIN EN 60909-0 (IEC 60909-0), DIN EN 61660 (IEC 61160), and DIN EN 60865 (IEC 60865) or treated according to DIN VDE 0101, DIN VDE 0102, and DIN VDE 0103. The calculation of short-circuit currents according to these standards does not take into account the temporal nature of the short-circuit current, but rather the individual parameters that are required for the design of the equipment and the network are decisive. The higher the voltage level at which a short-circuit occurs, the more loads are switched off or consume less power due to the voltage reduction. In the event of faults in the high voltage (HV system, the load reduction leads to the fact that the frequency in the network rises after only a few tenths of a second. During this period, however, the speed of the generators can be set as a constant, which is also a prerequisite for the short-circuit current calculation.

IEC 60909-0 is comprised of four main sections:

- 1: General
- 2: Properties of short-circuit currents. Method of calculation
- 3: Short-circuit impedances for electrical operational equipment
- 4: Calculation of short-circuit currents

Analysis and Design of Electrical Power Systems: A Practical Guide and Commentary on NEC and IEC 60364, First Edition. Ismail Kasikci.

The following points have been added to this standard:

- New classification into low-voltage and high-voltage networks.
- Supplementary sheets with examples and correction factors, supplementing the theoretical part.
- Low-voltage and high-voltage networks are treated in the same way. They differ only in the information listed in IEC 60364.
- The information regarding the calculation of three-pole and single-pole short-circuits applies equally to low-voltage and high-voltage networks.
- Voltage factors (c_{max} and c_{min}) are given.
- Impedance corrections, independent of the behavior of the short -circuit over time, have been introduced for generators and power Stations.
- μ and q curves have been newly included for $t_{min} = 0.02$ seconds in order to take account the decay of the AC component of the short-circuit current for asynchronous machines or motor groups.
- Equations for the calculation of μ and q factors have been newly included.
- In low-voltage networks a temperature rise from 20 to 80 °C (in future 160 °C for PVC-cable) is used for the single-pole short-circuit.
- Indices for symmetrical components (0, 1, 2) are internationally recognized.

The behavior of the initial symmetrical short-circuit current over time, from the beginning of the short-circuit to the end, is shown in Figure 8.1 beginning with the

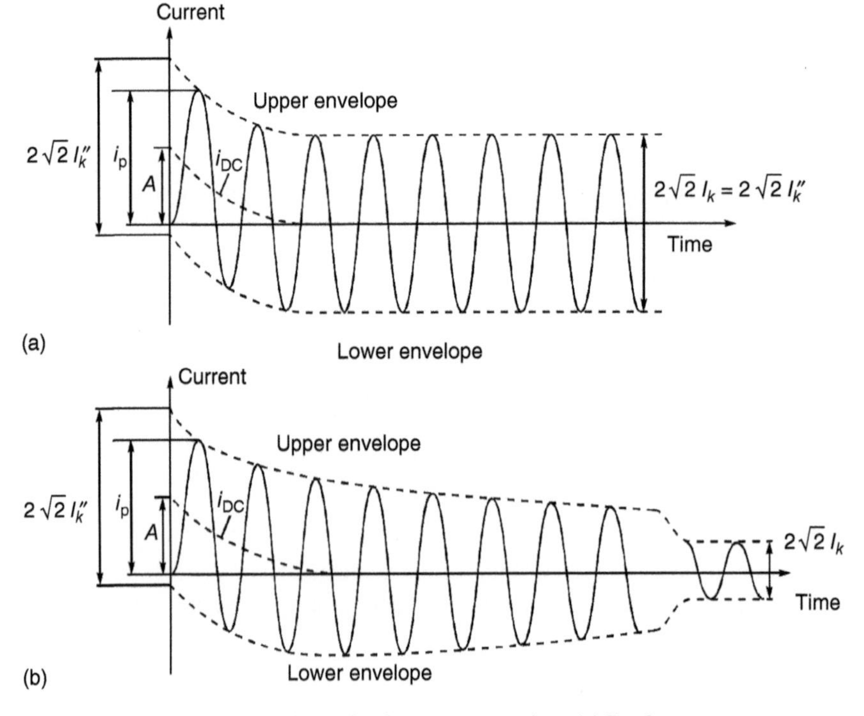

Figure 8.1 Behavior of the short-circuit current over time. (a) Far-from-generator short-circuit. (b) Near-to-generator short-circuit.

mΩ entry value of the voltage at the onset of the short-circuit. For a complete calculation of the short circuit it is necessary to consider this behavior. In most cases, the initial symmetrical short-circuit current and the peak short-circuit current must be calculated. The following pages will now deal with the fundamentals of calculating short circuit currents on the basis of this standard.

Here are the meanings of the symbols used in Figure 8.1:

I''_k Initial symmetrical short-circuit current
i_p Peak short-circuit current
I_k Steady-state short-circuit current
i_{DC} Decaying DC current component
A Initial value of DC current component i_{DC}

Different types of short-circuits (Figure 8.2) are calculated with the method of symmetrical components.

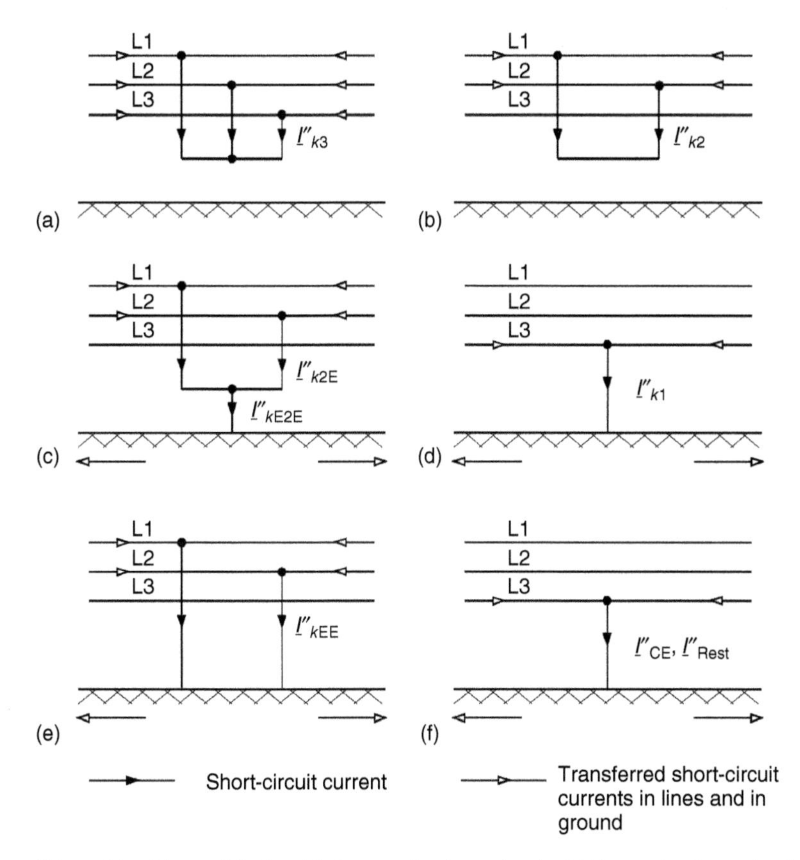

Figure 8.2 Types of faults. (a) Three-pole short-circuit. (b) Two-pole short-circuit without contact to ground. (c) Two-pole short-circuit with contact to ground. (d) Single-pole ground fault. (e) Double line-to-ground fault. (f) Capacitive ground fault current I_{CE}, residual ground fault current I_{Rest}.

8.2 Fault Types, Causes, and Designations

According to IEC 60909-0, a short circuit is the accidental or intentional conductive connection across a relatively low resistance or impedance between two or more points of a circuit that usually have different voltages. At a full short-circuit, the contact resistance is practically zero. With an arc short-circuit, the contact resistance represents a nonlinear resistance. If one or two live conductors are connected to earth, we speak of Shrouding. For networks with low-impedance transformer star earthing, the following is referred to as an earth short-circuit. Figure 8.3 shows the different short-circuit types and the designations of the so-called initial short-circuit AC currents.

Table 8.1 contains a compilation of the design criteria and short-circuit currents according to IEC 60909-0:2016-12 to be considered for the planning and project planning of electrical systems.

The cause of short-circuits or earth faults is the loss of operating insulation between the phases or between phases and earth, caused by the following:

- Insulation aging
- Pollution
- Internal (switching overvoltage) or external (lightning overvoltage) overvoltage
- External influences (e.g. excavators)

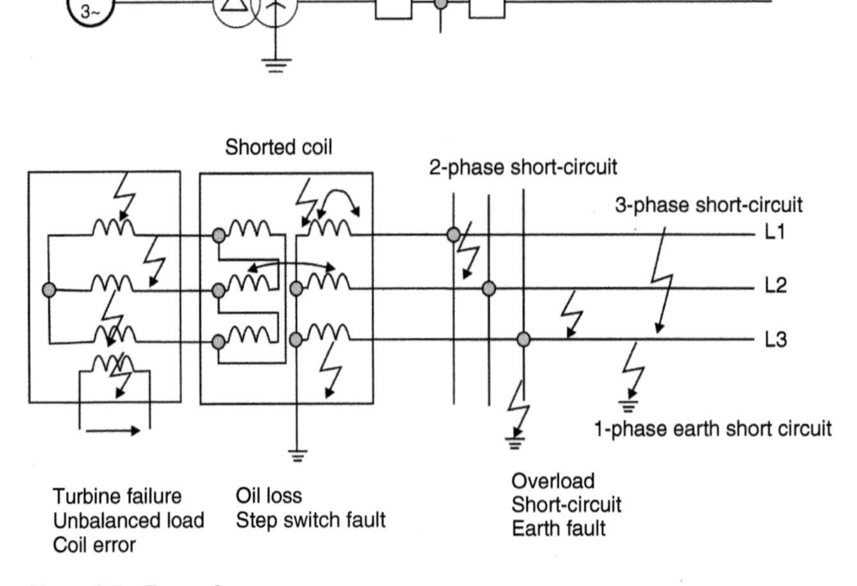

Figure 8.3 Types of errors.

Table 8.1 Selection via short-circuit currents.

Design criterion	Physical effect	Short-circuit stream	Limit
Mechanical stress	Forces F	Shock short current i_p	Instantaneous
Thermal stress	Heating	Continuous short-circuit current I_k	RMS value
Rated OFF or making capacity the OCP	Thermal stress I_{th}	Switch-off or making current I_a, I_e	RMS value
Protection setting	Protection measures	Initial short-circuit AC continuous short-circuit current I_k	Smallest short-circuit current $I''_{k1\,min}$

OCP: Overcurrent protection device, AC: Alternating current.

Electrical power supply networks must be planned in such a way that the systems and operating equipment can withstand the expected short-circuit currents. Short-circuit leads to the following effects in three-phase networks:

- Dynamic forces on current-carrying conductors (current forces)
- Thermal effect due to current heat ($\sim i^2 t$)
- Excitation of mechanical oscillations of the generators, and thus oscillations of the active and reactive power with the consequence of a possible loss of stability in the network
- Electromagnetic compensation processes in the network (e.g. travelling waves)

8.3 Short-circuit with *R*–*L* Network

The circuit with a short-circuit can be closed by a mesh with AC voltage source, reactances X, and resistances R (Figure 8.4). X and R replace all components such as cables and wires, transformers, generators, and motors.

In Figure 8.5 the switch-on processes are shown.

The following can be used to describe the short-circuit process differential equation:

$$i_k \cdot R_k + L_k \frac{di_k}{dt} = \hat{u} \cdot \sin(\omega t + \psi) \tag{8.1}$$

where ψ is the switching angle at the time of the short-circuit. It is presupposed that the current before the closing of S (short-circuit) is zero. The inhomogeneous differential equation of the first order can be described by determination of the homogeneous solution i_k and a particulate solution i''_k:

$$i_k = i''_{k\sim} + i_{k\_} \tag{8.2}$$

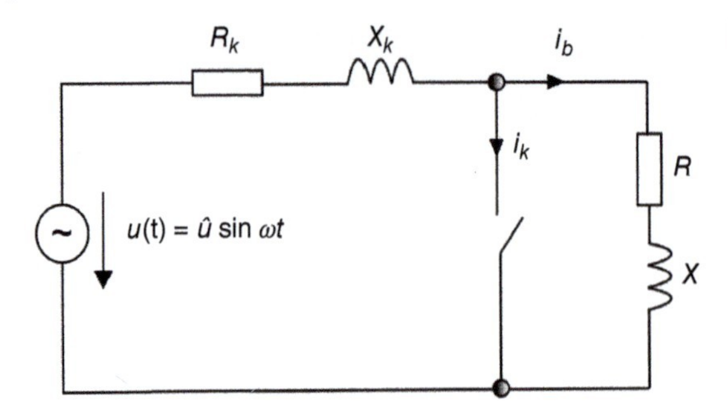

Figure 8.4 Short-circuit with R–L network.

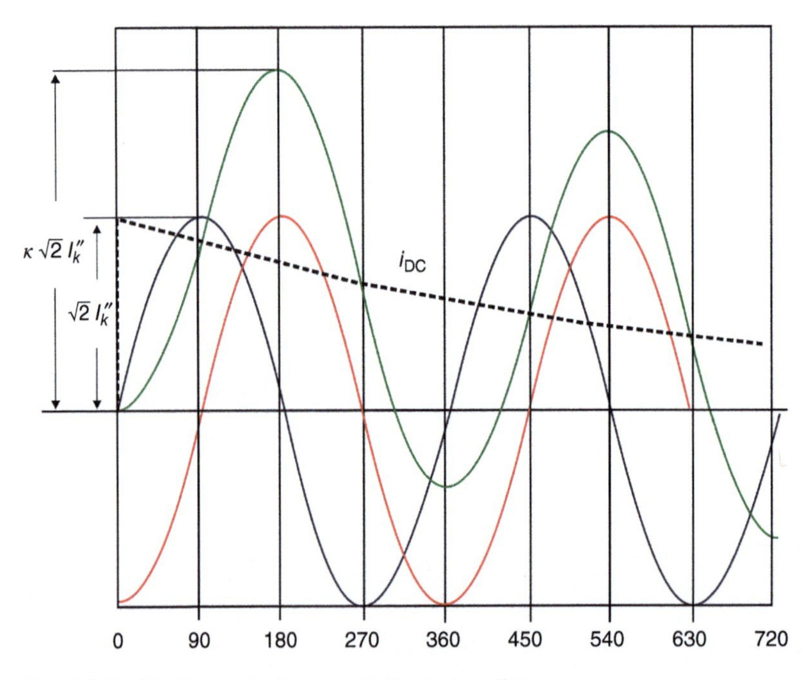

Figure 8.5 Short -circuit-current-switch-on operations.

The homogeneous solution with the time constant $\tau = L/R$ results in

$$i_k = \frac{-\hat{u}}{\sqrt{(R^2 + X^2)}} e^{\frac{t}{\tau}} \sin(\psi - \varphi_k) \tag{8.3}$$

For the particulate solution, we obtain or one gets:

$$i_k'' = \frac{\hat{u}}{\sqrt{(R^2 + X^2)}} \sin(\omega t + \psi - \varphi_k) \tag{8.4}$$

The total short-circuit current consists of the two components together:

$$i_k = \frac{\hat{u}}{\sqrt{(R^2 + X^2)}} [\sin(\omega t + \psi - \varphi_k) - e^{\frac{t}{\tau}} \sin(\psi - \varphi_k)] \tag{8.5}$$

The phase angle of the short-circuit current (short-circuit angle) is according to the above equation:

$$\varphi_k = \psi - \nu = \arctan\frac{X}{R} \tag{8.6}$$

$$\underline{Z} = R + jX = \sqrt{R^2 + (\omega L)^2} = \sqrt{R^2 + X^2} \tag{8.7}$$

The effective value of the current is of interest.

$$I_{\text{eff}}(t) = \sqrt{I_{\text{ac}}^2(t) + I_{\text{dc}}^2(t)} \tag{8.8}$$

$$I_{\text{eff}}(t) = \sqrt{I_{\text{ac}}^2(t) + [\sqrt{2} \cdot I_{\text{ac}} \cdot e^{-t/T}]^2} \tag{8.9}$$

$$I_{\text{eff}}(t) = I_{\text{ac}} \cdot \sqrt{1 + 2 \cdot e^{-2t/T}} \tag{8.10}$$

$$I_{\text{eff}}(\tau) = K \cdot (\tau) \cdot I_{\text{ac}} \tag{8.11}$$

whereby the asymmetrical factor

$$K(\tau) = \sqrt{1 + 2 \cdot e^{-4\pi\tau(X/R)}} \tag{8.12}$$

8.4 Calculation of the Stationary Continuous Short-circuit

In the case of a short-circuit far from the generator, the stationary continuous short-circuit current I_k is equal to the initial alternating current I''_k. With voltage factor c, $U''_Q = c \cdot U_{nQ}/\sqrt{3}$ applies to the driving mains voltage (sometimes called E''). The complex short-circuit current is calculated as follows:

$$\underline{I}_k = \frac{U''_Q}{\underline{Z}_k} = \frac{c \cdot U_{nQ}}{\sqrt{3} \cdot (R_k + jX_k)} = \frac{1.1 \cdot U_{nQ}}{\sqrt{3} \cdot \sqrt{R_k^2 + X_k^2}} \tag{8.13}$$

with

$$\varphi_k = \arctan\frac{X_k}{R_k} \tag{8.14}$$

with

$$u_Q(t) = \frac{\hat{u}_Q}{\sqrt{3}} \cdot \sin(\omega t + \varphi) \quad \text{with } \hat{u}_Q = \sqrt{2} \cdot 1.1 \cdot U_{nQ} \tag{8.15}$$

resulting in

$$i_{kp}(t) = \frac{\hat{u}_Q}{\sqrt{3} \cdot \sqrt{R_k^2 + X_k^2}} \cdot \sin(\omega t + \varphi - \varphi_k) \tag{8.16}$$

The continuous short-circuit current of the driving mains voltage rushes around the phase angle φ_k where this phase shift is also called the short-circuit angle. The short-circuit angle is determined by the impedances of the equipment in the short-circuit path. It amounts to

- $\approx 86°$ *endash* 89° in HV networks (higher overhead line share)
- $\approx 60°$ in MV and LV networks (higher cable portion)

8.5 Calculation of the Settling Process

From the simplified equivalent circuit diagram follows the differential equation

$$R_k \cdot i_k + L_k \frac{di_k}{dt} = \frac{\hat{u}}{\sqrt{3}} \cdot \sin(\omega t + \varphi) \tag{8.17}$$

whose solution is composed of the particulate and the homogeneous solution. The particulate solution $i_{kp}(t)$ corresponds to the permanent stationary solution. The homogeneous solution is

$$i_{kh}(t) = A \cdot e^{-\frac{t}{t}\tau} \quad \text{with} \quad \tau = \frac{L_k}{R_k} \tag{8.18}$$

where the constant A takes into account the initial condition for the current $i(t = 0)$. With $i(t = 0) = 0$ the total solution (= short-circuit current in the reference conductor) results in

$$i_k(t) = \frac{\hat{u}_Q}{\sqrt{3} \cdot \sqrt{R_k^2 + X_k^2}} \cdot \left(e^{-\frac{t}{\tau}} \cdot \sin(\varphi_k - \varphi) + \sin(\omega t + \varphi - \varphi_k)\right) \tag{8.19}$$

With the abbreviations

$$I_k = \frac{1.1 \cdot U_{nQ}}{\sqrt{3} \cdot \sqrt{R_k^2 + X_k^2}} \quad \text{and} \quad I_{kg} = \sqrt{2} \cdot I_k \cdot \sin(\varphi_k - \varphi) \tag{8.20}$$

results in

$$ik(t) = I_{kg} \cdot e^{-t/\tau} + \sqrt{2} \cdot I_k \cdot \sin(\omega t + \varphi - \varphi_k) \text{ and } I_{kg} = \sqrt{2} \cdot I_k \cdot \sin(\varphi_k - \varphi) \tag{8.21}$$

i.e. the short-circuit current consists of a decaying DC component and an AC component together.

Figure 8.6 shows the time course of the short-circuit current $i_k(t)$ and also the equal share $i_{kh}(t)$ and the exchange share $i_{kp}(t)$.

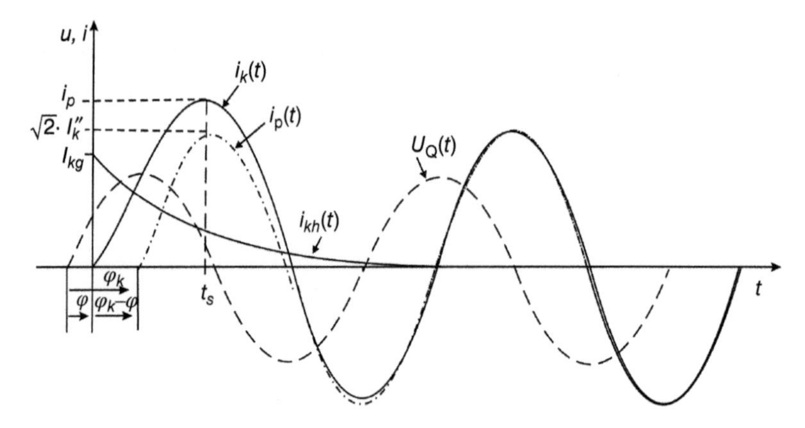

Figure 8.6 Short-circuit current over time.

8.6 Calculation of a Peak Short-Circuit Current

The largest value of the short-circuit current occurring, the so-called peak short-circuit current, is i_p. i_p is important for the mechanical design of network systems. The peak short-circuit current can be determined by an extreme value calculation, whereby it should be noted that i_k depends on both the time t and the phase angle of the voltage φ. The highest peak short-circuit current results when the short-circuit is initiated at zero crossing of the voltage and when $\varphi = 0$. The time t_s, at which the peak short-circuit current i_p occurs, results from the relationship

$$\sin(\omega t_s - \varphi_k) = 1 \tag{8.22}$$

because then the inhomogeneous portion becomes maximum. Thus, the calculation equation for the peak short-circuit current is

$$i_p(t = t_s) = \sqrt{2} \cdot I_k'' \cdot \left(1 + \sin\varphi_k \cdot e^{-\frac{t_s}{\tau}}\right) = \kappa\sqrt{2} \cdot I_k'' \tag{8.23}$$

with the factor κ for R/X:

$$\kappa = \left(1 + \sin\varphi_k \cdot e^{-\frac{t_s}{\tau}}\right) \tag{8.24}$$

The factor κ can be calculated accordingly from the equation

$$\kappa = 1.02 + 0.98 \cdot e^{-3\frac{R}{X}} \tag{8.25}$$

where R = R_k and X = X_k, i.e. resistance or reactance of the short-circuit path. The theoretical maximum value for the factor is $\kappa_{\text{max}} = 2$, i.e. the peak short-circuit current is twice the amplitude of the initial short-circuit AC current. This particularly unfavorable case occurs approximately in the event of short-circuits directly behind transformers.

8.6.1 Impact Factor for Branched Networks

In branched networks, i.e. with multiple short-circuits, the short-circuit impedance is determined for each supplying network branch i, and the corresponding factor κ_i is calculated from the R/X ratio.

In addition, the partial initial short-circuit AC currents I_k'' of the individual power supply branches are calculated. The total peak short-circuit current then results in

$$i_p = \sum_i \kappa_i \cdot \sqrt{2} \cdot I_{ki}'' \tag{8.26}$$

8.6.2 Impact Factor for Meshed Networks

In the case of a short-circuit in meshed networks, the peak short-circuit current at the fault location can no longer be determined from the superposition of partial short-circuit currents. Therefore, a shock factor κ is determined with which from the total initial short-circuit alternating current I_k'' of a network, the shock short-circuit current can be calculated. For the determination of κ different methods are available, two of which are mentioned here as examples:

- Uniform or lowest ratio R/X at the fault location: The shock factor κ is calculated from the smallest ratio R/X of all branches of the network that carry a partial short-circuit current, whereby the result is on the safe side if the R/X ratios of all network branches are relatively uniform.
- Ratio R/X of the short-circuit impedance at the short-circuit point: The shock factor κ is set from the ratio R/X of the calculated short-circuit current impedance at the short-circuit point with Eq. (8.25) and a safety factor of 1.15 is applied. In LV networks, the factor $1.15 \cdot \kappa$ is limited to the value 1.8 and in HV networks to the value 2.0.

8.7 Calculation of the Breaking Alternating Current

The short-circuit must be switched off as quickly as possible, but selectively. The reliable opening time is between 5 ms for fast fuses and 50–100 ms for circuit-breakers. The cut-off AC current I_b is the rms value of the short-circuit current flowing through the switch at the time of the first contact separation. It is calculated from the initial short-circuit alternating current I''_k with

$$I_b = \mu \cdot I''_k \tag{8.27}$$

where μ is the decay factor, which takes into account that the short-circuit current to the switch-off time, the so-called switching delay time t_{min}, has already elapsed in part, i.e. $\mu \leqslant 1.0$.

- *Short-circuit current of a near-from-generator*: The value of μ depends on the type of short-circuit, the fault location, the design of the feeding generator, and the switching delay time and can be found in tables or diagrams or calculated with the aid of approximate equations, e.g.

$$\mu = 0.62 + 0.72 \cdot e^{-0.32 \frac{I''_{kG}}{I_{rG}}} \quad \text{for} \quad t_{\text{min}} = 0.10 \text{ second} \tag{8.28}$$

- *Short-circuit current of a far-from-generator*: $\mu = 1$, i.e. $I_b = I''_k = I_k$.

8.8 Near-Generator Three-Phase Short-circuit

Figure 8.7 shows the basic time curve of the short-circuit current in the event of a generator-side short-circuit. Thus, I''_k is the initial short-circuit alternating current, I'_k the transition short-circuit alternating current, I_k the continuous short-circuit current, and i_p the peak short-circuit current. The closer a short-circuit occurs to the generator, the greater the short-circuit current that flows through the stator winding of the generator. The armature feedback increases to the same extent, so that the source voltage of the generator decreases and the short-circuit current decreases.

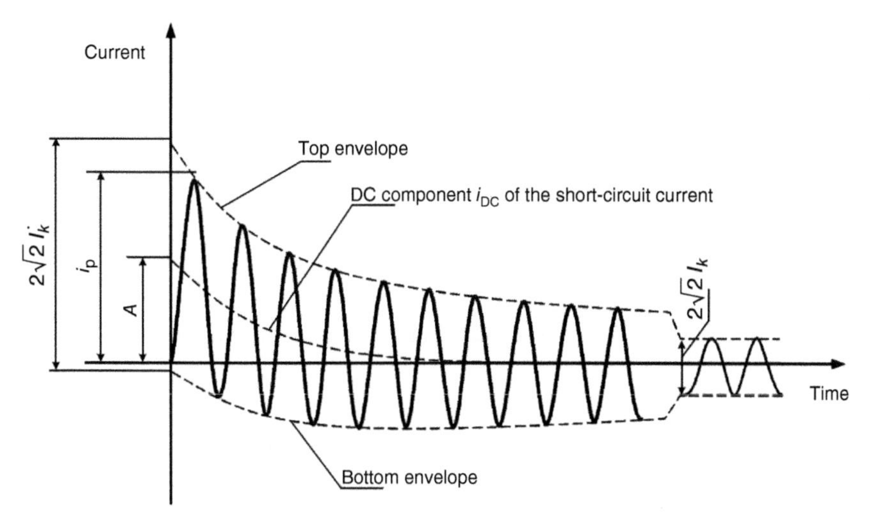

Figure 8.7 Near-generator three-pole short-circuit.

8.9 Calculation of the Initial Short-Circuit Alternating Current

The initial short-circuit alternating current I_k'' is calculated as for the generator distant short-circuit, whereby the subtransient longitudinal reactance X_d'' and, if necessary, the winding resistance of the generator must be used to calculate the short-circuit impedance.

Calculation of the Stationary Continuous Short-circuit Current:

The continuous short-circuit current I_k can be determined less precisely in the case of a short-circuit close to the generator. It depends on the following:

- Design of the synchronous generator
- Excitation device of the synchronous generator
- voltage regulation of the synchronous generator
- Voltage regulation of the variable transformers
- Signs of saturation
- Switching status of the network during the short-circuit period

The continuous short-circuit current I_k is calculated from the equation

$$I_{k\,\min} = \lambda_{\min} \cdot I_{\mathrm{rG}} \leqslant i_k \leqslant I_{k\,\max} = \lambda_{\max} \cdot I_{\mathrm{rG}}$$

with the so-called permanent factors $\lambda_{\min} \approx 0.4$–$1.0$ and $\lambda_{\max} \approx 1.7$–$5.5$, which can be taken from tables or diagrams.

8.10 Short-Circuit Power

The short-circuit power S''_k in the network is decisive for the voltage quality, e.g. when starting large motors. As a rule, the following shall apply:

$$S''_k \geqslant 40 \cdot S_{\text{Connection}} \tag{8.29}$$

The short-circuit power is calculated as follows:

$$S''_k = \sqrt{3} \cdot U_n \cdot I''_k \tag{8.30}$$

and is a fictitious power that does not occur at the fault location because the voltage at the short-circuit is zero.

8.11 Calculation of Short-Circuit Currents in Meshed Networks

The time curve of the total short-circuit current during a short-circuit in a meshed network corresponds to the curves, but the calculation is more difficult, because the total short-circuit current consists of several partial short-circuit currents. For the calculation of the total short-circuit current in a meshed network, the superposition method and the method of the equivalent voltage source are available.

8.11.1 Superposition Method

The superposition procedure presupposes that the load flow in the network before the short-circuit occurs is known, as well as all impedances and admittances of the equipment and the set voltage set-points of the feeding generators and the setting of the tap changers of the transformers. The procedure is based on the superposition of the network state before the short-circuit occurs with a change state caused by the short-circuit. To explain the procedure, a three-pole short-circuit is assumed, i.e. a single-phase equivalent circuit diagram of the network is sufficient. However, the method can also be used for other types of errors.

The following steps are necessary for calculating the short-circuit current:

1. *Calculation of the load flow before short-circuit occurrence*: Figure 8.8 shows the network state before short-circuit occurrence. The network is described by the admittance matrix. It consists of $i = 1 \cdots N$ load node and $j = 1 \cdots M$ generator node. The generator nodes are fictitious nodes within the network that only occur in the equivalent circuit diagram, since the generators described by their equivalent circuit diagram consist of voltage source and longitudinal impedance. The load nodes can be both loads and supply. In this case, the sign of the current must be reversed. Currents of load nodes $\underline{I}_{Li}$ are regarded as internal network currents and the current-free load nodes are regarded as external nodes. Only short-circuits at such nodes are considered. Cross impedances, i.e. impedances between network node and reference node 0 set in the admittance matrix, must be taken into account. Based on this mains replacement circuit diagram, a suitable

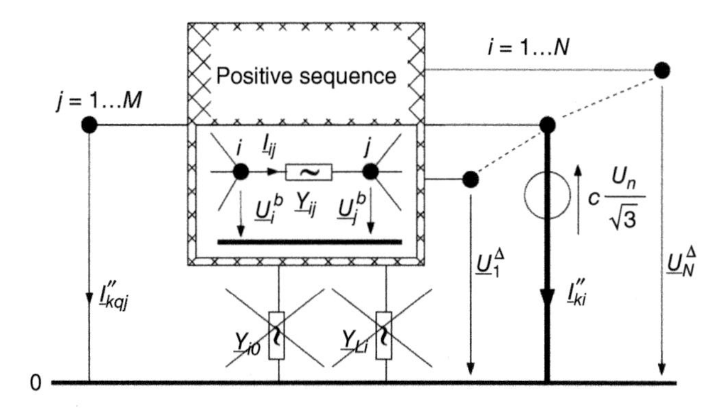

Figure 8.8 Superposition procedure. Source: Dietrich and Bernd Rüdiger [30].

load flow is calculated in the network before the short-circuit occurs. Thus all load currents $\underline{I}^b_{Li}$, all generator currents $\underline{I}^b_{qj}$, and all load node voltages $\underline{U}^b_{Li}$ are present. The index ^{b} indicates the condition before the short-circuit occurrence.

2. *Calculation of the state of change at a short-circuit*: Figure 8.8 shows the network state during the short-circuit, marked with the index ^{k}. The short-circuit occurs at load node i. The current flowing there $\underline{I}^k_i$ corresponds to the short-circuit current, because the current in the load node i was zero before the short-circuit occurred. The voltage at the generator node immediately after short-circuit occurrence corresponds to values before short-circuit occurrence $\underline{U}^b_{qj}$. Load currents $\underline{I}^k_{Li}$ and the generator currents $\underline{I}^k_{qj}$ have changed values. The network state during the short-circuit can also be described according to Figure 8.8, i.e. at node i affected by the short-circuit two voltages $\underline{U}^b_i$ of equal magnitude but directed against each other are introduced, so that the total voltage at the fault location is zero again. This allows the network state during the short-circuit to be described by superimposing the network state prior to the occurrence of the short-circuit and a so-called change state. Figure 8.8 shows the change state. The internal generator voltages all are set to zero and the voltage $-\underline{U}^b_i$ is effective only at the fault. Since the voltage acts in the opposite direction to its original direction, this condition is also known as reverse feed. For this change, state of the load flow of the network is calculated so that all load currents $\underline{I}^{\Delta}_{Li}$, all generator currents $\underline{I}^{\Delta}_{qj}$, all load node voltages $\underline{U}^{\Delta}_{Li}$ and the short-circuit current $\underline{I}^k_i$ are available for this state.
3. *Superposition of the network state before the short-circuit occurs with the change state*: If the load flow of the stationary network state is superimposed before the short-circuit entry with the load flow of the change state, the complete load flow during the short-circuit is obtained with the total short-circuit current at the short-circuit point and all partial short-circuit currents.

Disadvantage of the superposition method is that a complete load flow calculation of the network must be carried out several times, i.e. the procedure is very time-consuming. The advantage is that one can get in addition to the total short-circuit current at the fault location, all partial short-circuit currents as well as the voltages at all network nodes during the short-circuit.

8.11.2 Method of Equivalent Voltage Source

With the change state alone, the short-circuit current at the fault location can be fully calculated (Figure 8.9). Therefore, only the change state can be used for short-circuit current calculation. The advantages of this method are that the entire network is passive and a voltage source only occurs at the fault location. The disadvantage of this method is that the partial short-circuit currents in the individual network branches can only be calculated incompletely, because the stationary currents, i.e. the currents in the individual network branches before short-circuit occurrence, are missing in the individual network branches. However, since the stationary currents are usually small compared to the partial short-circuit currents, especially in the vicinity of the short-circuit point, the partial short-circuit currents are also determined with sufficient accuracy using these method. The accuracy of the calculation of the partial short-circuit currents is improved with the introduction of the so-called impedance correction factors, i.e. correction factors for calculating the impedances of individual equipment. This method is described in IEC 60909-0 as the alternative voltage source method at the fault source. To explain the procedure, a three-pole short-circuit is assumed again, i.e. a single-phase equivalent circuit diagram of the network is sufficient. However, the method can also be used for other types of errors.

Figure 8.10 shows the equivalent circuit diagram of the network for short-circuit current calculation. It is based on the following prerequisites:

- Introduction of backup voltage source $c \cdot U_n/\sqrt{3}$ at short-circuit instead of U_i^b
- Use of the impedance values of the transformers for the middle position of on-load tap changers or taps
- Conversion of impedances to other voltage levels with the design transmission ratios of the transformers $t_r = U_{\text{rTHV}}/U_{\text{rTLV}}$
- Neglect of the load impedances of non-motor loads
- Neglect of cross admittances (e.g. of line capacities) in the co-system (and negative sequence; take into account in the zero system)
- Introduction of correction factors for the impedances of certain equipment

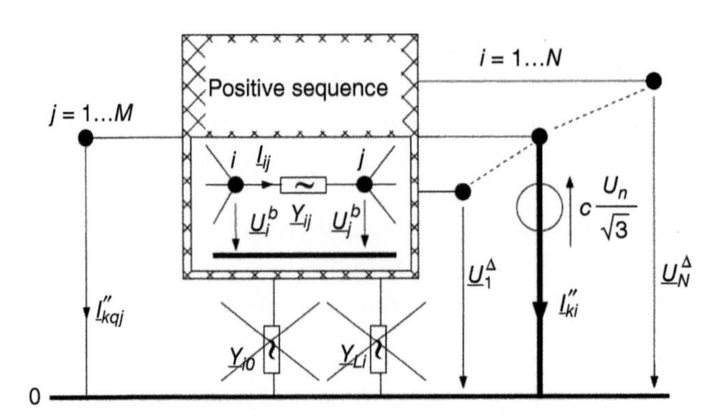

Figure 8.9 Voltage source. Source: Dietrich and Bernd Rüdiger [30].

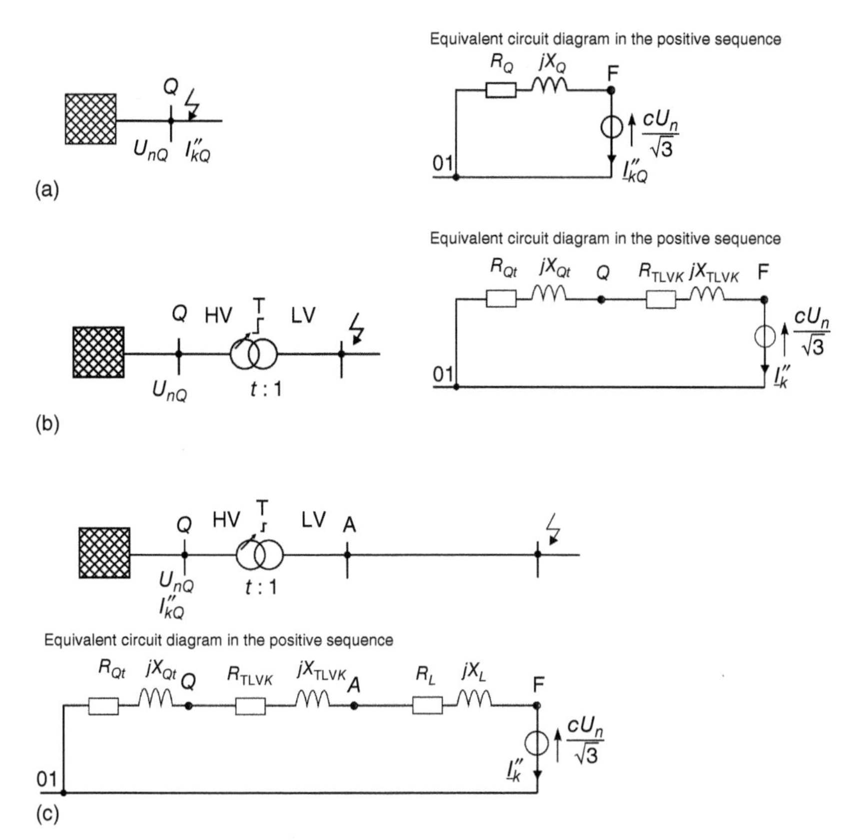

Figure 8.10 Power systems and equivalent circuit diagram of the network. (a) Structure of the mains circuit diagram without transformer. (b) Structure of the mains circuit diagram with transformer. (c) Structure of the network circuit diagram with supply, transformer, and cable.

For the short-circuit current calculation, the network is operated in accordance with the specified requirements. At the fault location the backup voltage source with voltage $c \cdot U_n/\sqrt{3}$ is inserted and all other voltage sources are short-circuited. Afterwards, the short-circuit current at the fault location calculated using the complex alternating current calculation to be. The short-circuit current calculated using the standby voltage source method is the initial short-circuit AC current I''_k. The values of the peak short-circuit current i_p, the breaking current I_b, and the continuous short-circuit current I_k can be calculated from this. The exact time curve of the short-circuit current is generally not required.

For the dimensioning of electrical systems to withstand thermal and mechanical stresses under fault conditions the three-pole initial symmetrical short-circuit current i''_k is of great importance. The fundamentals of the short-circuit calculation with the equivalent voltage source method at the fault location are found in IEC 60 909. This method represents a simplification, since the calculation is independent of the current operating state and future load flows. The electrical operational equipment parameters are taken into account in the calculation, and all network feeders, generators, and motors are short-circuited behind their internal reactances. The equivalent voltage source $c \cdot U_n/\sqrt{3}$ is then the only driving voltage in the network. The largest possible short-circuit current I''_{k3} is the value that determines the dimensioning of the operational equipment, and the smallest short-circuit current I''_{k1} is the value that determines the protective measure "protection by cut-off" and the setting for the network protection.

Except for the three-pole short-circuit current, all short-circuit types are calculated with the use of symmetrical components.

8.12 The Equivalent Voltage Source Method

The short-circuit current at the fault location F is calculated using the equivalent voltage source (Figure 8.11). This is the only voltage acting in the network at the fault location F. All other voltages for the operational equipment are short-circuited behind their internal impedances [29]. The voltage factor c is defined exactly in IEC 60909-0 and can be taken from Table 8.2.

The introduction of the voltage factor c is necessary for the following reasons:

- Different voltages within the network
- Subtransient behavior of generators, power stations, and motors
- Neglecting of loads and line capacitances
- Neglecting of the stationary operating state

8.13 Short-Circuit Impedances of Electrical Equipment

Power always begins with the external network. In general, the short-circuit power of the network is known and is given in MVA. Equipment found in the network includes generators, transformers, lines and cables, motors, and other loads. For the determination of short-circuit currents, we will discuss impedances and equivalent circuit diagrams for three-phase equipment in this section.

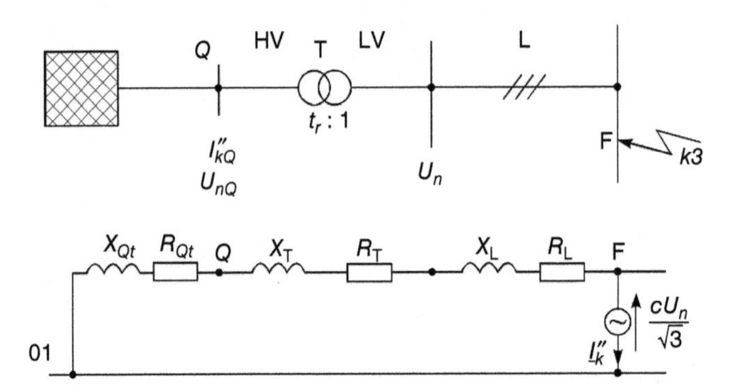

Figure 8.11 Equivalent voltage source.

Table 8.2 Voltage factor c in accordance with IEC 60909-0.

Rated voltage	Voltage factor c for calculation of the largest the smallest short-circuit current	
U_n	c_{max}	c_{min}
Low voltage		
100–1000 V	1.05	0.95
(IEC 60038, Table I)	1.10	0.9
Medium voltage		
>1–35 kV	1.10	1.00
(IEC 60038, Table III)		
High voltage		
>35 kV	1.10	1.00
(IEC 60038, Table IV)		

Note: cU_n must not exceed the highest voltage U_m for operational equipment in networks.

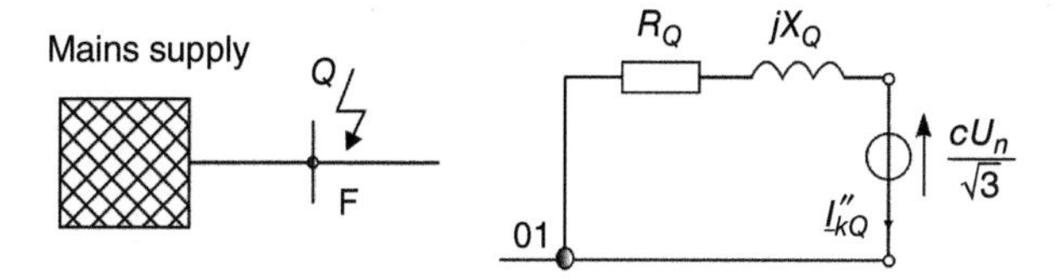

Figure 8.12 Power supply feeder.

8.13.1 Network Feeders

The internal impedance of the network Z_Q at the terminal Q is calculated from the initial symmetrical short-circuit current power S''_{kQ} or the initial symmetrical short-circuit current I''_{kQ} (Figure 8.12).

Internal impedance of the network Z_Q:

$$\underline{Z}_Q = R_Q + jX_Q \tag{8.31}$$

$$S''_{kQ} = \sqrt{3}U_{nQ}\, I''_{kQ} \tag{8.32}$$

$$Z_Q = \frac{cU_{nQ}}{\sqrt{3}\, I''_{kQ}} = \frac{c\, U^2_{nQ}}{S''_{kQ}} \tag{8.33}$$

The conversion of the internal impedance of the network to the lower voltage side of the transformer is given by

$$Z_{Qt} = \frac{c\, U_{nQ}^2}{S''_{kQ}} \frac{1}{t_r^2} \tag{8.34}$$

whereby

$$X_Q = \frac{Z_Q}{\sqrt{1 + \frac{R_Q}{X_Q}}} \tag{8.35}$$

When the resistance of the power supply feeder R_Q is not known, we can introduce in its place

$$X_Q = 0.995\, Z_Q \tag{8.36}$$

$$R_Q = 0.1\, X_Q \tag{8.37}$$

Here the symbols have the meanings:

U_{nQ}	Nominal voltage of the network at the terminal Q
S''_{kQ}	Initial symmetrical short-circuit apparent power
I''_{kQ}	Initial symmetrical short-circuit current
t_r	Rated transformation ratio at the tap changer in the main position
c	Voltage factor

8.13.2 Synchronous Machines

In general, for the calculation of the initial symmetrical short-circuit current I''_{kG} at the generator terminals we first find the subtransient part of the equivalent circuit E'' (Figure 8.13).

Synchronous generators are simulated by their longitudinal impedances and the subtransient reactance is given as reference.

$$\underline{Z}_{\mathrm{G}} = R_{\mathrm{G}} + j\, X''_d \tag{8.38}$$

$$x''_d = \frac{X''_d}{Z_{\mathrm{G}}}, \quad S_{\mathrm{rG}} = \frac{U_{\mathrm{rG}}^2}{Z_{\mathrm{G}}} \tag{8.39}$$

Then, the subtransient reactance is given by

$$X''_d = \frac{x''_d \cdot U_{\mathrm{rG}}^2}{S_{\mathrm{rG}}} \tag{8.40}$$

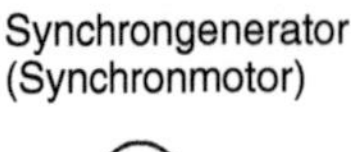

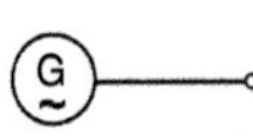

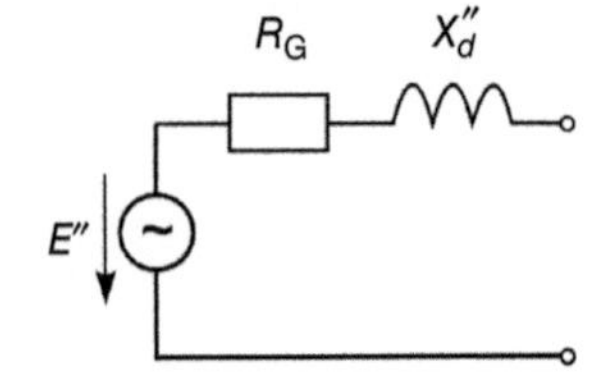

Figure 8.13 Synchronous machine.

The following approximate values can be used for R_G:

For low-voltage generators $U_{rG} < 1$ kV:

$$R_G = 0.15\, X''_d \tag{8.41}$$

For high-voltage generators $U_{rG} > 1$ kV
with $S_G \geq 100$ MVA:

$$R_G = 0.05\, X''_d \tag{8.42}$$

and with $S_G < 100$ MVA:

$$R_G = 0.07\, X''_d \tag{8.43}$$

Here the symbols have the meanings:

X''_d Subtransient reactance
X''_d Initial reactance in %
R_G Resistance of generator
Z_G Impedance of generator
U_{rG} Voltage of generator

8.13.3 Transformers

The short-circuit voltage is the primary voltage at which the transformer with short-circuited output winding takes up its primary current. It is a measure of the voltage change occurring under load. The positive sequence impedance of the transformer is as shown in Figure 8.14:

$$\underline{Z}_1 = \underline{Z}_T = R_T + jX_T \tag{8.44}$$

The positive-sequence impedance is

$$Z_T = \frac{u_{kr}\, U_{rT}^2}{100\%\; S_{rT}} \tag{8.45}$$

$$R_T = \frac{u_{Rr}\, U_{rT}^2}{100\%\; S_{rT}} = \frac{P_{krT}}{3\, I_{rT}^2} \tag{8.46}$$

$$X_T = \sqrt{Z_T^2 - R_T^2} \tag{8.47}$$

The characteristic values of three-phase distribution transformers can be found in DIN 42500, 42503, and 42511. For the calculation of the zero-sequence resistances, we can use Table 8.3.

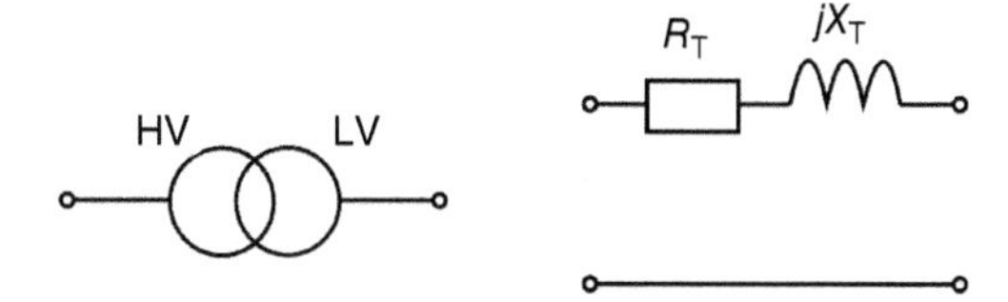

Figure 8.14 Transformer and its equivalent circuit.

Table 8.3 Zero-sequence resistances of transformers.

Zero-sequence resistances	Dy	Dz, Yy
R_{0T}	R_T	$0.4\,R_T$
X_{0T}	$0.95\,X_T$	$0.1\,X_T$

8.13.4 Consideration of Motors

In industry, mostly asynchronous motors are used (Figure 8.15). With the occurrence of a fault these contribute to the initial symmetrical short-circuit current, the peak short-circuit current, and the symmetrical breaking current, and for a two-pole short-circuit also to the steady-state short-circuit current, depending on the location of the motors and the fault [29]. The peak short-circuit component of the asynchronous motors must be taken into account. There is no difference in the calculations for squirrel cage rotor and slip-ring rotor motors, since the starting resistors of slip-ring rotor motors are short-circuited during operation.

The impedance $\underline{Z}_M$ of an asynchronous machine in the positive-sequence and negative-sequence systems is calculated from

$$Z_{\mathrm{M}} = \frac{1}{I_{an}} I_{\mathrm{rM}} \frac{U_{\mathrm{rM}}}{\sqrt{3}\, I_{\mathrm{rM}}} = \frac{1}{I_{an}} I_{\mathrm{rM}} \frac{U_{\mathrm{rM}}^2}{S_{\mathrm{rM}}} \tag{8.48}$$

$$I''_{\mathrm{kM}} = \frac{c\, U_n}{\sqrt{3}\, Z_{\mathrm{M}}} \tag{8.49}$$

Motors or groups of motors on a busbar following a three-pole short-circuit can be neglected, provided that

$$\sum I_{\mathrm{rM}} \leq 0.01\, I''_{kQ} \tag{8.50}$$

$$\frac{R_{\mathrm{M}}}{X_{\mathrm{M}}} = 0.42 \quad \text{and} \quad X_{\mathrm{M}} = 0.922\, Z_{\mathrm{M}} \tag{8.51}$$

High-voltage and low-voltage motors fed from a short-circuit at the terminal Q by way of transformers with two windings can also be neglected under the conditions $u_{\mathrm{kr}} = 6\%, \cos\varphi_{\mathrm{rM}} = 0.8$, and $I_{an}/I_{\mathrm{rM}} = 5$, so that

$$\frac{P_{\mathrm{rM}}}{S_{\mathrm{rT}}} \leq \frac{0.8}{\left| \frac{c\ 100\ S_{\mathrm{rT}}}{S''_{\mathrm{kQ}}} - 0.3 \right|} \tag{8.52}$$

Motor groups fed through transformers with different rated voltages can be calculated as follows:

$$\frac{\sum P_{\mathrm{rM}}}{\sum S_{\mathrm{rT}}} \leq \frac{\cos\varphi\ \eta_r}{\frac{I_{an}}{I_{\mathrm{rM}}}} \left(\frac{c \sum S_{\mathrm{rT}}}{0.05\ S''_{kQ}} - u_{kr} \right) \tag{8.53}$$

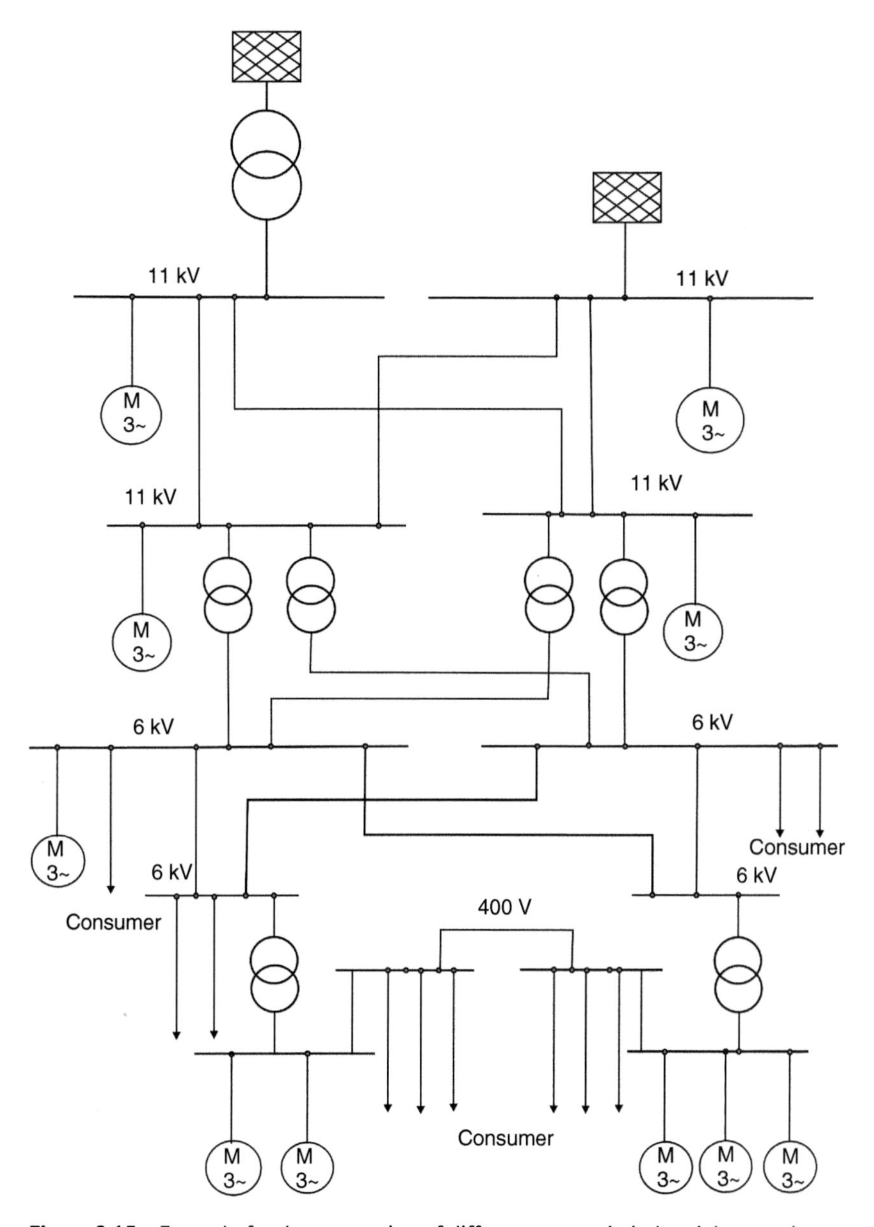

Figure 8.15 Example for the connection of different motors in industrial networks.

Here, the symbols have the meanings:

I_{an} Breakaway starting current

I_{rM} Rated current of motors in a group

U_{rM} Rated voltage of motor

I_{an}/I_{rM} Ratio of breakaway starting current to rated current for motor (lies between 4 and 8)

$\sum P_{rM}$	Sum of rated effective powers
$\sum S_{rT}$	Sum of rated apparent powers
S''_{kQ}	Initial symmetrical short-circuit apparent power
Z_M	Short-circuit impedance

Conclusion:
Under the following conditions asynchronous motors can be neglected for the Short-circuit calculation:

- Motors in public low-voltage networks
- Contributions from motors or groups of motors that represent less than 5% of the initial symmetrical short-circuit current without motors
- Motors, which due to interlocking or the type of process execution, do not run at the same time
- Motors fed from a short-circuit by way of transformers with two windings

8.13.5 Overhead Lines, Cables, and Lines

The short-circuit impedance in the positive-sequence system can be calculated from the line data, the tables, the cross sections, and the minimum spacing between the lines.

From Figure 8.16, the short-circuit impedance is

$$\underline{Z} = R_L + jX_L \tag{8.54}$$

Ohmic resistance:

$$R_L = R'_L \, l \tag{8.55}$$

Inductive reactance:

$$X_L = X'_L \, l \tag{8.56}$$

For the calculation of the single-pole short-circuit current in accordance with IEC 60909-0 a temperature rise of 80 and 160 °C is assumed:

$$R_{80\,°C} = 1.24 \frac{l}{\kappa \, S} \tag{8.57}$$

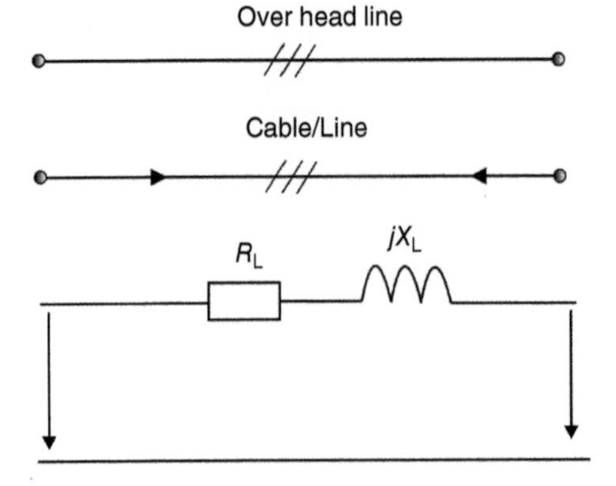

Figure 8.16 Cables and lines.

$$R_{160\,°C} = 1.56\,\frac{l}{\kappa\,S} \tag{8.58}$$

Zero-sequence resistances of lines:

$$R_{0L} = \text{Table value } R_L \tag{8.59}$$

$$X_{0L} = \text{Table value } X_L \tag{8.60}$$

The effective resistance R'_L for overhead lines can be calculated for a line temperature of 20 °C from

$$R'_L = \frac{1}{\kappa\,S} \tag{8.61}$$

The reactance per unit length X'_L for overhead lines can be calculated from

$$X'_L = 2\,\pi\,10^{-2}\left(0.25 + \ln\,\frac{d}{r}\right) \tag{8.62}$$

with

$$d = \sqrt[3]{d_{L1L2} \cdot d_{L2L3} \cdot d_{L3L1}}$$

The symbols have the meanings:

d	Average spacing between lines in mm
r	Radius of lines in mm
κ	Conductivity in $m/(\Omega \cdot mm^2)$
R'_L	Line resistance at 20 °C

8.13.6 Impedance Corrections

For the calculation of the three-pole initial symmetrical short-circuit current in networks with generators, with or without generator transformers, impedance corrections must be performed once (Figure 8.17). The correction factor K accounts for a voltage higher than E''.

1. *Generators*: For generators in industrial networks or in low-voltage networks with direct connection, we can use the following equation in the positive-sequence system:

$$\underline{Z}_G = R_G + jX''_d \tag{8.63}$$

$$\underline{Z}_{(GK)} = K_G\,\underline{Z}_G = K_G\,\left(R_G + jX''_d\right) \tag{8.64}$$

with the correction factor

$$K_G = \frac{U_n}{U_{rG}}\,\frac{c_{max}}{(1 + x''_d\,\sin\varphi_{rG})} \tag{8.65}$$

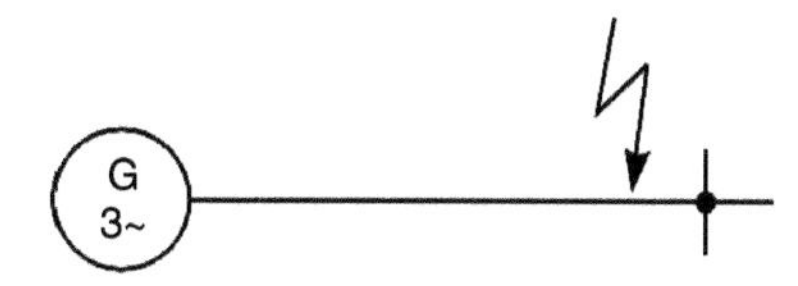

Figure 8.17 Impedance correction for generators.

Here the symbols have the meanings:

$c_{\max}$	Voltage factor
U_n	Nominal voltage of network
U_{rG}	Rated voltage of generator
$\underline{Z}_{\mathrm{GK}}$	Corrected impedance of generator
$\underline{Z}_{\mathrm{G}}$	Impedance of generator
X''_d	Subtransient reactance of generator
φ_{rG}	Phase angle between $U_{\mathrm{rG}}/\sqrt{3}$ and I_{rG}

2. *Impedance correction for a two-winding transformer*: Impedance correction for a two-winding transformer can be calculated with and without tap changer as follows:Short-circuit impedance of the transformer:

$$\underline{Z}_{\mathrm{T}} = R_{\mathrm{T}} + j\,X_{\mathrm{T}} \tag{8.66}$$

With the correction factor K_{T}:

$$\underline{Z}_{\mathrm{TK}} = K_{\mathrm{T}} \cdot \underline{Z}_{\mathrm{T}} \tag{8.67}$$

$$K_{\mathrm{T}} = 0.95\,\frac{c_{\max}}{1 + 0.6\,x_{\mathrm{T}}} \tag{8.68}$$

$$x_{\mathrm{T}} = \frac{X_{\mathrm{T}}}{(U_{\mathrm{rT}}^2/S_{\mathrm{rT}})} \tag{8.69}$$

Here the symbols have the meanings:

U_{rT}	Rated voltage of the transformer
I_{rT}	Rated current of the transformer
S_{rT}	Rated apparent power of the transformer
$P_{k\mathrm{rT}}$	Total loss of the transformer in the windings at rated current
u_{kr}	Rated short-circuit voltage in %
x_{T}	Related reactance of the transformer
u_{Rr}	Rated ohmic voltage in %

3. *Power station blocks with tap changer (Figure* 8.18*)*:

$$\underline{Z}_S = K_S(t_r^2\underline{Z}_{\mathrm{G}} + \underline{Z}_{\mathrm{THV}}) \tag{8.70}$$

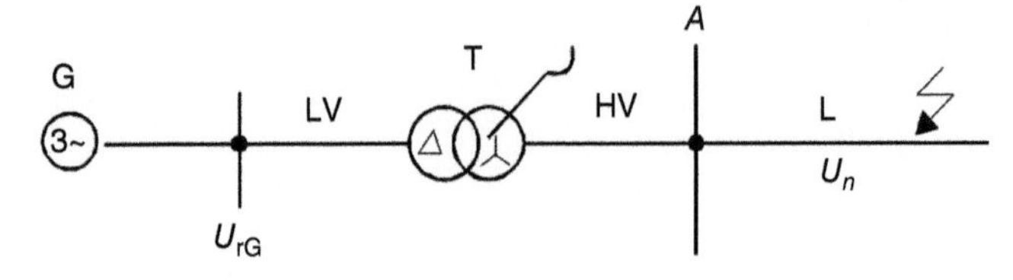

Figure 8.18 Impedance correction for power station transformer.

with the correction factor

$$K_S = \frac{U_{nQ}^2}{U_{rG}^2}\frac{U_{rTLV}^2}{U_{rTHV}^2}\frac{c_{\max}}{(|1 + x_d'' - x_T|\ \sin\varphi_{rG})} \tag{8.71}$$

4. *Power station blocks without tap changer*:
   The short-circuit impedance of power supply blocks on the higher voltage side of the block transformer is

$$\underline{Z}_{SO} = K_S(t_r^2\underline{Z}_G + \underline{Z}_{THV}) \tag{8.72}$$

with the correction factor

$$K_{SO} = \frac{U_{nQ}}{U_{rG}\ (1 + p_G)} \cdot \frac{U_{rTLV}}{U_{rTHV}} \cdot (1 \pm p_T) \cdot \frac{c_{\max}}{(1 + x_d''\ \sin\varphi_{rG})} \tag{8.73}$$

Here the symbols have the meanings:

Z_S, Z_{SO},	Corrected impedance with and without tap changer
Z_G	Impedance of generator
t_r	rated transformation ratio at which the tap changer at main position
Z_{TLV}	Impedance of transformer (lower voltage side)
Z_{THV}	Impedance of transformer (upper voltage side)
U_{nQ}	Nominal voltage of network
U_{rG}	Rated voltage of generator
X_d''	Subtransient reactance of generator
φ_{rG}	Phase angle between $U_{rG}/\sqrt{3}$ and I_{rG}

8.14 Calculation of Short-Circuit Currents

8.14.1 Three-Phase Short-circuits

In contrast with a single-pole short-circuit, a three-pole short circuit is a symmetrical fault, which is used for the evaluation of the rated breaking capacity of the overcurrent protection equipment (OPE) (Figure 8.19).

$$I_{k3}'' = \frac{c\ U_n}{\sqrt{3}\ Z_k} \tag{8.74}$$

$$Z_k = \sqrt{R_k^2 + X_k^2} \tag{8.75}$$

$$R_k = R_{Qt} + R_T + R_L \tag{8.76}$$

$$X_k = X_{Qt} + X_T + X_L \tag{8.77}$$

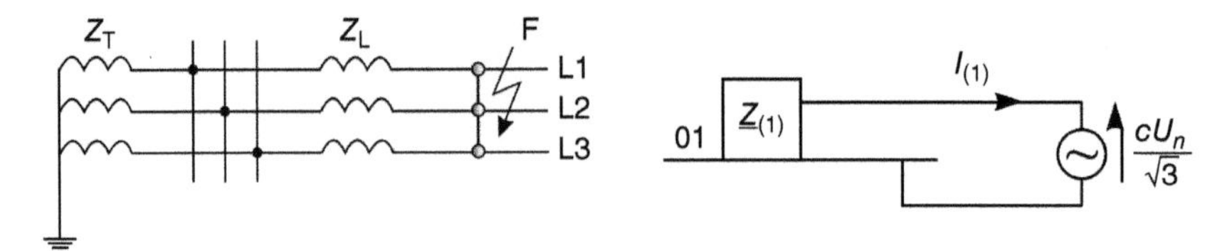

Figure 8.19 Network and equivalent circuit for a three-pole short-circuit current I''_{k3}.

Here, the meanings of the symbols are as follows:

R_k Sum of resistances in series

X_k Sum of reactances in series

Z_k Short-circuit impedance

c Voltage factor

8.14.2 Line-to-Line Short-circuit

In the case of a line-to-line short-circuit the initial short-circuit current shall be calculated by

$$I''_{k2} = \frac{c\,U_n}{2 \cdot |\underline{Z}_{(1)}|} = \frac{\sqrt{3}}{2} \cdot I''_{k3} \tag{8.78}$$

During the initial stage of the short-circuit, the negative impedance is approximately equal to the positive-sequence impedance, independent of whether the short-circuit is a near-to-generator or a far-from-generator short-circuit.

8.14.3 Single-Phase Short-circuits to Ground

The single-phase short-circuits to Ground is an asymmetrical fault, which is calculated on the basis of symmetrical components. This type of fault involves an external conductor (PEN) or a protective ground conductor (PE). In practice, however, a simpler approach is used, which entails an error of up to 20%. Since the OPE and the cross section of the conductor are matched to each other, tripping takes place within 0.2, 0.4, or 5 seconds. The initial symmetrical short-circuit current for a single-pole short-circuit to ground as in Figure 8.20 is calculated from

$\underline{Z}_{(1)} + \underline{Z}_{(0)}$

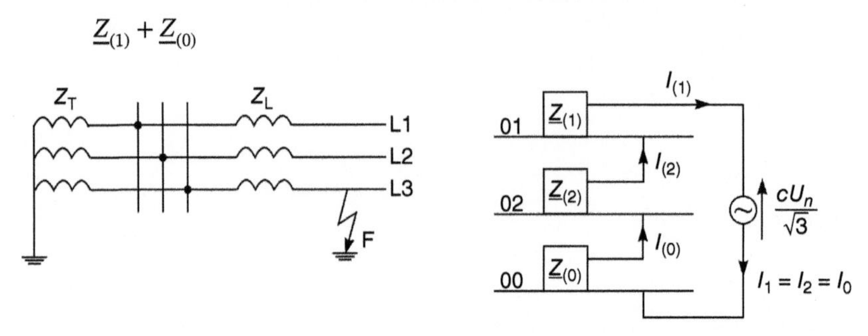

Figure 8.20 Network and equivalent circuit for a single-pole short-circuit current I''_{k1}.

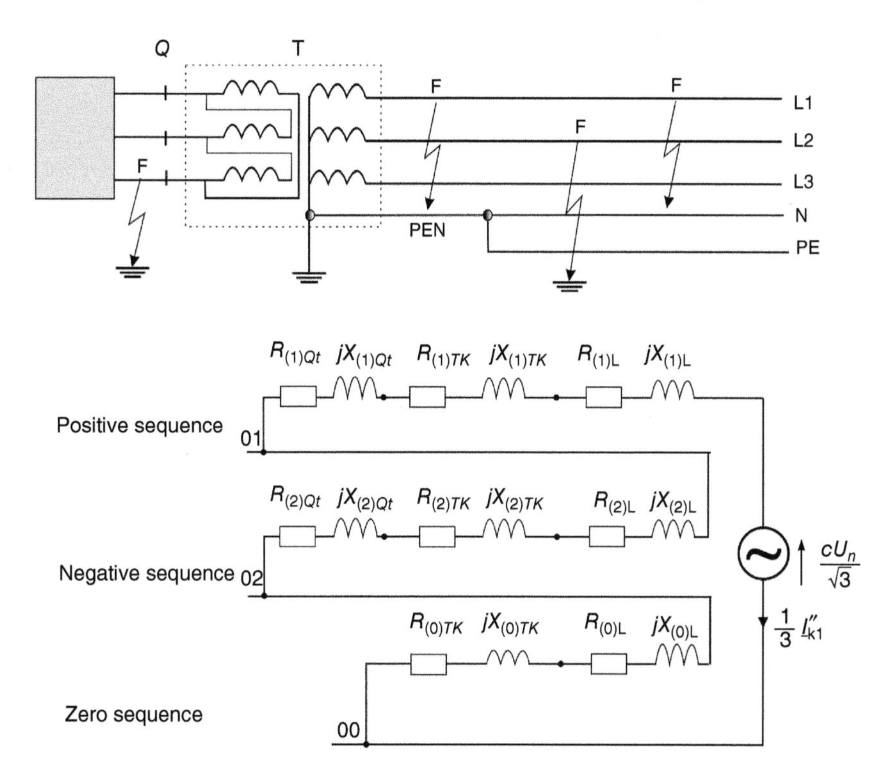

Figure 8.21 Network and equivalent circuit for a single-pole short-circuit with symmetrical components.

For the calculation of the single-pole short-circuit current the method of symmetrical components [28, 29] is used. This method requires the determination of three independent component systems (positive-sequence, zero-sequence, and negative-sequence systems). For the impedances of the operational equipment from Figure 8.21, it follows for the fault location that

$$I''_{k1} = \frac{3}{|\underline{Z}_{(1)} + \underline{Z}_{(2)} + \underline{Z}_{(0)}|} \cdot \frac{c\,U_n}{\sqrt{3}} \tag{8.79}$$

Here the symbols have the meanings:

$Z_{(1)}$ Positive-sequence impedance

$Z_{(0)}$ Negative-sequence impedance

c Voltage factor

8.14.4 Calculation of Loop Impedance

In low-voltage networks the loop impedance, which accounts for the effective resistances and for the reactances from the transformer to the fault location in the outward and return directions, is required in order to determine the cut-off conditions in accordance with IEC 60364, Parts 41 and 6. According to the installation of the system, this value can be determined very easily for the circuit assumed using measuring instruments. This value must not be greater than the values listed in Tables F.1 and F.2 in IEC 60364, Part 6. Figure 8.22 depicts the network, the equivalent circuit, and the related loop impedance.

$$\sum R_k = (2R_Q + 2R_T + 2R_K + 2R_{L1} + 2R_{L2} + R_{0T} + R_{0K} + R_{0L1} + R_{0L2})$$

$$\sum X_k = (2X_Q + 2X_T + 2X_K + 2X_{L1} + 2X_{L2} + X_{0T} + X_{0K} + X_{0L1} + X_{0L2})$$

$$Z_k = \sqrt{\sum R^2 + \sum X^2} \tag{8.80}$$

$$I''_{k1} = \frac{c_{\min}\, U_n}{\sqrt{3}\, Z_k} \tag{8.81}$$

The short-circuit impedance Z_k (loop impedance) includes the impedances of the supply network, the transformer, the external conductor, the PEN conductor, and the PE conductor. The result is on the safe side, since the short-circuit current that actually flows is greater than the value calculated. The source impedance or the single-pole short-circuit current is often known at the connection to the line, so that the maximum permissible circuit lengths and the cut-off conditions can then be determined (Figure 8.22).

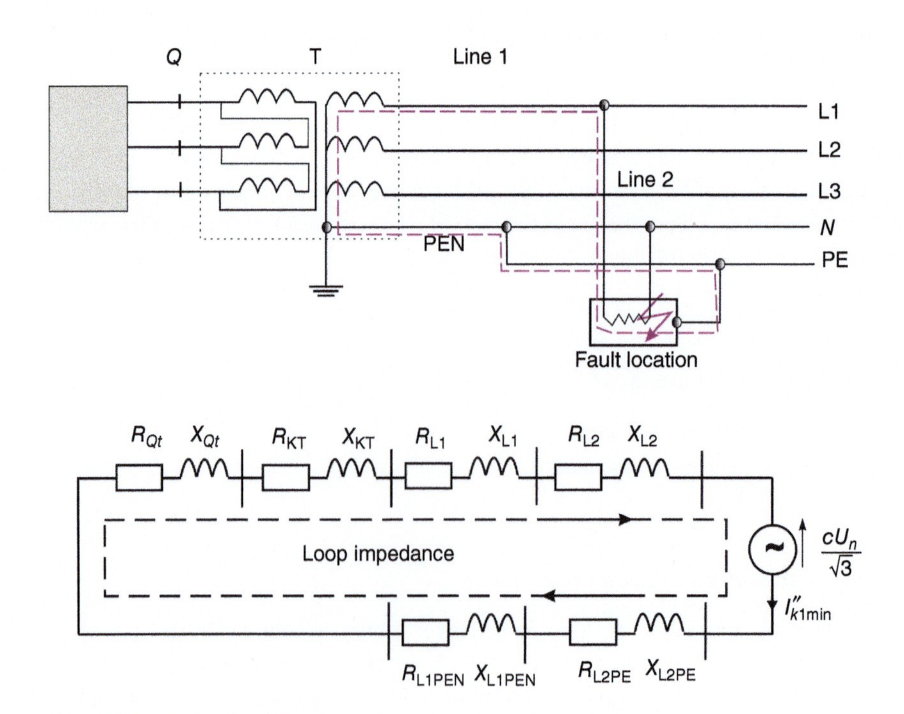

Figure 8.22 Loop impedance.

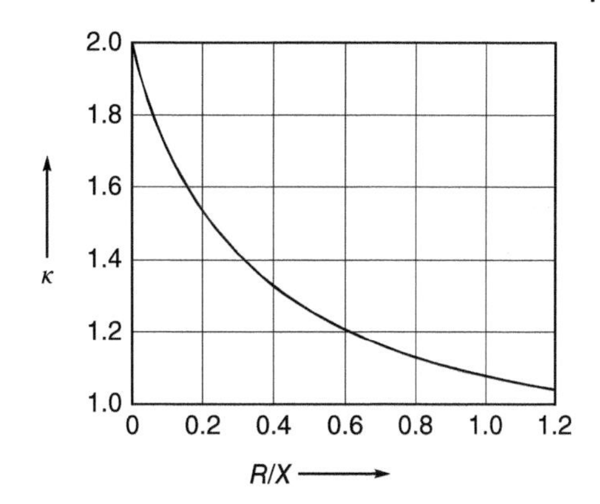

Figure 8.23 Factor κ for calculation of the peak short-circuit current i_p.

Here, the meanings of the symbols are as follows:

R_Q, X_Q	Resistance and inductive reactance of network feeder
R_T, X_T	Resistance and inductive reactance of transformer
R_K, X_K	Resistance and inductive reactance of cable
R_{L1}, X_{L1}	Resistance and inductive reactance of line 1
R_{L2}, X_{L2}	Resistance and inductive reactance of line 2
R_{0T}, X_{0T}	Zero-sequence resistance and inductive reactance of transformer
R_{0K}, X_{0K}	Zero-sequence resistance and inductive reactance of cable
R_{0L1}, X_{0L1}	Zero-sequence resistance and inductive reactance of line 1
R_{0L2}, X_{0L2}	Zero-sequence resistance and inductive reactance of line 2

8.14.5 Peak Short-Circuit Current

The calculation of the peak short-circuit current is important for

- the dynamic stressing of electrical systems,
- the making capacity of switchgear.

The following relationships apply:

$$i_p = \kappa \sqrt{2}\, I''_k \tag{8.82}$$

$$\kappa = f\left(\frac{R_k}{X_k}\right) \tag{8.83}$$

The factor κ is taken from Figure 8.23.

8.14.6 Symmetrical Breaking Current

The symmetrical breaking current is the effective value of the symmetrical AC component of the expected short-circuit current at the beginning of contact separation of the first pole, which quenches in the overcurrent protective device.

1. *For synchronous machines*:

$$I_a = \mu \, I''_{kG} \tag{8.84}$$

If $I_a = I''_k$, then $\mu = 1$, i.e. a far-from-generator short-circuit. For synchronous machines:

$$\frac{I''_{k3}}{I_{rG}} \leq 2 \tag{8.85}$$

If $I_a < I''_k$, i.e. a near-to-generator short-circuit, then,

$$\frac{I''_{k3}}{I_{rG}} \geq 2 \tag{8.86}$$

In practice, the minimum switching delay (for tripping an overcurrent protection device) is 0.1 seconds.

2. *For asynchronous machines*:

$$I_a = \mu \, q \, I''_{kM} \tag{8.87}$$

μ depends on the ratio I''_k / I_{rG} for the individual short-circuit sources and the minimum switching delay t_{min}, and q depends on the effective power of the pole pair (Figure 8.24).

3. *For power supply feeders*:

$$I_{aQ} = I''_{kQ} \tag{8.88}$$

The μ factor can be taken from Figure 8.25.

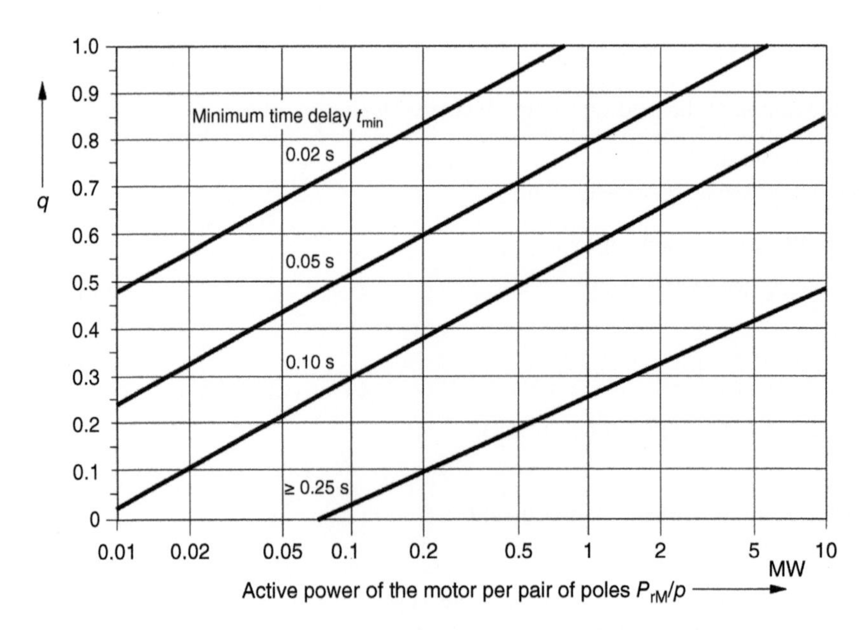

Figure 8.24 Factor q for calculation of the symmetrical breaking current of asynchronous machines.

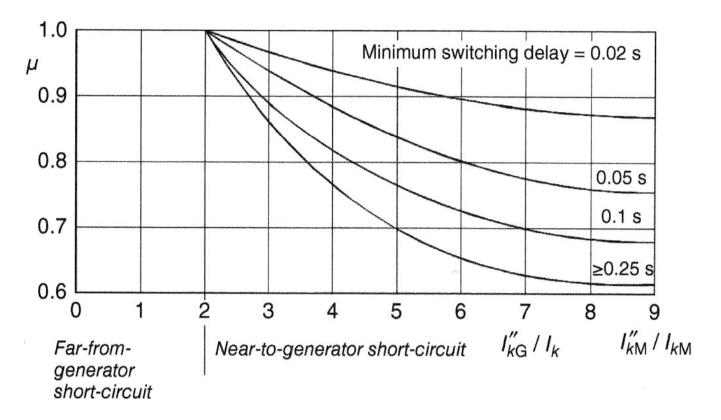

Figure 8.25 Factor μ for calculation of the symmetrical breaking current of asynchronous machines.

8.14.7 Steady-State Short-circuit Current

We distinguish between the maximum steady-state short-circuit current $I_{k\max}$ and the minimum current $I_{k\min}$, which applies for the maximum excitation voltage of the synchronous machine and arises for a constant unregulated no-load voltage. The factor λ depends on the ratio I''_{kG}/I_{rG}, the excitation, and the type of synchronous machine (Figure 8.26). λ can be taken from the figures for the upper and lower limit values. Furthermore, it is also necessary to consider the λ curves of the two generators for the excitation voltage in rated operation.

The following relationships hold true:

$$I_{k\max} = \lambda_{\max}\, I_{rG} \tag{8.89}$$

$$I_{k\min} = \lambda_{\min}\, I_{rG} \tag{8.90}$$

8.15 Thermal and Dynamic Short-circuit Strength

Electrical operational equipment such as busbars, overcurrent protection devices, cables, and lines are thermally and mechanically very strongly stressed in the event of a fault (IEC 893). For a short-circuit of duration T_K the short-circuit current I_{th} results, with the factor m for the temperature rise resulting from the DC aperiodic component and the factor n for the temperature rise resulting from the AC component:

$$I_{th} = I''_k\,\sqrt{m+n} \tag{8.91}$$

The factors m and n can be taken from Figure 8.27. The dynamic stressing of the systems through the short-circuit currents gives rise to forces that can destroy the systems and endanger the operating staff. The greatest possible force between the main supply lines with length l and spacing a is

$$F = 0.2\, i_p^2\, \frac{l}{a} \tag{8.92}$$

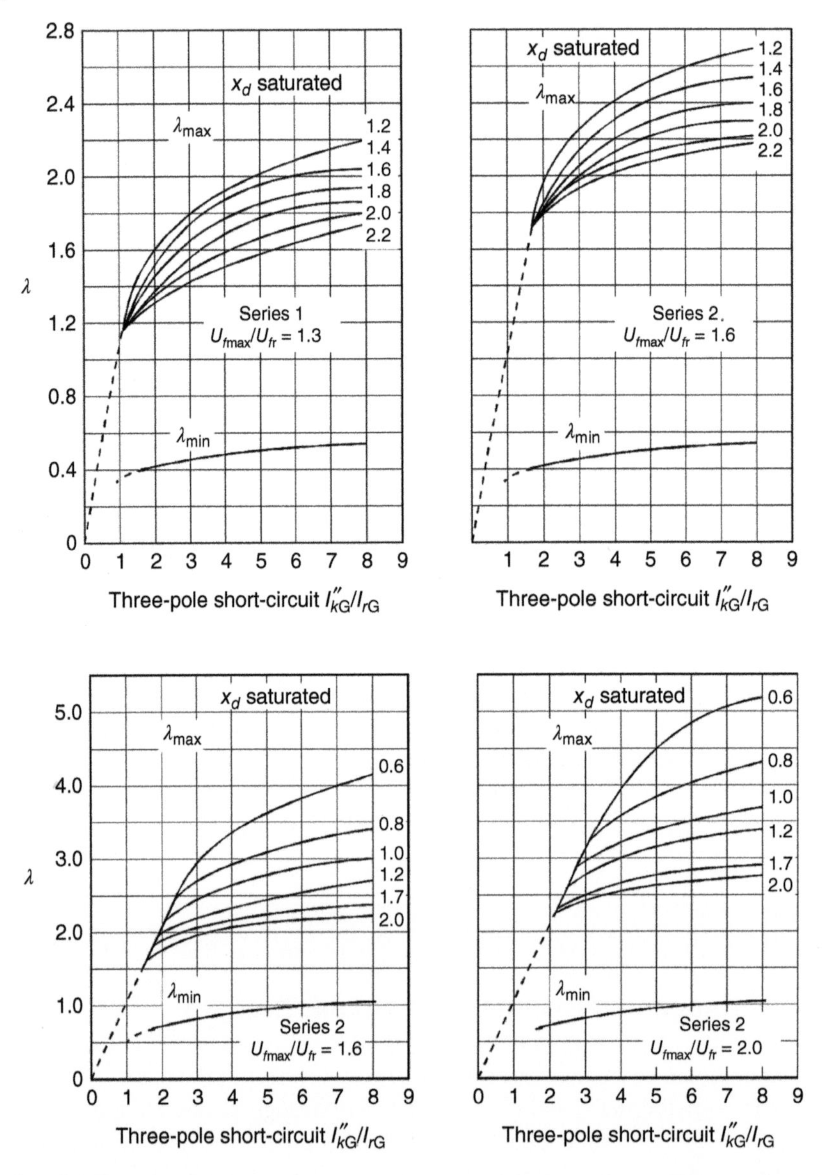

Figure 8.26 Factors λ_{max} and λ_{min} for calculation of the steady-state short-circuit current I_k.

Here the symbols have the meanings:

m Thermal effect of the DC aperiodic component for three-phase and single-phase AC current

n Thermal effect of the AC component for a three-pole short-circuit

F Electrodynamic force between lines

i_p Peak short-circuit current

I_{th} Short-time current

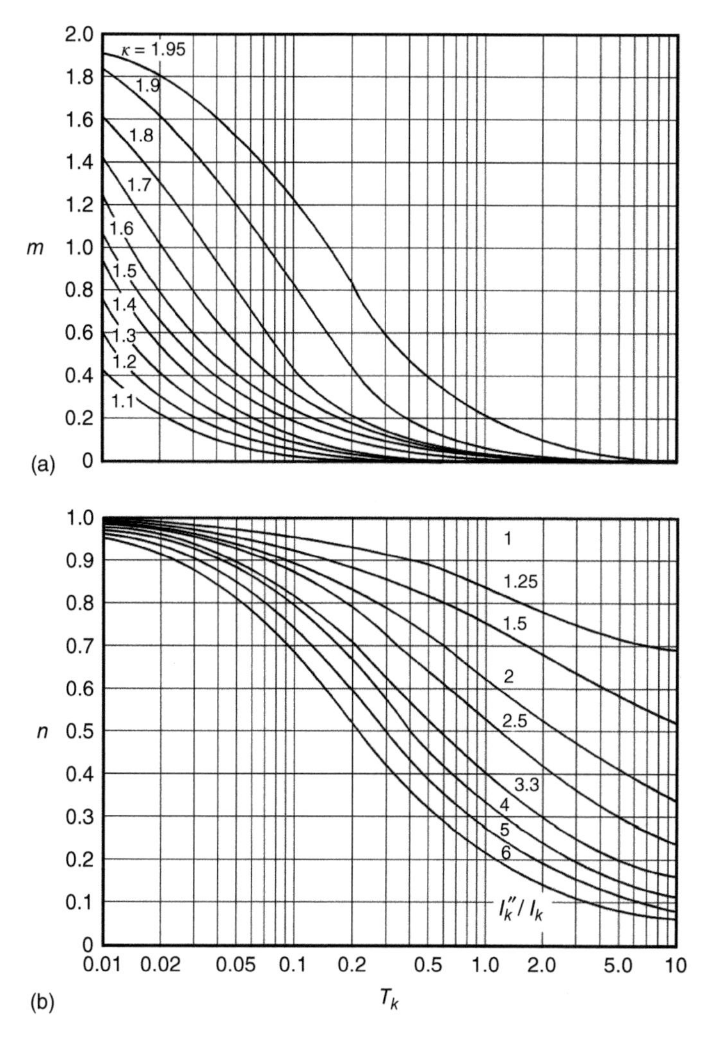

Figure 8.27 Factors m and n for calculation of the short-time current.

I''_k Initial symmetrical short-circuit current

l Length of line

a Spacing between lines

8.16 Examples for the Calculation of Short-Circuit Currents

8.16.1 Example 1: Calculation of the Short-Circuit Current in a DC System

With the occurrence of a fault, the operational equipment connected is not thermally protected. The sensitive components connected to the line (e.g. initiators or

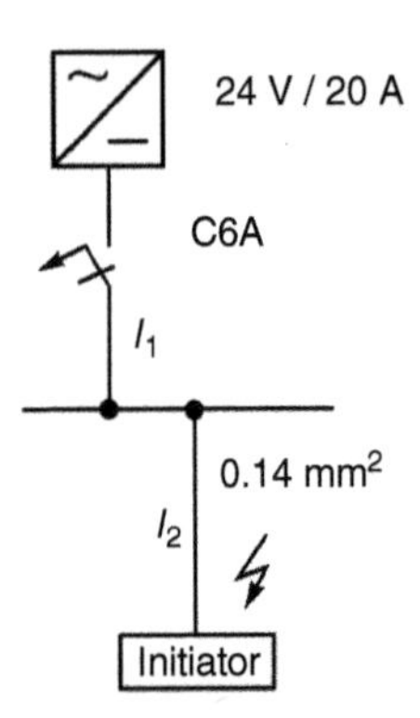

Figure 8.28 Wiring diagram.

printed circuit boards) are subjected to high Joule heat values. Given the system in Figure 8.28 with the following data:

Installation: $l_1 = 20$ m, $S_1 = 1.5$ mm^2 Cu
Initiator: $l_2 = 2$ m, $S_2 = 0.14$ mm^2 Cu

Check the thermal strength of the insulation for the 0.14-mm^2 Cu line in the event of a fault.

Calculation of the loop resistance with the use of a C6A (the internal resistances of the power supply [50 mΩ] and overcurrent protective device [55 mΩ] can be neglected) overcurrent protection device:

$$R_{\text{ges}(20°)\text{CC}} = 2(R_1 + R_2) = 2\left(\frac{l_1}{\kappa\, S_1} + \frac{l_2}{\kappa\, S_2}\right)$$

$$= 2\left(\frac{20\text{ m}}{56\frac{\text{m}}{\Omega\text{ mm}^2}\cdot 1.5\text{ mm}^2} + \frac{2\text{ m}}{56\frac{\text{m}}{\Omega\text{ mm}^2}\cdot 0.14\text{ mm}^2}\right)$$

$$= 0.986\ \Omega \approx 1\ \Omega$$

For a line temperature of 80 °C (as required by IEC 60 909):

$$R_{\text{ges}(80\,°\text{C})} = 1.24\ R_{\text{ges}(20\,°\text{C})} = 1.24 \cdot 1\ \Omega = 1.24\ \Omega$$

This results in a short-circuit current of

$$I''_{k1} = \frac{U_n}{R_{\text{ges}}} = \frac{24\text{ V}}{1.24\ \Omega} = 19.35\text{ A}$$

$$t_{k_{\text{zul.}}} = \left(\frac{k\cdot S}{I''_{k_1}}\right)^2 \text{ und } t_{k_{\text{zul.}}} > t_a$$

$$t_{k_{\text{zul.}}} = \left(\frac{115\cdot A\cdot\sqrt{s}\cdot 0.14\text{ mm}^2}{\text{mm}^2\cdot 19.35\text{ A}}\right)^2 = 0.69\text{ second}$$

$$\text{Factor } k = \frac{I''_{k_1}}{I_n} = \frac{19.35\text{ A}}{6\text{ A}} = 3.2$$

According to Figure 13.12, $t_a = 18$ seconds. Since $t_{k_{\text{zul.}}} < t_a$, the insulation strength is inadequate.

Solution: Replace C6A by Z6A.

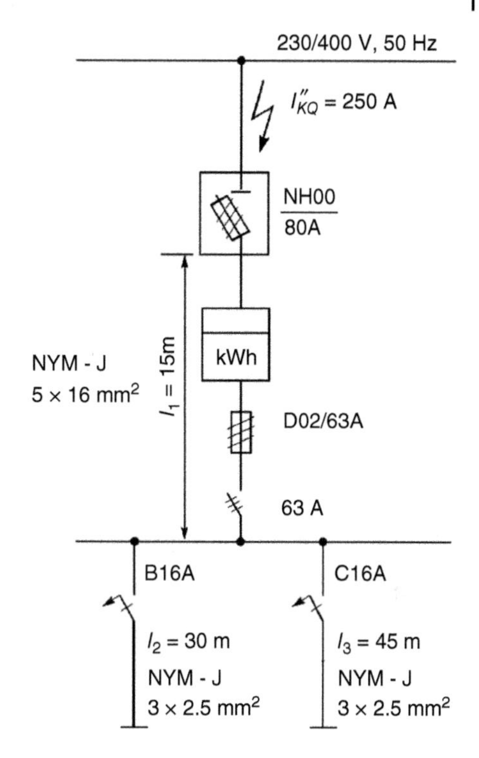

Figure 8.29 Building installation.

8.16.2 Example 2: Calculation of Short-Circuit Currents in a Building Electrical System

In the electrical installation of a building the following data are known (Figure 8.29). Calculate all required short-circuit currents and check the cut-off conditions.

1. Calculation of the single-pole short-circuit current

$$I''_{k1} = \frac{c\,U_n}{\sqrt{3}\,Z_k} \Rightarrow Z_k = \frac{c\,U_n}{\sqrt{3}I''_{k1}} = \frac{0.9 \cdot 400\ \text{V}}{\sqrt{3} \cdot 250\ \text{A}} = 0.831\ \Omega$$

$$Z_{\text{L1}} = 1.24\,\frac{2 \cdot l_1}{\kappa\,S} = 1.24 \cdot \frac{2 \cdot 15\ \text{m}}{56\frac{\text{m}}{\Omega\ \text{mm}^2} \cdot 16\ \text{mm}^2} = 0.0415\ \Omega$$

$$Z_{\text{G}} = Z_k + Z_{\text{L1}} = 0.872\ \Omega$$

$$I''_{k1} = \frac{c\,U_n}{\sqrt{3}\,Z_{\text{G}}}$$

$$I''_{k1} = \frac{0.9 \cdot 400\ \text{V}}{\sqrt{3} \cdot 0.919\,07\ \Omega} = 238.35\ \text{A}$$

$$Z_{\text{L2}} = Z_{\text{G}} + 1.24\,\frac{2\,l_2}{\kappa\,S}$$

$$Z_{\text{L2}} = 0.919\,07\ \Omega + 1.24 \cdot \frac{2 \cdot 30\ \text{m}}{56\frac{\text{m}}{\Omega\ \text{mm}^2} \cdot 2.5\ \text{mm}^2} = 1.451\ \Omega$$

$$I''_{k1} = \frac{c\,U_n}{\sqrt{3}\,Z_{\text{L2}}} = \frac{0.9 \cdot 400\ \text{V}}{\sqrt{3} \cdot 1.451\ \Omega} = 143.24\ \text{A}$$

$$Z_{L3} = Z_{L2} + 1.24\,\frac{2\,l_3}{\kappa\,S}$$

$$Z_{L3} = 1.451\ \Omega + 1.24\,\frac{2 \cdot 45\ \text{m}}{56\frac{\text{m}}{\Omega\ \text{mm}^2} \cdot 2.5\ \text{mm}^2} = 2.248\ \Omega$$

$$I''_{k1} = \frac{c\,U_n}{\sqrt{3}\,Z_{L3}} = \frac{0.9 \cdot 400\ \text{V}}{\sqrt{3} \cdot 2.248\ \Omega} = 92.45\ \text{A}$$

2. Evaluation of the cut-off conditions
   (a) Connection with B characteristic
       $I_a = 5\,I_n = 5 \cdot 16\ \text{A} = 80\ \text{A}$
       $I''_{k1} = 143.24\ \text{A}$
       $I''_{k1} > I_a$ satisfied
   (b) Connection with C characteristic
       $I_a = 10\,I_n = 10 \cdot 16\ \text{A} = 160\ \text{A}$
       $I''_{k1} = 92.45$
       Cut-off does not take place, since $I''_{k1} < I_a$.

   The following measures are possible:
   (a) Increasing the cross section
   (b) Installing an RCD
   (c) Limiting the line length

8.16.3 Example 3: Dimensioning of an Exit Cable

A load is connected through an overhead line having the following data to a transformer: Network data:

$S_{rT} = 400$ kVA
$u_{kr} = 4\%$
$S = 4 \times 50\ \text{mm}^2$ Cu
$l = 700$ m

1. Calculate the single-pole short-circuit current:

$$Z_T = \frac{u_{kr}}{100\%}\,\frac{U_n^2}{S_{rT}} = 0.016\ \Omega,\ Z_l = 0.565\ \Omega/\text{km}$$

$$Z_G = Z_T + 2\,l\,Z_l = 0.807\ \Omega$$

$$I''_{k1} = \frac{c\,U_n}{\sqrt{3}\,Z_G} = \frac{0.9 \cdot 400\ \text{V}}{\sqrt{3} \cdot 0.807\ \Omega} = 257.55\ \text{A}$$

2. Select the fuse according to overload conditions:

$$I_n \le \frac{1.45\,I_z}{1.6} = \frac{1.45 \cdot 250\ \text{A}}{1.6} = 226.5\ \text{A}$$

$I_n = 200$ A is selected.

3. Prove that the cut-off condition is satisfied:
   The cut-off current I_a for the 200 A fuse in 0.4 second is 2.1 kA (Figure 13.20). The following relationship must always hold true:

$$I''_{k_1} > I_a$$

In this practical case, however, it is found that

$$I''_{k_1} = 257.55 \text{ A} < I_a = 2.1 \text{ kA}$$

The cut-off current is greater than the single-pole short-circuit current. The cut-off condition is therefore not satisfied.

8.16.4 Example 4: Calculation of Short-Circuit Currents with Zero-Sequence Resistances

The single-pole short-circuit current is to be calculated at the end of the line, considering the zero-sequence resistances. The data are given in Figure 8.30.

Power supply feeder:

$$X_{Qt} = \frac{c\,U_{nQ}^2}{\sqrt{3}\,S''_{kQ}} = \frac{1.1 \cdot (400\text{ V})^2}{\sqrt{3} \cdot 350\text{ MVA}} = 0.290\text{ m}\Omega$$

$$R_{Qt} = 0.1 \cdot X_Q = 0.1 \cdot 0.290\text{ m}\Omega = 0.0290\text{ m}\Omega$$

Transformer:

$$R_{\text{TLV}} = \frac{u_{Rr}}{100\%} \cdot \frac{U_{\text{rTLV}}^2}{S_{\text{rT}}} = \frac{1.15\%}{100\%} \frac{(0.4\text{ kV})^2}{630\text{ kVA}} = 2.92\text{ m}\Omega$$

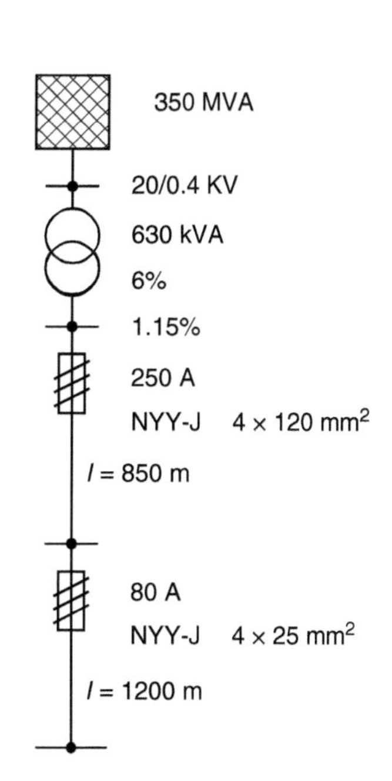

Figure 8.30 Power supply feeder.

$$Z_{\text{TLV}} = \frac{u_{kr}}{100\%} \cdot \frac{U_{\text{rTLV}}^2}{S_{rT}} = \frac{6\%}{100\%} \frac{(0.4\ \text{kV})^2}{630\ \text{kVA}} = 15\ \text{m}\Omega$$

$$X_{\text{TLV}} = \sqrt{Z_{\text{TLV}}^2 - R_{\text{TLV}}^2} = \sqrt{15^2 - 2.92^2}\ \text{m}\Omega = 14.71\ \text{m}\Omega$$

$$\frac{R_{\text{THV}}}{R_{\text{TLV}}} = 1 \Rightarrow R_{\text{THV}} = 1 \cdot 2.92\ \text{m}\Omega = 2.92\ \text{m}\Omega$$

$$\frac{X_{\text{THV}}}{X_{\text{TLV}}} = 0.96 \Rightarrow X_{\text{THV}} = 0.96 \cdot 14.71\ \text{m}\Omega = 14.12\ \text{m}\Omega$$

Cable:

$$R_l = R_l'\, l = 0.195\ \Omega/\text{km} \cdot 0.850\ \text{km} = 165.75\ \text{m}\Omega$$

$$X_l = X_l'\, l = 0.080\ \Omega/\text{km} \cdot 0.850\ \text{km} = 68\ \text{m}\Omega$$

$$\frac{R_{\text{0L}}}{R_l} = 4 \Rightarrow R_{\text{0L}} = 4 \cdot 165.75\ \text{m}\Omega = 663\ \text{m}\Omega$$

$$\frac{X_{\text{0L}}}{X_l} = 3.65 \Rightarrow X_{\text{0L}} = 3.65 \cdot 68\ \text{m}\Omega = 248.2\ \text{m}\Omega$$

Line:

$$R_l = R_l'\, l = 0.898\ \Omega/\text{km} \cdot 1.2\ \text{km} = 1077.6\ \text{m}\Omega$$

$$X_l = X_l'\, l = 0.086\ \Omega/\text{km} \cdot 1.2\ \text{km} = 103.2\ \text{m}\Omega$$

$$\frac{R_{\text{0L}}}{R_l} = 4 \Rightarrow R_{\text{0L}} = 4 \cdot 1077.6\ \text{m}\Omega = 4310.4\ \text{m}\Omega$$

$$\frac{X_{\text{0L}}}{X_l} = 4.13 \Rightarrow X_{\text{0L}} = 4.13 \cdot 103.2\ \text{m}\Omega = 426.21\ \text{m}\Omega$$

Calculation of the single-pole short-circuit current:

$$\sum R = (2R_{Qt} + 2R_{\text{T}} + 2R_{\text{K}} + 2R_l + R_{\text{0T}} + R_{\text{0K}} + R_{\text{0L}})$$

$$\sum X = (2X_{Qt} + 2X_{\text{T}} + 2X_{\text{K}} + 2X_l + X_{\text{0T}} + X_{\text{0K}} + X_{\text{0L}})$$

$$\sum R = (2 \cdot 0.0290 + 2 \cdot 2.92 + 2 \cdot 165.75 + 2 \cdot 1077.6 + 2.92 + 663 + 4310.4)\text{m}\Omega = 7.46\ \Omega$$

$$\sum X = (2 \cdot 0.290 + 2 \cdot 14.71 + 2 \cdot 14.12 + 2 \cdot 68 + 248.2 + 103.2 + 426.216)\text{m}\Omega = 0.971\ \Omega$$

$$I_{k1}'' = \frac{\sqrt{3}\, c\, U_n}{\sqrt{R^2 + X^2}} = \frac{\sqrt{3} \cdot 0.9 \cdot 400\ \text{V}}{\sqrt{(7.46\ \Omega)^2 + (0.971\ \Omega)^2}} = 82.88\ \text{A}$$

8.16.5 Example 5: Complex Calculation of Short-Circuit Currents

The complex calculation is shown with this example (Figure 8.31).

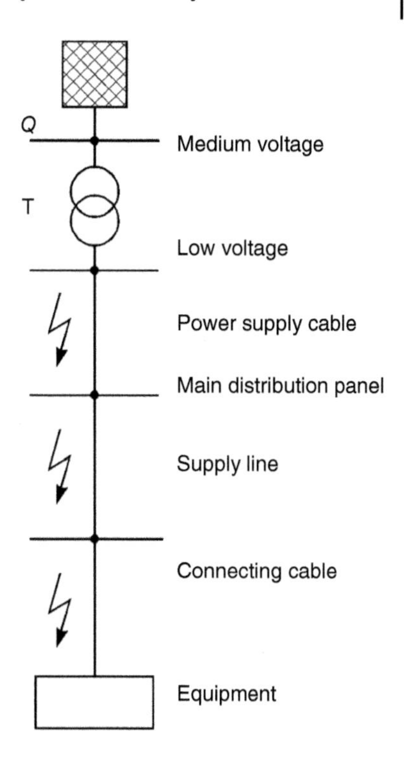

Figure 8.31 Complex calculation.

(a) *Calculation of primary distribution voltage*:

$$S''_{kQ} = 500\ \text{MVA}$$

$$Z_Q = \frac{c\,U_n^2}{S''_{kQ}} = \frac{1.1 \cdot (400\ \text{V})^2}{500\ \text{MVA}} = 0.352\ \text{m}\Omega$$

$$X_Q = 0.995\ Z_Q$$

$$X_Q = 0.995 \cdot 0.352\ \text{m}\Omega = 0.35\ \text{m}\Omega$$

$$R_Q = 0.1\ X_Q$$

$$R_Q = 0.1 \cdot 0.35\ \text{m}\Omega = 0.035\ \text{m}\Omega$$

(b) *Transformer*:

$$U_n = 400\ \text{V}$$

$$S_{\text{rT}} = 316\ \text{kVA}$$

$$u_{kr} = 6\%$$

$$u_{Rr} = 1\%$$

$$u_{xr} = \sqrt{u_{kr}^2 - u_{Rr}^2} = \sqrt{6^2 - 1}\% = 5.92\%$$

$$X_{\text{T}} = \frac{u_{xr}}{100\%} \cdot \frac{U_{\text{rT}}^2}{S_{\text{rT}}} = \frac{5.92\%}{100\%} \cdot \frac{(400\ \text{V})^2}{316\ \text{kVA}} = 29.9\ \text{m}\Omega$$

$$R_{\text{T}} = \frac{u_{Rr}}{100\%} \cdot \frac{u_{\text{rT}}^2}{S_{\text{rT}}} = \frac{1\%}{100\%} \cdot \frac{(400\ \text{V})^2}{316\ \text{kVA}} = 5.06\ \text{m}\Omega$$

(c) *Supply cable* $3 \times 185/95\ \text{mm}^2$:

$$R_1' = 0.105\ \text{m}\Omega/\text{m}$$
$$X_1' = 0.072\ \text{m}\Omega/\text{m}$$
$$l_1 = 30\ \text{m}$$
$$R_{\text{L1}} = R_1\ l = 0.105\ \text{m}\Omega/\text{m} \cdot 30\ \text{m} = 3.15\ \text{m}\Omega$$
$$X_{\text{L1}} = X_1\ l = 0.072\ \text{m}\Omega/\text{m} \cdot 30\ \text{m} = 2.16\ \text{m}\Omega$$

(d) *Main low-voltage distribution*:
Three-pole short-circuit:

$$Z_{\text{G}} = |(R_Q + R_{\text{T}} + R_{\text{L1}}) + j(X_Q + X_{\text{T}} + X_{\text{L1}})|$$
$$= |(0.035 + 5.06 + 3.15)\text{m}\Omega + j(0.35 + 29.9 + 2.16)\text{m}\Omega| = 33.44\ \text{m}\Omega$$
$$I''_{k3} = \frac{c\ U_n}{\sqrt{3}\ Z_{\text{G}}} = \frac{1.1 \cdot 400\ \text{V}}{\sqrt{3} \cdot 33.44\ \text{m}\Omega} = 7.6\ \text{kA}$$
$$i_{\text{p}} = \kappa\ \sqrt{2}\ I''_{k3} = 1.2 \cdot \sqrt{2} \cdot 7.6\ \text{kA} = 12.9\ \text{kA}$$

(e) *Supply line*:
Cable or line: $4 \times 35\ \text{mm}^2$

$$R_2 = 0.627\ \text{m}\Omega/\text{m}$$
$$X_2 = 0.083\ \text{m}\Omega/\text{m}$$
$$l_2 = 250\ \text{m}$$
$$R_{\text{L2}} = R_2'\ l_2 = 0.627\ \text{m}\Omega/\text{m} \cdot 250\ \text{m} = 156.75\ \text{m}\Omega$$
$$X_{\text{L2}} = X_2'\ l_2 = 0.089\ \text{m}\Omega/\text{m} \cdot 250\ \text{m} = 22.25\ \text{m}\Omega$$

(f) *Sub-distribution*:
Three-pole short-circuit:

$$Z_{\text{G}} = |(R_Q + R_{\text{T}} + R_{\text{L1}} + R_{\text{L2}})\text{m}\Omega + j(X_Q + X_{\text{T}} + X_{\text{L1}} + X_{\text{L2}})\text{m}\Omega|$$
$$Z_{\text{G}} = |(0.035 + 5.06 + 3.15 + 156.75)\text{m}\Omega + j(0.35 + 29.9 + 2.16 + 22.25)\text{m}\Omega|$$
$$Z_{\text{G}} = 173.81\ \text{m}\Omega$$
$$I''_{k3} = \frac{U_n}{\sqrt{3}\ Z_{\text{G}}} = \frac{1.1 \cdot 400\ \text{V}}{\sqrt{3} \cdot 173.81\ \text{m}\Omega} = 1.46\ \text{kA}$$
$$i_{\text{p}} = \kappa\ \sqrt{2}\ I''_{k3} = 2.47\ \text{kA}$$

(g) *Connection cable*:
Cable or line: $5 \times 2.5\ \text{mm}^2$

$$R_3' = 8.71\ \text{m}\Omega/\text{m}$$
$$X_3' = 0.11\ \text{m}\Omega/\text{m}$$
$$l_3 = 25\ \text{m}$$

$$R_{L3} = R'_3\, l_3 = 8.71\ \text{m}\Omega/\text{m} \cdot 25\ \text{m} = 217.75\ \text{m}\Omega$$
$$X_{L3} = X'_3\, l_3 = 0.11\ \text{m}\Omega/\text{m} \cdot 25\ \text{m} = 2.75\ \text{m}\Omega$$

(h) *Equipment*:
Three-pole short-circuit

$$Z_G = |(R_Q + R_T + R_{L1} + R_{L2} + R_{L3}) + jk(X_Q + X_T + X_{L1} + X_{L2} + X_{L3})|$$
$$Z_G = (0.035 + 5.06 + 3.15 + 156.75 + 217.75)\text{m}\Omega$$
$$+ jk(0.35 + 29.9 + 2.16 + 22.25 + 2.75)\text{m}\Omega$$
$$Z_G = 387.03\ \text{m}\Omega$$
$$I''_{k3} = \frac{c\, U_n}{\sqrt{3}\, Z_G} = \frac{1.1 \cdot 400\ \text{V}}{\sqrt{3} \cdot 387.03\ \text{m}\Omega} = 0.656\ \text{kA}$$
$$i_p = \kappa\, \sqrt{2}\, I''_{k3} = 1.41 \cdot \sqrt{2} \cdot 0.656\ \text{kA} = 1.3\ \text{kA}$$

8.16.6 Example 6: Calculation with Effective Power and Reactive Power

Find the following for the electrical system shown in Figure 8.32:

1. Calculate the value $\cos\varphi$ for the system.
2. Select the overcurrent protection devices.
3. Calculate the voltage drops.
4. Dimension the transformer.
5. Design the reactive current compensation for a value $\cos\varphi = 0.95$.
6. Calculate the single-pole and three-pole short-circuit currents.
7. Calculate the single-pole and three-pole short-circuit currents with the simplified method.
8. Determine the permissible lengths.

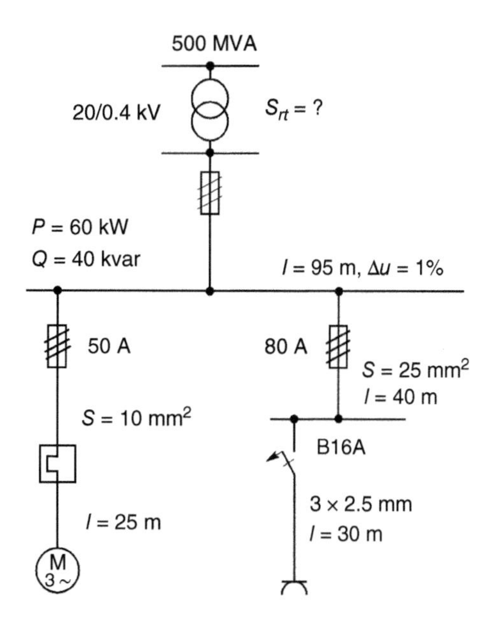

Figure 8.32 Calculation with effective and reactive power.

Solution:

1. $\cos\varphi$ *for the system*:

$$\tan\varphi = \frac{Q}{P} = \frac{40\ \text{kvar}}{60\ \text{kW}} = 0.66 \Rightarrow \cos\varphi = 0.83$$

2. *Selection of the overcurrent protective equipment*:

$$I_B = \frac{P}{\sqrt{3}\ U_n\ \cos\varphi} = \frac{60\ \text{kW}}{\sqrt{3}\cdot 400\ \text{V}\cdot 0.83} = 104.46\ \text{A}$$

$$I_B \leq I_n \leq I_z$$

$$104.4\ \text{A} \leq 125\ \text{A} \leq 128\ \text{A}$$

$$I_n = 125\ \text{A} \quad \text{for } 25\ \text{mm}^2$$

NYY–J$4\times 25\ \text{mm}^2$ underground installation,

from Table 14.18 we obtain $I_z = 133$ A

$$I_r = \frac{104.46\ \text{A}}{0.91} = 115\ \text{A},$$

$f_1 = 0.91$ (DIN VDE 0276 Part 1000 Table 4)

$$I_z' = I_z\, f_1 = 133\ \text{A}\cdot 0.91 = 121.03\ \text{A}$$

$$S = 35\ \text{mm}^2 \Rightarrow I_z = 159\ \text{A}\cdot 0.91 = 144.69\ \text{A}$$

$$I_B \leq I_n \leq I_z$$

$$104.5\ \text{A} \leq 125\ \text{A} \leq 144.69\ \text{A}$$

$$S = \frac{\sqrt{3}\cdot 125\ \text{A}\cdot 95\ \text{m}\cdot 0.83}{56\frac{\text{m}}{\Omega\ \text{mm}^2}\cdot 4\ \text{V}} = 76\ \text{mm}^2$$

Selected : $4\times 95\ \text{mm}^2$

$$I_z' = 280\ \text{A}\cdot 0.91 = 254.8\ \text{A} \Rightarrow I_n = 160\ \text{A}$$

3. *Actual voltage drop: in accordance with DIN 18015*:

$$\Delta U = \frac{\sqrt{3}\, l\, I_n\ \cos\varphi}{\kappa\ S} = \frac{\sqrt{3}\cdot 95\ \text{m}\cdot 160\ \text{A}\cdot 0.83}{56\frac{\text{m}}{\Omega\ \text{mm}^2}\cdot 95\ \text{mm}^2} = 4.1\ \text{V}$$

$$\Delta u = \frac{\Delta U}{U_n}\cdot 100\% = \frac{4.1\ \text{V}}{400\ \text{V}}\cdot 100\% = 1\%$$

Voltage drop calculation for the main line

$$\Delta U = \frac{\Delta u}{100\%}\ U_n = \frac{1\%}{100\%}\cdot 400\ \text{V} = 4\ \text{V}$$

$$R_\text{L}'\ \cos\varphi + X_\text{L}'\ \sin\varphi = \frac{\Delta U}{\sqrt{3}\, l\, I_B}$$

$$R_\text{L}'\ \cos\varphi + X_\text{L}'\ \sin\varphi = \frac{4\ \text{V}}{\sqrt{3}\cdot 0.095\ \text{km}\cdot 104.46\ \text{A}} = 0.233\ \Omega/\text{km}$$

4mm

From Table 16.5, the effective resistance per unit length for $4\times 95\ \text{mm}^2$ $0.232\Omega/\text{km}$. The actual voltage drop is therefore

$$\Delta U = \sqrt{3}\, l\, I\ (R_l'\ \cos\varphi + X_\text{L}'\ \sin\varphi)$$

$$\Delta U = \sqrt{3}\cdot 0.095\ \text{km}\cdot 104.46\ \text{A}\cdot 0.232\ \Omega/\text{km} = 4\ \text{V}$$

This corresponds to

$$\Delta u = \frac{\Delta U}{U_n} \cdot 100\% = \frac{4\ \text{V}}{400\ \text{V}} \cdot 100\% = 1\%$$

Motor: $$\Delta u = \frac{\sqrt{3}\, l\, I_r\ \cos\varphi \cdot 100\%}{\kappa\ S\ U_n}$$

$$\Delta u = \frac{\sqrt{3} \cdot 25\ \text{m} \cdot 50\ \text{A} \cdot 0.8 \cdot 100\%}{56 \frac{\text{m}}{\Omega\ \text{mm}^2} \cdot 10\ \text{mm}^2 \cdot 400\ \text{V}} = 0.77\%$$

Sub-distribution: $$\Delta U = \frac{\sqrt{3} \cdot 40\ \text{m} \cdot 80\ \text{A} \cdot 0.8 \cdot 100\%}{56 \frac{\text{m}}{\Omega\ \text{mm}^2} \cdot 25\ \text{mm}^2 \cdot 400\ \text{V}} = 0.79\%$$

Electrical outlet: $$\Delta U = \frac{2 \cdot 30\ \text{m} \cdot 16\ \text{A} \cdot 1 \cdot 100\%}{56 \frac{\text{m}}{\Omega\ \text{mm}^2} \cdot 2.5\ \text{mm}^2 \cdot 230\ \text{V}} = 2.98\%$$

4. *Dimensioning of transformer*:

$$S_{\text{rT}} = \frac{P_{\text{ges}}}{\cos\varphi} = \frac{60\ \text{kW}}{0.83} = 72.28\ \text{kVA} \quad \text{or}$$

$$S_{\text{rT}} = \sqrt{3}\ U_n\ I_n = \sqrt{3} \cdot 400\ \text{V} \cdot 160\ \text{A} = 110.89\ \text{kVA}$$

Selected: 100 kVA with $I_{\text{rT}} = 144A$, $u_{Rr} = 1.75\%$, $u_{kr} = 3.6\%\ (4\%)$

$$R_{\text{T}} = \frac{u_{Rr}\ U_n^2}{100\%\ S_{\text{rT}}} = \frac{1.75\% \cdot (400\ \text{V})^2}{100\% \cdot 100\ \text{kVA}} = 28\ \text{m}\Omega$$

$$Z_{\text{T}} = \frac{u_{kr}\ U_n^2}{100\%\ S_{\text{rT}}} = \frac{4\% \cdot (400\ \text{V})^2}{100\% \cdot 100\ \text{kVA}} = 64\ \text{m}\Omega$$

$$X_{\text{T}} = \sqrt{Z_{\text{T}}^2 - R_{\text{T}}^2} = 57.55\ \text{m}\Omega$$

5. *Reactive current compensation*:

$$Q_{\text{G}} = P \cdot (\tan\varphi_1 - \tan\varphi_2)$$

$$Q_{\text{G}} = 60\ \text{kW} \cdot (0.672 - 0.2029) = 28.146\ \text{kVar}$$

$$C_{\text{ges}} = \frac{Q_{\text{G}}}{w\ U_n^2} = \frac{28.146\ \text{kVar}}{314\ \text{s}^{-1} \cdot (400\ \text{V})^2} = 560.23\ \mu\text{F}$$

6. *Calculation of the single-pole short-circuit current*:
Impedance in the main distribution system

$$Z_{\text{MDP}} = 64\ \text{m}\Omega + \frac{1.24 \cdot 2 \cdot 95\ \text{m}}{56 \frac{\text{m}}{\Omega\ \text{mm}^2} \cdot 95\ \text{mm}^2} = 108.28\ \text{m}\Omega$$

$$I''_{k1} = \frac{c\ U_n}{\sqrt{3}\ Z_{\text{MDP}}} = \frac{0.9 \cdot 400\ \text{V}}{\sqrt{3} \cdot 108.28\ \text{m}\Omega} = 1.91\ \text{kA}$$

With $I_n = 160$ A, $I_a = 1$ kA, read at five seconds
$I''_{k1} > I_a$, so that the cut-off condition is satisfied.

Main distribution: $$I''_{k1} = \frac{c\ U_n}{\sqrt{3}\ Z_{\text{T}}} = \frac{0.9 \cdot 400\ \text{V}}{\sqrt{3} \cdot 64\ \text{m}\Omega} = 3.24\ \text{kA}$$

Motor: $$Z_L = \frac{1.24 \cdot 2 \cdot l}{\kappa\, S} = \frac{1.24 \cdot 2 \cdot 25\ \text{m}}{56 \frac{\text{m}}{\Omega\ \text{mm}^2} \cdot 10\ \text{mm}^2} = 0.11\ \Omega$$

$$Z_G = Z_{MDP} + Z_L = 108.28 + 110\ \text{m}\Omega = 218.28\ \text{m}\Omega$$

$$I''_{k1} = \frac{0.9 \cdot 400\ \text{V}}{\sqrt{3} \cdot 218.28\ \text{m}\Omega} = 0.952\ \text{A}$$

Sub-distribution: $$Z_l = \frac{1.24 \cdot 2 \cdot l}{\kappa\, S} = \frac{1.24 \cdot 2 \cdot 40\ \text{m}}{56 \frac{\text{m}}{\Omega\ \text{mm}^2} \cdot 25\ \text{mm}^2} = 70.85\ \text{m}\Omega$$

$$Z_G = Z_{MDP} + Z_l = 108.28\ \text{m}\Omega + 70.85\ \text{m}\Omega = 179.13\ \text{m}\Omega$$

$$I''_{k1} = \frac{0.9 \cdot 400\ \text{V}}{\sqrt{3} \cdot 179.13\ \text{m}\Omega} = 1.16\ \text{kA}$$

Electrical outlet: $$Z_L = \frac{1.24 \cdot 2 \cdot l}{\kappa\, S} = \frac{1.24 \cdot 2 \cdot 30\ \text{m}}{56 \frac{\text{m}}{\Omega\ \text{mm}^2} \cdot 2.5\ \text{mm}^2} = 531.42\ \text{m}\Omega$$

$$Z_G = Z_{MDP} + Z_L = 108.28\ \text{m}\Omega + 531.42\ \text{m}\Omega$$
$$= 639.7\ \text{m}\Omega$$

$$I''_{k1} = \frac{0.9 \cdot 400\ \text{V}}{\sqrt{3} \cdot 639.7\ \text{m}\Omega} = 325\ \text{A}$$

7. *Simplified method with the specified impedance values:*

Transformer $$Z_T = \frac{u_{kr}}{100\%} \frac{(U_n)^2}{S_{rT}} = \frac{4\%}{100\%} \cdot \frac{(400\ \text{V})^2}{100\ \text{kVA}} = 64\ \text{m}\Omega$$

Main distribution feeder

$$Z_l = 2\, z\, l = 2 \cdot 0.257\ \Omega/\text{km} \cdot 0.095\ \text{km} = 48.83\ \text{m}\Omega$$

$$I''_{k1} = \frac{c\, U_n}{\sqrt{3}\, Z_G} = \frac{0.95 \cdot 400\ \text{V}}{\sqrt{3} \cdot (64 + 48.83)\text{m}\Omega} = 1.944\ \text{kA}$$

Motor feeder

$$Z_{LM} = 2\, z\, l = 2 \cdot 2.246\ \Omega/\text{km} \cdot 0.025\ \text{km} = 112.3\ \text{m}\Omega$$

$$Z_G = Z_T + Z_l + Z_{LM} = 225.13\ \text{m}\Omega$$

$$I''_{k1} = \frac{0.9 \cdot 400\ \text{V}}{\sqrt{3}\, Z_G} = \frac{0.95 \cdot 400\ \text{V}}{\sqrt{3} \cdot 225.13\ \text{m}\Omega} = 923.22\ \text{A}$$

Sub-distribution feeder

$$Z_{SD} = 2\, z\, l = 2 \cdot 0.902\ \Omega/\text{km} \cdot 0.040\ \text{km} = 72.16\ \text{m}\Omega$$

$$Z_G = Z_T + Z_l + Z_{SD} = 184.99\ \text{m}\Omega$$

$$I''_{k1} = \frac{0.9 \cdot 400\ \text{V}}{\sqrt{3} \cdot 184.99\ \text{m}\Omega} = 1.123\ \text{kA}$$

Electrical outlet

$$Z_{SD} = 2\, z\, l = 2 \cdot 8.77\ \text{m}\Omega/\text{m} \cdot 0.030\ \text{km} = 526.2\ \text{m}\Omega$$

$$Z_G = Z_T + Z_l + Z_{SD} + Z_{LM} + Z_{SD} = 823.49\ \text{m}\Omega$$

$$I''_{k1} = 252.4\ \text{A}$$

8. *Determination of the maximum length:*
   The permissible lengths are

$$l = \frac{\frac{c\,U_n}{\sqrt{3}\,I_a} - Z_V}{2\,z}$$

Main distribution $$l = \frac{\frac{0.9 \cdot 400\ \text{V}}{\sqrt{3} \cdot 950\ \text{A}} - 64\ \text{m}\Omega}{2 \cdot 0.257\ \Omega/\text{km}} = 301\ \text{m} > 95\ \text{m}$$

Motor $$l = \frac{\frac{0.9 \cdot 400\ \text{V}}{\sqrt{3} \cdot 250\ \text{A}} - 79.82\ \text{m}\Omega}{2 \cdot 2.246\ \Omega/\text{km}} = 167.31\ \text{m} > 25\ \text{m}$$

Sub-distribution $$l = \frac{\frac{0.9 \cdot 400\ \text{V}}{\sqrt{3} \cdot 450\ \text{A}} - 79.82\ \text{m}\Omega}{2 \cdot 0.902\ \Omega/\text{km}} = 211.78\ \text{m} > 40\ \text{m}$$

Electrical outlet $$l = \frac{\frac{0.9 \cdot 400\ \text{V}}{\sqrt{3} \cdot 80\ \text{A}} - 102.94\ \text{m}\Omega}{2 \cdot 8.770\ \Omega/\text{km}} = 142.25\ \text{m} > 30\ \text{m}$$

The circuits are now correctly dimensioned.

8.16.7 Example 7: Complete Calculation for a System

The following calculations are for an electrical system as shown in Figure 8.33:

1. Determine the feeders, fuses, and motor circuit-breakers for all motors for
   (a) direct starting
   (b) Y/Δ starting.
2. Determine I''_{k1} and I''_{k3} for
   (a) main distribution 1
   (b) sub-distribution 1
   (c) sub-distribution 2
   (d) motor 1
   (e) circuit no. 20.
3. Determine the voltage drops for:
   (a) main distribution 1
   (b) sub-distribution 1
   (c) sub-distribution 2
   (d) circuit no. 20, connected to sub-distribution 3, with $I_B = 8$ A.
4. Select all circuit breakers and fuses.
5. Demonstrate the selectivity.
6. Determine the transformer data.
7. Determine the compensation power.
8. Determine the cut-off conditions for all circuits.

The network voltage is 400/230 V, 50 Hz. As a protective measure, the TN system is required.

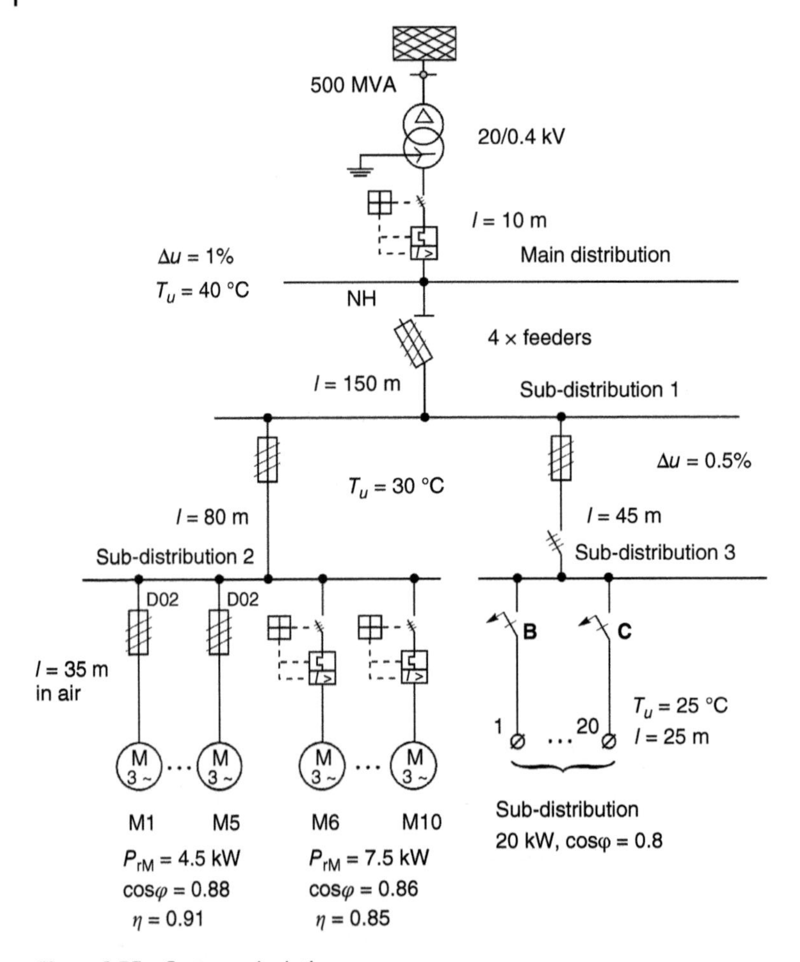

Figure 8.33 System calculation.

Solution:

1. *Power balance for the system*:
   (a) Sub-distribution (SD) 3
   Given: $P_{\text{SD3}} = 20\ \text{kW},\ \cos\varphi = 0.8, U_n = 400\ \text{V}$

$$I_{\text{SD3}} = \frac{P_{\text{SD3}}}{\sqrt{3}\,U_n \cos\varphi} = \frac{20\ \text{kW}}{\sqrt{3} \cdot 400\ \text{V} \cdot 0.8} = 36.12\ \text{A}$$

$$S_{\text{SD3}} = \frac{P_{\text{SD3}}}{\cos\varphi} = \frac{20\ \text{kW}}{0.8} = 25\ \text{kVA}$$

   (b) Sub-distribution 2
   Given:

Motors 1–5	Motors 6–10
$P_2 = 4.5\ \text{kW}$	$P_2 = 7.5\ \text{kW}$
$\eta_1 = 0.91$	$\eta_6 = 0.85$
$\cos\varphi_1 = 0.88$	$\cos\varphi_6 = 0.86$

$$I_{1-5} = \frac{4500\ \mathrm{W}}{\sqrt{3} \cdot 400\ \mathrm{V} \cdot 0.91 \cdot 0.88} = 8.11\ \mathrm{A}$$

$$I_{6-10} = \frac{7500\ \mathrm{W}}{\sqrt{3} \cdot 400\ \mathrm{V} \cdot 0.85 \cdot 0.86} = 14.82\ \mathrm{A}$$

$$P_1 = \frac{P_1}{\eta} = \frac{4.5\ \mathrm{kW}}{0.91} = 4.95\ \mathrm{kW}$$

$$S_{1-5} = \frac{5\ P_1}{\cos\varphi_1} = 28.125\ \mathrm{kVA}$$

$$P_6 = \frac{P_6}{\eta} = \frac{7.5\ \mathrm{kW}}{0.85} = 8.82\ \mathrm{kW}$$

$$S_{6-10} = \frac{5\ P_{6zu}}{\cos\varphi_6} = 51.28\ \mathrm{kW}$$

$$P_{\mathrm{SD2}} = 5\ P_1 + 5\ P_6 = 5 \cdot 4.95\ \mathrm{kW} + 5 \cdot 8.82\ \mathrm{kW} = 68.85\ \mathrm{kW}$$

$$S_{\mathrm{SD2}} = S_{1-5} + S_{6-10} = 79.4\ \mathrm{kVA}$$

$$I_{\mathrm{SD2}} = \frac{S_{\mathrm{SD2}}}{\sqrt{3}\ U_n} = \frac{79.4\ \mathrm{kVA}}{\sqrt{3} \cdot 400\ \mathrm{V}} = 114.6\ \mathrm{A}$$

$$\cos\varphi = \frac{P_{\mathrm{SD2}}}{S_{\mathrm{SD2}}} = \frac{68.86\ \mathrm{kW}}{79.4\ \mathrm{kVA}} = 0.867$$

(c) Sub-distribution 1

$$P_{\mathrm{SD1}} = P_{\mathrm{SD3}} + P_{\mathrm{SD2}} = 20\ \mathrm{kW} + 68.86\ \mathrm{kW} = 88.86\ \mathrm{kW}$$

$$S_{\mathrm{SD1}} = S_{\mathrm{SD3}} + S_{\mathrm{SD2}} = 25\ \mathrm{kVA} + 79.4\ \mathrm{kVA} = 104.4\ \mathrm{kVA}$$

$$\cos\varphi_{\mathrm{SD1}} = \frac{P_{\mathrm{SD1}}}{S_{\mathrm{SD1}}} = \frac{88.86\ \mathrm{kW}}{104.4\ \mathrm{kVA}} = 0.85$$

$$I_{\mathrm{SD1}} = \frac{S_{\mathrm{SD1}}}{\sqrt{3}\ U_n} = \frac{104.4\ \mathrm{kVA}}{\sqrt{3} \cdot 400\ \mathrm{V}} = 150.7\ \mathrm{A}$$

For $I_n = 35$ A the cross section is chosen: NYY 5×4 mm^2.
From Table 14.20, the current-carrying capacity is 34 A.
$I_z = f_1\ I_{z\mathrm{Tab.}} = 1.07 \cdot 34\ \mathrm{A} = 36.38\ \mathrm{A}$
From the rule for nominal currents
$14.81\ \mathrm{A} \leq 16\ \mathrm{A} \leq 36.38\ \mathrm{A}$

(d) The motor protection is placed before the network protection

$$I_n = 16\ \mathrm{A}, I_r = 18.5\ \mathrm{A}, \text{from Table 14.20} \rightarrow \text{NYY } 5 \times 4\ \mathrm{mm}^2$$

$$I_z = I_r\ f_1 = 19.5\ \mathrm{A} \cdot 1.07 = 20.865\ \mathrm{A}$$

$$I_B \leq I_n \leq I_z \rightarrow 14.81\ \mathrm{A} \leq 16\ \mathrm{A} \leq 20.865\ \mathrm{A}$$

2. *Selection of lines (except for motors)*:
   (a) Feeder to main distribution 1 ⟶ power circuit -breaker in transformer station

$$I_B = 150.7\ \mathrm{A},\ I_n = 250\ \mathrm{A},\ \text{ambient temperature } T_U = 30\,^\circ\mathrm{C},$$
underground installation.

$$I_z = 280\ \text{A, from Table 14.18}$$
$$I_B \leq I_n \leq I_z$$
$$150.7\ \text{A} \leq 250\ \text{A} \leq 280\ \text{A}$$

Selected: NYCWY 4 × 95 mm^2

(b) Feeder to sub-distribution 1

$$I_B = 150.7\ \text{A},\ I_n = 160\ \text{A, ambient temperature } 40\,^\circ\text{C}$$
free installation, perforated cable rack,
grouping $f_2 = 4$ lines.
$$I_z = 202\ \text{A from Table 14.20}$$
$$f_1 = 0.87\ \text{from Table 14.24 at } 40\,^\circ\text{C}$$
$$f_2 = 0.95\ \text{from Table 14.22}$$
$$I_z = I_r\, f_1\, f_2$$
$$= 202\ \text{A} \cdot 0.87 \cdot 0.95 = 166.9\ \text{A}$$
$$I_B \leq I_n \leq I_z \rightarrow 150.7\ \text{A} \leq 160\ \text{A} \leq 166.9\ \text{A from Table 14.19}$$

Selected: NYY 4 ×50 mm^2

3. *Selection of protective equipment (backup fuses for distribution)*:
   (a) Backup fuse for sub-distribution 3

$$I_B = 36.12\ \text{A} \Rightarrow I_n = 50\ \text{A (D02)}$$
$$\text{RCD} = 63\ \text{A}/0.5\ \text{A}$$

   (b) Backup fuse for sub-distribution 2

$$I_B = 114.68\ \text{A} \Rightarrow I_n = 125\ \text{A/(NH1)}$$
$$\text{RCD} = 125\ \text{A}/0.3\ \text{A}$$

   (c) Backup fuse for sub-distribution 1

$$I_B = 150.7\ \text{A} \Rightarrow I_n = 160\ \text{A (NH1)}$$

   Because of the selectivity

$$125\ \text{A (sub-distribution 2)} \cdot 1.6 = 200\ \text{A (NH1)}$$

   Feeders, fuses for all motors M1-M10

   (d) Directly connected
   Given: M1–M5 $P_{ab} = 4.5$ kW, $I = 8.11$ A
   The fuse is chosen with $I_n = 20$ A
   The line cross section follows from Table 14.20
   $I_r = 25$ A, $f_1 = 1.06$ (25°) from Table 14.24
   $I_z = f_1\, I_r = 1.06 \cdot 25\ \text{A} = 26.5\ \text{A} \Rightarrow$ NYY 4 × 2.5 mm^2
   $I_B \leq I_n \leq I_z$
   $8.11\ \text{A} \leq 20\ \text{A} \leq 26.5\ \text{A} \Rightarrow$ The requirement is satisfied.

(e) in $Y/\Delta = 10$ A

(I) Motor protection installed before network protection $I_e = 1 \cdot I_B$

Line cross section, from Table 14.20:

$I_r = 19.5$ A, $f_1 = 1.06$ (25 °C) from Table 14.24

$I_z = I_r\, f_1 = 1.06 \cdot 19.5\text{ A} = 20.67\text{ A} \Rightarrow$ NYY $4 \times 1.5\text{ mm}^2$

$I_B \le I_n \le I_z$

$8.11\text{ A} \le 10\text{ A} \le 20.67\text{ A} \Rightarrow$ The requirement is satisfied.

(II) Motor protection installed after network protection $I_e = 0.58 \cdot I_B = 4.7$ A

$I_r = 19.5$ A, $f_1 = 1.06$, $f_2 = 0.65$ from Table 14.25

$I_z = f_1\, f_2\, I_r = 1.06 \cdot 0.65 \cdot 19.5\text{ A} = 13.43\text{ A} \Rightarrow$ NYY $7 \times 1.5\text{ mm}^2$

(f) Feeder to sub-distribution 2

$I_B = 114.68$ A, $I_n = 125$ A,

ambient temperature $T_U = 30°$, $\Rightarrow f_1 = 1$, free installation, cable rack, unperforated, no grouping

$I_r = 129$ A from Table 14.20

$f_2 = 0.97$ from Table 14.22

$I_z = f_2\, I_r = 0.97 \cdot 129\text{ A} = 125.13\text{ A} \Rightarrow I_z$ value too small, increase cross section.

$I_r = 157$ A from Table 14.20

$I_2 = f_2\, I_r = 0.97 \cdot 157\text{ A} = 152.29$ A

$I_B \le I_r \le I_z$

$114.6\text{ A} \le 125\text{ A} \le 152.29\text{ A} \Rightarrow$ The requirement is satisfied.

Selected: NYY $4 \times 50\text{ mm}^2$

(g) Feeder to sub-distribution 3

$I_B = 36.12$ A, $I_n = 50$ A

ambient temperature $T_U = 25°$, free installation, cable rack, unperforated, no grouping

$I_r = 59$ A from Table 14.20

$f_1 = 1.06$ from Table 14.24 at 25 °C

$f_2 = 0.97$ from Table 14.22, for one cable tray = 1

$I_z = f_1\, f_2\, I_r = 1.06 \cdot 0.97 \cdot 59\text{ A} = 60.66$ A

$I_B \le I_r \le I_2$

$36.1\text{ A} \le 50\text{ A} \le 60.66\text{ A} \Rightarrow$ The requirement is satisfied.

Selected: NYY $4 \times 10\text{ mm}^2$

(h) Feeder to circuit no. 20

$I_B = 8$ A, $I_n = 16$ A

ambient temperature $T_U = 25$ °C

installation type B2, no grouping

$I_z = 16.5$ A from Table 14.9

$I_B \le I_n \le I_z$

$8\text{ A} \le 16\text{ A} \le 16.5\text{ A} \Rightarrow$ The requirement is satisfied.

Selected: NYY $3 \times 1.5\text{ mm}^2$

4. *Determination of the voltage drops in accordance with IEC 60 364, with the operating current I_B Determination of cable or line:*

(a) Main distribution 1

Given: $I_B = 150.7\,\text{A}, S = 95\ \text{mm}^2, l = 10\,\text{m}, \Delta u_{\text{max}} = 1\%$

$$\Delta u = \frac{\sqrt{3}\, I_B\, l\, 100\%}{U_n}\left(\frac{1}{\kappa\, S}\cos\varphi + X_1'\sin\varphi\right)$$

$$= \frac{\sqrt{3}\cdot 150.7A\cdot 10\ \text{m}\cdot 100\%}{400\ \text{V}} \cdot\left(\frac{1}{56\frac{\text{m}}{\Omega\ \text{mm}^2}\cdot 95\ \text{mm}^2}\cdot 0.9787 + 0.08\cdot 10^{-3}\frac{\Omega}{\text{m}}\cdot 0.199\right)$$

$\Delta u = 0.13\%$, so that the requirement is satisfied.

Selected: NYCWY $4\times 95\,\text{mm}^2$

(b) Sub-distribution 1

Given: $I_B = 150.7\,\text{A}, S = 150\ \text{mm}^2, l = 150\ \text{m}, \Delta u_{\text{max}} = 1\%$

$$\Delta u = \frac{\sqrt{3}\, I_B\, l\, 100\%}{U_n}\left(\frac{1}{\kappa\, S}\cos\varphi + X_l'\sin\varphi\right)$$

$$\Delta u = \frac{\sqrt{3}\cdot 150.7\ \text{A}\cdot 150\ \text{m}\cdot 100\%}{400\ \text{V}} \cdot\left(\frac{1}{56\frac{\text{m}}{\Omega\ \text{mm}^2}\cdot 150\ \text{mm}^2}\cdot 0.98 + 0.08\cdot 10^{-3}\frac{\Omega}{\text{m}}\cdot 0.199\right)$$

$\Delta u = 1.3\%$

⇒ The value is too high, increase S

$$S = \frac{l\cos\varphi}{\kappa\left(\frac{\Delta U_{\text{max}}}{\sqrt{3}\, I_B} - X_l'\, l\sin\varphi\right)}$$

$$= \frac{150\ \text{m}\cdot 0.98}{56\frac{\text{m}}{\Omega\ \text{mm}^2}\left(\frac{4\ \text{V}}{\sqrt{3}\cdot 150.7\ \text{A}} - 0.08\cdot 10^{-3}\frac{\Omega}{\text{m}}\cdot 150\ \text{m}\cdot 0.199\right)}$$

$S = 202.9\ \text{mm}^2$

⇒ next standard cross section will be taken = 240 mm²

$\Delta u_{actual} = 0.87\%$

Selected: NYY $4\times 240\ \text{mm}^2$

(c) Sub-distribution 2

Given: $I_B = 114.6\ \text{A}, S = 50\ \text{mm}^2, l = 80\ \text{m}, \Delta u_{\text{max}} = 0.5\%, \cos\varphi = 0.867$

$$\Delta u = \frac{\sqrt{3}\cdot 114.6\ \text{A}\cdot 80\ \text{m}\cdot 100\%}{400\ \text{V}} \cdot\left(\frac{1}{56\frac{\Omega}{\text{m}\ \text{mm}^2}\cdot 50\ \text{mm}^2}\cdot 0.867 + 0.08\cdot 10^{-3}\frac{\Omega}{\text{m}}\cdot 0.489\right)$$

$\Delta u = 1.4\%$

⇒ The value is too high, increase S.

$$S = \frac{80\text{ m} \cdot 0.867}{56\frac{\text{m}}{\Omega\text{ mm}^2} \cdot (\frac{2V}{\sqrt{3}\cdot 114.6\text{ A}} - 0.08 \cdot 10^{-3}\frac{\Omega}{\text{m}} \cdot 80\text{ m} \cdot 0.498)}$$

$$S = 179.8\text{ mm}^2 \Rightarrow$$

next standard cross section will be taken = 185 mm^2

$$\Delta u_{actual} = 0.42\%$$

Selected: NYY 4 × 185 mm^2

(d) Sub-distribution 3

Given: $I_B = 36.1$ A, $S = 10$ mm^2, $l = 45$ m, $\Delta u_{\text{max}} = 0.5\%$, $\cos\varphi = 0.8$

$$\Delta u = \frac{\sqrt{3}\, l\, I_B \cos\varphi\, 100\%}{\kappa\, S\, U_n} = \frac{\sqrt{3} \cdot 45\text{ m} \cdot 36.12\text{ A} \cdot 0.8 \cdot 100\%}{56\frac{\text{m}}{\Omega\text{ mm}^2} \cdot 10\text{ mm}^2 \cdot 400\text{ V}} = 1\%$$

The value is too high, increase S.

$$S = \frac{45\text{ m} \cdot 0.8}{56\frac{\text{m}}{\Omega\text{ mm}^2}\,(\frac{2\text{ V}}{\sqrt{3}\cdot 36.1\text{ A}} - 0.08 \cdot 10^{-3}\frac{\Omega}{\text{m}} \cdot 45\text{ m} \cdot 0.6)} = 21.6\text{ mm}^2$$

The next standard cross section $S = 25$ mm^2 ⇒ $\Delta u = 0.44\%$

Selected: NYY 4 × 25 mm^2

(e) Circuit no. 20

In accordance with IEC 60 364, Part 52, the maximum permissible voltage drop in the system is 4%.

Main distribution = 0.13%, Sub-distribution 1= 0.87%, Sub-distribution 3 = 0.44%

The maximum calculated voltage drop from the main distribution, sub-distribution 1, and sub-distribution 3 is 1.44%.

⇒ $u_{\text{max}} = 4\% - 1.44\% = 2.56\%$

$I_B = 8$ A, $S = 1.5$ mm^2, $l = 25$ m, $\Delta u_{\text{max}} = 2.54\%$, $\cos\varphi = 0.8$, $U = 230$ V

$$\Delta u = \frac{2\, l\, I_B\, \cos\varphi\, 100\%}{\kappa\, S\, U_n} = \frac{2 \cdot 25\text{ m} \cdot 8\text{ A} \cdot 0.8 \cdot 100\%}{56\frac{\text{m}}{\Omega\text{ mm}^2} \cdot 1.5\text{ mm}^2 \cdot 230\text{ V}} = 1.65\%$$

⇒ The requirement is satisfied.

Selected: NYM 3 × 1.5 mm^2

5. *Determination of the transformer data*:
   Given: $u_{kr} = 4\%$, $U_{\text{THV}}/U_{\text{TLV}} = 20$ kV/0.4 kV, $S''_{kQ} = 500$ MVA
   Power balance of the system: $S_{\text{G}} = 104.4$ kVA
   The rated power of the transformer is selected: $S_{rT} \Rightarrow 160$ kVA
6. *Short-circuit current calculation*:

$$Z_{QT} = \frac{c\, U_n^2}{S''_k} = \frac{1.1 \cdot (0.4\text{ kV})^2}{500\text{ MVA}} = 0.352\text{ m}\Omega$$

$$X_{QT} = 0.995\, Z_{QT} = 0.35\text{ m}\Omega$$

$$R_{QT} = 0.1\, X_{QT} = 0.035\text{ m}\Omega$$

$$R_{\text{T}} = 15\text{ m}\Omega$$

$$X_{\mathrm{T}} = 37\ \mathrm{m\Omega}$$

$$Z_{\mathrm{T}} = \frac{u_{kr}\,U_{rT}^2}{100\%S_{rT}} = \frac{4\% \cdot (400\ \mathrm{V})^2}{100\% \cdot 160\ \mathrm{kVA}} = 40\ \mathrm{m\Omega}$$

(a) Main distribution 1 – I''_{k3}

$$R_{\mathrm{LHV}} = \frac{l}{\kappa\ S} = \frac{10\ \mathrm{m}}{56\frac{\mathrm{m}}{\Omega\ \mathrm{mm}^2} \cdot 95\ \mathrm{mm}^2} = 1.88\ \mathrm{m\Omega}$$

$$X_{\mathrm{LHV}} = l\,X'_l = 10\ \mathrm{m} \cdot 0.08\frac{\mathrm{m\Omega}}{\mathrm{m}} = 0.8\ \mathrm{m\Omega}$$

Summary of resistances and reactances:

	R	X
Network	0.035 mΩ	0.35 mΩ
Transformer	15 mΩ	37 mΩ
R_{LHV}	1.88 mΩ	0.80 mΩ
$R_{\mathrm{MDP}} =$	16.92 mΩ,	$X_{\mathrm{MDP}} = 38.15$ mΩ

$$Z_{\mathrm{MDP}} = \sqrt{R_{\mathrm{MDP}}^2 + X_{\mathrm{MDP}}^2}$$

$$Z_{\mathrm{MDP}} = \sqrt{(16.92\ \mathrm{m\Omega})^2 + (38.15\ \mathrm{m\Omega})^2} = 41.73\ \mathrm{m\Omega}$$

$$I''_{k3} = \frac{c\,U_n}{\sqrt{3}\,Z_{\mathrm{MDP}}} = \frac{1.1 \cdot 400\ \mathrm{V}}{\sqrt{3} \cdot 41.73\ \mathrm{m\Omega}} = 7\ \mathrm{kA}$$

(b) Main distribution 1 – I''_{k1}

$$R_{\mathrm{LHV}} = 1.24\frac{2\,l}{\kappa\ S} = 1.24 \cdot \frac{2 \cdot 10\ \mathrm{m} \cdot 1000}{56\frac{\mathrm{m}}{\Omega\ \mathrm{mm}^2} \cdot 95\ \mathrm{mm}^2} = 4.66\ \mathrm{m\Omega}$$

$$X_{\mathrm{LHV}} = 2\,l\,X_{\mathrm{L}} = 2 \cdot 10\ \mathrm{m} \cdot 0.08\frac{\mathrm{m\Omega}}{\mathrm{m}} = 1.6\ \mathrm{m\Omega}$$

Summary of resistances and reactances:

	R	X
Network	0.035 mΩ	0.35 mΩ
Transformer	15 mΩ	37 mΩ
R_{LHV}	4.66 mΩ	1.6 mΩ
R_{MDP}	19.695 mΩ,	$X_{\mathrm{MDP}} = 39.95$ mΩ

$$Z_{\mathrm{MDP}} = \sqrt{(19.695\ \mathrm{m\Omega})^2 + (39.95\ \mathrm{m\Omega})^2} = 43.64\ \mathrm{m\Omega}$$

$$I''_{k1} = \frac{c\,U_n}{\sqrt{3}\,Z_{\mathrm{MDP}}} = \frac{0.9 \cdot 400\ \mathrm{V}}{\sqrt{3} \cdot 43.64\ \mathrm{m\Omega}} = 4.76\ \mathrm{kA}$$

(c) Main distribution 1 – I''_{k3}

$$R_{\mathrm{LSD1}} = \frac{150\ \mathrm{m} \cdot 1000}{56\frac{\mathrm{m}}{\Omega\ \mathrm{mm}^2} \cdot 240\ \mathrm{mm}^2} = 11.16\ \mathrm{m\Omega}$$

$$X_{\text{SD1}} = 2 \cdot 150\ \text{m} \cdot 0.08\frac{\text{m}\Omega}{\text{m}} = 24\ \text{m}\Omega$$
$$R_{\text{SD1}} = R_{\text{LSD1}} + R_{\text{MDP}} = 11.16\ \text{m}\Omega + 16.97\ \text{m}\Omega = 28.13\ \text{m}\Omega$$
$$X_{\text{SD1}} = X_{\text{LSD1}} + X_{\text{MDP}} = 24\ \text{m}\Omega + 38.76\ \text{m}\Omega = 62.76\ \text{m}\Omega$$
$$Z_{\text{SD1}} = \sqrt{R_{\text{SD1}}^2 + X_{\text{SD1}}^2} = \sqrt{(28.13\ \text{m}\Omega)^2 + (62.676\ \text{m}\Omega)^2} = 68.77\ \text{m}\Omega$$
$$I''_{k3} = \frac{1.1 \cdot 400\ \text{V}}{\sqrt{3} \cdot 68.77\ \text{m}\Omega} = 3.7\ \text{kA}$$

(d) Sub-distribution 1 – I''_{k1}

$$R_{\text{LSD1}} = 1.24 \cdot \frac{2 \cdot 150\ \text{m}}{56\frac{\text{m}}{\Omega\ \text{mm}^2} \cdot 240\ \text{mm}^2} = 27.68\ \text{m}\Omega$$
$$X_{\text{LSD1}} = 2 \cdot 150\ \text{m} \cdot 0.08\frac{\text{m}\Omega}{\text{m}} = 24\ \text{m}\Omega$$
$$R_{\text{SD1}} = R_{\text{LSD1}} + R_{\text{LHV}} = 27.68\ \text{m}\Omega + 20.54\ \text{m}\Omega = 48.22\ \text{m}\Omega$$
$$X_{\text{SD1}} = X_{\text{LSD1}} + R_{\text{LHV}} = 24\ \text{m}\Omega + 39.476\ \text{m}\Omega = 63.476\ \text{m}\Omega$$
$$Z_{\text{SD1}} = \sqrt{(48.22\ \text{m}\Omega)^2 + (63.476\ \text{m}\Omega)^2} = 79.71\ \text{m}\Omega$$
$$I''_{k1} = \frac{0.9 \cdot 400\ \text{V}}{\sqrt{3} \cdot 79.71\ \text{m}\Omega} = 2.6\ \text{kA}$$

(e) Sub-distribution 2 – I''_{k3}

$$R_{\text{LSD2}} = \frac{80\ \text{m} \cdot 1000}{56\frac{\text{m}}{\Omega\ \text{mm}^2} \cdot 185\ \text{mm}^2} = 7.72\ \text{m}\Omega$$
$$X_{\text{LSD2}} = 80\ \text{m} \cdot 0.08\frac{\text{m}\Omega}{\text{m}} = 6.4\ \text{m}\Omega$$
$$R_{\text{SD2}} = R_{\text{LSD1}} + R_{\text{SD1}} = 7.72\ \text{m}\Omega + 28.13\ \text{m}\Omega = 35.85\ \text{m}\Omega$$
$$X_{\text{SD2}} = X_{\text{LSD2}} + X_{\text{SD1}} = 6.4\ \text{m}\Omega + 62.68\ \text{m}\Omega = 69\ \text{m}\Omega$$
$$Z_{\text{SD2}} = \sqrt{(35.85\ \text{m}\Omega)^2 + (69\ \text{m}\Omega)^2} = 77.75\ \text{m}\Omega$$
$$I''_{k3} = \frac{1.1 \cdot 400\ \text{V}}{\sqrt{3} \cdot 77.75\ \text{m}\Omega} = 3.26\ \text{kA}$$

(f) Sub-distribution 2 – I''_{k1}

$$R_{\text{LSD2}} = 1.24 \cdot \frac{2 \cdot 80\ \text{m} \cdot 1000}{56\frac{\text{m}}{\Omega\ \text{mm}^2} \cdot 185\ \text{mm}^2} = 19.15\ \text{m}\Omega$$
$$X_{\text{LSD2}} = 2 \cdot 80\ \text{m} \cdot 0.08\frac{\text{m}\Omega}{\text{m}} = 12.8\ \text{m}\Omega$$
$$R_{\text{LSD2}} = R_{\text{LSD2}} + R_{\text{SD1}} = 19.15\ \text{m}\Omega + 48.22\ \text{m}\Omega = 67.37\ \text{m}\Omega$$
$$X_{\text{SD2}} = X_{\text{LSD2}} + X_{\text{SD1}} = 12.8\ \text{m}\Omega + 63.48\ \text{m}\Omega = 76.28\ \text{m}\Omega$$
$$Z_{\text{SD2}} = \sqrt{(67.37\ \text{m}\Omega)^2 + (76.28\ \text{m}\Omega)^2} = 101.77\ \text{m}\Omega$$
$$I''_{k1} = \frac{1.1 \cdot 400\ \text{V}}{\sqrt{3} \cdot 0.101\ 77\ \Omega} = 2.5\ \text{kA}$$

(g) Motor 1 – I''_{k3}

$$R_{\mathrm{LM1}} = \frac{35\ \mathrm{m} \cdot 1000}{56\frac{\mathrm{m}}{\Omega\ \mathrm{mm}^2} \cdot 1.5\ \mathrm{mm}^2} = 416.7\ \mathrm{m\Omega}$$

$$R_{\mathrm{M1}} = Z_{\mathrm{M1}} = R_{\mathrm{LM1}} + R_{\mathrm{SD2}} = 416.7\ \mathrm{m\Omega} + 35.85\ \mathrm{m\Omega}$$
$$= 452.5\ \mathrm{m\Omega}$$

$$I''_{k3} = \frac{1.1 \cdot 400\ \mathrm{V}}{\sqrt{3} \cdot 452.5\ \mathrm{m\Omega}} = 561.4\ \mathrm{A}$$

(h) Motor 1 – I''_{k1}

$$R_{\mathrm{LM1}} = 1.24 \cdot \frac{2 \cdot 35\ \mathrm{m} \cdot 1000}{56\frac{\mathrm{m}}{\Omega\ \mathrm{mm}^2} \cdot 1.5\ \mathrm{mm}^2}$$
$$= 1033\ \mathrm{m\Omega}$$

$$R_{\mathrm{M1}} = R_{\mathrm{LM1}} + R_{\mathrm{SD2}} = 1033\ \mathrm{m\Omega} + 67.33\ \mathrm{m\Omega}$$
$$= 1100.33\ \mathrm{m\Omega}$$

$$I''_{k1} = \frac{1.1 \cdot 400\ \mathrm{V}}{\sqrt{3} \cdot 1100.33\ \mathrm{m\Omega}} = 230.87\ \mathrm{A}$$

(i) Circuit no. 20 – I''_{k3}

$$R_{\mathrm{L20}} = \frac{25\ \mathrm{m} \cdot 1000}{56\frac{\mathrm{m}}{\Omega\ \mathrm{mm}^2} \cdot 1.5\ \mathrm{mm}^2} = 297.6\ \mathrm{m\Omega}$$

$$R_{\mathrm{LSD3}} = \frac{45\ \mathrm{m} \cdot 1000}{56\frac{\mathrm{m}}{\Omega\ \mathrm{mm}^2} \cdot 25\ \mathrm{mm}^2} = 32.14\ \mathrm{m\Omega}$$

$$X_{\mathrm{LSD3}} = 45\ \mathrm{m} \cdot 0.08\frac{\mathrm{m\Omega}}{\mathrm{m}} = 3.6\ \mathrm{m\Omega}$$

$$R_{20} = R_{\mathrm{L20}} + R_{\mathrm{LSD3}} + R_{\mathrm{SD1}}$$
$$= 297.6\ \mathrm{m\Omega} + 32.14\ \mathrm{m\Omega} + 28.13\ \mathrm{m\Omega} = 357.8\ \mathrm{m\Omega}$$

$$X_{20} = X_{\mathrm{LSD3}} + X_{\mathrm{SD1}} = 3.6\ \mathrm{m\Omega} + 62.68\ \mathrm{m\Omega} = 66.28\ \mathrm{m\Omega}$$

$$Z_{20} = \sqrt{(357.87\ \mathrm{m\Omega})^2 + (66.28\ \mathrm{m\Omega})^2} = 364\ \mathrm{m\Omega}$$

$$I''_{k3} = \frac{1.1 \cdot 400\ \mathrm{V}}{\sqrt{3} \cdot 0.364\ \Omega} = 697.89\ \mathrm{A}$$

(j) Circuit no. 20 – I''_{k1}

$$R_{\mathrm{L20}} = 1.24\frac{2 \cdot 25\ \mathrm{m} \cdot 1000}{56\frac{\mathrm{m}}{\Omega\ \mathrm{mm}^2} \cdot 1.5\ \mathrm{mm}^2} = 738.1\ \mathrm{m\Omega}$$

$$R_{\mathrm{LSD3}} = 1.24 \cdot 2 \cdot R_{\mathrm{LSD3}} = 1.24 \cdot 2 \cdot 32.14\ \mathrm{m\Omega} = 79.71\ \mathrm{m\Omega}$$

$$X_{\mathrm{LSD3}} = 2 \cdot 45\ \mathrm{m} \cdot 0.08\frac{\mathrm{m\Omega}}{\mathrm{m}} = 7.2\ \mathrm{m\Omega}$$

$$R_{20} = R_{\mathrm{L20}} + R_{\mathrm{LSD3}} + R_{\mathrm{SD1}}$$
$$= 738.1\ \mathrm{m\Omega} + 79.71\ \mathrm{m\Omega} + 48.22\ \mathrm{m\Omega} = 866.03\ \mathrm{m\Omega}$$

$$X_{20} = X_{\mathrm{SD3}} + X_{\mathrm{SD1}} = 7.2\ \mathrm{m\Omega} + 63.48\ \mathrm{m\Omega} = 70.68\ \mathrm{m\Omega}$$

$$Z_{20} = \sqrt{(866.03\ \mathrm{m\Omega})^2 + (70.68\ \mathrm{m\Omega})^2} = 869\ \mathrm{m\Omega}$$

$$I''_{k1} = \frac{0.9 \cdot 400\ \mathrm{V}}{\sqrt{3} \cdot 0.869\ \Omega} = 239.17\ \mathrm{A}$$

7. *Calculation of the compensation system*:
   Given: $\cos\varphi_1 = 0.85$, $\cos\varphi_2 = 0.95$, $P = 88.86$ kW

$$Q_c = P(\tan\varphi_1 - \tan\varphi_2) = 88.86\ \text{kW} \cdot (0.62 - 0.32) = 26.65\ \text{kVar}$$

$$C = \frac{Q_c}{U_n^2\ 2\ \pi\ f} = \frac{26.65\ \text{kVar}}{(400\ \text{V})^2 \cdot 2 \cdot \pi \cdot 50\ \text{s}^{-1}} = 530.34\ \mu\text{F}$$

$$I_{\text{neu}} = \frac{P}{\sqrt{3}\ U\ \cos\varphi^2} = \frac{88.86\ \text{kW}}{\sqrt{3} \cdot 400\ \text{V} \cdot 0.95} = 135\ \text{A}$$

8. *Determination the cut-of conditions for all circuits*:
   (a) Main distribution 1 ⇒ $I_r = 250$ A, $I''_{k1} = 3.7$ kA, $t_a = 0.015$ second
   (b) Sub-distribution 1 ⇒ $I_r = 200$ A, $I''_{k1} = 2.6$ kA, $t_a = 0.3$ second (this satisfies the requirement of maximum five seconds)
   (c) Sub-distribution 2 ⇒ $I_r = 125$ A, $I''_{k1} = 2.5$ kA, $t_a = 0.06$ second
   (d) Circuit no. 20 ⇒ $I_r = 16$ A, $I''_{k1} = 239.17$ A, $t_a = 0.01$ second (maximum 0.2 second)

8.16.8 Example 8: Calculation of Short-Circuit Currents with Impedance Corrections

Given a 220 kV network with the operating data in Figure 8.34 calculate the short-circuit currents in Q and A with impedance corrections:

Network:

$$Z_Q = \frac{c\ U_{nQ}^2}{S''_{kQ}} = \frac{1.1 \cdot (220\ \text{kV})^2}{6000\ \text{MVA}} = 8.873\ \Omega$$

Generator:

$$Z_{\text{G}} = \frac{x''_d\ U_{\text{rG}}^2}{100\%\ S_{\text{rG}}} = \frac{28\% \cdot (21\ \text{kV})^2}{100\% \cdot 880\ \text{MVA}} = 0.14\ \Omega$$

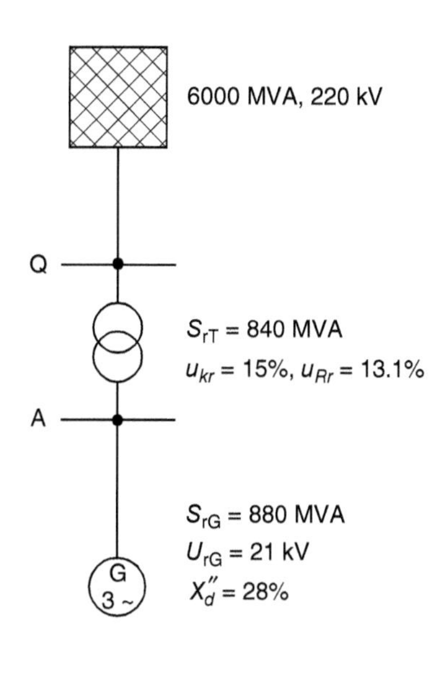

Figure 8.34 Short-circuit currents with impedance corrections.

Correction factor:

$$K_{G,KW} = \frac{c}{1 + x''_d \sin\varphi_{rG}} = \frac{1.1}{1 + 0.28 \cdot 0.557} = 0.951$$

Corrected generator impedance:

$$Z_{G,KW} = K_{G,KW}\, Z_G = 0.951 \cdot 0.14\Omega = 0.133\ \Omega$$

Block transformer:

$$Z_{THV} = \frac{u_{kr}}{100\%} \cdot \frac{U^2_{rTHV}}{S_{rT}} = \frac{15\%}{100\%} \cdot \frac{(220\ \text{kV})^2}{840\ \text{MVA}} = 8.64\ \Omega$$

$$Z_{TLV} = \frac{u_{kr}}{100\%} \cdot \frac{U^2_{rTLV}}{S_{rT}} = \frac{13.1\%}{100\%} \cdot \frac{(21\ \text{kV})^2}{840\ \text{MVA}} = 0.068\ 77\ \Omega$$

$$Z_{T,KW} = c\, Z_{TLV} = 1.1 \cdot 0.068\ 77\ \Omega = 0.0756\ \Omega$$

Calculation of the short-circuit currents at the network power supply feeder Q:

$$I''_k = I''_{kQ} + I''_{kKW}$$

$$I''_{kQ} = \frac{c\, U_{nQ}}{\sqrt{3}\, Z_Q} = \frac{1.1 \cdot 220\ \text{kV}}{\sqrt{3} \cdot 8.873\ \Omega} = 15.76\ \text{kA}$$

$$K_{KW} = \left(\frac{"t_f}{"t_r}\right)^2 \cdot \frac{c}{1 + (x''_d - x_T)\sin\varphi_{rG}}$$

$$= \left(\frac{220\ \text{kV}}{21\ \text{kV}}\right)^2 \cdot \left(\frac{21\ \text{kV}}{233\ \text{kV}}\right)^2 \cdot \frac{1.1}{1 + (0.28 - 0.131) \cdot 0.63}$$

$$= 0.8965$$

$$Z_{KW} = K_{KW}\,("t_r^2\, Z_G + Z_{THV})$$

$$= 0.8965\ \left[\left(\frac{220\ \text{kV}}{21\ \text{kV}}\right)^2 \cdot 0.14 + 8.64\Omega\right] = 21.52\ \Omega$$

$$I''_{kKW} = \frac{1.1 \cdot 220\ \text{kV}}{\sqrt{3} \cdot 21.52\ \Omega} = 6.49\ \text{kA}$$

$$I''_k = 15.76\ \text{kA} + 6.49\ \text{kA} = 22.25\ \text{kA}$$

Calculation of the short-circuit currents in A:

$$I''_k = I''_{kG} + I''_{kT}$$

$$I''_{kG} = \frac{c\, U_{rG}}{\sqrt{3}\, Z_{G,KW}} = \frac{1.1 \cdot 21\ \text{kV}}{\sqrt{3} \cdot 0.133\ \Omega} = 100.4\ \text{kA}$$

$$I''_{kT} = \frac{1.1\, U_{rG}}{\sqrt{3}\,(Z_{T,KW} + \frac{1}{\ddot{u}_r^2}\, Z_Q)}$$

$$I''_{kT} = \frac{1.1 \cdot 21\ \text{kV}}{\sqrt{3} \cdot (0.0756\ \Omega + (\frac{21\ \text{kV}}{220\ \text{kV}})^2 \cdot 8.873\ \Omega)} = 85.24\ \text{kA}$$

$$I''_k = 100.4\ \text{kA} + 85.24\ \text{kA} = 185.64\ \text{kA}$$

8.16.9 Example: Load Voltage and Zero Impedance

An insulated four-core cable that is 300 m long is connected to a rigid network with 400 V. The cable constants are $R_L = 0.852\frac{\Omega}{\text{km}}, X_L = 0$. The cable is loaded at the end between phase R and the neutral conductor with an effective resistance of $R_V = 352\ \Omega$. What is the consumer voltage, if the zero impedance is greater than the positive sequence impedance by a factor of 4? How large would the load voltage be if all would be burdened?

The star voltage of the network is (reference value!)

$$\underline{U}_{1l} = \frac{400\ \text{V}}{\sqrt{3}} = 230.5\ \text{V}$$

The impedances of the line are

$$\underline{Z}_1 = \underline{Z}_2 = R_l' = 0.24\ \Omega \quad \underline{Z}_0 = 4 \cdot \underline{Z}_1 = 0.96\ \Omega$$

It must apply to the consumer:

$$\underline{I}_S = \underline{I}_T = 0 \quad \underline{U}_R = R_V \cdot \underline{I}_R$$

It follows from this that

$$\underline{I}_0 + a^2 \cdot \underline{I}_1 + a \cdot \underline{I}_2 = 0$$
$$\underline{I}_0 + a \cdot \underline{I}_1 + a^2 \cdot \underline{I}_2 = 0$$
$$\underline{U}_0 + \underline{U}_1 + \underline{U}_2 = R_V \cdot (\underline{I}_0 + \underline{I}_1 + \underline{I}_2)$$

From the first two equations one gets by addition and subtraction

$$\underline{I}_1 = \underline{I}_2 = \underline{I}_0$$

In the third equation, the connection equations of the symmetrical components are inserted among themselves:

$$-\underline{Z}_0 \cdot \underline{I}_0 + \underline{U}_{1l} - \underline{Z}_1 \cdot \underline{I}_1 - \underline{Z}_2 \cdot \underline{I}_2 = R_V \cdot (\underline{I}_0 + \underline{I}_1 + \underline{I}_2)$$

or

$$U_{1l} - \underline{I}_1 \cdot (\underline{Z}_0 + \underline{Z}_1 + \underline{Z}_2) = 3 \cdot R_V \cdot \underline{I}_1$$

This results in

$$\underline{I}_1 = \frac{U_{1l}}{\underline{Z}_0 + \underline{Z}_1 + \underline{Z}_2 + 3 \cdot R_V}$$
$$= \frac{3 \cdot 230.5\ \text{V}}{(0.96 + 0.24 + 0.24 + 9)}\ \Omega = \frac{692.5\ \text{A}}{10.44} = 66.2\ \text{A}$$
$$\underline{U}_R = \underline{I}_R \cdot R_V = 66.2\ \text{A} \cdot 3\ \Omega = 198.6\ \text{V}$$

Since there are no reactances in the circuit, a symmetrical load with the simple voltage divider rule results in

$$\frac{U_R}{U_{1l}} = \frac{R_V}{\underline{Z}_1 + R_V} = \frac{3}{3.24} = 0.926$$
$$U_R = 0.926 \cdot 230.5\ \text{V} = 213.5\ \text{V}$$

This is considerably more than with single-phase load.
The maximum star voltage is

$$U_{1l} = \frac{1.04 \cdot 400\ \text{V}}{\sqrt{3}} = 240\ \text{V}$$

The positive, negative, and zero impedances of the transformer are

$$R_\text{T} = \frac{u_r U_B^2}{S_\text{rT}} = \frac{0.0215 \cdot 400^2\ \text{V}^2}{10^5\ \text{VA}} = 34.4\ \text{m}\Omega = R_1 = R_2$$

$$X_\text{T} = \frac{u_s U_B^2}{S_\text{rT}} = \frac{0.0337 \cdot 400^2\ \text{V}^2}{10^5\ \text{VA}} = 54.0\ \text{m}\Omega = X_1 = X_2$$

$$R_0 = R_1 = 34.4\ \text{m}\Omega$$

$$X_0 = 8 \cdot X_1 = 432\ \text{m}\Omega$$

(a) Single-pole short-circuit

$$\underline{I}_R = \frac{3 \cdot U_{1l}}{\underline{Z}_0 + \underline{Z}_1 + \underline{Z}_2} = \frac{3 \cdot 240\ \text{V}}{(34.4 + j432 + 68.8 + j108)}\ \text{m}\Omega$$

$$= \frac{720\ \text{kA}}{(103 + j540)} = \frac{720\ \text{kA}}{550 \cdot e^{j79.2°}} = 1.31\ \text{kA} \cdot e^{-j79.2°}$$

$$\underline{U}_S = \frac{-j\sqrt{3} \cdot U_{1l} \cdot (\underline{Z}_2 - a \cdot \underline{Z}_0)}{\underline{Z}_0 + \underline{Z}_1 + \underline{Z}_2}$$

$$= \frac{-j\sqrt{3} \cdot 240\ \text{V} \cdot (34.4 + j54 - (-0.5 + j0.866) \cdot (34.4 + j432))}{550 \cdot e^{j79.2°}}$$

$$= \frac{-j416\ \text{V} \cdot (34.4 + j54 + 17.2 + j216 - j29.8 + 374)}{550 \cdot e^{j79.2°}}$$

$$= \frac{-j416\ V \cdot (425.6 + j240.2)}{550 \cdot e^{j79.2°}}$$

$$= \frac{416\ \text{V} \cdot e^{-j90°} \cdot 489 \cdot e^{j29.4°}}{550 \cdot e^{j79.2°}} = 370\ \text{V} \cdot e^{-j139.8°}$$

$$\underline{U}_\text{T} = \frac{j\sqrt{3} \cdot U_{1l} \cdot (\underline{Z}_2 - a^2 \cdot \underline{Z}_0)}{\underline{Z}_0 + \underline{Z}_1 + \underline{Z}_2}$$

$$= \frac{j\sqrt{3} \cdot 240\ \text{V} \cdot (34.4 + j54 - (-0.5 - j0.866) \cdot (34.4 + j432))}{550 \cdot e^{j79.2°}}$$

$$= \frac{j416\ \text{V} \cdot (34.4 + j54 + 17.2 + j216 + j29.8 - 374)}{550 \cdot e^{j79.2°}}$$

$$= \frac{j416\ \text{V} \cdot (-322.4 + j299.8)}{550 \cdot e^{j79.2°}}$$

$$= \frac{416\ \text{V} \cdot e^{-j90°} \cdot 440 \cdot e^{j137.1°}}{550 \cdot e^{j79.2°}} = 333\ \text{V} \cdot e^{-j147.9°}$$

(b) Two-pole short-circuit without ground contact

$$\underline{I}_S = \frac{-j\sqrt{3} \cdot U_{1l}}{(\underline{Z}_1 + \underline{Z}_2)} = \frac{-j\sqrt{3} \cdot 240\ \text{V}}{68.8 + j108}\ \text{m}\Omega$$

$$= \frac{416\ \text{kA} \cdot e^{-j90°}}{128 \cdot e^{j57.5°}} = 3.25\ \text{kA} \cdot e^{-j147.5°}$$

$$\underline{U}_R = \frac{2 \cdot \underline{Z}_2 \cdot U_{1l}}{(\underline{Z}_1 + \underline{Z}_2)} = U_{1l} = 240\ \text{V}$$

$$\underline{U}_S = \underline{U}_T = \frac{-\underline{Z}_2 \cdot U_{1l}}{(\underline{Z}_1 + \underline{Z}_2)} = -0.5 \cdot U_{1l} = -120\ \text{V}$$

(c) Two-pole short-circuit with ground contact

$$\underline{I}_S = \frac{-j\sqrt{3} \cdot U_{1l} \cdot (\underline{Z}_0 - a \cdot \underline{Z}_2)}{\underline{Z}_0 \cdot \underline{Z}_1 + \underline{Z}_0 \cdot \underline{Z}_2 + \underline{Z}_1 \cdot \underline{Z}_2}$$

$$= \frac{-j\sqrt{3} \cdot 240\ \text{V} \cdot (34.4 + j432 - (-0.5 + j0.866) \cdot (34.4 + j54))}{(34.4 + j54) \cdot (103 + j918)\ \text{m}\Omega}$$

$$= \frac{-j416\ \text{kA}(34.4 + j432 + 17.2 + j27 - j29.8 + 46.7)}{3.540 + j31.600 + j5.560 - 49.500}$$

$$= \frac{-j416\ \text{kA} \cdot (98.3 + j358.3)}{-45.960 + j37.160}$$

$$= \frac{416\ \text{kA} \cdot e^{-j90°} \cdot 372 \cdot e^{j74.7°}}{59.000 \cdot e^{j129°}} = 2.62\ \text{kA} \cdot e^{-j144.3°}$$

$$\underline{I}_T = \frac{j\sqrt{3} \cdot U_{1l} \cdot (\underline{Z}_0 - a^2 \cdot \underline{Z}_2)}{\underline{Z}_0 \cdot \underline{Z}_1 + \underline{Z}_0 \cdot \underline{Z}_2 + \underline{Z}_1 \cdot \underline{Z}_2}$$

$$= \frac{j\sqrt{3} \cdot 240\ \text{V} \cdot (34.4 + j432 - (-0.5 - j0.866) \cdot (34.4 + j54))}{59.000 \cdot e^{j129°}}$$

$$= \frac{j416\ \text{kA}(34.4 + j432 + 17.2 + j27 + j29.8 - 46.7)}{59.000 \cdot e^{j129°}}$$

$$= \frac{j416\ \text{kA} \cdot (4.9 + j434.8)}{59.000 \cdot e^{j129°}}$$

$$= \frac{416\ \text{kA} \cdot e^{-j90°} \cdot 435 \cdot e^{j89.4°}}{59.000 \cdot e^{j129°}} = 3.07\ \text{kA} \cdot e^{j50.4°}$$

$$\underline{I}_M = \frac{3 \cdot U_{1l} \cdot \underline{Z}_0 \cdot \underline{Z}_2}{\underline{Z}_0 \cdot \underline{Z}_1 + \underline{Z}_0 \cdot \underline{Z}_2 + \underline{Z}_1 \cdot \underline{Z}_2}$$

$$= \frac{-3 \cdot U_{1l} \cdot \underline{Z}_2 \cdot (34.4 + j54)}{59.000 \cdot e^{j129°}\ \text{m}\Omega}$$

$$= \frac{-720\ \text{kA} \cdot 64 \cdot e^{j57.5°}}{59.000 \cdot e^{j129°}} = 0.78\ \text{kA} \cdot e^{j108.5°}$$

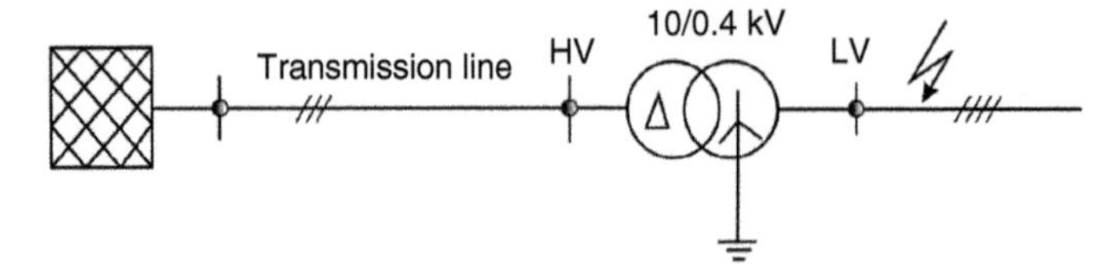

Figure 8.35 Power system.

$$\underline{U}_R = \frac{3 \cdot U_{1l} \cdot \underline{Z}_0 \cdot \underline{Z}_2}{\underline{Z}_0 \cdot \underline{Z}_1 + \underline{Z}_0 \cdot \underline{Z}_2 + \underline{Z}_1 \cdot \underline{Z}_2}$$
$$= \frac{-3 \cdot 240\ \text{V} \cdot (34.4 + j54)}{59.000 \cdot e^{j129^\circ}}$$
$$= \frac{720\ \text{V} \cdot (1.180 + j14.850 + j1.860 - 23.350)}{59.000 \cdot e^{j129^\circ}}$$
$$= \frac{720\ \text{V} \cdot (-22.170 + j16.710)}{59.000 \cdot e^{j129^\circ}}$$
$$= \frac{720\ \text{V} \cdot 27.750 \cdot e^{-j143^\circ}}{59.000 \cdot e^{j129^\circ}} = 341\ \text{V} \cdot e^{-j14^\circ}$$

(d) Three-phase short-circuit

$$\underline{I}_R = \frac{U}{\underline{Z}_1} = \frac{240\ \text{V}}{(34.4 + j54)\ \text{m}\Omega} = \frac{240\ \text{kA}}{64 \cdot e^{j57.5^\circ}}$$
$$= 3.75\ \text{kA} \cdot e^{-j57.5^\circ}$$

8.16.10 Example: Power Transmission

Given is a high-voltage power system with an overhead line energy transfer according to Figure 8.35:

Grid: 10.4 kV, 250 MVA, $\frac{R_N}{X_N} = 0.1$
Overhead line: 6 km long, $R_L' = 0.28 \frac{\Omega}{\text{km}}, X_L' = 0.32\ \frac{\Omega}{\text{km}}$
Transformer: 3.15 MVA, 10/0.4 kV, $u_{kr} = 0.9\ \%$, $u_s = 6\%$

How large is the three-phase continuous short-circuit current on the secondary side of the transformer? All resistors are calculated or converted to the reference voltage of 0.4 kV (star voltage 0.231 kV).

$$Z_N = X_N = \frac{1.1 \cdot U_B^2}{S_k} = \frac{1.1 \cdot 0.4^2 \cdot 10^6\ \text{V}^2}{250 \cdot 10^6\ \text{VA}} = 0.705\ \text{m}\Omega$$
$$R_N = 0.1 \cdot X_N = 0.07\ \text{m}\Omega$$
$$R_L = R_L' \cdot l \cdot \frac{U_B}{U_N}^2 = 0.28 \frac{\Omega}{\text{km}} \cdot 6\ \text{km} \cdot \frac{0.4}{10}^2 = 2.69\ \text{m}\Omega$$
$$X_L = X_L' \cdot l \cdot \frac{U_B}{U_N}^2 = 0.32 \frac{\Omega}{\text{km}} \cdot 6\ \text{km} \cdot \frac{0.4}{10}^2 = 3.07\ \text{m}\Omega$$

$$R_{\mathrm{T}} = u_{kr} \cdot \frac{U_B^2}{S_{\mathrm{rT}}} = \frac{0.009 \cdot 0.4^2 \cdot 10^6\ \mathrm{V}^2}{3.15 \cdot 10^6\ \mathrm{VA}} = 0.457\ \mathrm{m\Omega}$$

$$X_{\mathrm{T}} = u_s \cdot \frac{U_B^2}{S_{\mathrm{rT}}} = \frac{0.06 \cdot 0.4^2 \cdot 10^6\ V^2}{3.15 \cdot 10^6\ \mathrm{VA}} = 3.04\ \mathrm{m\Omega}$$

$$R_{\mathrm{ges}} = 3.22\ \mathrm{m\Omega}$$

$$X_{\mathrm{ges}} = 6.82\ \mathrm{m\Omega}$$

$$Z_{\mathrm{ges}} = 7.54\ \mathrm{m\Omega}$$

When calculating the short-circuit current, it must be taken into account that the voltage at the beginning of the overhead line (to cover the voltage drops on the line) is 4% higher than the primary rated voltage of the transformer. For further voltage drops in the network and possible voltage increases, the factor 1.1 is also used. This results in

$$I_k = \frac{1.04 \cdot 1.1 \cdot U}{Z_{\mathrm{ges}}} = \frac{1.04 \cdot 1.1 \cdot 231\ \mathrm{V}}{7.54\ \mathrm{m\Omega}} = 35\ \mathrm{kA}$$

9
Relays

9.1 Terms and Definitions

- *Relay*: A device that is influenced by a change in the operating variable in the drive system and that actuates switching elements.
- *Distance time*: Scale time: The command time specified for a certain area of the characteristic size (zone).
- *Primary relays*: Protective devices whose windings are in or on the monitored circuit, to which the measured variable is fed directly, and which mechanically initiate an unlocking process.
- *Secondary relays*: Protective devices whose windings are connected to the monitored circuit via current transformers or voltage transformers, to which the measured variable is indirectly supplied, and which electrically control further devices.
- *Command time*: The time from the occurrence of a certain value of the input variable until the output circuit is switched.
- *Protection system*: The entirety of the protective devices and other devices, which are used for the realization of a certain protection principle.

9.2 Introduction

When planning a switchgear system, tasks, functions, and influencing variables must be harmonized and an economical solution found. In principle, the switchgear should have a high degree of security, so that both the protection of personnel and trouble-free operation of the network are guaranteed. It must meet the requirements for basic protection and fault protection and eliminate the possibility of incorrect operation. However, if an error does occur, its effect on the place of origin should be limited and should not result in personal injury [45].

The switchboards can be assigned to the primary or secondary distribution level (Figure 9.1). Characteristics of the primary distribution level are high load and short-circuit currents as well as a high level of secondary equipment of the switchgear with regard to protection, measurement, and (remote) control. In the primary distribution level is the transformer station (substation), in which the

Analysis and Design of Electrical Power Systems: A Practical Guide and Commentary on NEC and IEC 60364, First Edition. Ismail Kasikci.

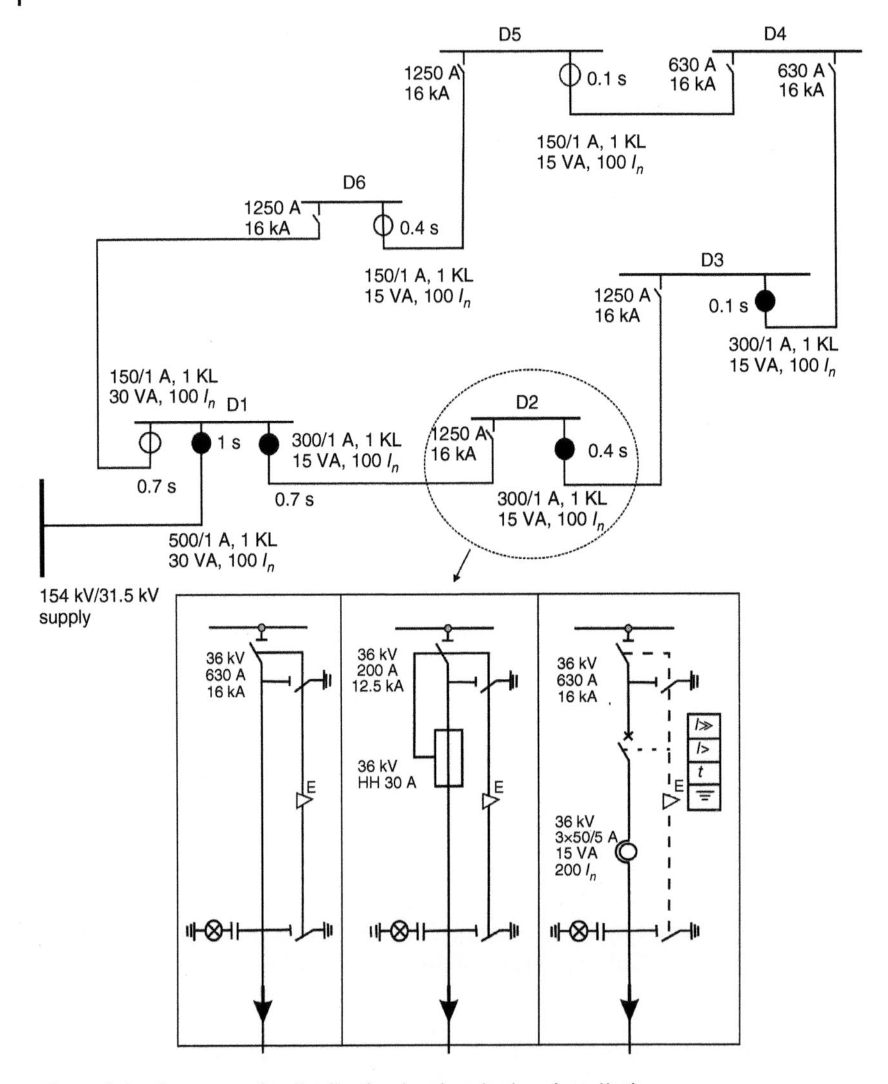

Figure 9.1 Structure of a distribution level and relays installation.

energy with higher voltage is transformed to medium voltage. The switchgear is almost completely equipped with circuit-breakers. They switch large consumers, mostly in industrial plants, or cable rings, which in turn supply switchgear at the secondary distribution level.

In the secondary distribution level, the switchgear has load-break switches or a mixture of load- and circuit-breakers, with the proportion of circuit-breakers clearly predominating. The currents are smaller, and short-circuit protection is often provided by the assigned circuit-breaker at the primary distribution level. The following section describes the protection of electrical systems and their parts against electrical faults by protection technology, in particular the assessment of the usefulness of the most important protective devices. Technical details on the design and function of the protective relays and the associated modules can be found in the relevant publications of the manufacturers.

9.3 Requirements

The built-in protection devices (relays, selective protection devices, protective devices, or protective relays) cannot prevent the occurrence of faults. Protective devices only become active after a fault has occurred and, as already mentioned, have the task of controlling the effects of the fault, of the error for humans, animals, and equipment.

The following requirements apply to protective devices:

- Selectivity (only disconnect faulty plant section or equipment ($I - t$))
- Speed (as fast as necessary or technically possible)
- Availability (no shutdown in case of under- or over-operation or faulty messages)
- Sensitivity (distinction of normal operation or overload)
- Economic efficiency (maintenance, repair, investments)

9.4 Protective Devices for Electric Networks

Protective devices have the task of providing a short-circuit in the electrical network as quickly as possible and switching off selectively. The components of the network and the consumers should be be subjected to short-circuit currents and voltage dips for as short a time as possible. After breaker tripping by the protection, either all consumers should continue to be supplied without restriction, or as few consumers as possible should be switched off, these should be resupplied immediately after fault localization and fault supplied again immediately after the fault has been localized and rectified. Protective devices for electrical power distribution provide the following:

- Overcurrent time protection
- Line differential protection
- Transformer differential protection
- Machine protection
- Busbar differential protection
- Distance protection

For the connection and operation of the protective devices, three current transformers and, if necessary, three voltage transformers on the busbars and the circuit-breakers are required.

9.5 Type of Relays

9.5.1 Electromechanical Protective Relays

The so-called first generation relays are for electromechanical protection. Characteristic of this is the use of hinged and rotary armature relays. A spring was used as counter-torque, at which the response value or time delay of any existing timepiece can be set. Electromechanical relays are robust and do not tend to overfunction (unintentional releases). The disadvantages are, however, the high Energy requirement in the measuring circuit, unfavorable release ratio, long staggered times, and considerable planning effort and space requirements. In addition comes the lack of self-monitoring. The error is displayed via drop flap relays or trailing indicators, which are activated on site and have to be reset manually.

9.5.2 Static Protection Relays

With the advent of transistor technology it was possible to reduce the movable and maintenance-intensive parts due to electronic components replacement. The protection device, also called analog-electronic relays, had an improved speed and accuracy compared to the electromechanical protection relay. Nevertheless, not all expectations were met with the introduction of the user to this 2nd generation of protection at the beginning of 1970s. In particular, the high DC demand in idle mode, the greater inclination to hyperfunction, the necessity to take electromagnetic combability (EMC) measures, as well as a large documentation effort had an unfavorable effect. In contrast, the number of electromechanical relays are significantly higher than the number of individual components, leading to an increased need for repairs and therefore to a decrease in availability and reliability of static relays.

9.5.3 Numeric Protection Relays

The rapid development of microprocessor technology in the 1970s limited the introduction of static relays. At the end of the 1980s the first microprocessor-controlled protection relays were available. Correspondingly, developments in the highly dynamic PC sector have been delayed as also the developments in protection technology and in device technology.

The positive experience with the third generation of protective devices led to the fact that today almost exclusively numerical protection devices are used. Some advantages of this technology are as follows:

1. *Multifunctionality*: Several protection tasks (e.g. differential and overcurrent protection) can be implemented in a device such that integration of control and monitoring functions is possible.
2. *Reliability*: Owing to the omission of almost all movable, mechanical parts and by implementing extensive self-monitoring functions the reliability is significantly increased.
3. *Speed and accuracy*: This results in lower scale and accuracy and trigger times.

4. *Adaptability*: Excitation and triggering characteristics (as provided by software) can be adapted to the physical properties of the equipment optimally.
5. *Cost-effectiveness*: Lower planning, installation, commissioning, and maintenance costs are required; the lower maintenance costs are partly compensated by the need to perform software updates when necessary.

Digital protection relays operate on the basis of digital measuring principles. The analog measured current and voltage values are measured galvanically from the secondary circuit of the system via input transmitters. After an analog filter process, the measurement is carried out and conversion of analog to digital measurements is performed. For certain devices (e.g. generator protection) the sampling rate is continuously adjusted depending on the actual sampling rate.

Mains frequency. The protection principle is based on a cyclic calculation algorithm using the analog measured current and voltage values. The error detection, which is determined by this process, must be determined in several consecutive calculations before a protective reaction can take place. The computer transmits a trigger command to the command relay.

9.6 Selective Protection Concepts

The central task of network protection is to selectively disconnect disturbed, i.e. overloaded or faulty, equipment from the power supply network at all poles and on all sides. The supply of electrical energy to other consumers is then ensured without interruption. Selective means here that only the faulty equipment is switched off. Since the energy supply network in Germany is built according to the $n-1$ principle a device can be switched off selectively without the need to interrupt the power supply. The concept of selective protection is therefore an important prerequisite for implementing the $n-1$ principle. To switch off the operating equipment, the circuit-breaker is tripped by the mains protection.

With the concept of selective protection, all equipment, e.g. generators, transformers, lines, or busbars, form a protection zone assigned to the protection zone, which includes the equipment itself and the circuit-breakers, as well as the voltage and current transformers.

The protection zone is protected by a mains protection device, through which the main Protection is monitored and protected. The protection zones of a network overlap so that if the breakers of one protection zone fail, the breakers of the following protection zone can take over their task. The mode of operation of the mains protection device of the following protection zone corresponds to that of a reserve protection (backup protection), which operates downstream of the main protection.

1. *Main protection*: The main protection takes over the protective function for a protection zone in terms of a selection section. Faults within the assigned protection zone are recorded in the fastest possible (i.e. minimum) command time and the affected equipment is selectively disconnected from the network.

2. *Reserve protection*: If the main protection could not clear a fault in its protection zone, e.g. due to failure of the main protective device or the associated circuit-breaker, in order to control the incident, the reserve protection must be effective. The reliability of the entire system is guaranteed by the reserve protection. The protection system consists of main and reserve protection against failing protection. False tripping, i.e. unauthorized tripping, cannot be avoided. The reserve protection can be designed as local or remote reserve protection.

Figure 9.2 shows the high integration density of functions in numeric protective devices.

9.7 Overcurrent Protection

Simple overcurrent protection can be achieved with the so-called independent maximum time (IMT) relays, or definite time overcurrent protection can be set up. IMT relays open if the measured phase current exceeds an adjustable excitation threshold $I >$ when an adjustable delay time t is exceeded.

The phase currents are measured by a current transformer, which monitors the primary phase current $i_{\text{prim}}(t)$ into the secondary phase current $i_{\text{sek}}(t)$. Usually magnetic current transformers are used. Primary and secondary windings are galvanically isolated. The overcurrent stage uses either the base of the measured current $i_{\text{sek}}(t)$ or its peaks. Figure 9.3 shows the one-level overcurrent protection and the scale plan.

In order to achieve selectivity in the radiation network with only one-sided network feed-in, a selectivity with regard to the fault location, the delay time tI must be $>$ in the direction of the mains supply and can be incrementally staggered. The IMT relay farthest from the mains supply is set to a delay time $tI > 100$ ms, with all further delay times starting from 0.5 second in steps of 0.5 second each. The use of numerical (microprocessor-controlled) protective devices can considerably reduce the staggered times. The respective opening times of the upstream switches (= time between the OFF command on the trip coil and the extinguishing of the arc in the switching path) must be taken into account. In addition, the intrinsic time of the relay must be observed. In the beam output, all IMT relays that are connected between the mains supply and the short-circuit have the same short-circuit current. If the excitation thresholds $I >$ of all IMT relays are set to the same value, all IMT relays detect simultaneously if the excitation threshold relay $I >$ is exceeded by the short-circuit current.

If the excitation threshold $I >$ is exceeded, a time step for measuring the delay time $tI >$ is started. Since the IMT relay has the switch open when the measured current has reached the complete delay time $tI \gg$, which is greater than the set excitation threshold $I >$, this triggers a Short-circuit in the nearest IMT relay first. The disadvantage of the concept of time scaling is that the longer the shutdown time, the closer the fault location is to the mains supply. Since the short-circuit current also becomes larger the closer the fault location is to the grid feed-in, high-current short-circuits last longer than low-current short-circuits.

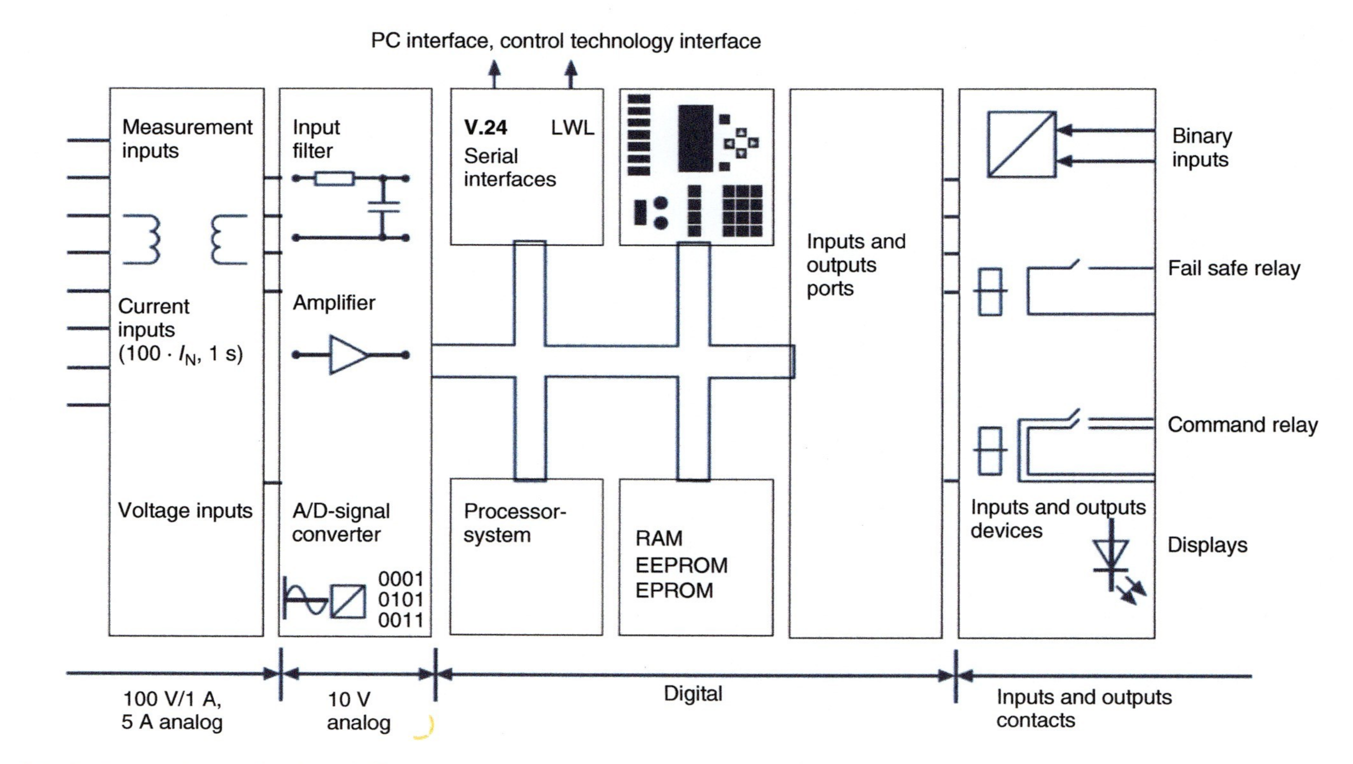

Figure 9.2 Built-in protection functions [45].

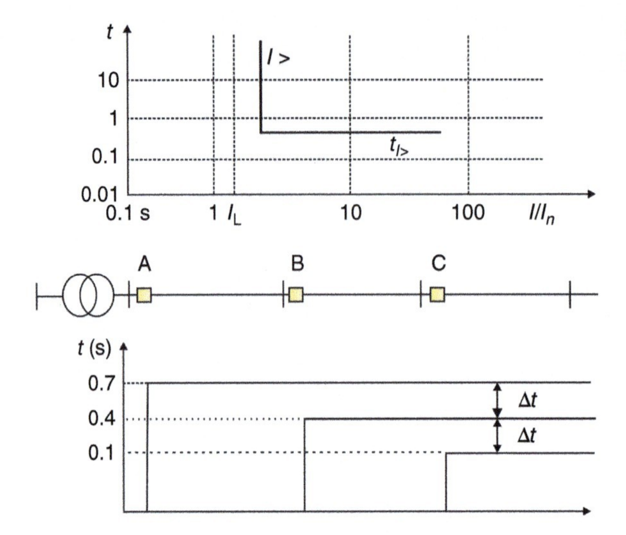

Figure 9.3 Overcurrent protection.

9.7.1 Examples for Independent Time Relays

Time-overcurrent protection devices detect a fault on account of its amperage and clear the fault after a certain delay time has elapsed. In Figure 9.4, 9.5, and 9.6 two practical examples are shown. The graduation starts with the smallest time step and is increased toward the source. The disadvantage of these devices is that the largest short-circuit current with the largest time step occurs at the feeding source.

9.8 Reserve Protection for IMT Relays with Time Staggering

When using IMT relays in time staggering, all IMT relays located upstream of an IMT relay automatically operate as backup protection. Upstream here means closer to the mains supply. In the example, IMT relays a and b act as backup protection for IMT relay c, and IMT relay a acts as backup protection for IMT relay B (Figure 9.3). Nevertheless, the disadvantage remains that, due to the time staggering, high-current faults near the mains supply are present for longer than low-current faults. The hazard potential for the equipment therefore increases the closer the equipment is to the mains supply.

9.9 Overcurrent Protection with Direction

Overcurrent time protection devices can also be used for cables with two-sided power supply when an IMT relay with additional short-circuit direction detection is used. With the help of this type of IMT relays, selectivity can be adjusted for one-sided as well as for two-sided supply (Figure 9.7).

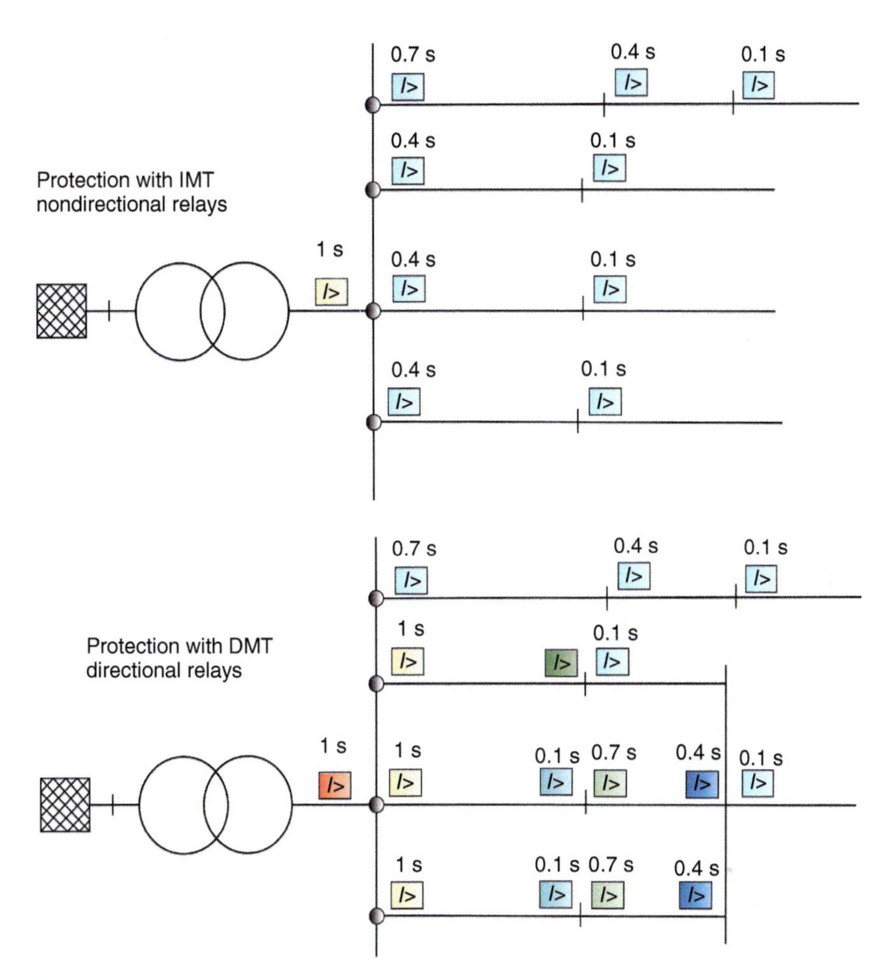

Figure 9.4 Examples for IMT.

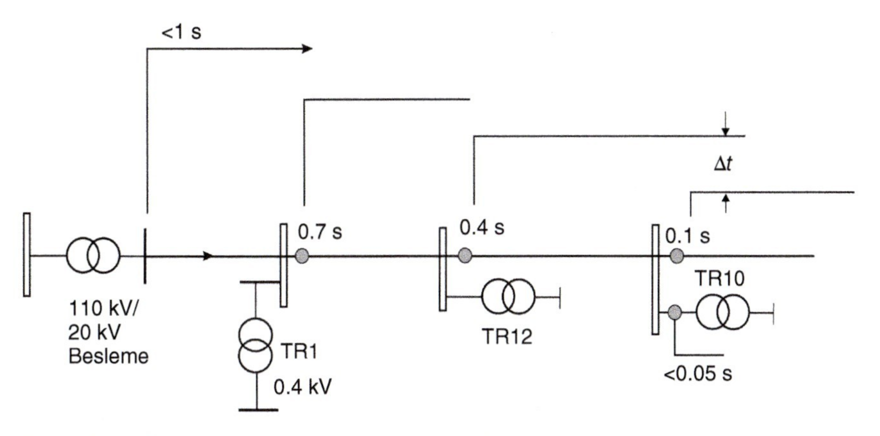

Figure 9.5 Principal circuit diagram of definite time relays protection.

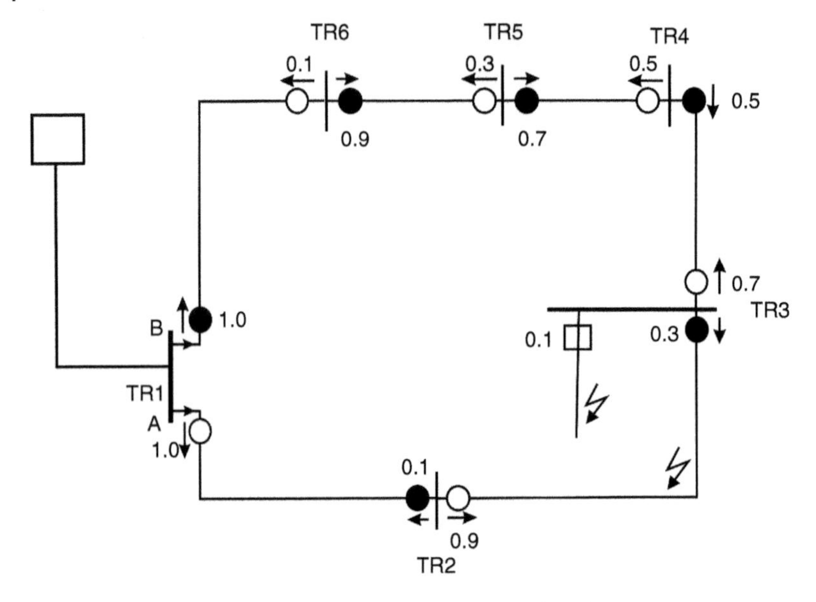

Figure 9.6 Directional overcurrent protection: general settings.

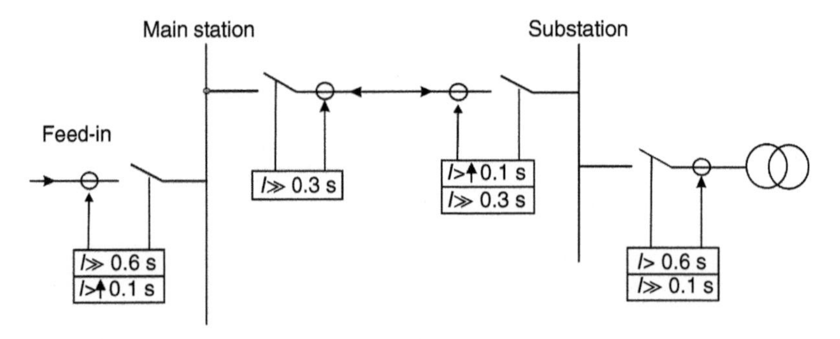

Figure 9.7 Overcurrent protection with direction [3].

IMT relays with short-circuit direction detection open the corresponding switch if the measured phase current provides an adjustable excitation threshold $I >$ for an adjustable delay time $tI >$ and the measured short-circuit direction corresponds to the set short-circuit direction (forward or reverse).

To determine the short-circuit direction, the IMT direction relay must be measured both for the conductor-earth voltages and the conductor currents. In today's standard power supply networks the short-circuit current is inductive as seen from the grid feed-in, if a generator counting arrow system is assumed.

To selectively switch off a fault on one of the cable systems, there are two possibilities:

Line start and end are protected with overcurrent time protection, with the one on the end designed as directional overcurrent protection.

Both cable systems are protected by a cable differential protection. Since differential protection only switches off the fault within its protection range, it must be used for faults on the busbar of the subordinate. Further protection is to be provided for the switchgear, usually a separate overcurrent protection in the outputs of the double switches or a function of the overcurrent time protection within the differential protection.

9.10 Dependent Overcurrent Time Protection (DMT)

As already mentioned, the use of an independent overcurrent protection (IMT relay) with or without short-circuit direction detection has the disadvantage that high-current faults are close to the supply and can be switched off for longer staggered times than current-weak errors, away from the power supply. This disadvantage can be avoided by using dependent maximum time relays or inverse definite maximum time (DMT relays). Unlike IMT relays, DMT relays have no fixed delay time, $tI >$, which is independent of the measured current amplitude. DMT relays determine the current-dependent delay time $tI >$ with the aid of characteristic curves of type $tI >= f(I_{\text{Measure}}/I_B)$. The characteristic curves are defined in such a way that the tripping time of the DMT relay becomes smaller with increasing current, i.e. the tripping time runs inverse to the course of the current as with the fuses. Therefore, DMT relays are usually referred to as inverse time relays.

The tripping time t of the Relay is calculated from the characteristic curve according to the following equation:

$$t = \frac{\beta}{\left(\frac{I}{I_p}\right)^{\alpha} - 1} \cdot T_p \tag{9.1}$$

The meanings of the symbols are

t	Tripping time
T_p	Setting value of the time multiplier
I	Fault current
I_p	Setting value of the current

All characteristics are described by the formulae (9.1). At the same time, there are also distinctions between characteristics (Table 9.1).

Figure 9.8 shows DMT levels for all common characteristic curves according to IEC and ANSI/IEEE.

The tripping characteristics commonly used in DMT relays today vary by continent, country, or even region differently defined and viewed from a historical point of view. For DMT relay of the electromechanical or analog-static generation, the characteristic curves were realized with the aid of mechanical or electronic components. As a result, the realization possibilities were more or less restrictive. In numerical protection technology, these characteristic curves are determined by mathematical

Table 9.1 Values of α and β.

	α	β
Normal inverse	0.02	0.14
Very inverse	1.0	13.5
Extremely inverse	2.0	80.0
Long time inverse	1.0	120.0

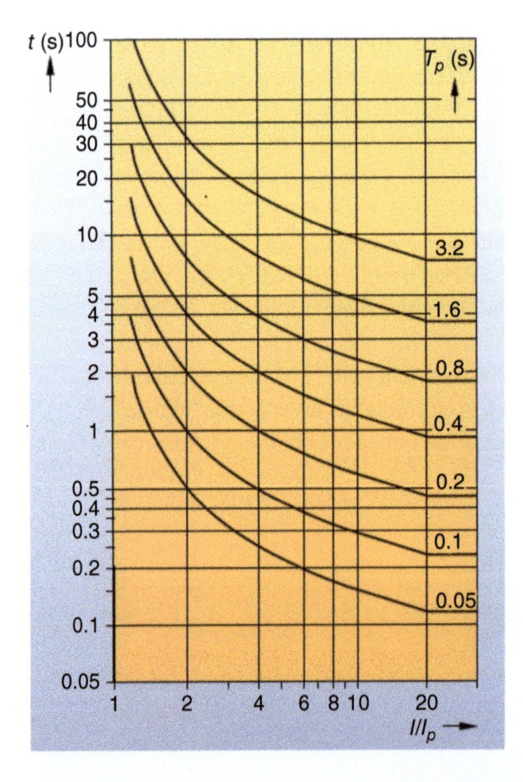

Figure 9.8 Dependent overcurrent protection [45].

equations and implemented in the device firmware. Many DMT characteristics are defined in internationally valid standards such as IEC255-3, IEEE C37.112, or ANSI. Besides manufacturer-specific inverse characteristics are still often used.

The selectivity, which can be achieved with IMT directional relays in simple network structures, can be used for distance protection or differential protection with increasing mesh ratio of the networks. With a high degree of meshing, e.g. annular structures, overlapping of staggered times can occur quickly. The selectivity can be lost very quickly. Even reserve protection concepts can no longer be implemented.

Protection concepts implemented with the simple IMT relay enable no or only a very low selectivity, even with the use of time staggering. This disadvantage has been eliminated by the introduction of short-circuit direction detection. Sufficient selectivity can be achieved with both one-sided and two-sided supply. By using the dependent maximum time relay (DMT relay), high-current (near) faults are switched off faster than low-current (remote) faults compared to the IMT relay.

9.11 Differential Relays

In order to minimize the thermal and mechanical damage to equipment caused by short-circuits, the short-circuit current must be switched off as quickly as possible. The previously considered concept of overcurrent time protection with time staggering has the disadvantage that the tripping is generally delayed. In addition, high-current faults with the longest delay time are switched off.

The differential protection concept, on the other hand, offers the possibility of switching off a short-circuit within the protection zone without further time delay, i.e. in a fast time of 0.1 second (Figures 9.9 and 9.10).

The fast time is the shortest possible time that a protective device starting with the physical error occurrence is required to output the OFF command to the corresponding switch. The fast time is often also called the intrinsic time of the protective device. The inherent time of a protective device depends on the structure of the protective device and the algorithms used: Measured value acquisition, determination of current and voltage indicators, evaluation methods, etc. The characteristic times of protective devices of the same type but from different manufacturers can therefore differ considerably.

The protection concept of differential protection is based on Kirchhoff's node equation. The sum of the currents flowing to and from a node must be zero. With a faultless line or transformer, power is transported through the equipment. For generally constant nominal voltage of the networks, this means a current flow $I(t)$ into and out of the equipment proportional to the power transport. If the equipment is regarded as an N-port, the sum of the currents flowing into and out of the N-ports must be zero in the error-free state ($D_i = 0$). If a current difference $Di > 0$ results, an internal fault of the equipment is present.

The principle of differential protection can also be applied to lines, cables, and generators. For line or cable differential protection, the measuring points of the phase currents are a few kilometers apart. The transmission of the measured variables then requires auxiliary connections over the entire distance. With increasing distance,

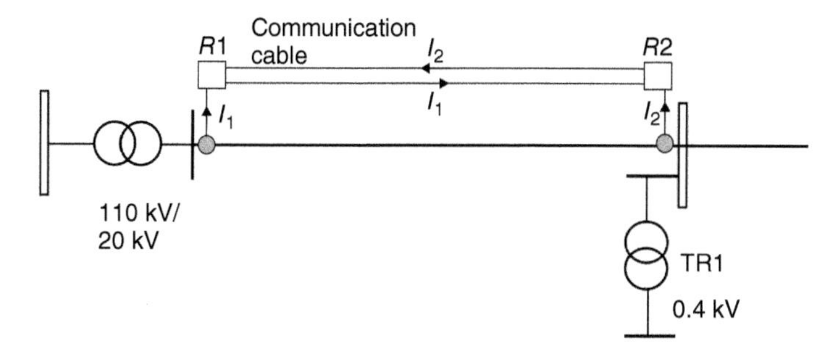

Figure 9.9 Principal circuit diagram of line differential protection.

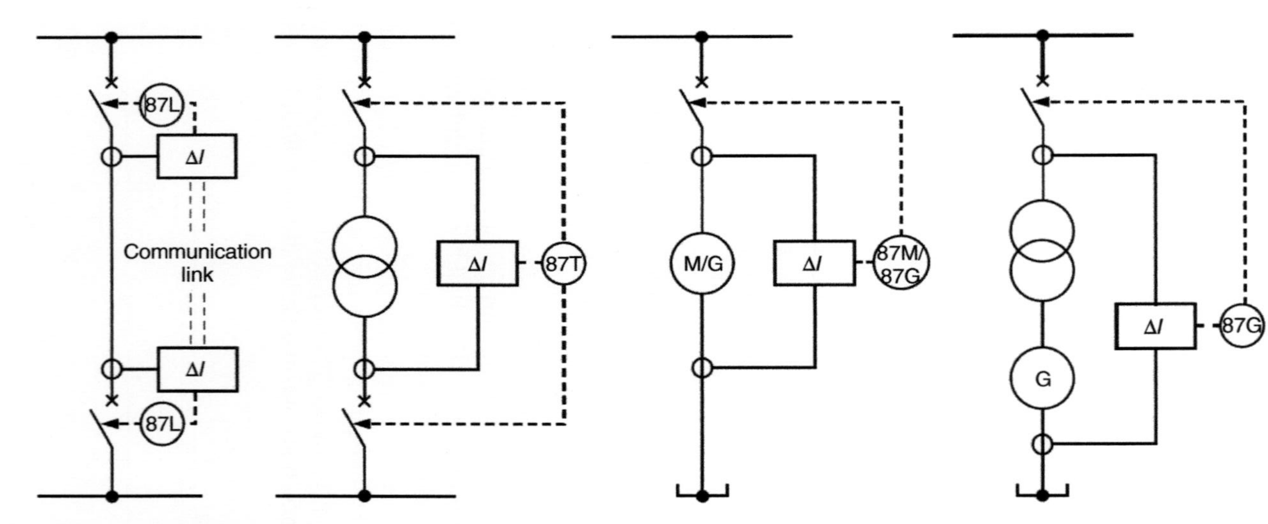

Figure 9.10 Principle of differential protection.

larger and larger cross sections must be used for the auxiliary wires to minimize power dissipation. Regardless of the equipment to be protected, all protection methods based on the concept of differential protection have one problem in common: the effects of saturation of current transformers.

With cable differential protection, each section of the ring cable is protected with a differential protection. One failure within the range will cause both ends to shut down simultaneously due to the cable differential protection. An error outside the differential range is not recognized as an error. Therefore, for possible faults within the stations, protection can be provided with overcurrent time. Communication between the device pairs for cable differential protection takes place in modern devices generally via optical fiber, but communication via copper cable is also possible. Another advantage of the differential protection principle is the simpler structure and strict selectivity.

9.12 Distance Protection

The first distance protection devices were developed around 1920. Until today they are also called distance protection or distance relays for short. The task of distance protection is to calculate the fault distance from the measured variables, current and voltage. The error distance here is the shortest distance between the installation location (measuring location) of the distance protection and the physical fault location (short-circuit location) and is understood in terms of wiring. **F**igures 9.11 and 9.12 show the schematic.

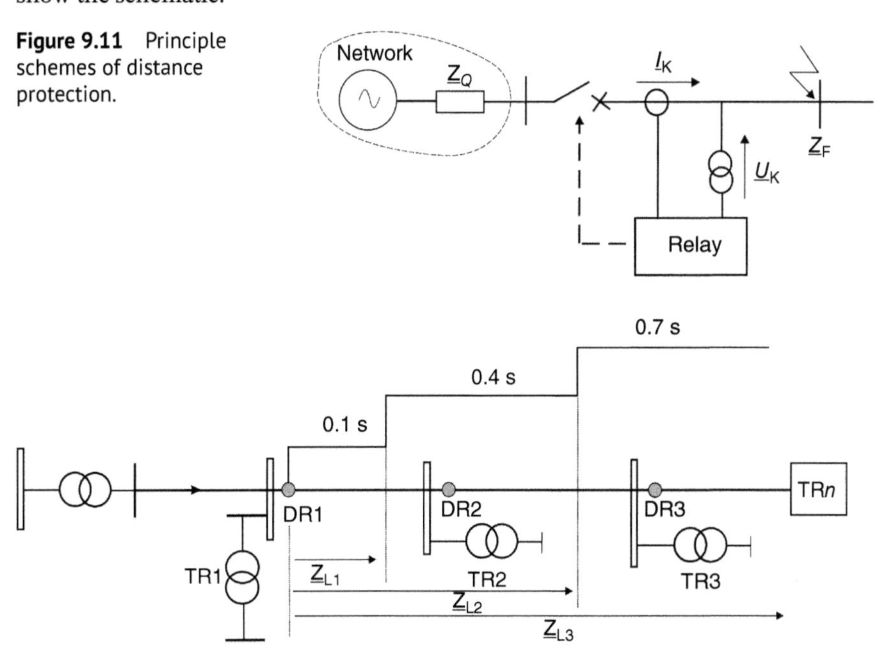

Figure 9.11 Principle schemes of distance protection.

Figure 9.12 Distance protection characteristics, general settings.

The range of an impedance zone is defined by the reactance X, not by the resistance R. The resistance X of a line is practically not subject to weather-related influences such as the temperature and not even by the resistance of the arc to the arc fault additively superimposed. The figure shows the equivalent circuit diagram of those affected by the short-circuit Line in the co-system. The arc can be used in very good approximation as resistance R_{Light}. The distance protection is calculated from the measured variables, voltage $\underline{U}$ and current $\underline{I}$, and the error loop impedance $\underline{Z}_{\text{F}}$:

$$\underline{Z}_{\text{F}} = \frac{U}{I} = (R_1 + R_{\text{Light}} + jX_1) \tag{9.2}$$

It can be seen that the calculated resistance is derived from line resistance and arc resistance and therefore no reliable measure exists for the error distance. The calculated line reactance, on the other hand, can be used as a measure of the error distance.

For monitoring, the relay is set as follows:

Immediate triggering (fast step)

$$Z < 0.85 \cdot Z_{\text{L}} \tag{9.3}$$

Trip with delay by one relay time ($t_{\text{v}} = 0.3 \cdots 0.5$ second)

$$Z > 0.85 \cdot Z_{\text{L}} \tag{9.4}$$

In practice, new settings have proved their worth due to the low impedance distances between the stages of adjacent relays.

First level

$$Z_1 = 0.85 \cdot Z_{\text{AB}} \tag{9.5}$$

Second level

$$Z_2 = 0.85 \cdot Z_{\text{AB}} + 0.72 \cdot Z_{\text{BC}} \tag{9.6}$$

Third level

$$Z_3 = 0.85 \cdot Z_{\text{AB}} + 0.72 \cdot Z_{\text{BC}} + 0.61 \cdot Z_{\text{CD}} \tag{9.7}$$

Fourth level

$$Z_4 = 0.85 \cdot Z_{\text{AB}} + 0.72 \cdot Z_{\text{BC}} + 0.61 \cdot Z_{\text{CD}} + 0.52 \cdot Z_{\text{CD}} \tag{9.8}$$

Current and voltage are measured between all outer conductors and the impedance is formed from them. Depending on the measured impedance, the output command is given to the protection device with the set time. The distance protection device provides reserve protection for further line sections.

$$\underline{I}_{\text{L1}} \cdot \underline{Z}_1' l_{\text{L}} - \underline{I}_{\text{L2}} \cdot \underline{Z}_1' l_{\text{L}} = \underline{U}_{\text{L1}} - \underline{U}_{\text{L2}} \tag{9.9}$$

Impedance of the line:

$$\underline{Z}_{\text{L}} = \frac{\underline{U}_{\text{L1}} - \underline{U}_{\text{L2}}}{\underline{I}_{\text{L1}} - \underline{I}_{\text{L2}}} = \underline{Z}_1' l_{\text{L}} = R' l_{\text{L}} + jX' l_{\text{L}} \tag{9.10}$$

On earth fault (example: L1 earth):

$$\underline{I}_{L1} \cdot \underline{Z}'_{l} l_{\mathrm{L}} - I_{\mathrm{E}} \cdot \underline{Z}_{\mathrm{E}} \cdot l_{\mathrm{L}} = \underline{Z}_{l1} l_{\mathrm{L}} \left(\underline{I}_{\mathrm{L1}} - \underline{I}_{\mathrm{E}} \cdot \frac{\underline{Z}'_{\mathrm{E}}}{\underline{Z}'_{1}} \right) = \underline{U}_{\mathrm{L1}} \tag{9.11}$$

$$\frac{\underline{U}_{\mathrm{L1}}}{\underline{I}_{1}} l_{\mathrm{L}} - I_{\mathrm{E}} \frac{\underline{Z}'_{\mathrm{E}}}{\underline{Z}_{1}} = \frac{\underline{U}_{\mathrm{L1}}}{\underline{I}_{\mathrm{L1}}} - I_{\mathrm{E}} \cdot k_{\mathrm{E}} = \underline{Z}'_{1} l_{\mathrm{L}} = R' \cdot l_{\mathrm{L}} + jX'_{l_{\mathrm{L}}} \tag{9.12}$$

$$k_{\mathrm{E}} = \frac{\underline{Z}'_{\mathrm{E}}}{\underline{Z}'_{1}} = \frac{\underline{Z}'_{0} - \underline{Z}'_{1}}{3\underline{Z}'_{1}} \tag{9.13}$$

The arc resistance can be estimated with a safety factor of $s = 2 \cdots 4$.

$$R_{\mathrm{Lb}} = \frac{1800 \cdot l_{\mathrm{Lb}}}{I_{\mathrm{kmin}}} \cdot s \tag{9.14}$$

9.12.1 Method of Distance Protection

Distance protection measures the conductor-earth voltages and conductor currents at the installation location. Voltage and current transformers must therefore be installed at the relay location. From the voltages and currents, the distance protection calculates the fault impedance between the relay installation location and the fault location. The division of the error distance by the impedance coating results in the error distance in kilometers. The impedance layer of a line, for example, indicates the line impedance Z'_{L} in Ω/km related to 1 km.

9.12.2 Distance Protection Zones

Up to now, only the error impedance (= co-impedance between measuring location of the protective device and the fault location) has been calculated in Ω or the fault distance in km. Only a comparison of the fault impedance with a tripping characteristic makes it possible to assess whether the short-circuit is in the protection zone to be monitored and thus leads to an OFF command. Similar to overcurrent protection, distance protection is provided by a scale characteristic (scale plan), which is defined however, not according to time but staggered according to the error distance. The individual sections of the grading curve are called impedance zones or simply zones. For distance protection, a distinction is made between setting values.

9.12.3 Relay Plan

Figure 9.13 shows a scale plan for the protection of two lines, if cable ends distance protection is used. The distance protection A uses three directional impedance zones 1–3, all staggered forward. In the example, the two lines to be protected are to be staggered, each having a length of 100 km. The squadron plan is now used for distance protection. Here, a radiant network without intermediate supply was assumed. The tripping characteristics of the distance protection in the R/X level are characterized by the forward–backward characteristic and the adjustable distance steps, zone 1, zone 1B for restarting, zone 2, and zone 3.

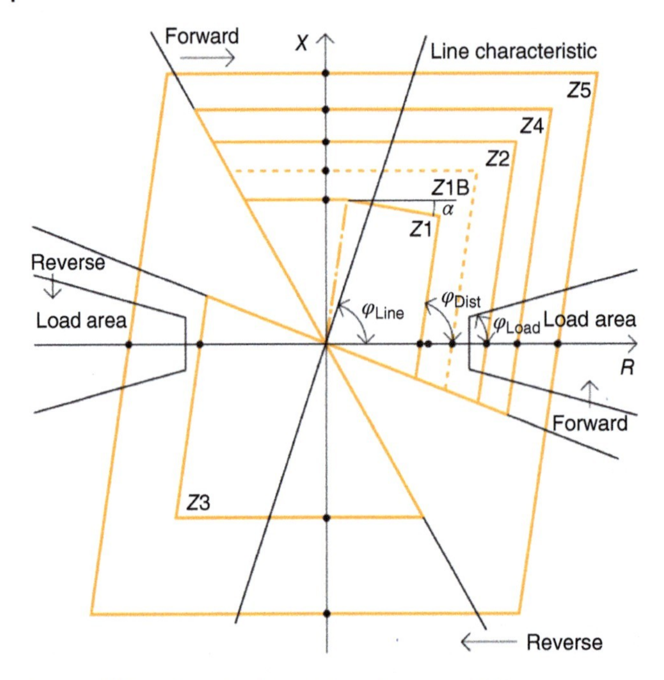

Figure 9.13 Characteristics of the relay zones [45].

The fast time of distance protection A protects approximately 85% of line 1. When determining the range of the impedance zones, make sure that the fault impedance is determined by the protective device with the positive impedance Z_1 of the line. The manually calculated co-impedance of the management is based on conditions, which in reality are only approximately fulfilled or not at all:

1. With regard to the geometric dimensions (mast diagram), the cable shown as *a* is assumed to be completely symmetrical. Weather-induced influences on the conductor cable temperature and the conductor sag are neglected.
2. The cable is considered with regard to the resistive, capacitive, and inductive coating ($R1', C1'$ and $X1'$ in Ω/km) as homogeneous over the entire cable length – this has never been thought of before, i.e. the impedance coating of the line is constant.
3. Any nonlinearities or frequency dependencies are avoided, especially the impedance of the earth return line is neglected. To avoid nonselective false tripping due to over-ranges, wires or cables are valued at a maximum of $5 \cdots 90\%$ of their wire length in fast time protected zone. The range of the fast time zone is displayed as grading factor *fs*. It is decisive for the design of the further relay zones.

Figure 9.14 shows an example with distance protection devices.

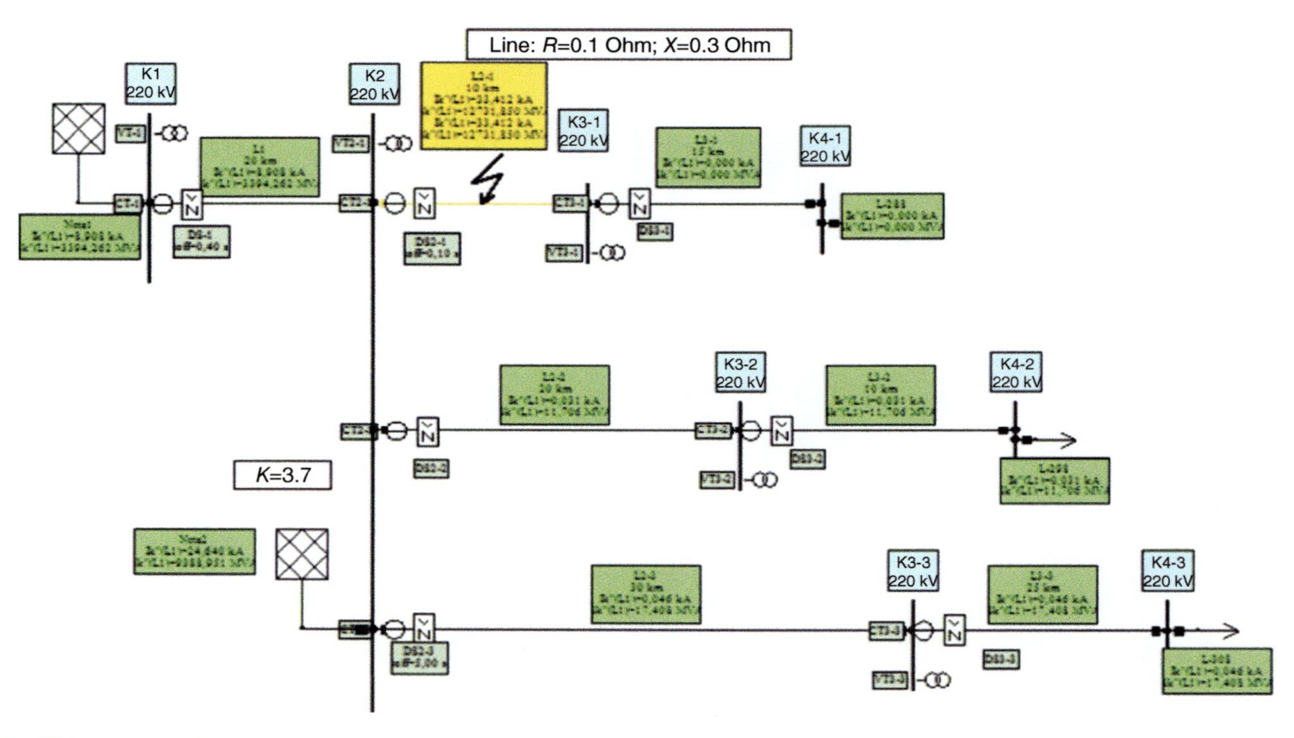

Figure 9.14 Distance protection.

9.13 Motor Protection

Special protective functions are available for protecting medium-voltage motors, such as the overload protection function. The temperatures of critical points of the motor (e.g. bearings) are recorded and monitored, and can be controlled by the so-called "thermoboxes." In particular, this can reduce the sensitivity of the thermal overload protection. The short-circuit and overload protection of the supply lines must be ensured either by fuses and circuit-breakers or by fuses alone.

9.14 Busbar Protection

Busbar faults within switchgear systems nowadays due to the construction of the facilities are very unlikely, but not impossible. Usually such errors are caused by an upstream overcurrent time protection, detected, and switched off. For this method, however, the time to shutdown depends on the grading times as they result from the selective design of the network. It is to be used for busbar faults to achieve shorter shutdown times and thus reduce the extent of damage as far as possible; a higher level of protection can be achieved either when a special busbar differential protection is used or a backward locking by the directional overcurrent time protection.

Figures 9.15 and 9.16 show the possibility of backward locking at a network station that is connected in a closed ring via directional comparison protection or is integrated via cable differential protection.

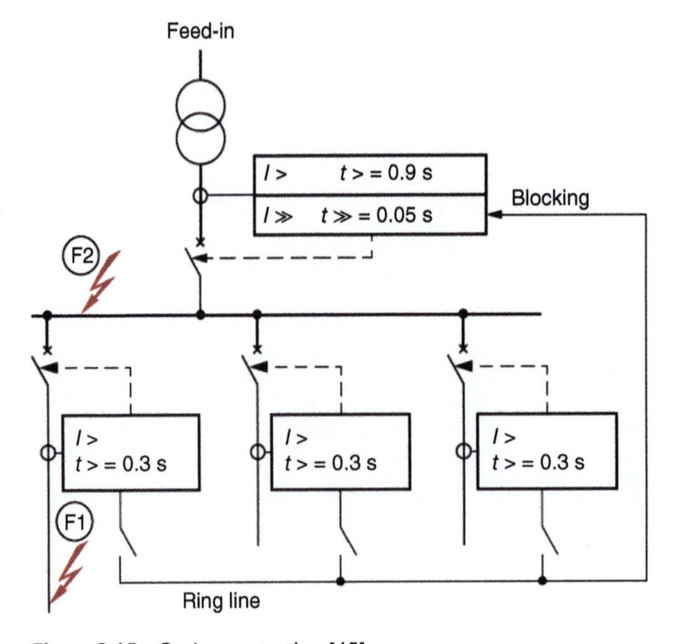

Figure 9.15 Busbar protection [45].

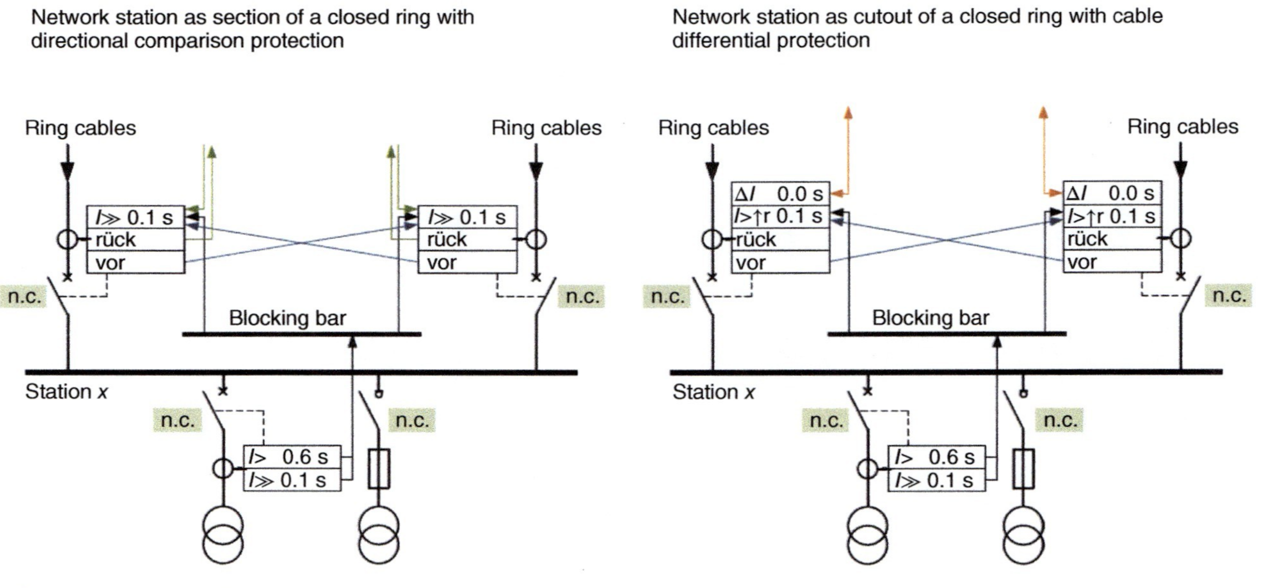

Figure 9.16 Busbar protection [45].

In order to increase the protection level for faults within metal-enclosed switchgear, arcing faults, which are naturally associated with a pressure rise within the encapsulated system, can be detected and switched off by means of pressure switches. In the event of such a fault, the pressure switch acts like a busbar protection.

Depending on the complexity of the busbar system busbar protection can become very complex. For simpler systems, the principle of the backward locking is used. In doing so XMZ protection of the feed-in is independent of staggered time in fast time, provided that its fast trip level is not blocked in a drain by the short-circuit or earth fault current excitation. In the output devices, the excitation is transferred to an extra contact and all excitation contacts are parameterized and connected in parallel. A protective excitation in an outgoing feeder field (fault F1) means that the pending error is not in the area of the busbar. Through the excitation of the output protection, the undelayed trip $I \gg$ of the feed-in protection ($t \gg= 50$ ms in) can be blocked via the binary input. The feed-in protection acts as reserve protection, with $t >$. If an error occurs on the busbar (Fault F2) only the protective device of the power supply is excited with $t >$ and $t \gg$; thus blocking is missing and the busbar fault is triggered in fast time. The quick shutdown reduces the load due to the error.

9.15 Saturation of Current Transformers

Current transformers transform the very high primary currents in the event of a short-circuit into smaller secondary currents with smaller conductor cross sections that can be led to the protective devices. They also realize the galvanic isolation of the protective devices from the high voltage side.

The transformation ratio of the current transformers can be taken from defined standards and manufacturers. Primary nominal currents are, e.g. 200 or 600 A for medium voltage, and 1200 or 2000 A for the maximum voltage, as secondary rated currents are 5 A for medium-voltage networks, and 1 A for high and extra-high voltage networks. In the secondary circuit, the current flowing in the primary circuit has an influence on the secondary current corresponding to the transmission ratio.

Current transformers are connected to the terminals on the secondary side with the load impedance completed. The load impedance consists of the internal load (winding resistance) and the external load. The external burden consists of the resistance of the supply lines to the protective device and the input resistance of the protective device. Note that the supply lines have a return route. Current transformers must never be operated without load, i.e. open. In in this case, the current transformer is destroyed. Therefore, the secondary side must be short-circuited. In general, for all differential protection applications, it must be noted that the resistances of all auxiliary conductors etc. between the terminals of the secondary circuit of the current transformer and the input terminals of the protection device are included in the rating of the current transformers. In particular, considerable impedance values can be achieved here with line or cable differential protection.

The load on the current transformer, i.e. the value of the resulting load impedance, influences the transmission behavior of the current transformer considerably. The higher the current transformer is loaded (i.e. the higher the current at the current transformer terminals is switched on load resistance), the sooner the current transformer is in saturation. An uneven charging of otherwise identical current transformers means that the higher loaded current transformer saturates earlier than the less burdened current transformer. Through asymmetric saturation of both the current transformers, even with external faults, there is a differential current $DI > 0$, which is used to excite the differential protection and the Output of an OFF command to the devices assigned to the switch. A nonselective and also unauthorized shutdown of fault-free equipment can be the result here.

The cause for the saturation of current transformers and the resulting nonlinear transfer behavior is the nonlinear magnetization behavior of the iron core of the current transformers.

9.16 Summary

High and extra-high voltage networks serve to transport electrical energy over medium to long distances. The electrical energy is transported from the power plants to the medium-voltage distribution networks and to large industrial customers. The nominal voltages of these networks are usually 110–380 kV in Europe. In countries with longer distances between power plants and consumer centers, such as North America or Russia, nominal voltages up to 1000 kV are used in the transport network. In contrast to medium-voltage networks, the high-voltage and extra-high-voltage grids used are generally meshed in order to ensure that, in the event of a failure of one of the equipment (e.g. line or transformer), the energy transport to the subordinate medium-voltage networks occurs without interruption ($(n-1)$ principle).

The time-dependent protection commonly used in medium-voltage networks cannot be used in high-voltage and extra-high-voltage networks because, due to the meshing, there are generally two-sided supplies and, due to the significantly higher short-circuit ratings, the time scale protection is not applicable. In these networks, equipment affected by faults must generally be switched on and off at high speed, in order to prevent destruction of the equipment and to avoid endangering plants and personnel. That is why distance protection systems and differential protection systems are used in high and extra-high voltage networks as line protection.

Differential protection systems are used as already discussed for transformers, short power or cable connections, and busbars. Distance protection systems are used for longer lines and cables but also as reserve protection systems for transformers and busbars.

The use of distance protection devices is not limited to high and extra-high voltage networks. Even in meshed medium-voltage networks, selectivity can also be improved by the use of definite time relays. With short-circuit direction detection no longer being guaranteed, distance protection devices are used.

10

Power Flow in Three-Phase Network

10.1 Terms and Definitions

The aim of the load flow calculation is to determine the complex stresses in all network nodes and to calculate the power flows on the connecting elements between the nodes, i.e. the overhead lines, cables, and transformers. Load flow calculations are determined for the two load states, heavy load and light load.

Three different network nodes are distinguished (Figure 10.1):

- *Slack node*: At this node, only the voltage (amplitude, phase control) is specified. The slack node covers the power difference, since only after the current and voltage distribution mentioned results.
- *Load node*: As a rule, the complex power is specified, which is constant during the calculation (P, Q node). In addition, the behavior of the loads at a changed voltage must be considered (constant current, constant impedance).
- *Generator node*: In general, the amplitude of the voltage and the active power are fixed (P, V-knot). In many cases, the limits of the reactive power result from the operating diagram of the generator.

10.2 Introduction

Power flow calculation (load flow) includes the stationary current and voltage distribution in a network and covers the following aspects:

- Power distribution in the network (optimization)
- Voltages at the network nodes
- Overload of equipment
- Behavior of the grid in the event of a generator or consumer failure
- Loss minimization
- Transformer stage setting

Analysis and Design of Electrical Power Systems: A Practical Guide and Commentary on NEC and IEC 60364, First Edition. Ismail Kasikci.

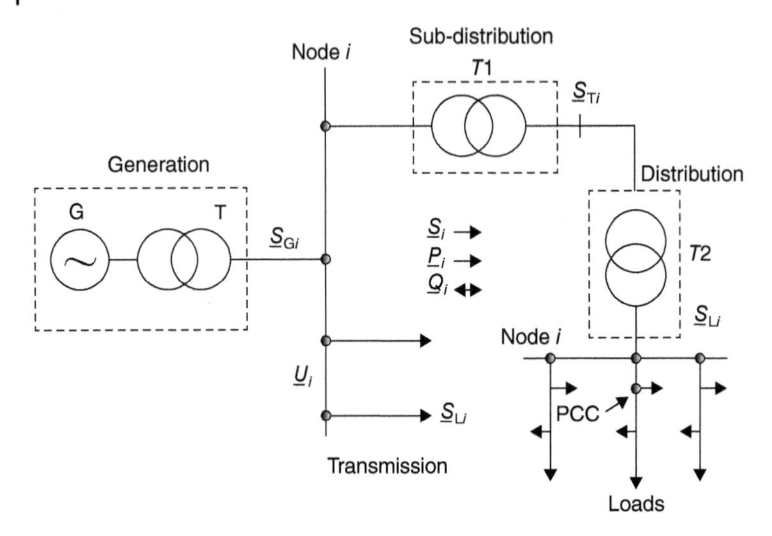

Figure 10.1 Power flow network nodes.

The analytical calculation of three-phase networks described so far was based on simplifications, which led to a real calculation. This procedure is permitted in low-voltage (LV) networks, partly also in medium voltage (MV) networks. In high-voltage (HV) networks and to a large extent also in MS networks, the networks, however, are more strongly meshed with strongly different line impedances between the knots. For this reason, complex calculations usually have to be carried out here. The calculation a three-phase network with the aim of determining the distribution of active and reactive power on the individual lines and nodes as well as the determination of the node stresses is called load flow calculation.

An important characteristic of a meshed network with L lines and N nodes (without the reference node) is the degree of meshing. It applies:

$$\nu = \frac{L}{N} \tag{10.1}$$

The maximum degree of meshing exists when the nodes are connected to each other, and is $\nu_{\max} = N(N-1)/2$. Actual meshing degrees are in the range of 1–2, in HV networks usually even under 2. For the load flow calculation, specifications must be made at the individual network nodes regarding the type of load. For this purpose, the nodes are divided into load or load nodes and supply or generator nodes. At least one node must exist where the stress is specified according to amount and angle, the so-called slack node. P and Q adjust themselves in such a way that there is a balance in the whole network between the services fed in and the services taken off. The slack nodes must be capable of taking over the current account balance assigned to them. Therefore it is usually a matter of grid feed-ins or the strongest power plant. The following table shows the different Node types with the specification of the given and the required sizes:

Designation of nodes	Given sizes	Required sizes	Remarks
Slack nodes	U, φ_u	P, Q	Reference node (grid supply)
Feed-in node	U, P	φ_u, Q	Power plant feed-ins (generator node)
	P, Q	U, φ_u	Negative consumer node
Consumer nodes	P, Q	U, φ_u	Consumer with voltage-dependent power

Only if a load flow calculation without iteration is specified for all nodes in the network except for slack node I and φ_i, i.e. for all supply and load nodes, is a load flow calculation possible. In all other cases, an iteration process must be used to calculate the load flow of the nonlinear system of equations. The system of equations is nonlinear because the power depends quadratically on the voltage.

10.3 Node Procedure

The following values are given or searched for in the node method:

- *Founded*:
  - Input and load power (P, Q) of all nodes except the slack node. In general, the node performance depends on the respective node voltage and must therefore be calculated iteratively. Only if the node currents instead of the node powers are given, i.e. if the voltage dependency is not applicable, no iteration necessary.
  - Voltage of the slack node (amount U and angle φ_u).
- *Sought*:
  - Node current of the slack node or active and reactive power (P, Q) of the slack node.
  - Node voltages at all nodes (magnitude gU and angle φ_u) except the slack node.

First, the sum of all currents occurring at each node is calculated. At the same time, incoming currents are counted positive, and outgoing currents are counted negative. In addition, each node i is assigned a node voltage $\underline{U}_i$ as star voltage. By means of Kirchhoff's laws the current sums of the nodes can be connected to the node voltages $\underline{U}_i$ via the admittances of the network.

$$\begin{matrix} \underline{Y}_{11} & -\underline{Y}_{12} & \cdots & -\underline{Y}_{1i} & \cdots & -\underline{Y}_{1N} \\ -\underline{Y}_{21} & \underline{Y}_{22} & \cdots & -\underline{Y}_{2i} & \cdots & -\underline{Y}_{2N} \\ \vdots & \vdots & \vdots & \vdots & \vdots & \\ -\underline{Y}_{i1} & -\underline{Y}_{i2} & \cdots & \underline{Y}_{ii} & \cdots & -\underline{Y}_{iN} \\ \vdots & \vdots & \vdots & \vdots & \vdots & \\ -\underline{Y}_{N1} & -\underline{Y}_{N2} & \cdots & -\underline{Y}_{Ni} & \cdots & -\underline{Y}_{NN} \end{matrix} \cdot \begin{matrix} \underline{U}1 \\ \underline{U}_2 \\ \vdots \\ \underline{U}_i \\ \vdots \\ \underline{U}_N \end{matrix} = \begin{matrix} \underline{I}1 \\ \underline{I}_2 \\ \vdots \\ \underline{I}_i \\ \vdots \\ \underline{I}_N \end{matrix} \tag{10.2}$$

or in matrix notation

$$[\underline{Y}] \cdot [\underline{U}] = [\underline{I}]$$

with $[\underline{Y}]$ as the so-called node admittance matrix.

The node admittance matrix is always square in the form $N \times N$ and is always symmetrical. If there are branch admittances between two nodes i and j in a network, then the associated total conductance $\underline{Y}_{ij}$ is entered with a negative sign in the fields i,j and j,i. If there is no branch admittance between the two nodes i and j, zero is entered in the fields i,j and j,i. On the main diagonal (elements $\underline{Y}_{ij}$) stands the negative sum of all secondary diagonal elements of the associated row plus cross admittances, i.e. admittances between node i and reference node 0 ($\underline{Y}_{ij} = \underline{Y}_{i0} + \sum j, j = 1 \cdots N, j \neq i$). The reference node is usually Earth. The system of equations can only be solved when a node stress is specified, namely the stress at the slack node. For example, if the Nth node is intended as a slack node, the system of equations is reduced by the Nth equation and reads as follows:

$$\begin{bmatrix} \underline{Y}_{11} & -\underline{Y}_{12} & \cdots & -\underline{Y}_{1i} & \cdots & -\underline{Y}_{1N} \\ -\underline{Y}_{21} & \underline{Y}_{22} & \cdots & -\underline{Y}_{2i} & \cdots & -\underline{Y}_{2N} \\ \vdots & \vdots & \vdots & \vdots & \vdots & \\ -\underline{Y}_{i1} & -\underline{Y}_{i2} & \cdots & \underline{Y}_{ii} & \cdots & -\underline{Y}_{iN} \\ \vdots & \vdots & \vdots & \vdots & \vdots & \\ -\underline{Y}_{N1} & -\underline{Y}_{N2} & \cdots & -\underline{Y}_{Ni} & \cdots & -\underline{Y}_{NN} \end{bmatrix} \cdot \begin{bmatrix} \underline{U}1 \\ \underline{U}_2 \\ \vdots \\ \underline{U}_i \\ \vdots \\ \underline{U}_N \end{bmatrix} = \begin{bmatrix} \underline{I}1 \\ \underline{I}_2 \\ \vdots \\ \underline{I}_i \\ \vdots \\ \underline{I}_N \end{bmatrix} + \begin{bmatrix} \underline{Y}1N \\ \underline{Y}_{2N} \\ \vdots \\ \underline{Y}_{iN} \\ \vdots \\ \underline{Y}_{N-1,N} \end{bmatrix} \underline{U}_N \tag{10.3}$$

or in matrix notation

$$\underbrace{[\underline{Y}]_{\text{LK}}}_{\text{known}} \cdot \underbrace{[\underline{U}]_{\text{LK}}}_{\text{sought}} = \underbrace{[\underline{I}] + [\underline{Y}]_{\text{SK}} \cdot \underline{U}_N}_{\text{known}} \tag{10.4}$$

with $[\underline{Y}]_{\text{LK}}$ as the load node admittance matrix, $[\underline{U}]_{\text{LK}}$ as the vector of unknown load node voltages, and $[\underline{Y}]_{\text{LSK}}$ as the feed node admittance matrix (Note: At only one given voltage [slack node], $[\underline{Y}]_{\text{LSK}}$ is a vector). Note that the load currents $\underline{I}_i$ must be counted negatively because they flow out of the load node i. The $N - 1$ load node voltages and the current at the slack node are required. All branch currents, power flows, network losses, and the reactive power requirements of the network can then be calculated. The node currents on the right side of the equation system (10.3) are either given or must be calculated for each subsequent iteration step from the given node powers with the current approximate values for the node stresses:

$$\underline{I}_i = \frac{\underline{S}_i^*(\underline{U}_i)}{3U_i^*} = \frac{P_i(U_i) - jQ_i(U_i)}{3\underline{U}_i^*} = \frac{P_i(U_i) - jQ_i(U_i)}{3\underline{U}^2}\underline{U}_i \tag{10.5}$$

For this purpose the voltage dependence of performances must be known and the sign of $\underline{I}_i$ must be observed. At the beginning of the iteration calculation there are no approximate values of the individual node voltages. In order that initial values for the node currents are available, the nominal voltages $U_n/\sqrt{3}$ are used as initial values of the node voltages and derived with Eq. (10.5); initial values of node streams are calculated. The node current at the slack node is calculated according to the solution of Eq. (10.3) calculated from:

$$\underline{I}_N = \begin{bmatrix} -\underline{Y}_{N1} & -\underline{Y}_{N2} & \cdots & -\underline{Y}_{Ni} & \cdots & \cdots & -\underline{Y}_{N,N-1} & \underline{Y}_{NN} \end{bmatrix} [\underline{U}] \tag{10.6}$$

Equation (10.3) is solved on diagonal elements

$$\begin{bmatrix} \underline{Y}_{11}\underline{U}_1 \\ \underline{Y}_{22}\underline{U}_2 \\ \vdots \\ \underline{Y}_{ii}\underline{U}_i \\ \vdots \\ \underline{Y}_{N-1,N-1}\underline{U}_{N-1} \end{bmatrix} = \begin{bmatrix} \underline{I}_1 \\ \underline{I}_2 \\ \vdots \\ \underline{I}_i \\ \vdots \\ \underline{I}_{N-1} \end{bmatrix} - \begin{bmatrix} 0 & -\underline{Y}_{12} & \cdots & -\underline{Y}_{1i} & \cdots & \underline{-1, N-1} \\ -\underline{Y}_{21} & 0 & \cdots & -\underline{Y}_{2i} & \cdots & -\underline{Y}_{2,N-1} \\ \vdots & \vdots & \cdots & \vdots & \cdots & \vdots \\ -\underline{Y}_{i1} & -\underline{Y}_{i2} & \cdots & 0 & \cdots & -\underline{Y}_{i,N-1} \\ \vdots & \vdots & \cdots & \vdots & \cdots & \vdots \\ -\underline{Y}_{N-1,1} & -\underline{Y}_{N-1,2} & \cdots & -\underline{Y}_{N-1,i} & \cdots & 0 \end{bmatrix}$$

$$\cdot \begin{bmatrix} \underline{U}_1 \\ \underline{U}_2 \\ \vdots \\ \underline{U}_i \\ \vdots \\ \underline{U}_{N-1} \end{bmatrix} + \begin{bmatrix} \underline{Y}_{1N} \\ \underline{Y}_{2N} \\ \vdots \\ \underline{Y}_{iN} \\ \vdots \\ \underline{Y}_{N-1,N} \end{bmatrix} \underline{U}_N \tag{10.7}$$

and can then be solved with methods of numerical calculation. In the so-called single-step procedure, for example, the rule for calculating the stress of node i in iteration step $(v + 1)$ is U_i.

$$\underline{U}_i^{v+1} = \frac{1}{\underline{Y}_{ii}} \left(\underline{I}_i^{(v)} + \underline{Y}_{iN}\underline{U}_N + \sum_{\substack{j=1 \\ j \neq 1}}^{i-1} \underline{Y}_{ij}\underline{U}_j^{(v+1)} + \sum_{j=i+1}^{N-1} \underline{Y}_{ij}\underline{U}_j^{(v)} \right) \tag{10.8}$$

with $\underline{I}_i^v$ as the (v)th approximation of the impressed current in node i, $\underline{U}_j^{(v+1)}$ as the $(v + 1)$th approximation of stresses in node j with $j < i$, $\underline{U}_j^{(v)}$ as the (v)th approximation of stresses in node j with $j > I$, and $\underline{U}_N$ as the impressed voltage of the slack node. The node method as previously presented has two limitations:

- *Limitation to an impressed voltage*: Up to now it was assumed that there is only one known impressed voltage in the network. All other voltages inevitably result from the iteration calculation. If there are several impressed voltages in the net, then the method can be suitably extended by using the method described in Eq. (10.3), the vector $[\underline{U}_{SK}]$ of known tensions at table knots.

$$\underbrace{[\underline{Y}]_{LK}}_{\text{known}} \cdot \underbrace{[\underline{U}]_{LK}}_{\text{sought}} = \underbrace{[\underline{I}] + [\underline{Y}]_{SK} \cdot [\underline{U}]_{SK}}_{\text{known}} \tag{10.9}$$

The node admittance matrix $[\underline{Y}]_{LK}$ links the nodes to the load nodes. This matrix has the form (number of load nodes) × (number of supply nodes). With Eq. (10.9) unknown stresses at the load node $[\underline{U}]_{LK}$ can be iteratively calculated. The currents of the feed nodes can then be calculated.

- *Limitation to grid feed-ins*: Up to now it was assumed that the grid is fed from a grid feed-in, i.e. from a higher level grid, and that the voltage is known in terms of magnitude and angle at the node of the grid feed-in. If the three-phase network is not supplied by higher level networks but by power plant feed-ins, then other conditions exist. At the nodes of the power plant feed-ins, the voltage is known in terms of magnitude but not in terms of angle. The power plant regulation regulates the voltage amount U_{Gi} = const. and the active power P_{Gi} = const. These node conditions can be suitable for integration into the intersection method. However, the more power plant feed-ins to be considered, the more convergence difficulties of the iteration calculation occur.

10.4 Simplified Node Procedure

Figure 10.2 shows an example of a mesh network with four load nodes and three feed-in nodes.

The simplified node method can be used for the calculation of LV networks and partly of MV networks. It is based on the following simplifying assumptions:

- A symmetrical three-phase network is assumed, which can be described by a single-phase equivalent circuit diagram.
- The phase angles of all phase voltages occurring in the network should be approximately equal. In addition, the voltages at the individual network nodes are generally different, but in each case close to the nominal voltage, i.e. the line angle is neglected and the invoice becomes real.
- cross admittances are neglected, i.e. only longitudinal admittances are calculated.
- The power factors of all loads on the network should differ only slightly, so that the average power factor $\cos\varphi_{mi}$ can be expected.

The following terms also apply:

- Load nodes are nodes in which at least three lines converge, and/or are connected to the loads. Load node voltages are measured with $U_1, U_2 \cdots$ notations.

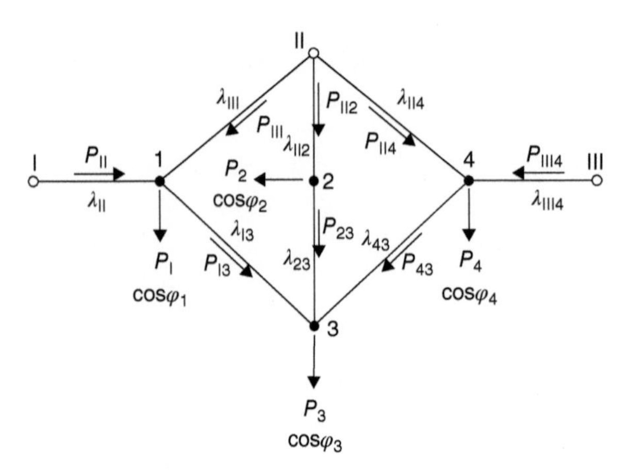

Figure 10.2 Example network for simplified node procedure: Load nodes 1, 2, 3, and 4, Feed-in Points I, II, and III.

- Power supply nodes are nodes for which the phase voltage is specified for the calculation. Nodal points into which generators feed a given power are regarded as load nodes with negative power consumption. Feeding point voltages are marked with $U_{\mathrm{I}}, U_{\mathrm{II}}, \dots$

Equation (10.10) results in the case that on a line between two nodes i and k no loads are present, i.e. with $\sum_{P_\nu L_\nu}^{n}$, the compensation flowing from i to k

$$P_{ik} = (U_i - U_k)U_n\lambda_{ik} \tag{10.10}$$

This makes it possible to set up the following system of equations for the example network in Figure 8.6:

$$\begin{aligned}
P_{\mathrm{I}1} &= (U_{\mathrm{I}} - U_1)U_n\lambda_{\mathrm{I}1} \quad ; \quad P_{23} = (U_2 - U_3)U_n\lambda_{23}\\
P_{\mathrm{III}1} &= (U_{\mathrm{II}} - U_1)U_n\lambda_{\mathrm{III}1} \quad ; \quad P_{\mathrm{II}4} = (U_{\mathrm{II}} - U_4)U_n\lambda_{\mathrm{II}4}\\
P_{13} &= (U_1 - U_3)U_n\lambda_{13} \quad ; \quad P_{43} = (U_4 - U_3)U_n\lambda_{43}\\
P_{\mathrm{II}2} &= (U_{\mathrm{II}} - U_2)U_n\lambda_{\mathrm{II}2} \quad ; \quad P_{\mathrm{III}4} = (U_{\mathrm{III}} - U_4)U_n\lambda_{\mathrm{III}4}
\end{aligned}$$

The following applies to the current account balance of the individual nodes:

$$\begin{aligned}
P_1 &= P_{\mathrm{I}1} + P_{\mathrm{III}1} - P_{13} = U_n[(U_I - U_1)\lambda_{\mathrm{I}1} + (U_{\mathrm{II}} - U_1)\lambda_{\mathrm{III}1} - (U_1 - U_3)\lambda_{13}]\\
P_2 &= P_{\mathrm{I}1} - P_{23} = U_n[(U_{\mathrm{II}} - U_2)\lambda_{\mathrm{II}2} - (U_2 - U_3)\lambda_{23}]\\
P_3 &= P_{13} + P_{23} + P_{43} = U_n[(U_1 - U_3)\lambda_{13} + (U_2 - U_3)\lambda_{23} + (U_4 - U_3)\lambda_{43}]\\
P_4 &= P_{\mathrm{II}4} + P_{\mathrm{III}4} - P_{43} = U_n[(U_{\mathrm{II}} - U_4)\lambda_{\mathrm{II}4} + (U_{\mathrm{III}} - U_4)\lambda_{\mathrm{III}4} - (U_4 - U_3)\lambda_{43}]
\end{aligned}$$

Conversion of the equations is done in such a way that the unknown load node voltages are on one side and all known quantities are on the other side of the system of equations; the result is

$$\begin{aligned}
(\lambda_{\mathrm{I}1} + \lambda_{\mathrm{III}1} + \lambda_{13})U_1 - \lambda_{13}U_{13} &= \lambda_{\mathrm{I}1}U_{\mathrm{I}} + \lambda_{\mathrm{III}1}U_{\mathrm{II}} - (P_1/U_n)\\
(\lambda_{\mathrm{II}2} + \lambda_{23})U_2 - \lambda_{23}U_3 &= \lambda_{\mathrm{II}2}U_{\mathrm{II}} - (P_2/U_n)\\
-\lambda_{13}U_1 - \lambda_{23}U_2 + (\lambda_{13} + \lambda_{23} + \lambda_{43})U_3 - \lambda_{43}U_4 &= -(P_3/U_n)\\
-\lambda_{43}U_3 + (\lambda_{\mathrm{II}4} + \lambda_{\mathrm{III}4} + \lambda_{43})U_4 &= \lambda_{\mathrm{II}4}U_{\mathrm{II}} + \lambda_{\mathrm{III}4}U_{\mathrm{III}} - (P_4/U_n)
\end{aligned}$$

or in matrix notation:

$$\begin{bmatrix} (\lambda_{\mathrm{I}1} + \lambda_{\mathrm{III}1} + \lambda_{13}) & 0 & -\lambda_{13} & 0 \\ 0 & (\lambda_{\mathrm{II}2} + \lambda_{23}) & -\lambda_{23} & 0 \\ -\lambda_{13} & -\lambda_{23} & (\lambda_{13} + \lambda_{23} + \lambda_{43}) & -\lambda_{43} \\ 0 & 0 & -\lambda_{43} & (\lambda_{\mathrm{II}4} + \lambda_{\mathrm{III}4} + \lambda_{43}) \end{bmatrix} \cdot \begin{bmatrix} U_1 \\ U_2 \\ U_3 \\ U_4 \end{bmatrix} =$$

$$\begin{bmatrix} \lambda_{\mathrm{I}1} & \lambda_{\mathrm{III}1} & 0 \\ 0 & \lambda_{\mathrm{II}2} & 0 \\ 0 & 0 & 0 \\ 0 & \lambda_{\mathrm{II}4} & \lambda_{\mathrm{III}4} \end{bmatrix} \cdot \begin{bmatrix} U_{\mathrm{I}} \\ U_{\mathrm{II}} \\ U_{\mathrm{III}} \end{bmatrix} - \frac{1}{U_n} \begin{bmatrix} P_1 \\ P_2 \\ P_3 \\ P_4 \end{bmatrix}$$

In general, the following system of equations results for N-load nodes and M-supply nodes and the nominal voltage U_n.

$$\begin{bmatrix} \lambda 11 & -\lambda_{12} & \cdots & -\lambda_{1i} & \cdots & -\lambda_{1N} \\ -\lambda_{21} & \lambda_{22} & \cdots & -\lambda_{2i} & \cdots & -\lambda_{2N} \\ \vdots & \vdots & \vdots & \vdots & \vdots & \vdots \\ -\lambda_{i1} & -\lambda_{i2} & \cdots & -\lambda_{ii} & \cdots & \lambda_{iN} \\ \vdots & \vdots & \vdots & \vdots & \vdots & \vdots \\ -\lambda_{N1} & -\lambda_{N2} & \cdots & -\lambda_{Ni} & \cdots & \lambda_{NN} \end{bmatrix} \cdot \begin{bmatrix} U_1 \\ U_2 \\ \vdots \\ U_i \\ \vdots \\ U_N \end{bmatrix}$$
$$= \begin{bmatrix} \lambda 1\mathrm{I} & \lambda_{1\mathrm{II}} & \cdots & -\lambda_{1M} \\ \lambda_{2\mathrm{I}} & \lambda_{2\mathrm{II}} & \cdots & -\lambda_{2M} \\ \vdots & \vdots & \vdots & \vdots \\ \lambda_{i\mathrm{I}} & \lambda_{i\mathrm{II}} & \cdots & -\lambda_{iM} \\ \vdots & \vdots & \vdots & \vdots \\ \lambda_{N\mathrm{I}} & -\lambda_{N\mathrm{II}} & \cdots & -\lambda_{NM} \end{bmatrix} \cdot \begin{bmatrix} U_{\mathrm{I}} \\ U_{\mathrm{II}} \\ \vdots \\ U_M \end{bmatrix} - \frac{1}{U_n} \begin{bmatrix} P_1 \\ P_2 \\ \vdots \\ P_i \\ \vdots \\ P_N \end{bmatrix} \tag{10.11}$$

or abbreviated in matrix notation as

$$[\lambda]_{\mathrm{LK}} \cdots [U]_{\mathrm{LK}} = [\lambda]_{\mathrm{SK}} \cdot [U]_{\mathrm{SK}} - \frac{1}{U_n}[P]_{\mathrm{LK}} \tag{10.12}$$

with $[\lambda]_{\mathrm{LK}}$ as the load node conductance matrix, $[U]_{\mathrm{LK}}$ as the vector of unknown load node voltages, $[\lambda]_{\mathrm{SK}}$ as the supply node conductance matrix, $[\lambda]_{\mathrm{SK}}$ as the vector of known supply node voltages, and $[P]_{\mathrm{LK}}$ as the vector of known loads of the load nodes.

On the main diagonals of the load node conductance matrix $[\lambda]_{\mathrm{LK}}$ are the sums of the route conductance values of all network branches leading into the respective load node. On the secondary diagonals are the negative conductance values $-\lambda_{ik}$ of the network branches connecting two load nodes i and k. The feed node admittance matrix $[\lambda]_{\mathrm{SK}}$ links the feed nodes with the load nodes. This matrix has the form (number of load nodes) × (number of supply nodes)($N \times M$). The elements of this matrix are the positive conductance values λ_{ix}, which connect a load node i with a supply node X. The resulting system of equations is linear and can be calculated using the methods for solving linear systems of equations. The solution of the equation system delivers the unknown load node voltages $U_1, U_2, \ldots, U_N$. The line capacities of the individual lines can be calculated from this, so that the load distribution in the entire network is known.

Load shift from network branches: To calculate the number of linear equations in Eq. (10.12), the number of load nodes must be small. The number of load nodes can be reduced if the load of load nodes where only two lines meet and to which a load is connected is meet and to which a load is connected, is transformed, i.e. shifted, to neighboring network nodes, i.e. nodes with at least three connected lines, i.e. relocated. In Figure 10.3 a network branch is given between nodes i and k, which is connected to the Debits P_1, P_2, P_3 and P_4, which are loaded. The sum of these two services is equal to the sum of all purchase services along the given network branch, i.e. equal to $\sum_{v=1}^{4} P_v$ in the example from Figure 10.3. The compensatory payment P_{ik}

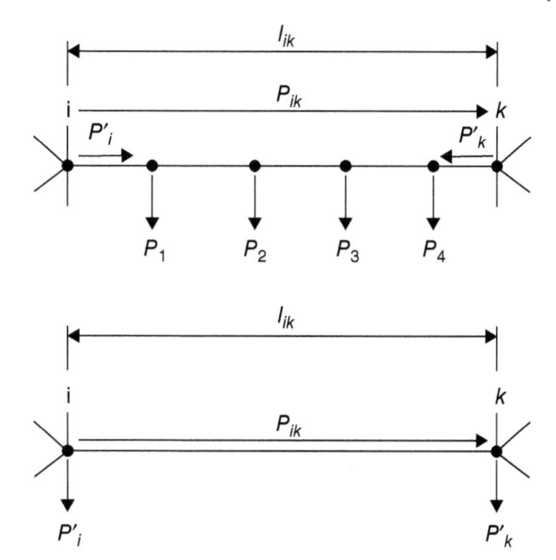

Figure 10.3 Example network for the simplified node method: Load nodes 1, 2, 3, and 4, feed nodes I, II, and III.

overlays the acceptance services that flow from node i into node k due to the voltage difference $U_i - U_k$ and which is calculated with Eq. (10.10).

Now the services $P_i^{'}$ and $P_k^{'}$ can be interpreted as acceptance services in nodes i and k, whereby the load flow in the network is not changed (see Figure 10.3). The compensation power P_{ik} is calculated with the simplified load flow calculation. The actual power that flows from node i into the network branch then results in $P_{ik} + P_i^{'}$, while the actual power resulting from node k flows to $P_{ik} - P_k^{'}$.

Special case: Parallel network branches are created after the load shifting of parallel, load-free network branches, and if these network branches have identical line structure, i.e. identical related series resistances, then the distribution of the balancing power to the individual network branches can be easily calculated. Two parallel network branches 1 and 2 are given between nodes i and k with the route guide values λ_{ik_2} and λ_{ik_1} and cable lengths l_{ik_1} and l_{ik_2}. The compensation payments P_{ik_1} and P_{ik_2}, each with Eq. (10.10), can be calculated:

$$P_{ik_1} = (U_i - U_k)U_n\lambda_{ik_1} \quad With\ \lambda_{ik_1} = \frac{1}{l_{ik_1}\Psi} \tag{10.13}$$

$$P_{ik_2} = (U_i - U_k)U_n\lambda_{ik_2} \quad With\ \lambda_{ik_2} = \frac{1}{l_{ik_2}\Psi} \tag{10.14}$$

Division of the two equations results in

$$\frac{P_{ik_1}}{P_{ik_2}} = \frac{\lambda_{ik_1}}{\lambda_{ik_2}} = \frac{l_{ik_1}}{l_{ik_2}} \tag{10.15}$$

10.5 Newton–Raphson Procedure

The Newton–Raphson method is based on the application of Kirchhoff's node rule to power instead of current. Then, in each node i, the sum of the inflowing and outgoing services always results in the value zero, that is, the following must apply:

$$\underline{S}_i = P_i + jQ_i = \underline{S}_{i0} + \sum_{k=1}^{N} \underline{S}_{ik} \tag{10.16}$$

with $\underline{S}_i = P_i + jQ_i$ as the load of node i, $\underline{S}_{ik}$ as the outgoing power against the reference node (earth), i.e. power of node i, and $\underline{S}_{ik}$ as the power flowing off from node i via a line to node k. If there is no load in a node i, or if node i is not connected to the reference node (earth), or if nodes i and k are not connected, the power concerned is zero.

The starting point is the complex node admittance matrix of a three-phase network with N nodes (without the reference node) $[\underline{Y}]$, according to Eq. (10.3), with

$$\underline{S} = 3\,\underline{U}\underline{I}^* = 3\,\underline{U} \cdot \underline{Y}^* \cdot \underline{U}^* \tag{10.17}$$

which is

$$\begin{bmatrix} \underline{S}_1 \\ \underline{S}_2 \\ \vdots \\ \underline{S}_i \\ \underline{S}_N \end{bmatrix} = 3 \begin{bmatrix} \underline{U}_1 \\ \underline{U}_2 \\ \vdots \\ \underline{U}_i \\ \underline{U}_N \end{bmatrix} \cdot \begin{bmatrix} \underline{Y}_{11}^* & -\underline{Y}_{12}^* & \cdots & -\underline{Y}_{1i}^* & \cdots & -\underline{Y}_{1N}^* \\ -\underline{Y}_{21}^* & \underline{Y}_{22}^* & \cdots & -\underline{Y}_{2i}^* & \cdots & -\underline{Y}_{2N}^* \\ \vdots & \vdots & \vdots & \vdots & \vdots & \vdots \\ -\underline{Y}_{i1}^* & -\underline{Y}_{i2}^* & \cdots & -\underline{Y}_{ii}^* & \cdots & \underline{Y}_{iN}^* \\ \vdots & \vdots & \vdots & \vdots & \vdots & \vdots \\ -\underline{Y}_{N1}^* & -\underline{Y}_{N2}^* & \cdots & -\underline{Y}_{Ni}^* & \cdots & \underline{Y}_{NN}^* \end{bmatrix} \cdot \begin{bmatrix} \underline{U}_1^* \\ \underline{U}_2^* \\ \vdots \\ \underline{U}_i^* \\ \underline{U}_N^* \end{bmatrix} \tag{10.18}$$

or in matrix notation

$$\underbrace{[\underline{S}]}_{\text{known}} = 3\,\underbrace{[\underline{U}]}_{\text{searched}} \cdot \underbrace{[\underline{Y}^*]}_{\text{known}} \cdot \underbrace{[\underline{U}^*]}_{\text{searched}} \tag{10.19}$$

The splitting of the apparent power $[S]$ into active and reactive power results in

$$3\,[\underline{U}][\underline{Y}^*][\underline{U}^*] = [\underline{S}] = ([P] + j[Q]) \tag{10.20}$$

The system of equations is changed so that there is a zero vector on the right side:

$$3\,[\underline{U}][\underline{Y}^*][\underline{U}^*]([P] + j[Q]) = [\Delta P] + j[\Delta Q] = [0] + j[0] \tag{10.21}$$

In this system of equations the node voltages $[\underline{U}]$ are unknown. The load flow calculation problem thus consists in the determination of the zero of this system of equations. The equation system is generally nonlinear and can only be solved iteratively. It is created using the Newton–Raphson process where $[\Delta P] + j[\Delta Q]$ are the deviations of the results after an iteration step from the final result. The iteration is aborted if $[\Delta P]$ and $[\Delta Q]$ are less than a defined value, the so-called abort criterion. The system of equations is reduced by one, since at the slack-node, the voltage is known by magnitude and angle or real and imaginary part, i.e. a system of equations with $(N - 1)$ equations is to be solved. The slack node is the node N. The system of equations is divided into real and imaginary parts.

$$\mathrm{Re}\{3[\underline{U}][\underline{Y}^*][\underline{U}^*]\} - [P] = [\overline{P}] - [P] = [\Delta P] = 0 \tag{10.22}$$

$$\mathrm{Re}\{3[\underline{U}][\underline{Y}^*][\underline{U}^*]\} - [Q] = [\overline{Q}] - [Q] = [\Delta P] = 0 \tag{10.23}$$

with $[\overline{P}]$ or $[\overline{Q}]$ as the iteratively calculated value of the active and reactive power in the individual network nodes and $[P]$ or $[Q]$ as the given voltage-dependent active and reactive power loads of the individual network nodes. It must be specified whether Cartesian coordinates (real and imaginary part or longitudinal and transverse stresses) or polar coordinates (magnitude and angle) are used. If Cartesian coordinates are selected, the following applies to the stresses or admittances:

$$\underline{U}_i = U_{i_l} + jU_{i_q} \tag{10.24}$$

$$\underline{Y}_i = G_{i_k} + jB_{i_k} \tag{10.25}$$

with a longitudinal voltage U_{i_l}, a transverse voltage U_{i_q}, an effective conductance (conductance) G_{i_k}, and the susceptance B_{i_k}. If one takes out the ith line of the system of equations according to Eq. (10.18), this line reads as follows:

$$\underline{S}_i = P_i + jQ_i = \underbrace{3 \cdot \underline{U}_i \sum_{k=1}^{N-1} \underline{Y}_{ik}^* \cdot \underline{U}_k^*}_{\overline{P}_i + j\overline{Q}_i}$$

respectively with (10.24) and (10.25).

$$\underline{S}_i = P_i + jQ_i = \underbrace{3 \cdot (\underline{U}_{i_l} + \underline{U}_{i_q}) \cdot \sum_{k=1}^{N-1} (G_{i_k} + jB_{i_k})^* \cdot (U_{k_l} + jU_{k_q})^*}_{\overline{P}_i + j\overline{Q}_i} \tag{10.26}$$

It follows that

$$\overline{P}_i = 3 \cdot \sum_{k=1}^{N-1} U_{i_l}(U_{k_l}G_{i_k} - U_{k_q}B_{i_k}) + U_{i_q}(U_{k_q}G_{i_k} - U_{k_l}B_{i_k}) \tag{10.27}$$

$$\overline{Q}_i = 3 \cdot \sum_{k=1}^{N-1} U_{i_{ql}}(U_{k_l}G_{i_k} - U_{k_q}B_{i_k}) + U_{i_l}(U_{k_q}G_{i_k} - U_{k_l}B_{i_k}) \tag{10.28}$$

The outputs are nonlinearly dependent on the node voltages and are determined using the Taylor series development linearized by one operating point (AP). The linearization is done by Eqs. (10.27) and (10.28), which are partially derived according to the longitudinal and transverse stresses, and then the voltage values assumed at the operating point are used, i.e. the voltage values of the previous iteration step. This results in the following system of equations:

$$\begin{bmatrix} [\Delta P] \\ \cdots \\ [\Delta Q] \end{bmatrix} = \underbrace{\begin{bmatrix} \dfrac{\partial P}{\partial U_{q\,AP}} & \vdots & \dfrac{\partial P}{\partial U_{l\,AP}} \\ \cdots & \cdots & \cdots \\ \dfrac{\partial Q}{\partial U_{q\,AP}} & \vdots & \dfrac{\partial Q}{\partial U_{l\,AP}} \end{bmatrix}}_{\text{Jacobi matrix}} \cdot \begin{bmatrix} [\Delta U_q] \\ \cdots \\ [\Delta U_l] \end{bmatrix} \tag{10.29}$$

In this equation $[\Delta P]$ and $[\Delta Q]$ and the Jacobi matrix are known, while the voltages$[\Delta U_l]$ and $[\Delta U_q]$ are unknown. The Jacobi matrix depends on the operating point and must therefore be recalculated for each iteration step. Equation (10.29) represents an inhomogeneous linear system of equations, which must be solved for each iteration step, $[\Delta U_l]$ and $[\Delta U_q]$; to solve the system of equations, the node voltages $[U_l]$ and $[U_q]$ are tracked and the new operating point is calculated.

The iteration process is specified as follows:

1) *Start values (ν = 0)*: Since there are no initial values for the unknown node voltages at the start of the iteration calculation, the voltage at the Slack node U_N is used.

$$U_{i_l}^{\nu=0} = U_N \quad U_{i_q}^{\nu=0} = 0 \quad i = 1 \cdots (N-1)$$

where the node N is the slack node.

2) *(ν+1)-ter iteration step*: With the values for the stresses from the νth iteration step, i.e. with the stresses $[U_l]^{(\nu)}$ and $[U_q]^{(\nu)}$, the node powers $[P]^{(\nu)}$ and $[P]^{(\nu)}$ can be calculated. At the same time, Eqs. (10.27) and (10.28) are used to calculate the powers $[\overline{P}]^{(\nu)}$ and $[\overline{P}]^{(\nu)}$, respectively. This allows the power deviations to be calculated:

$$[\Delta P]^{(\nu+1)} = [\overline{P}]^{\nu} - [P]^{(\nu)}$$
$$[\Delta PQ]^{(\nu+1)} = [\overline{Q}]^{\nu} - [Q]^{(\nu)}$$

At the same time, for the current operating point, i.e. for the stresses $[U_l]^{(\nu)}$ and $[U_q]^{(\nu)}$, the Jacobi matrix must be calculated. Then the system of equations according to Eq. (10.29) is solved and returns the new correction values $[\Delta U_l]$ and $[\Delta U_q]$, with which the searched node voltages of the current iteration step can be calculated:

$$[U_l]^{(\nu+1)} = [[U_l]^{\nu} - [\Delta U_l]^{(\nu+1)}$$
$$[U_q]^{(\nu+1)} = [[U_q]^{\nu} - [\Delta U_q]^{(\nu+1)}$$

This concludes the current iteration step. The next step is whether or not to abort the iteration.

3) *End of iteration*: The iteration calculation can be terminated if the deviations between two consecutive iteration steps exceed a specified error limit.

After performing the iteration calculation, the required node voltages are available. From this, the still unknown branch currents and branch powers can be calculated so that the complete load flow of the network is available. The advantage of the Newton–Raphson method is the fast convergence, i.e. usually only a few iteration steps are required. In addition, power plant feed-ins (generator nodes) can be easily integrated. The disadvantage of the method is the sensitivity to poor starting values, i.e. if the node voltages in the network are very different, i.e. the assumption of the voltage at the slack node as starting values deviates strongly from the actual voltages, then the method converges only badly or not at all. In addition, for each iteration step the Jacobi matrix, which is dependent on the operating point, must be recalculated.

11

Substation Earthing

This chapter explains the requirements for the design, construction, testing, and maintenance of earthing of power installations exceeding 1 kV AC in order to ensure that they are effective under all conditions and that the safety of persons is ensured in any place of the installation [43, 44].

Earthing systems serve to protect people, animals, and property in the event of short-circuits, lightning, and switching operations. They must be dimensioned in such a way that the permissible step and touch voltages and certain earthing resistances are not exceeded. The earthing of neutral star point or conductor to earth is for the purpose of controlling the voltage to earth within the required limits that must be in compliance with EN 50522.

11.1 Terms and Definitions

The most important terms are compiled here (see also IEV 195-02-10) (Figures 11.1–11.5).

- *Earth*: Part of the Earth which is in electric touch with an earth electrode and the electric potential of which is not necessarily equal to zero.
- *Reference earth (remote earth)*: Part of the Earth considered as conductive, the electric potential of which is conventionally taken as zero, being outside the zone of influence of the relevant earthing arrangement.
- *Earth electrode*: Conductive part, which may be embedded in a specific conductive medium, e.g. in concrete or coke, in electric touch with the Earth.
- *Earthing conductor*: Conductor which provides a conductive path, or part of the conductive path, between a given point in a system or in an installation or in equipment and an earth electrode.
- *Earthing system*: It is the set of electrical connections and devices used to ground a network, an installation, or a device.
- *Earth resistance*: It is the effective resistance between the earth and the reference earth. The resistance of an earth electrode to propagation depends on the specific earth resistance as well as on the dimensions and arrangement of the earth electrode. It is mainly of the length of the earth electrode, less of its cross section.

Analysis and Design of Electrical Power Systems: A Practical Guide and Commentary on NEC and IEC 60364, First Edition. Ismail Kasikci.

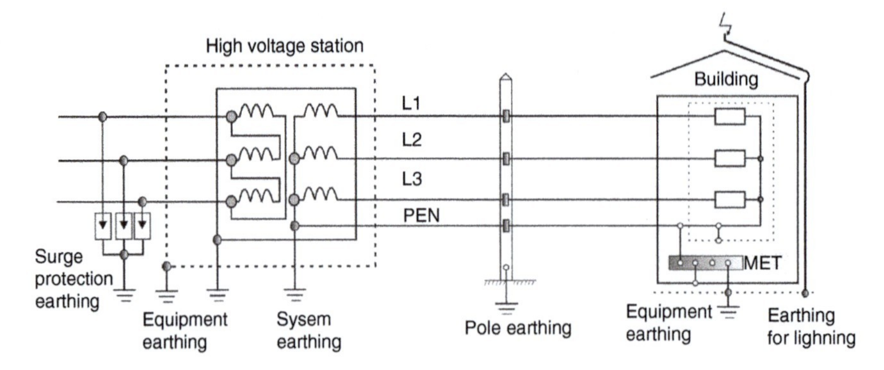

Figure 11.1 Terms and definitions.

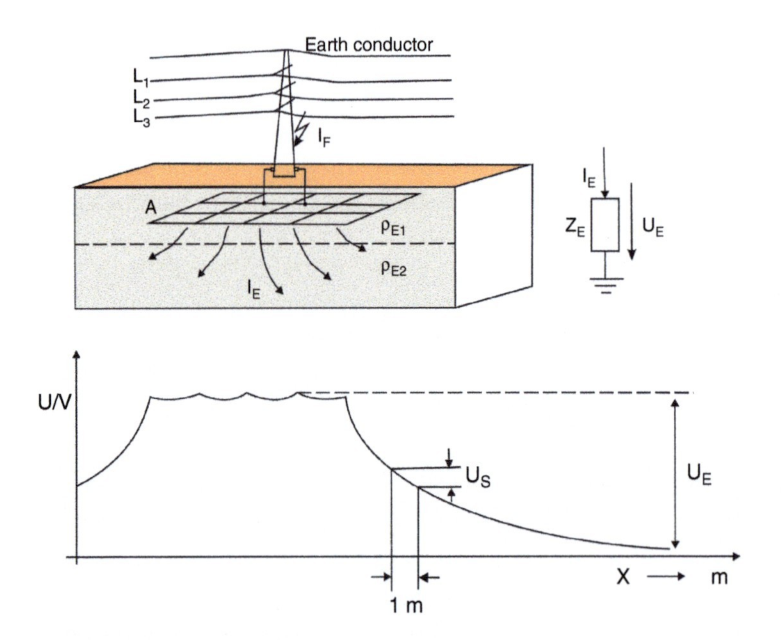

Figure 11.2 Terms and definitions.

- *Potential control*: It is the influence of earth potential through the earth electrode.
- *Earth short-circuit current*: It is generated in networks with low-resistance neutral earthing.
- *Earth fault current*: It is the current passing from the operating circuit to earth or to earthed parts at the earth fault point.
- *Capacitive earth short-circuit current*: It is generated in networks with an isolated star point.
- *Residual earth fault current*: It arises in networks with earth fault compensation.
- *Main earthing bar*: Main earthing terminal, main earthing connection point: connection point or bar which is part of the earthing system of a system and enables the electrical connection of several conductors for earthing purposes.

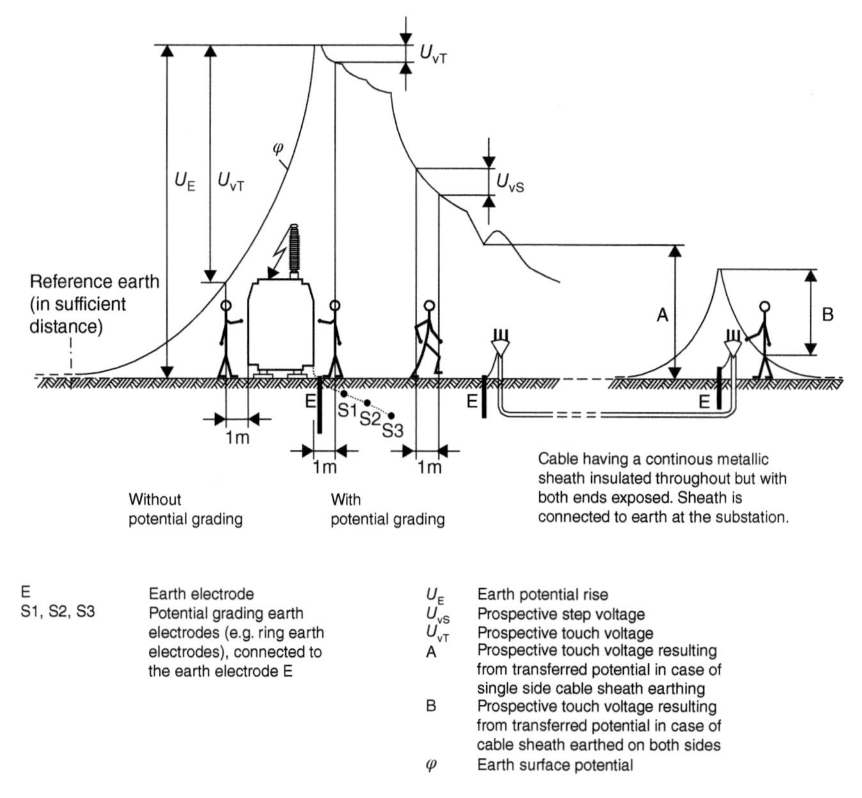

Figure 11.3 Prospective touch voltage U_{Tp}, ground voltage U_{E}, Step voltage U_{S} [44].

- *PE conductor (designation PE)*: Conductors for safety purposes, for example, to protect against electric shock.
- *Foundation earthing*: Conductive part embedded under a building foundation in the ground or preferably in the concrete of a building foundation, generally as a closed ring.
- *Low voltage* (LV) earthing: It is the earthing of the neutral necessary to operate the LV grid. It is also used to protect people against excessive touch voltage.
- *High-voltage protective earth*: It is the direct earthing of a conductive part of the equipment or system that does not belong to the operating circuit for the protection of persons against excessive touch and step voltage.
- *Substation*: It is a part of a power system, concentrated in a given place, including mainly the terminations of transmission or distribution lines, switchgear, and housing and which may also include transformers. It generally includes facilities necessary for system security and control (e.g. the protective devices).

Note that the EN 50522 uses the term **earthing** rather than **grounding**, which is used in IEEE Std. 80.

A main distinction can be made between protective earthing, operational earthing, and lightning protection earthing. Figure 11.1 shows examples of major earthing methods.

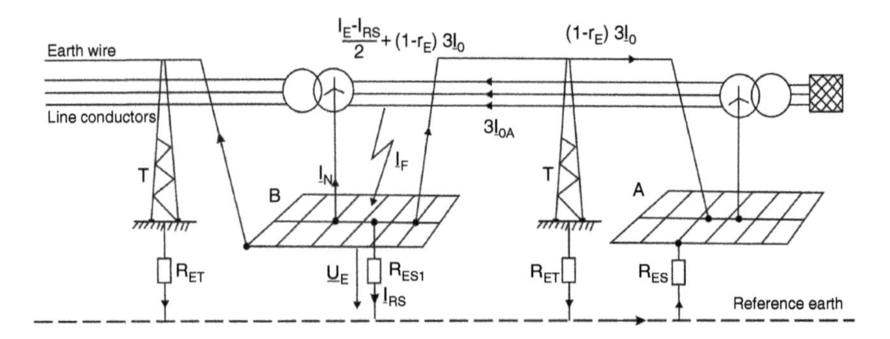

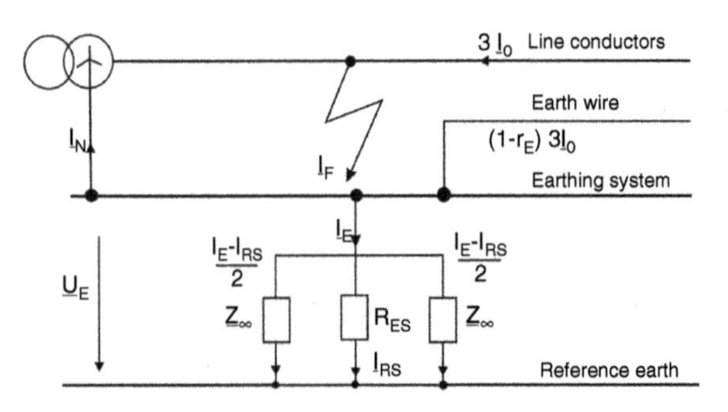

Figure 11.4 Partial short-circuit currents in the event of an earth short-circuit in system B of a network with low-impedance Star point grounding [44].

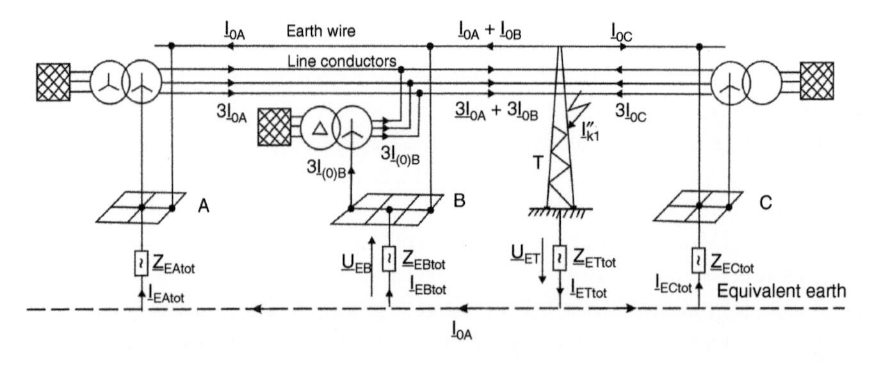

Figure 11.5 Partial short-circuit currents in the event of an earth short-circuit in system B of a network with low-impedance Star point grounding [44].

The currents flowing through earthing systems and earth can be caused by earth faults or double earth short-circuits in networks with an isolated neutral point or with earth fault compensation. In networks with low-resistance neutral earthing, currents over earth occur as a result of single-pole earth short-circuits or two-pole short-circuits with earth contact.

For $X_0/X_1 > 1, I''_{k1}$ is larger than I''_{kE2E}, so that for the design of the earthing systems mostly I''_{k1} can be used as a basis.

Table 11.6 shows a summary of the currents that are decisive for the earth, step, and touch voltages on the one hand, and the thermal load on the earth electrodes, earth collector lines, and earthing cables on the other.

In networks with low-resistance star point earthing, the earth short-circuit current I''_{k1} flows only partially over the total earth impedance Z_E.

The earth voltage U_E is therefore only caused by the earth current I''_E (part of the earth short-circuit current I''_{k1} at the short-circuit point) that flows over the total ground impedance Z_E (e.g. propagation resistance of a grounding system including the connected spur groundings in the form of connected ground wires, ground ropes, cable sheaths, pipes).

Figure 11.5 shows an example of an earth short-circuit in a substation B with outgoing overhead lines to determine the partial short-circuit currents.

In this case, the earth short-circuit current I''_{k1} is composed of the three components of three times the zero-sequence current:

$$I''_{k1} = 3I_{0A} + 3I_{0B} + 3I_{0C} \tag{11.1}$$

To determine the earth voltage, special considerations must therefore be made and the currents that cause the touch and step voltages must be calculated. The total ground impedance of the system consists of the parallel connection of the ground resistance R_B of the mesh grounding with the chain cable impedances Z_P, grounding cable, masts, all outgoing overhead lines, and the characteristic impedances of all connected cable sheaths and Piping (see for more information IEC 60909-0 Part 3). It means

$\underline{I}_0$ Zero current,
$\underline{I}_{Tr}$ Current flowing through the transformer,
$\underline{I}_F$ Fault current,
$\underline{I}_E$ Ground current,
I_{RS} Current flowing through the mesh,
r_E Reduction factor,
R_{ES} Spreading resistance of the mesh,
R_{ET} Spreading resistance of the mast,
$\underline{Z}\infty$ Chain impedance,
$\underline{Z}_E$ Ground impedance,
$\underline{U}_E$ Ground voltage,
n Number of lines leaving the system.

The following parameters are important for dimensioning the earthing system:

- Value of the fault current, magnitude of the current (depending on the type of neutral point treatment)
- Fault duration
- Soil characteristics

The following applies to the earthing current in the transformer station:

$$\underline{I}_E = r_E \cdot (\underline{I}_F - \underline{I}_N) \tag{11.2}$$

The residual current can be combined with

$$\underline{I}_F = 3 \cdot \underline{I}_0 + \underline{I}_N \tag{11.3}$$

With the earthing voltage

$$U_E \leq 2 \cdot U_{Tp} \tag{11.4}$$

$$\underline{U}_E = \underline{I}_E \cdot \underline{Z}_E \tag{11.5}$$

being followed by the ground impedance

$$\underline{Z}_E = \frac{1}{\frac{1}{R_{ES}} + n \cdot \frac{1}{\underline{Z}_\infty}} \tag{11.6}$$

The earthing resistance can also be calculated in a simplified way if the minimum short-circuit current is taken as the reference value.

$$R_E = \frac{U_E}{I''_{k1\,min}} \tag{11.7}$$

$$R_E \leq 2 \cdot \frac{U_{Tp}}{I''_{k1\,min}}, \quad \text{with } I''_{k1\,min} = \frac{\sqrt{3} \cdot c_{min} \cdot U_n}{2\underline{Z}_1 + Z_0} \tag{11.8}$$

11.2 Methods of Neutral Earthing

In symmetrically operated three-phase systems, the currents of the three phases in the star points of the equipment always add up to zero.

The most frequently occurring faults are the single-pole short-circuit and the earth fault. The earth fault is a conductive connection between a network point belonging to the operating circuit and earth. Earth faults account for 80–90% of the faults in the grounded system.

If the short-circuit currents for the three-phase short-circuit in all three conductors are equal, then the error is balanced. In all other cases, the short-circuit currents in the three conductors are different. These faults are unbalanced. In addition, various so-called transverse faults are possible in the three-phase system. In addition to the cross faults, line interruptions can also occur. In this case, longitudinal faults occur, but these are of no significance for short-circuit calculations.

In the case of earth faults and earth short-circuits, the magnitude of the short-circuit current is decisively dependent on how the neutral point of the network is connected to earth.

The short-circuit currents are determined by the voltage sources present in the network (generators and motors) and Mains impedances. The demand for an optimal and inexpensive network can lead to different star point treatments [44].

Furthermore, the neutral earthing method is important with regard to the following:

- Selection of insulation level
- Characteristics of overvoltage limiting devices such as spark gaps or surge arresters

Type of high voltage system		Relevant for thermal loading		Releant for earth potential rise and touch voltages
		Earth electrode	Earthing conductor	
Systems with isolated neutral				
		I''_{KEE}	I''_{KEE}	$I_E = r \cdot I_C$
System with resonant earthing Includes short time earthing for detection				
	Substations without arc-suppression coils	I''_{KEE}	I''_{KEE}	$I_E = r \cdot I_{RES}$
	Substations with arc- suppression coils	I''_{KEE}	I''_{KEE}	$I_E = r \cdot \sqrt{I_L^2 + I_{RES}^2}$
Systems with low-impedance neutral earthing Includes short time earthing for tripping				
	Substation without neutral earthing	I''_{K1}	I''_{K1}	$I_E = r \cdot I''_{k1}$
	Substation with neutral earthing	I''_{K1}	I''_{K1}	$I_E = r \cdot (I''_{k1} - I_N)$

Figure 11.6 Relevant currents for the design of earthing systems.

- Selection of protective relays
- Design of earthing system
- Relevant currents for the design of earthing systems are shown in Figure 11.6.

  Legend:

I_C Calculated or measured capacitive earth fault current.

I_{RES} Earth fault residual current.
If the exact value is not available,
10% of I_C may be assumed.

I_L Sum of the rated currents of the parallel
arc-suppression coils in the relevant substation.

I''_{kEE} Double earth fault current calculated
in accordance with IEC 60909-0.
For I''_{kEE} 85% of the initial symmetrical
short-circuit current may be used as a maximum value,

I''_{k1} Initial symmetrical short-circuit current
for a line-to-earth short-circuit, calculated in accordance with IEC 60909-0.

I_E Current to earth.

I_N Current via neutral earthing of the transformer.

The choice of the type of neutral earthing is normally based on the following criteria:

- Local regulations (if any)
- Continuity of supply required for the network
- Limitation of damage to equipment caused by earth faults
- Selective elimination of faulty sections of the network
- Detection of fault location

- Touch and step voltages
- Inductive interference
- Operation and maintenance aspects

Methods of neutral earthing and essential components of earth fault currents in high-voltage power systems will be explained and are shown in following pages. The method of neutral earthing strongly influences the fault current level and the fault current duration.

11.2.1 Isolated Earthing

Isolated earthing is used in small cable networks with a nominal voltage of 10(20) kV. In balanced networks with isolated star points, the star points and the earthed environment have practically the same potential φ_E, i.e. earth potential. This is due to the symmetrical earth capacitance C_E between the phase conductors and earth. Since in symmetrical three-phase systems the three phase voltages also always complement each other to zero, the neutral point N also has the potential $\varphi_N = \varphi_E = 0$. This means that even without a conductive connection between neutral point and earth, all insulation against earth is permanently loaded with star voltage. The isolated neutral earthing is shown in Figure 11.7.

For a fault of a phase conductor to earth, in the absence of a low-impedance return line to the star point N of the feeder, no short-circuit current can occur. Only a relatively small capacitive earth fault current I_{CE} flows via earth and the earth capacitances C_E to the two unaffected conductors $L2$ and $L3$. This is called an earth fault.

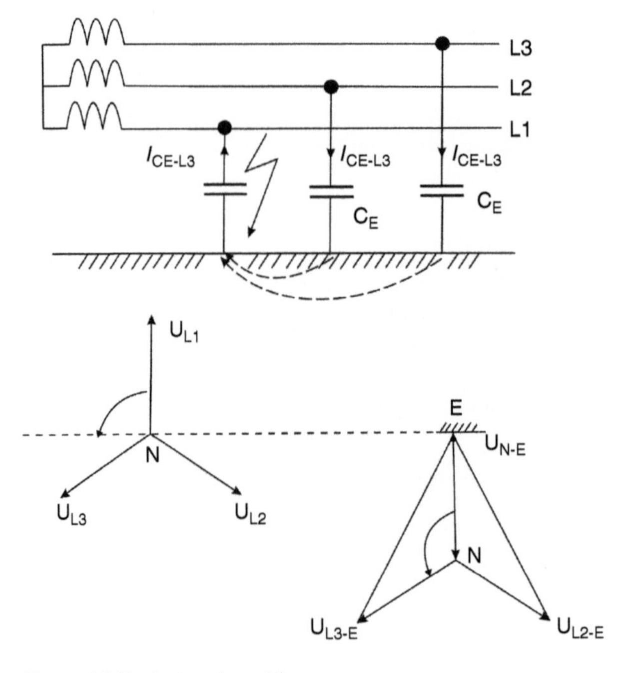

Figure 11.7 Isolated earthing.

Table 11.1 Advantages and disadvantages of networks with isolated, free neutral point.

Advantages	Disadvantages
No interruption of supply	Voltage increase
Self-extinguishing	of the healthy line
Mains operation possible	High-current arc
Low requirements for earthing systems	Limited to small networks
Voltage triangle on the primary side	Difficult fault location
maintains	Full isolation in the network
	Star point isolation
	Danger of subsequent errors

The limit of the earth fault current is 35 A and lasts over one hour. The height of the earth fault I_{CE} depends on the size of the earth capacitances C_E and thus after the expansion of the network,

$$I_{CE} = 3 \cdot \omega \cdot C_E \frac{c \cdot U_n}{\sqrt{3}} \quad (11.9)$$

It means
I_{CE} Capacitive earth fault current
C_E Earth leakage capacitance
U_n Rated voltage

Table 11.1 shows advantages and disadvantages of networks with isolated, free neutral point.

11.2.2 Resonant Earthing

With increasing network size or earth capacitance, the residual current in the event of an earth fault is no longer self-extinguishing. The occurrence of a double fault is becoming increasingly probable. In order to be able to maintain the advantages of isolated operation, earth fault compensation is used. Here, the high capacitive earth fault residual current is compensated by one or several chokes, the so-called coils, connected to the star points of the transformers. After compensation, a small residual current I_{Rest} remains, which extinguishes automatically as with isolated star points.

According to DIN VDE 0228, the maximum permissible residual current is 60 A in 20(10) kV networks and 130 A in 110 kV networks. As long as these currents are not exceeded, the self-resetting function is maintained.

The resonant earthing is shown in Figure 11.8.

$$\omega \cdot L_E = \frac{1}{3 \cdot \omega \cdot C_E} \quad (11.10)$$

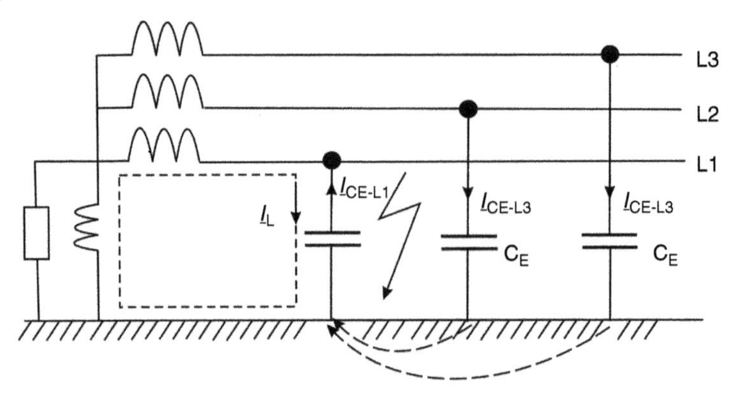

Figure 11.8 Resonant earthing.

Table 11.2 Advantages and disadvantages of networks with ground fault compensation.

Advantages	Disadvantages
No supply interruption	Voltage increase
Self-extinguishing	of the healthy line
Mains operation during earth fault possible	High-current arc
Low requirements for earthing systems	Limited to small networks
Compensation prevents	One coil in each network area
	Earth-fault arcing necessary
Cancellation if current and voltage	Traveling wave hazard
are in phase	Difficult fault location
Voltage triangle remains	in cable networks explosion hazard
Residual current at the fault location	Mains size limited by residual current
	Low earth fault current, disadvantageous
	on location
	Risk of resonance
	Short side voltage reduction
	Danger of multiple errors
	High insulation stress
	Additional costs for transformer insulation
	Investments for coils

Table 11.2 shows the advantages and disadvantages of networks with earth fault compensation.

11.2.3 Double Earth Fault

Double earth fault currents occur in isolated and compensated networks, which must always be calculated for the design of the earthing systems. Double earth fault

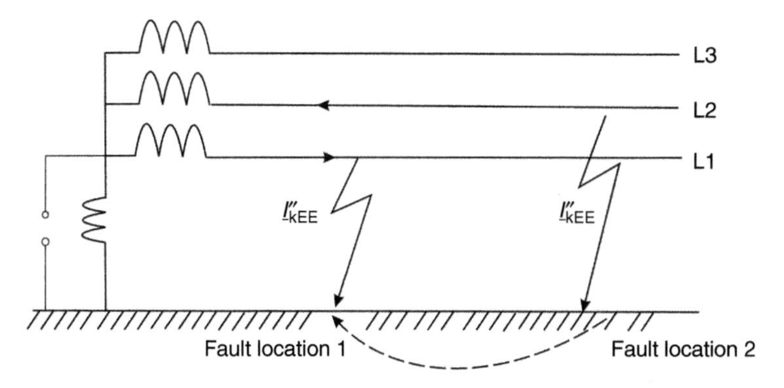

Figure 11.9 Solid (low impedance) earthing.

I''_{kEE} is shown in Figure 11.9. For the thermal stress of an earthing conductor or earthing busbars, the worst case of a double ground short-circuit shall be calculated.

For the double earth short-circuit current I''_{kEE}, the following applies according to EN 50522.

$$I''_{kEE} = 85\% \cdot I''_{k3} \tag{11.11}$$

It is further considered that the earth short-circuit current is distributed via the earth wire and protection potential equalization conductor to the transformer in the mesh system. With sufficient accuracy, a current splitting factor for the distribution can be taken for the earthing system. Then we can calculate the earth short-circuit current for the dimensioning of the earth conductor. The cross section of a conductor can be determined from the material and the disconnection time. The Norm EN 50522 specifies the maximum short-circuit current density G (A/mm^2) for different materials (Table 11.3).

$$I''_{kEE-Design} = 65\% \cdot I''_{kEE} \tag{11.12}$$

The cross section of a conductor is

$$A = \frac{I''_{kEE-Design}}{G} \tag{11.13}$$

Table 11.3 Short-circuit current density G (maximum temperature of 200 °C).

Time (s)	St/tZn (A/mm^2)	Copper (A/mm^2)	(V4A) (A/mm^2)
0.3	129	355	70
0.5	100	275	55
1	70	195	37
3	41	112	21
5	31	87	17

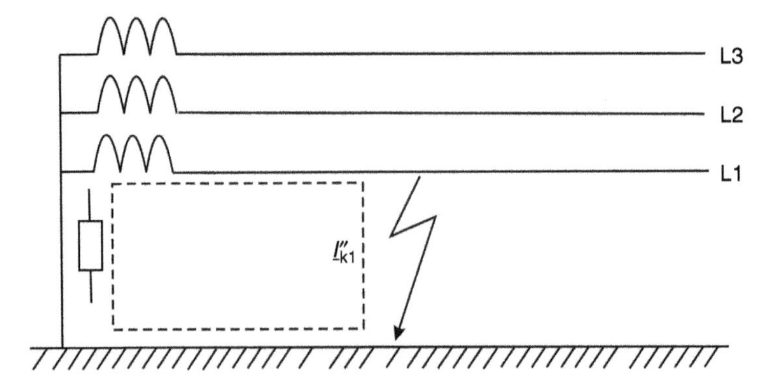

Figure 11.10 Solid (low impedance) earthing.

11.2.4 Solid (Low-Impedance) Earthing

Directly earthed networks or networks earthed via a resistor are used in 110(220) kV networks where isolated and compensated networks are not common. Networks with earthed star points also are used in low-voltage power systems under 1 kV. Compensated networks also have their limits. Through a network expansion, the respective residual current limit is approached; it is necessary to use earthed star points. Solid (low impedance) earthing is shown in Figure 11.10. These earthed neutral points generate very high single-pole earth short-circuit currents that must be immediately disconnected by fuses or circuit-breakers. The effectiveness of the neutral point earthing with regard to the reoccurring neutral point displacement is characterized by the earth fault factor [43, 44].

$$\underline{I}''_{k1} = \frac{\sqrt{3} \cdot c_{\min} \cdot U_n}{\underline{Z}_1 + \underline{Z}_2 + \underline{Z}_0} \tag{11.14}$$

It means

$\underline{Z}_1$ Positive impedance
$\underline{Z}_2$ Negative impedance
$\underline{Z}_0$ Zero impedance
$\underline{I}''_{k1}$ One-pole Short-circuit current

In Table 11.4 the advantages and disadvantages of networks with low-resistance neutral point treatment are summarized.

11.3 Examples for the Treatment of the Neutral Point

11.3.1 Example: Earth Fault Current When Operating with Free Neutral Point

Behind the block transformer of a generator, three overhead lines of 110 kV with a length of 70 km each go off. How large would the earth fault current be in operation with free neutral point, if the earth capacitance of the overhead lines is 5 nF/km?

Table 11.4 Advantages and disadvantages of networks with low-resistance neutral point treatment.

Advantages	Disadvantages
Selective disconnection of single pole fault	Each short-circuit leads to a shutdown
Low overvoltage	Fault current over ground
No limitation of the network expansion	High stress on the system
No travelling wave hazards	Earth fault is a short-circuit
High transmission voltages possible	High breaking capacities
Low insulation levels	Three-phase mains protection
Cost saving	High demands on earth systems

Total line length: $l = 3 \cdot 70\,\text{km} = 210\,\text{km}$
Total earth capacity: $C_\text{E} = C'_\text{E} \cdot l = 210\,\text{km} \cdot 5\,\text{nF/km} = 1.05\,\text{pF}$
The earth fault current is then

$$I_\text{E} = \frac{3 \cdot \omega \cdot C_\text{E} \cdot U = 3 \cdot 314\frac{1}{s} \cdot 1.05 \cdot 10^{-6}\,s \cdot 110 \times 10^3\,\text{V}}{\sqrt{3}} = 63\,\text{A}$$

11.3.2 Example: Calculation of Earth Fault Currents

A 10.4 kV network district consists of three overhead lines, each 10 km long, and three overhead lines, each 15 km long. How large is the earth fault current during operation with a free neutral point, if the lines have an earth capacitance of 6 nF/km?

Total line length: $\text{l} = 3 \cdot 10\,\text{km} + 3 \cdot 15\,\text{km} = 75\,\text{km}$
Total earth capacity: $C_\text{E} = C'_\text{E} \cdot l = 75\,\text{km} \cdot 6\,\text{nF/km} = 0.45\,\mu F$
The earth fault current is then

$$I_\text{E} = \frac{3 \cdot \omega \cdot C_\text{E} \cdot U = 3 \cdot 314\frac{1}{s} \cdot 0.45 \cdot 10^{-6}\,Ss \cdot 10.4 \cdot 10^3\,\text{V}}{\sqrt{3}} = 2.55\,\text{A}$$

This current is so small that an earth leakage quenching coil is not required.

11.3.3 Example: Ground Fault Current of a Cable

What would be the earth fault current of a cable 110 kV, 6 km long, earth capacitance 280 nF/km in operation with a free neutral point?

The earth fault current is

$$I_\text{E} = 3 \cdot \omega \cdot C'_\text{E} \cdot l \cdot U = \frac{3 \cdot 314\frac{1}{s} \cdot 0.28 \times 10^{-6}\frac{\mu\text{F}}{km} \cdot 6\,\text{km} \cdot 110 \times 10^3\,\text{V}}{\sqrt{3}} = 100\,\text{A}$$

11.3.4 Example: Earth Leakage Coil

Calculate the data of the earth fault extinction coil (ideal choke) that would be necessary to fully compensate the earth fault current according to task 1.

Earth fault current and choke current must be equal. This results in the following:

$$I_\mathrm{D} = I_\mathrm{E} = 63\ \mathrm{A}$$

$$U_\mathrm{D} = \frac{U_\mathrm{N}}{\sqrt{3}} = 63.5\ \mathrm{kV}$$

$$X_\mathrm{D} = \frac{U_\mathrm{D}}{I_\mathrm{D}} = \frac{63.5\ \mathrm{kV}}{63\ \mathrm{A}} = 1.01\ \mathrm{k\Omega}$$

$$Q_\mathrm{D} = U_\mathrm{D} \cdot I_\mathrm{D} = 63.5\ \mathrm{kV} \cdot 63\ \mathrm{A} = 4\ \mathrm{MVAr}$$

11.3.5 Example: Arc Suppression Coil

Calculate the data of the earth fault clearance coil for the overhead lines according to task 1. For the choke, a ratio $\frac{R_\mathrm{D}}{X_\mathrm{D}} = 0.1$ can be achieved. How large would the residual current be if any harmonics and the leakage resistances were neglected?

For the residual current one obtains with U_TO as reference value:

$$\underline{I}_\mathrm{R} = \underline{I}_\mathrm{E} - \underline{I}_\mathrm{D} = j \cdot 3 \cdot \omega \cdot C_\mathrm{E} \cdot U_\mathrm{TO} + \frac{U_\mathrm{TO}}{(R_\mathrm{D} + j\omega L_\mathrm{D})}$$

$$= j \cdot \omega \cdot C_\mathrm{E} \cdot U_\mathrm{TO} + U_\mathrm{TO} \cdot \frac{(R_\mathrm{D} - j\omega \cdot L_\mathrm{D})}{R_\mathrm{D}^2 + \omega^2 + L_\mathrm{D}^2}$$

The reactive components must compensate each other:

$$3 \cdot \omega \cdot C_\mathrm{E} \cdot U_\mathrm{TO} - \frac{X_\mathrm{D} \cdot U_\mathrm{TO}}{R_\mathrm{D}^2 + X_\mathrm{D}^2} = 63\ \mathrm{A} - \frac{X_\mathrm{D} \cdot 63.5\ \mathrm{kV}}{1.01 \cdot X_\mathrm{D}^2} = 0$$

$$X_\mathrm{D} = \frac{63.5\ \mathrm{kV}}{1.01 \cdot 63\ \mathrm{A}} = 1\ \mathrm{k\Omega}$$

$$R_\mathrm{D} = 0.1 \cdot X_\mathrm{D} = 100\ \Omega$$

This gives for the residual current

$$I_\mathrm{R} = \frac{U_\mathrm{TO} \cdot R_\mathrm{D}}{1.01 \cdot X_\mathrm{D}^2} = \frac{63.5\ \mathrm{kV} \cdot 100\ \Omega}{1.01 \times 10^6}\ \Omega^2$$

$$= 6.3\ \mathrm{A}$$

This is 10% of the earth leakage current.

11.4 Dimensioning of Thermal Strength

The currents to be taken into account for earthing conductors and earth electrodes are specified in Figure 11.6. The fault current is often subdivided in the earth electrode system; it is, therefore, feasible to dimension each electrode and earthing conductor for only a fraction of the fault current.

Final temperatures involved in the design and to which reference is made in Annex D of EN 50522 shall be chosen in order to avoid reduction of the material strength and to avoid damage of the material surrounding, for example, concrete or insulating materials. No permissible temperature rise of the soil surrounding the earth electrodes is given in this standard because experience shows that soil temperature rise is usually not significant.

The calculation of the cross section of the earthing conductors or earth electrodes depending on the value and the duration of the fault current is given in EN 50522. There is discrimination between fault duration lower than five seconds (adiabatic temperature rise) and greater than five seconds. The final temperature is to be chosen taking into account the material and the surroundings.

11.5 Methods of Calculating Permissible Touch Voltages

Very important definitions concerning design and safety measures against electric shock will be given here.

- *Earth potential rise* (EPR): U_E voltage between an earthing system and reference earth.
- *Touch voltage* U_T: Effective voltage that occurs on the human body or on the body of a farm animal when the touch current flows through it (Figure 11.11). The touch voltage is measured for people with a voltmeter with an internal resistance of 1000 Ω and for farm animals with an internal resistance of 500 Ω.

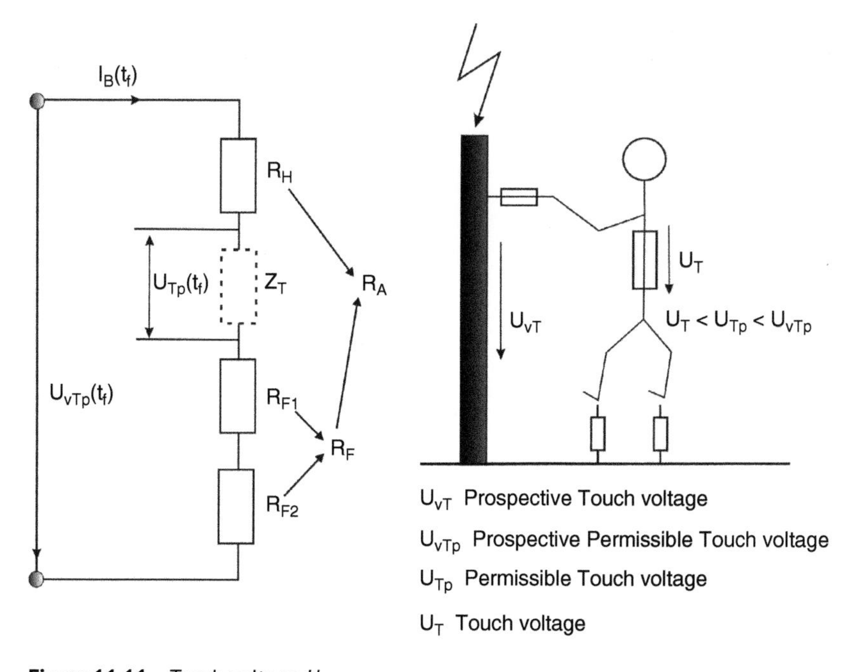

Figure 11.11 Touch voltage U_T.

- *Permissible touch voltage, U_{Tp}*: Limit value of touch voltage U_T.
- *Prospective touch voltage, U_{vT}*: Voltage between simultaneously accessible conductive parts when those conductive parts are not being touched.
- *Prospective permissible touch voltage, U_{vTp}*: Limit value of prospective touch voltage U_{vT}.
- *Transferred potential*: Potential rise of an earthing system caused by a current to earth transferred by means of a connected conductor (for example, a metallic cable sheath, PEN conductor, pipeline, rail) into areas 359 with low or no potential rise relative to reference earth, resulting in a potential difference occurring between the conductor and its surroundings.
- *Step voltage, U_S*: That part of the ground voltage which can be bridged by a person with a step of 1 m length, the current path over the human body from foot to foot (Figure 11.3).
- *Earth current*: It is the total current flowing into the ground via the ground impedance.

Touch voltage occurs on the human body or on the body of a farm animal when the touch current flows through it. The touch voltage is measured for people with a voltmeter with an internal resistance of 1000 Ω and for farm animals with an internal resistance of 500 Ω. The relationship between uninfluenced touch voltage U_{Tp} and touch voltage U_T is shown in Figure 11.11.

Figure 11.12 shows the touch voltage curve and calculation method according to EN 50522 and EN 61936-1. The values of the permissible touch voltage U_{Tp} can be read depending on the switch-off time.

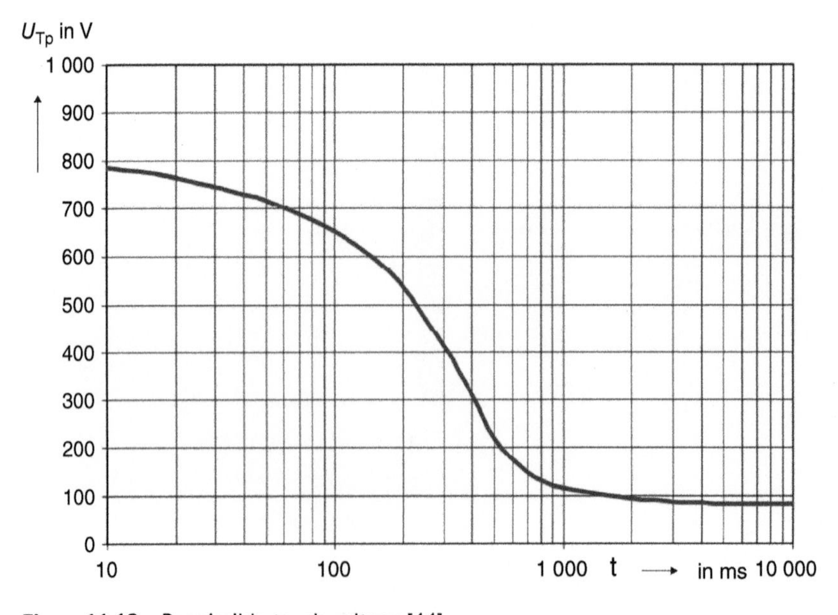

Figure 11.12 Permissible touch voltage [44].

Different touching conditions lead to different tolerable touch voltages. The limit values of the permissible touch voltages U_{Tp} of this standard (Figure 11.12) are based on the average of weighted values taken from four different permissible touch voltage calculations with Formula (11.16), corresponding to four different touching conditions. These four conditions are as follows:

- Touch voltage, U_{Tp1}, left hand to feet (weighted 1.0)
- Touch voltage, U_{Tp2}, right hand to feet (weighted 1.0)
- Touch voltage, U_{Tp3}, both hand to feet (weighted 1.0)
- Touch voltage, U_{Tp4}, hand to hand (weighted 0.7)

$$U_{Tp} = \frac{(U_{Tp1} + U_{Tp2} + U_{Tp3} + U_{Tp4} \cdot 0.7)}{4} \tag{11.15}$$

Equation (11.16) helps evaluate the permissible touch voltage (U_{Tp}) for any configuration depending on four specific conditions:

1) The fault duration (t_f), which determines the body current limit $I_B(t_f)$
2) The sensitivity of human heart to fibrillate (HF = heart factor)
3) The body impedance (Z_T), which depends on the touch voltage (U_T)
4) The touching condition (e.g. hand-hand, hand-to-feet; considering the body factor B_F)

Formula 1:

$$U_{Tp} = I_B(t_f) \cdot \frac{1}{\mathrm{HF}} \cdot Z_T(U_T) \cdot \mathrm{BF} \tag{11.16}$$

Formula 2:

Taking into account additional resistances, the formula for the open circuit touch voltage is

$$U_{vTp} = I_B(t_f) \cdot \frac{1}{\mathrm{HF}} \cdot (Z_T(U_T) \cdot \mathrm{BF} + R_H + R_F) \tag{11.17}$$

Another suitable formula to calculate the prospective permissible touch voltage is Formula 3, where $I_B(t_f)$ is the body current for a current path from left hand to the feet (heart current factor HF = 1) and $U_{Tp}(t_f)$ is the permissible touch voltage according to Figure 11.12.

Formula 3:

$$U_{vTp}(t_f) = U_{Tp}(t_f) + \frac{I_B(t_f)}{\mathrm{HF}} \cdot (R_H + R_F) \tag{11.18}$$

It means

I_B	Permissible body current (depending on the fault duration), EN 50522
U_{Tp}	Permissible touch voltage EN 50522 Figure B.4
U_{vTp}	Permissible open-circuit touch voltage
R_H	Additional hand resistance
R_F	Additional foot resistance

t_f Fault duration
I_{RS} Current flowing through the mesh,
Z_T Body impedance EN 50522 Table B.2
BF Body impedance correction factor
I_T Contact current through a person
R_{F1} Additional resistance through clothing, e.g. footwear
R_{F2} Additional resistance by the location, e.g. foot spreading resistance ambient series resistance (walls, floors, etc.)
U_{Tp} Uninfluenced touch voltage (partial fault voltage in the bridging range)
U_T Touch voltage
ΔU_F ($Z1$) Partial fault voltage (clothing)
($Z2$) Partial fault voltage (location, building)

11.6 Methods of Calculating Permissible Step Voltages

Touch and step voltages shall always be considered. As a general rule, the touch voltage requirements satisfy the step voltage requirements, because the tolerable step voltage limits are much higher than the touch voltage limits due to the different current path through the body (heart current factor HF = 0.04). In special cases with high U_E (EPR) step voltage limits have to be considered. In this case, the permissible step voltage is derived similarly to the method taking into account Formula 1 but without weighing. Step voltage should be considered at the boundaries of the earthing system when e.g. $U_E > 20 \cdot U_{Tp}$ depending on fault duration.

11.7 Current Injunction in the Ground

When earthing, a part of the current path passes through the ground, whereby the current filaments diverge from the ground and flow evenly through a large cross section of the ground. Since the earth electrodes are always in the surface area, a hemisphere embedded in the surface of the earth can be assumed for observation (Figure 11.13).

$$U = I_E \cdot R \tag{11.19}$$

with the resistance

$$R = \frac{\rho_E \cdot l}{A} \tag{11.20}$$

we get the step voltage:

$$U_S = \frac{I_E \cdot \rho_E \cdot l}{A} \tag{11.21}$$

In the hemisphere with $A = 2 \cdot \pi \cdot r^2$

$$U_S = \frac{I_E \cdot \rho_E \cdot l}{2 \cdot \pi \cdot r^2} \tag{11.22}$$

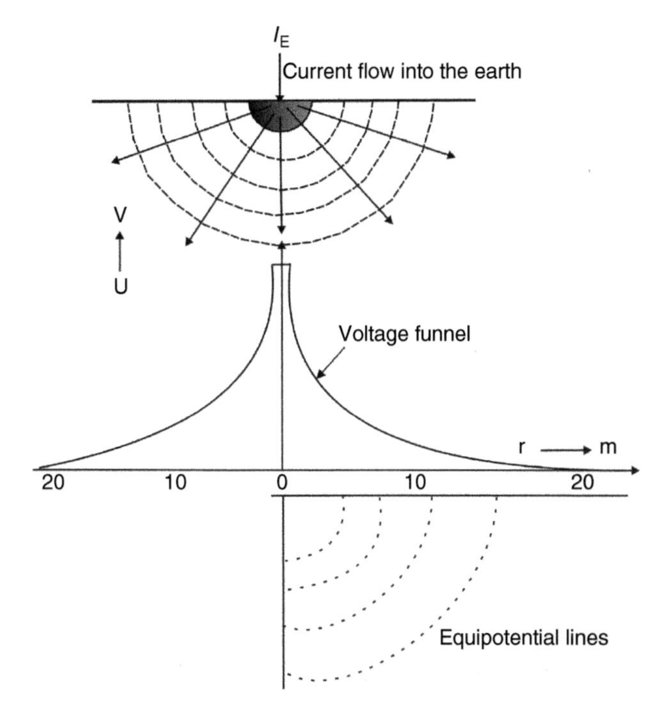

Figure 11.13 Half sphere with tension funnel.

11.8 Design of Earthing Systems

Figure 11.14 shows a flowchart for the design of the earthing system if not part of a global earthing system with regard to permissible touch voltage U_{Tp} by checking the EPR U_E or the touch voltage U_T and without regarding the transferred potential. The individual steps in designing an earthing system according to EN 50522 are explained in more detail in the following part.

- *Data collection*: Earth fault current determined by calculating or simulating the various short-circuits in the faulty power supply of the equipment to be earthed.
- *Initial design*: Initial design of the earthing system, taking into account the functional requirements such as earth fault current, fault duration, and the premises.
- *Determination of soil properties*: Determination of the specific earth resistance varies greatly depending on the type of soil, grain size, density, and moisture. Depending on the depth, changes in the moisture content can also result in temporal fluctuations in the specific earth resistance.
- *Determination of the earth sum current*: The earth sum current is the earth fault current component that enters the earth in the area of the earthing system. It depends on the type of neutral earthing, the occurring fault current, and the characteristic values of the surrounding network.
- *Determination of the total impedance to earth*: Determination of the total impedance to earth is based on the plant design, soil characteristics, and the

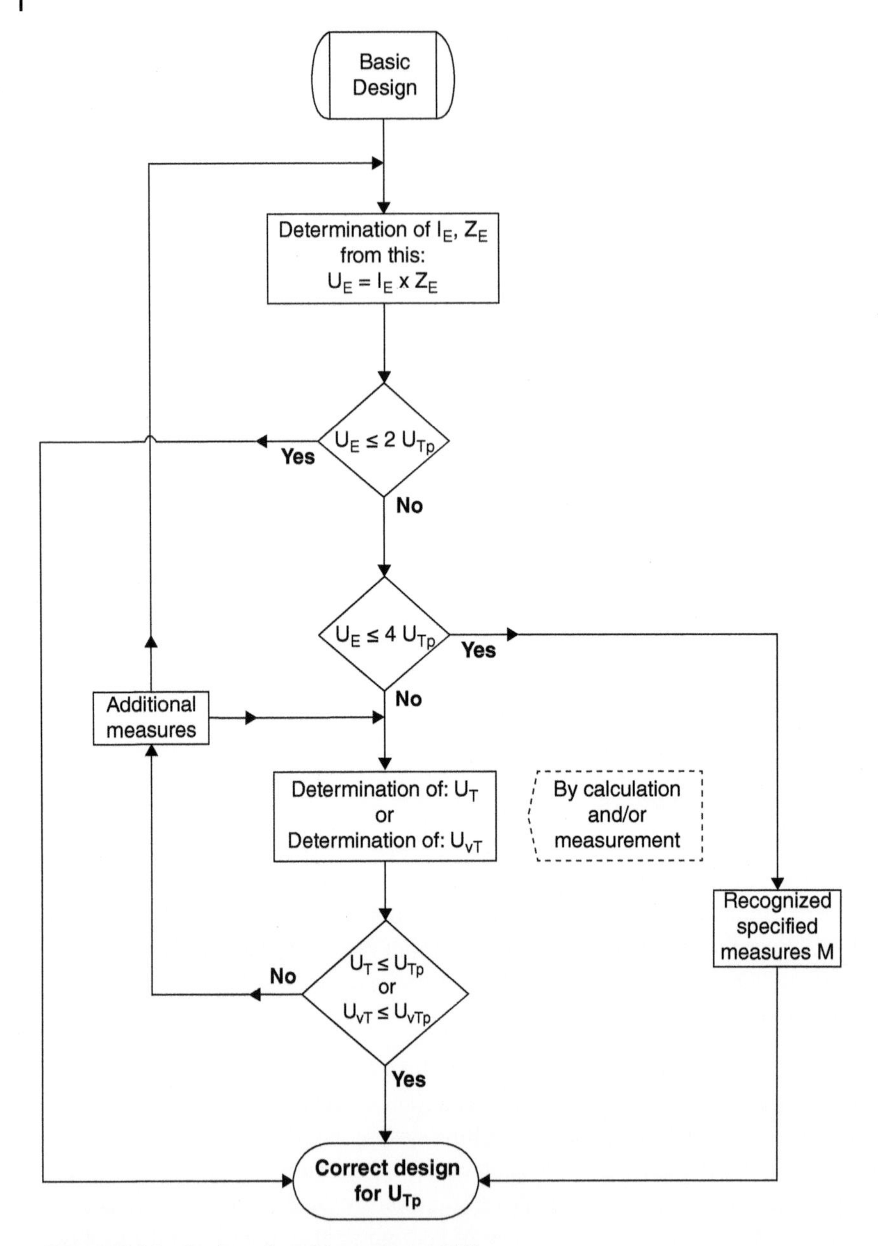

Figure 11.14 Design of earthing systems [44].

parallel earth electrodes. With the determination of the specific earth resistance and the previously designed earthing system, the resistance of the earthing system can now be calculated in advance.

- *Determination of the earthing voltage U_E (EPR)*: The earthing voltage is the voltage between an earthing system and the surrounding reference earth, which is generated when the earth flows through the earthing system into the ground and thus raises its potential.

- *Determination of the permissible touch voltages*: EN 50522 defines how permissible touch voltages are to be dimensioned. For the dimensioning of U_{Tp} and U_{vTp} the estimated fault duration t_{f} (s) is decisive.
- *Compliance with the permissible touch voltages*: Application of the essential requirements of EN 50522 general, electrical, and functional requirements results in the basic design of the earthing system. This must be checked with regard to the touch voltages.
  The requirements regarding touch voltages are considered to be fulfilled if one of the following points is met.
  - If the installation concerned is part of a Global Earthing System.
  - The earthing voltage determined by measurement or calculation does not exceed twice the permissible touch voltage.
  - If the appropriate approved specified measures M have been taken depending on the level of the earth voltage and the duration of the fault.

11.9 Types of Earth Rods

Figure 11.15 shows the most important types of earth rods. Earth electrodes can be erected individually or together. If the required earthing resistance is not achieved, other measures must be provided. The calculation equations of the resistance to propagation are given approximately below.

11.9.1 Deep Rod

If $x \geq$ is d, then the ground resistance is valid:

$$R_{\mathrm{E}} = \frac{\rho_E}{2\pi \cdot l} \cdot \ln \frac{4 \cdot l}{d} \tag{11.23}$$

Potential distribution of the earth rod:

$$\varphi_x = \frac{I_{\mathrm{E}} \cdot \rho_{\mathrm{E}}}{2\pi \cdot l} \cdot \ln\left[\frac{l}{x}x + \sqrt{1 + \left(\frac{l}{x}\right)^2}\right] \tag{11.24}$$

11.9.2 Earthing Strip

Grounding resistance:

$$R_{\mathrm{E}} = \frac{\rho_{\mathrm{E}}}{\pi \cdot l} \cdot \ln \frac{2 \cdot l}{d} \tag{11.25}$$

Potential distribution of the band earth electrode:

$$\varphi_x = \frac{I_{\mathrm{E}} \cdot \rho_{\mathrm{E}}}{\pi \cdot l} \cdot \ln\left[\frac{l}{2\sqrt{h^2 + x^2}} + \sqrt{1 + \left(\frac{l}{2\sqrt{h^2 + x^2}}\right)^2}\right] \tag{11.26}$$

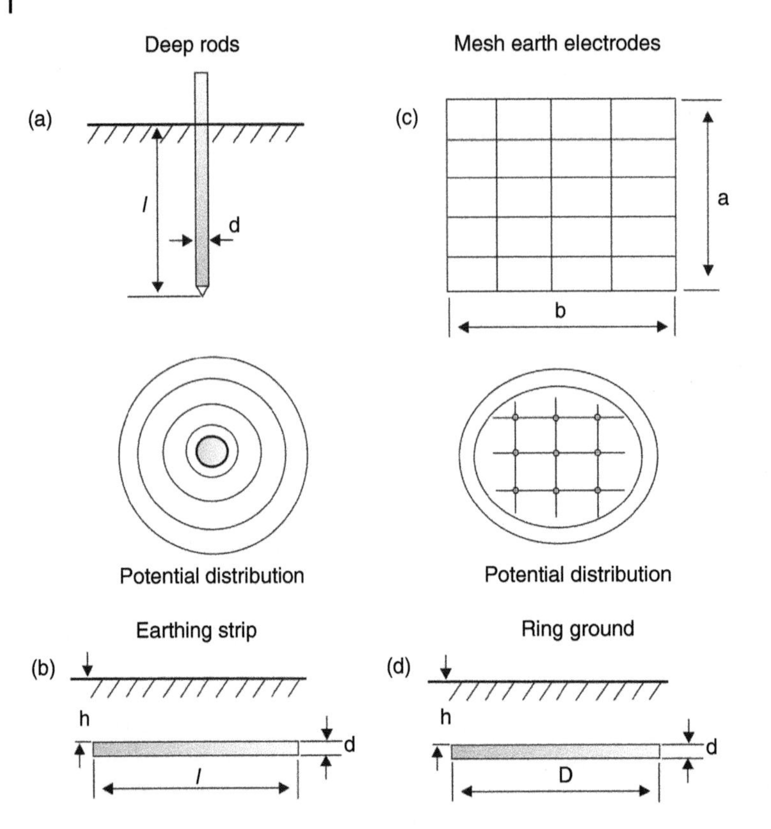

Figure 11.15 Types of earth rods. (a) Deep rods. (b) Earthing strip. (c) Mesh earth electrodes. (d) Ring ground.

simplified with $l \leq 10$ m results in

$$R_A = \frac{2 \cdot \rho_E}{l} \tag{11.27}$$

If $l > 10$m, then

$$R_A = \frac{3 \cdot \rho_E}{l} \tag{11.28}$$

11.9.3 Mesh Earth

Grounding resistance:

$$R_E \approx \frac{\rho_E}{2 \cdot D} + \frac{\rho_E}{l_{sum}} \tag{11.29}$$

With the diameter of the spare radius in ring form:

$$D = \sqrt{\frac{4 \cdot b \cdot l}{\pi}} \tag{11.30}$$

Potential distribution of the earth mesh:

$$\varphi_x = \frac{I_E \cdot \rho_E}{\pi \cdot \sqrt{A}} \cdot \arcsin \frac{\sqrt{A}}{2 \cdot x} \tag{11.31}$$

11.9.4 Ring Earth Electrode

If $x > D/2$, then

$$R_E = \frac{\rho_E}{15 \cdot D} \ln \frac{8 \cdot D}{d} \tag{11.32}$$

Potential distribution of the ring rode:

$$\varphi_x = \frac{I_E \cdot \rho_E}{\pi \cdot D} \cdot \arcsin \frac{D}{2 \cdot x} \tag{11.33}$$

11.9.5 Foundation Earthing

Earthing resistance:

$$R_E = \frac{2 \cdot \rho_E}{\pi \cdot D} \tag{11.34}$$

With the diameter of the spare radius in ring form:

$$D = \sqrt{\frac{4 \cdot b \cdot l}{\pi}} \tag{11.35}$$

11.10 Calculation of the Earthing Conductors and Earth Electrodes

For fault currents that are interrupted in less than five seconds the cross section of the earthing conductor or earth electrode according to the EN 50522 shall be calculated from the following formula:

$$A = \frac{I}{K} \cdot \sqrt{\frac{t_f}{\ln \frac{\Theta_f + \beta}{\Theta_i + \beta}}} \tag{11.36}$$

where

- A is the cross section in mm^2
- I is the conductor current in amperes (RMS value)
- t_f is the duration of the fault current in seconds
- K is a constant depending on the material of the current-carrying component (for Copper 226 $A \cdot \sqrt{s}/mm^2$);
  Table D.1 in EN 50522 provides values for the most common materials assuming an initial temperature of 20 °C

β is the reciprocal of the temperature coefficient of resistance of the current-carrying component at 0 °C (for Copper 234.5 °C)

Θ_i is the initial temperature in degrees Celsius. Values may be taken from IEC 60287-3-1. If no value is laid down in the national tables, 20 °C as ambient ground temperature at a depth of 1 m should be adopted.

Θ_f is the final temperature in degrees Celsius (for Copper 300 °C).

11.11 Substation Grounding IEEE Std. 80

In this chapter, the American Standard IEEE Std. 80-2013 (IEEE Guide for Safety in AC Substation Grounding Guide for Safety in AC Earthing Systems) will be introduced briefly [31].

11.11.1 Tolerable Body Current

The magnitude and duration of the current through the human body at 50 Hz should be less than the value that can cause ventricular fibrillation of the heart. Figure 11.16 shows the relationship between the critical flow and the body weight for several animal species (calves, dogs, sheep, and pigs) and a common threshold region of 0.5% for mammals.

Ferris et al. [35] proposed 100 mA as the fibrillation threshold derived from extensive experiments. In the experiments, animals with body and heart weights

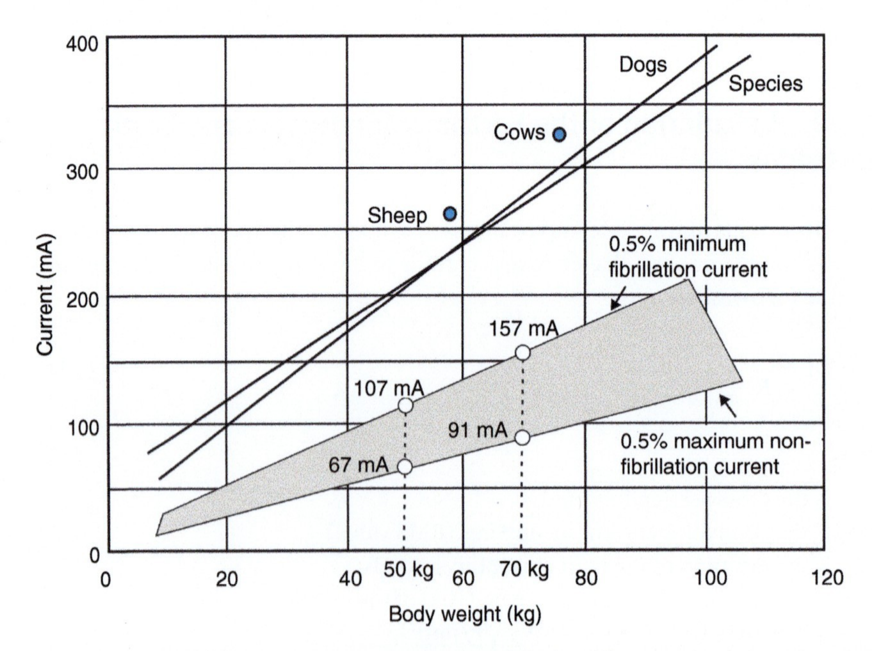

Figure 11.16 Fibrillation current against body weight for different animals based on three seconds duration of electric shock. Source: IEEE [35].

comparable with humans were exposed to a maximum shock duration of three seconds. Some of the more recent experiments suggest the existence of two different thresholds: one in which the shock duration is shorter than a heartbeat period and another for the current duration, which is longer than a heartbeat. For a 50 kg adult Biegelmeier beat [32] the threshold values are at 500 or 50 mA.

The tolerable body current equation developed by Dalziel [36] is the basis for the derivation of tolerable voltages used in the guideline Std. 80. It is assumed that 99% of all persons will safely withstand the passage of a current of size and duration without ventricular fibrillation, which can be determined by the following formula:

$$I_{\mathrm{B}} = \frac{k}{\sqrt{t_{\mathrm{s}}}} \quad \text{with} \quad k = \sqrt{S_{\mathrm{B}}} \tag{11.37}$$

The factor k can be (Figure 11.16)

$$k_{50} = \frac{I_{\mathrm{B}}}{\sqrt{3}} = \frac{0.067\,\mathrm{A}}{\sqrt{3}\ \text{seconds}} = 0.116$$

$$k_{70} = \frac{I_{\mathrm{B}}}{\sqrt{3}} = \frac{0.091\,\mathrm{A}}{\sqrt{3}\ \text{seconds}} = 0.157$$

Body current, e.g. for 50 kg person:

$$I_{\mathrm{B}} = \frac{0.116}{\sqrt{t_{\mathrm{s}}}} \tag{11.38}$$

Body current, e.g. for a 70 kg person:

$$I_{\mathrm{B}} = \frac{0.157}{\sqrt{t_{\mathrm{s}}}} \tag{11.39}$$

11.11.2 Permissible Touch Voltages

When calculating the touch voltage according to IEEE Std. 80, a distinction is made between 50 and 70 kg people.

Calculation of the permissible touch voltage:

$$U_{\mathrm{T50}} = \frac{(1000 + 1.5\ C_{\mathrm{s}}\ \rho_{\mathrm{s}})\ 0.116}{\sqrt{t_{\mathrm{F}}}} \tag{11.40}$$

$$U_{\mathrm{T70}} = \frac{(1000 + 1.5\ C_{\mathrm{s}}\ \rho_{\mathrm{s}})\ 0.157}{\sqrt{t_{\mathrm{F}}}} \tag{11.41}$$

Calculation of the permissible step voltage:

$$U_{\mathrm{S50}} = \frac{(1000 + 6\ C_{\mathrm{s}}\ \rho_{\mathrm{s}})\ 0.116}{\sqrt{t_{\mathrm{F}}}} \tag{11.42}$$

$$U_{\mathrm{S70}} = \frac{(1000 + 6\ C_{\mathrm{s}}\ \rho_{\mathrm{s}})\ 0.157}{\sqrt{t_{\mathrm{F}}}} \tag{11.43}$$

Calculation of the maximum mesh tension:

$$E_{\mathrm{m}} = \frac{\rho_{\mathrm{E}} \cdot K_{\mathrm{m}} \cdot K_{\mathrm{i}} \cdot I_{\mathrm{G}}}{L_{\mathrm{g}} + 1.15 \cdot L_{\mathrm{r}} \cdot N_{\mathrm{r}}} \tag{11.44}$$

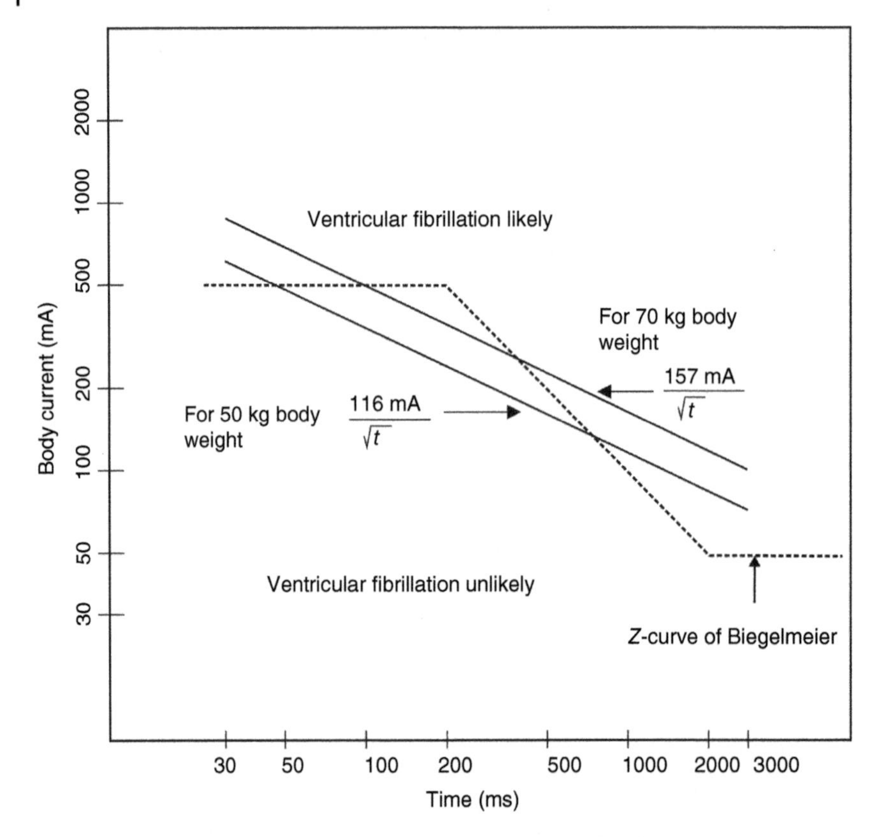

Figure 11.17 Body current–time characteristic. Source: IEEE [34].

Calculation of the maximum step stress:

$$E_S = \frac{\rho_E \cdot K_s \cdot K_i \cdot I_G}{L_g + 1.15 \cdot L_r \cdot N_r} \tag{11.45}$$

The comparison of Eq. (11.38), Eq. (11.39) published by Biegelmeier and Lee, and the *Z*-shaped curve of the body current against time developed by Biegelmeier is shown in Figure 11.17.

The *Z*-curve has a limit of 500 mA for short periods of up to 0.2 second and then drops to 50 mA at 2.0 second and beyond. Using Eq. (11.38) the tolerable body current for times from 0.06 to 0.7 second is smaller than the Biegemeier *Z* curve.

11.11.3 Calculation of the Conductor Cross Section

The required conductor cross section according to the IEEE Std. 80 can be calculated as a function of a conductor current with the short-term temperature rise in a conductor with the formula (11.47).

$$A = \frac{I}{\sqrt{\left(\frac{t_c\ \alpha_r\ \rho_r \times 10^{-4}}{\mathrm{TCAP}}\right)\ \ln\left[1 + \frac{K_0 + T_m}{K_0 + T_a}\right]}} \tag{11.46}$$

$$I = A\sqrt{\left(\frac{\text{TCAP} \times 10^{-4}}{t_c\ \alpha_r\ \rho_r}\right)\ \ln\left(\frac{K_0 + T_m}{K_0 + T_a}\right)} \tag{11.47}$$

11.11.4 Calculation of the Maximum Mesh Residual Current

Symmetrical current with a loop earth electrode:

$$I_g = S_f\ I_f \tag{11.48}$$

Maximum mesh current with decrement factor (Table 11.5):

$$I_G = D_f\ I_g \tag{11.49}$$

Division factor of the residual current in a loop earth electrode:

$$S_f = \frac{I_g}{3\ I_0} \tag{11.50}$$

Effectively nonsymmetrical residual current:

$$I_F = \frac{I_g}{3\ I_0} \tag{11.51}$$

Grounding current for a loop earth electrode:

$$I_E = D_f\ I_f \tag{11.52}$$

Zero current between conductor and earth:

$$I_0 = \frac{E}{3R_f + j(R_1 + R_2 + R_0) + j(X_1 + X_2 + X_0)} \tag{11.53}$$

It means

G Short-circuit current density in A/mm^2
M Material constant
t_F Error time in s
k Coefficient of materials
ϑ_e Permissible final temperature in °C
ϑ_0 Material constant in °C
ϑ_a Output temperature in °C

Table 11.5 Decrement factors.

Fault current duration t_F (s)	**decrement factor D_F**
0.008	1.65
0.1	1.25
0.25	1.10
>0.5	1.0

U_E	Earthing voltage in V
U_T	Touch voltage in V
R_E	Earthing resistance in Ω
I_E	Rest current in A
I_C	Capacitive current in A
w	Expectation factor ($w = 0.7$)
C_E	Earth capacity in μ*F*
X_L	Inductive resistance in Ω
R_B	Total earth resistance Ω
U_0	Conductor-earth voltage in V
I^2_{Esp}	Sum of the rated currents of the earth fault coils connected in parallel in A
I^2_{Rest}	Rest earth fault current in A
r	Reduction factor
I''_{k1}	Single pole short-circuit current in A
C_s	Reduction factor when using a ballast layer
ρ_s	Specific soil resistance at gravel layer in Ωm
r	Reduction factor
D	Decrement factor for error duration $t \geq 0.5$ second$D = 1$
I_0	Zero current in A
TCAP	Heat capacity per unit volume in $\text{J}/(\text{cm}^3 \cdot °\text{C})$
T_m	Maximum allowed temperature in °C
T_a	Ambient temperature in °C
X_1	Positif sequence reactance in Ω
X_2	Negatif sequence reactance in Ω
X_0	Zero reactance in Ω
R_1	Positif sequence resistance in Ω
R_2	Negatif sequence resistance in Ω
R_0	Zero resistance in Ω

11.12 Soil Resistivity Measurement

Geoelectric methods are used to measure the specific resistance of the earth. The resistivity is measured in Ohmmeter and is the reciprocal of the electrical conductivity, which is a measure of how well a material can conduct electricity. The resistance can vary by many orders of magnitude and show high contrasts; therefore, it is often a very good diagnostic parameter.

Geoelectrics is part of applied geophysics and includes methods for exploring the earth by measuring electrical voltage and current at the earth's surface.

These include the following:

Direct current methods and alternating current methods, in which electrodes supply artificial currents to the ground. Artificial current methods often use four-point arrangements (two electrodes A, B for current supply, two probes C, D for potential measurement), since only in this way can the contact resistance occurring at the electrodes be eliminated. When arranging the electrodes in a line (e.g. current supply through the outer electrodes, measurement at the inner electrodes = probes), there are various possibilities, e.g. according to the Wenner method: All electrodes have the same distance to each other, and the Schlumberger method: The current electrodes have a greater distance than the potential probes.

Soil resistivity is related to geology, soil temperature, electrolyte salt content, and moisture level, all of which can vary considerably across a site area both laterally and with depth. Boring test samples and other geological investigations often provide useful information on the presence of various layers and the nature of soil material, leading at least to some ideas as to the range of resistivity at the site. It is most likely, however, that a series of soil resistivity measurements will be required at design stage for determining both the general soil composition and degree of homogeneity.

Determining the soil resistivity structure is fundamental in the earthing system design process and represents a relatively constant parameter over the lifetime of an installation. It is advisable to make as thorough a soil resistivity investigation of the site as can be justified. It may also be likely that future developments adjacent to the site may mean that access to nearby open areas to take soil resistivity measurements may be limited.

There are different methods that can be applied for soil resistivity measurements. e.g. Schlumberger method for single location and Wenner method for large areas. The Wenner method uses four probes, which are driven into the earth along a straight line, at equal distances a apart, driven to a depth b. The voltage between the two inner (potential) electrodes is then measured and divided by the current between the two outer (current) electrodes to give the value of resistance R (Figure 11.18).

$$\rho_{\mathrm{E}} = k \cdot \frac{\Delta U}{I} \tag{11.54}$$

The constant of proportionality k is the geometry factor that depends on the configuration of the current and voltage electrodes. The geometry factor of the Wenner measurement is $k = 2 \cdot \pi \cdot a$.

The techniques for measuring soil resistivity are essentially the same for most measurements. The interpretation of the recorded data however can vary considerably, especially where soils with nonuniform resistivities are encountered. The added complexity caused by nonuniform soils is common, and in only a few cases are the soil resistivities constant with increasing depth.

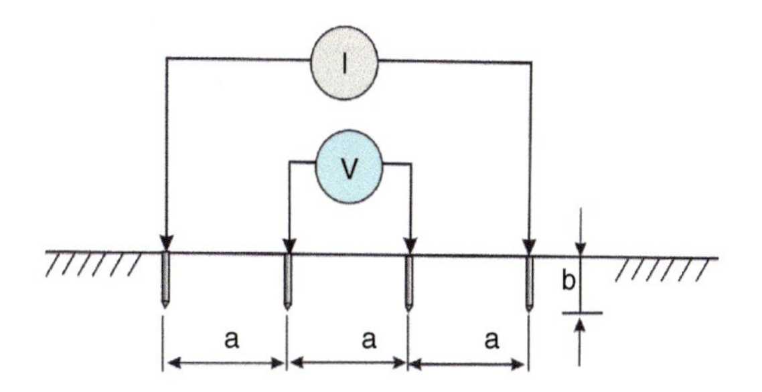

Figure 11.18 Wenner four-pin method.

The resistances and impedances may be determined in different ways. Which method is suitable depends on the extent of the earthing system and the degree of interference. The following are examples of suitable methods of measurements and types of instruments:

- *Fall-of-potential method with the earth tester*: This method is used for earth electrodes and earthing systems of small or medium extent, for example, single-rod earth electrodes, strip earth electrodes, earth electrodes of overhead line towers with lifted off or attached earth wires, medium voltage earthing systems, and separation of the low-voltage earthing systems. The frequency of the used alternating voltage should not exceed 150 Hz.
- *Heavy-current injection method*: This method is used particularly for the measurement of the impedance to earth of large earthing 1927 systems.

11.13 Measurement of Resistances and Impedances to Earth

In principle, the resistance of an earthing system can be measured using the current–voltage method. In order to determine the impedance of an earthing system, a current is injected into the system, which ideally does not flow to an opposite earth, but instead spreads evenly on all sides into the surrounding soil. The voltage increase of the earthing system then becomes the neutral zone, area in which the gradient of the voltage funnel is zero, i.e. at an infinite distance from the earthing system.

To form an electric circuit, a counter-ground, at which the current flowing into the electricity fed into the ground current emerges from the ground again and from there flows back via a line to the source, is used (Figure 11.19).

In a 110/20 kV sub-distribution system, the earthing resistance, potential rise, and touch voltage were measured. Measurement results are given in Table 11.6.

11.14 Example: Calculation of a TR Station

For a transformer station the earthing system will be calculated (Figure 11.20). The specific earth resistance is $\rho_E = 150\ \Omega\text{m}$. The diameter of the earth conductor is $d = 0.02$ m. Four additional earth electrodes with a length of 1.5 m are provided. Calculate the resistance of the earthing system.

Diameter of the replacement earth electrode in ring form for the foundation earth electrode is

$$D = \sqrt{\frac{4 \cdot a \cdot b}{\pi}} = \sqrt{\frac{4 \cdot 24\ \text{m} \cdot 10\ \text{m}}{\pi}} = 17.48\ \text{m}$$

According to Laurent, the earthing resistance is

$$R_T = \frac{\rho_E}{2 \cdot D} + \frac{\rho_E}{L} = \frac{150\ \Omega\text{m}}{2 \cdot 17.48\ \text{m}} + \frac{150\ \Omega\text{m}}{68\ \text{m}} = 6.49\ \Omega$$

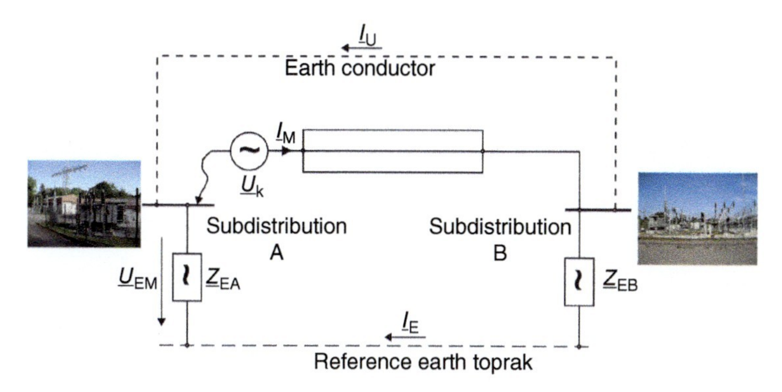

Figure 11.19 Measurement of earth resistance.

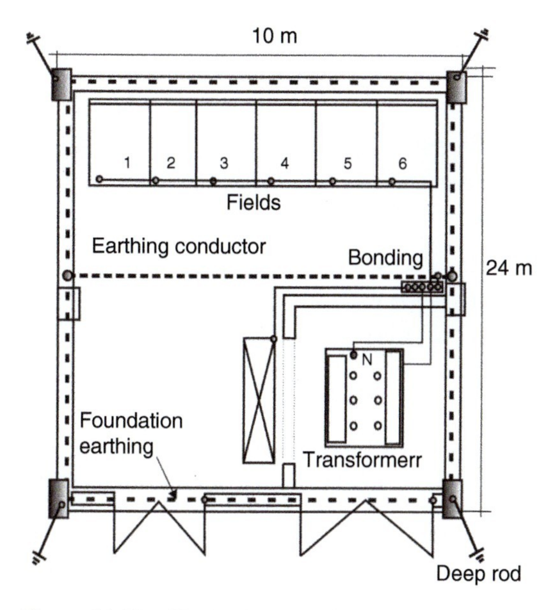

Figure 11.20 Distribution transformer.

With the formula of the foundation earth electrode

$$R_T \approx \frac{2 \cdot \rho_E}{\pi \cdot D} \approx \frac{2 \cdot 150\ \Omega\text{m}}{\pi \cdot 17.48} = 5.46\ \Omega$$

Resistance of the earth rod:

$$R_s = \frac{\rho_E}{2 \cdot \pi \cdot n \cdot l} \cdot \ln \frac{4 \cdot L}{d} = \frac{150\ \Omega\text{m}}{2 \cdot \pi \cdot 4 \cdot 1.5\ \text{m}} \cdot \ln \frac{4 \cdot 1.5\ \text{m}}{0.02}$$

Total resistance to propagation:

$$R_A = \frac{R_H \cdot R_d}{R_T + R_d} = \frac{5.46\ \Omega \cdot 22.7\ \Omega}{5.46\ \Omega + 22.7\Omega} = 4.4\ \Omega$$

Table 11.6 Measurement results.

Voltage U_a	226 V	
Voltage U_b	230 V	
Voltage U_K	2286 V	
Current I_M	124.78 A	
Reduction factor r_{EM}	0.64	
Earth voltage U_{EM}	8.71 V	
Earthing impedance Z_E	0.108 Ω	$\lvert\underline{Z}_E\rvert = \frac{\lvert U_{EM}\rvert}{\lvert r_E\rvert \cdot I_M}$
Earth faults		
380 kV	I_E= 12.42 kA	
110 kV	I_E= 84.48 A	
20 kV	I_E= 870.32 A	
Earth voltage		
380 kV	U_E = 1353.34 V	$\lvert\underline{U}_E\rvert = I_E \cdot Z_E$
110 kV	U_E = 9.21 V	
20 kV	U_E = 94.87 V	
Touch voltage		
380 kV	U_T = 333.46 V	
110 kV	U_T = 2.27 V	
20 kV	U_T = 23.37 V	
Permissible touch voltage $t_F \leq 0.10$ s		
U_{Tp}	683.3 V	

11.15 Example: Earthing Resistance of a Building

Figure 11.21 shows the earthing system of a building, which consists of a meshed combination of a foundation earth electrode, two ring earth electrodes, and four depth earth electrodes. The specific resistance of the earth was assumed to be $\rho_E = 45$ Ωm. Calculate the total earth resistance.

11.15.1 Foundation Earthing R_{EF}

Calculation of the earthing resistance R_{EF} of the foundation earth electrode with 10 m × 20 m.

$$D_{EF} = \sqrt{\frac{4 \cdot a \cdot b}{\pi}} = \sqrt{\frac{4 \cdot 20\ \text{m} \cdot 10\ \text{m}}{\pi}} = 15.96\ \text{m}$$

$$R_{EF} = \frac{2 \cdot \rho_E}{3 \cdot D} = \frac{2 \cdot 45\ \Omega\text{m}}{3 \cdot 15.96\ \text{m}} = 1.88\ \Omega$$

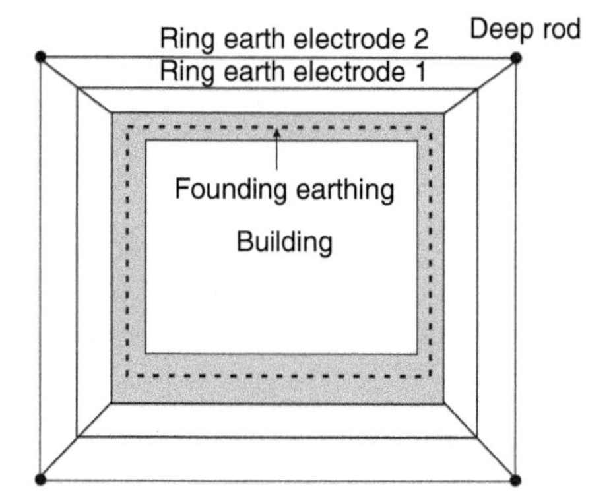

Figure 11.21 Design of the earthing system.

11.15.2 Ring Earth Electrode 1 R_{ER1}

Calculation of the ring resistance R_{ER1} of the inner ring earth electrode with 12 m × 22 m.

$$D_{ER1} = \sqrt{\frac{4 \cdot a \cdot b}{\pi}} = \sqrt{\frac{4 \cdot 22\ \text{m} \cdot 12\ \text{m}}{\pi}} = 18.3\ \text{m}$$

$$\begin{aligned} R_{ER1} &= \frac{\rho_E}{\pi^2 \cdot D_{ER1}} \cdot \ln\left(\frac{2 \cdot \pi \cdot D_{ER1}}{d}\right) \\ &= \frac{45\ \Omega\text{m}}{\pi^2 \cdot 18.3\ \text{m}} \cdot \ln\left(\frac{2 \cdot \pi \cdot 18.3\ \text{m}}{0.015\ \text{m}}\right) = 2.23\ \Omega \end{aligned}$$

11.15.3 Ring Earth Electrode 2 R_{ER2}

Calculation of the ring resistance R_{ER1} of the inner ring earth electrode with 16 m × 26 m.

$$D_{ER2} = \sqrt{\frac{4 \cdot a \cdot b}{\pi}} = \sqrt{\frac{4 \cdot 26\ \text{m} \cdot 16\ \text{m}}{\pi}} = 23.01\ \text{m}$$

$$\begin{aligned} R_{ER2} &= \frac{\rho_E}{\pi^2 \cdot D_{ER2}} \cdot \ln\left(\frac{2 \cdot \pi \cdot D_{ER2}}{d}\right) \\ &= \frac{45\ \Omega\text{m}}{\pi^2 \cdot 23.01\ \text{m}} \cdot \ln\left(\frac{2 \cdot \pi \cdot 23.01\ \text{m}}{0.015\ \text{m}}\right) = 1.82\ \Omega \end{aligned}$$

11.15.4 Deep Earth Electrode R_{ET}

Calculation of the ring resistance R_{ET} of the deep earth electrode with $d = 20$ mm and $L = 2$ m.

$$R_{ET} = \frac{\rho_E}{2 \cdot \pi \cdot l} \cdot \ln\left(\frac{4 \cdot l}{d}\right) = \frac{45\ \Omega\text{m}}{2 \cdot \pi \cdot 2\ \text{m}} \cdot \ln\left(\frac{4 \cdot 2\ \text{m}}{0.02\ \text{m}}\right) = 21.46\ \Omega$$

11.15.5 Total Earthing Resistance R_{ETotal}

Owing to the meshing, the differently meshed earth electrodes form a parallel connection. Using Eq. (11.55), the total earthing resistance R_{ETotal} can be calculated.

$$R_{\mathrm{ETotal}} = \left(\sum_{j=1}^{4} \frac{1}{R_{\mathrm{E}j}} \right)^{-1} \tag{11.55}$$

$$= \left(\frac{1}{1.88\ \Omega} + \frac{1}{2.23\ \Omega} + \frac{1}{1.82\ \Omega} + \frac{4}{21.46\ \Omega} \right)^{-1} = 0.58\ \Omega$$

11.16 Example: Cross-Sectional Analysis

With a HV switchgear the mains short-circuit capacity is 350 MVA and the nominal voltage 20 kV. The transformer has the data: 630 kVA, 6%, 400 V. Calculate the cross section of the earthing cable.

First the earth fault current is calculated on the HV side.

$$I''_{\mathrm{k2E}} = \frac{S''_{kQ}}{2 \cdot U_n} = \frac{350\ \mathrm{MVA}}{2 \cdot 20\ \mathrm{kV}} = 8.75\ \mathrm{kA}$$

According to EN 50522, the cross section of the earthing cable with the short-circuit current density (G) according to Figure 11.22 for a fault time of one second:

$$S = \frac{I''_{\mathrm{k2E}}}{G} = \frac{8.75\ \mathrm{kA}}{180\ (\mathrm{A/mm^2})} = 48.61\ \mathrm{mm}^2$$

Selected: 50 mm^2.

The earth fault current on the LV side is calculated by

$$Z_Q = \frac{c \cdot U_n^2}{S''_{kQ}} = \frac{1.1 \cdot 400\mathrm{V}^2}{350\ \mathrm{MVA}} = 0.5\ \mathrm{m\Omega}$$

The impedance of the transformer:

$$Z_{\mathrm{T}} = \frac{u_{\mathrm{kr}} \cdot U_{\mathrm{rT}}^2}{S_{\mathrm{rT}}} = \frac{6\% \cdot 400\ \mathrm{V}^2}{630\ \mathrm{kVA}} = 15.2\ \mathrm{m\Omega}$$

This means that the single-pole earth short-circuit:

$$I''_{\mathrm{k1E}} = \frac{c \cdot U_n}{\sqrt{3} \cdot (2 \cdot Z_Q + Z_{\mathrm{T}})} = \frac{1 \cdot 400\ \mathrm{V}}{\sqrt{3} \cdot (2 \cdot 0.5\ \mathrm{m\Omega} + 15.2\ \mathrm{m\Omega})} = 14.25\ \mathrm{kA}$$

According to EN 50522, the cross section of the earthing cable with the short-circuit current density (G) (Figure 11.22) for a fault time of 0.5 second:

$$S = \frac{I''_{\mathrm{k2E}}}{G} = \frac{14.25\ \mathrm{kA}}{270\ (\mathrm{A/mm^2})} = 52.77\ \mathrm{mm}^2$$

Selected: 70 mm^2.

In the case of double earth faults, the residual current is

$$I''_{\mathrm{kEE}} = 0.85 \cdot 14.25\ \mathrm{kA} = 12.11\ \mathrm{kA}$$

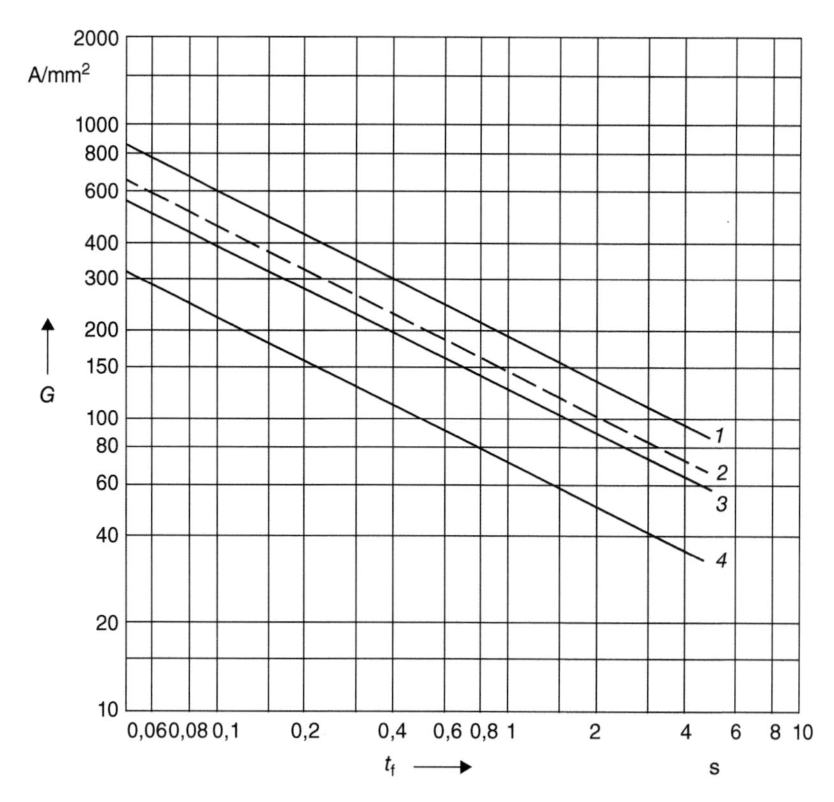

Figure 11.22 Short-circuit current density G for earthing conductors and earth electrodes relative to the duration of the fault current t_f [EN 50522]. 1 Copper, bare or zinc-coated; 2 Copper, tin-coated or with lead sheath; 3 Aluminum, only earthing conductors; 4 Galvanized steel.

With the reduction factor for underground cables you get

$$I''_{\mathrm{kEE}} = 0.65 \cdot 12.11\ \mathrm{kA} = 7.87\ \mathrm{kA}$$

This current is decisive for the dimensioning of the cross section of the earthing cable.

11.17 Example: Cross-Sectional Analysis of the Earthing Conductor

We distinguish between two earthing versions.

a) The operating earthing resistance and the protective earthing resistance are laid separately.
   In this case, the earth fault current flows over the two earth resistances that we want to assume: $R_B = 2\ \Omega$ and $R_A = 20\ \Omega$.

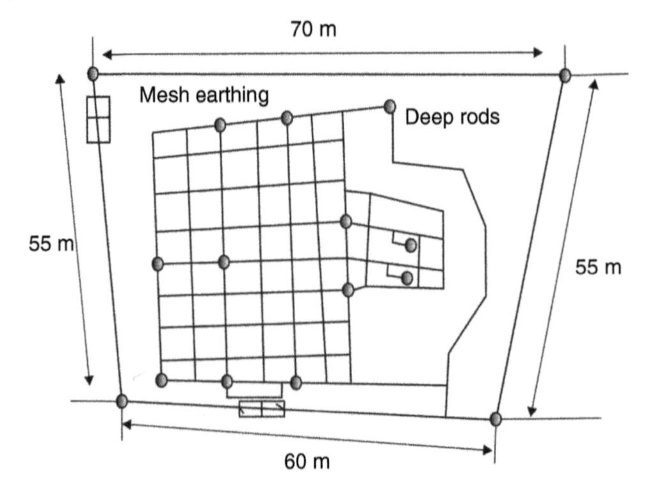

Figure 11.23 Substation.

This means that the residual current is

$$I_F \approx \frac{230\ \text{V}}{R_A + R_B} = \frac{230\ \text{V}}{2\ \Omega + 20\ \Omega} = 10.45\ \Omega$$

Selected: 50 mm².

b) The operating earthing resistance and protective earthing resistance are combined. In this case, the earth fault current flows through the neutral point of the transformer. The residual current of a 630 kV transformer is 22 kA.
   This gives the cross section

$$S = \frac{I \cdot \sqrt{s}}{k} = \frac{22\ \text{kA} \cdot \sqrt{0.5\ \text{second}}}{159 \frac{\text{A}\sqrt{s}}{\text{mm}^2}} = 107\ \text{mm}^2$$

   Selected: 3 × 40 mm copper.

11.18 Example: Grounding Resistance According to IEEE Std. 80

Given is a substation (figure 11.23) with 912.64 m² and two 1000 MVA transformers. The voltage levels are 154/33.6 kV.

Data of the substation:

1) 154 kV grid
   - Three-pole short-circuit current: 31.5 kA
   - Unipolar short-circuit current: 25 kA
   - Switch-off time: One second
   - Frequency: 50 Hz
   - Star point is grounded
   - Reduction factor: 0.45

2) 33.6 kV grid
   - Three-polar short-circuit current: 31.5 kA
   - Single pole short-circuit current: 1 kA
   - Switch-off time: One second
   - Frequency: 50 Hz
   - Star point is grounded
   - Reduction factor: 0.45
3) Specific earth resistance $\rho_E = 11.93\ \Omega$.
4) Cable cross section for the 154 kV system
   Data of the sizes in the formula:

I_{max}	Short-circuit current: 25 kA (85 kA for IEEE Std. 80),
T_m	For copper 300 °C, according to IEEE Std. 80., 1084 °C is used for thermal welding,
T_a	30 °C,
K_0	Material index 242 °C, $\frac{1}{\alpha_0}$,
t_c	Error duration in s,
TCAP	Thermal capacity 3.42 J/(cm³ °C),
α_r	Resistance index 0.003 81 1/°C,
ρ_r	Resistance for earthing conductor 1.78 mΩ/cm.

With these data, the cable average cross section:

$$A = I \cdot \frac{1}{\sqrt{\left(\frac{\text{TCAP} \times 10^{-4}}{t_c \cdot \alpha_r \cdot \rho_r}\right) \cdot \ln\left(\frac{K_0 + T_m}{K_0 + T_a}\right)}} = 146\ \text{mm}^2$$

5) cable cross section for the 33.6 kV system

$$A = I \cdot \frac{1}{\sqrt{\left(\frac{\text{TCAP} \times 10^{-4}}{t_c \cdot \alpha_r \cdot \rho_r}\right) \cdot \ln\left(\frac{K_0 + T_m}{K_0 + T_a}\right)}} = 6\ \text{mm}^2$$

6) Calculations for earthing mesh
   Given:
   $\rho_E = 11.93\ \Omega\text{m}, A = 912.64\ \text{m}^2$,
   $L = 200\ \text{m}, h = 0.5\ \text{m}$.
   Total resistance to propagation:

$$R_g = \rho_E \cdot \left[\frac{1}{L_T} + \frac{1}{\sqrt{20 \cdot A}} \cdot \left(1 + \frac{1}{1} 1 + h \cdot \sqrt{\frac{20}{A}}\right)\right] = 0.23\ \Omega$$

Calculation according to EN 50522:

$$R_h = \frac{\rho_E}{2 \cdot \sqrt{A}} = \frac{11.93\ \Omega\text{m}}{2 \cdot \sqrt{912.64\ \text{m}^2}} = 0.197\ \Omega$$

7) Ground impedance

$$Z_g = \frac{1}{\left[\frac{1}{R_g} + \left(\frac{6}{Z}\right)\right]} = 0.171\ \Omega$$

8) Calculation of the current for the ground strap
Data:
$n = 3$, $U_0 = 46.1$ kV, $Z_0 = 25.64\,\Omega$.

$$U_0 = \frac{c \cdot U_n}{\sqrt{3} \cdot \frac{2}{3}\left(\frac{I_{k1}}{I_{k3}}\right)} = 46.1 \text{ kV}$$

$$Z_p = \frac{1}{\left(\frac{6}{Z}\right)}$$

$$I_n = n \cdot 3 \cdot \frac{U_0}{Z_0} = 1.62 \text{ kA}$$

$$I_g = (I_{k1} - I_{k3}) \cdot r_E \cdot \left(\frac{Z_p}{R_g + Z_p}\right) = 1.63 \text{ kA}$$

9) Increase of the earthing potential

$$U_E = I_g \cdot R_g = 376.2 \text{ V}$$

10) Calculation of the touch voltage
Critical touch voltage:

$$E_t = [1000 + 1.5 \cdot C_s \cdot (h_s \cdot k) \cdot \rho_s] \cdot \frac{k}{\sqrt{t_s}} = 160 \text{ V}$$

For $C_s = 1$ was used.
Calculated touch voltage:
Data:
Ladder spacing: $D = 6$ m, Diameter of the conductor: $d = 0.014$ m, Number of conductors on the long side: $n_a = 6$, Number of conductors on the shorter side: $n_b = 4$, K_m = (Formula 69 in IEEE Std. 80), $K_{1m} = 1.4986$ (formula 69 in IEEE Std. 80).

$$E_m = K_m \cdot K_{1m} \cdot \rho_E \cdot \frac{I_g}{L} = 114.9 \text{ V}$$

Calculated touch voltage is less than the permissible $E_m < E_t$.

11) Calculation of the step voltage
Data:
Earth surface resistance: $\rho_s = 11.93\,\Omega$m, with error duration $t_s = 1$ second, $K_s \Rightarrow$ (formula 74 in IEEE Std. 80), Correction factor for surface resistance $C_s = 1.0$.
Step voltage limit E_{step}:

$$E_{step} = [1000 + 6 \cdot C_s \cdot (h_s \cdot k) \cdot \cdot \rho_s] \cdot \frac{k}{\sqrt{t_s}} = 168.2 \text{ V}$$

Calculated step voltage:
Data:
Conductor spacing $D = 6$ m, minimum conductor diameter $d = 0.014$ m, number of parallel conductors $n_a = 6$.

$$E_s = K_s \cdot K_{1s} \cdot \rho_E \cdot \frac{I_g}{L} = 68.6 \text{ V}$$

Calculated step voltage is smaller than the limit value $E_s < E_{step}$.

11.19 Example: Comparison of IEEE Std. 80 and EN 50522

A transformer station is installed on a floor space of 16 m × 26 m. The specific earth resistance was measured at 45 Ωm. Calculate the earth resistance of the earthing system and compare both results with EN 50522 and IEEE Std. 80.

The following data were used as a basis.

- $\rho = 45\ \Omega\text{m}$
- $h = 0.5$ m, is the depth of the grid in m
- $A = 416\ \text{m}^2$

1) Calculation of the total earthing resistance of the earthing system according to IEEE Std. 80
   We consider Eq. (57) in IEEE Std.:80-2013 IEEE Guide for Safety in AC Substation Grounding

$$R_g = \rho \cdot \left[\frac{1}{L_T} + \frac{1}{\sqrt{20 \cdot A}} \cdot \left(1 + \frac{1}{1 + h \cdot \sqrt{\frac{20}{A}}} \right) \right] \tag{11.56}$$

$$= 45\ \Omega\text{m} \cdot \left[\frac{1}{482\ \text{m}} + \frac{1}{\sqrt{20 \cdot 416\ \text{m}^2}} \cdot \left(1 + \frac{1}{1 + 0.5 \cdot \sqrt{\frac{20}{416\ \text{m}^2}}} \right) \right] = 1.03\ \Omega$$

2) Calculation of the total earthing resistance of the earthing system according to EN 50522
   The resistance to earth R_E of an earth electrode depends on the soil resistivity as well as on the dimensions and the arrangement of the earth electrode. It depends mainly on the length of the earth electrode, and less on the cross section.
   The resistance to earth of a meshed earth electrode is approximately

$$R_{EM} = \frac{\rho_E}{2 \cdot \sqrt{D}} \tag{11.57}$$

$$D = \sqrt{\frac{4 \cdot a \cdot b}{\pi}} = \sqrt{\frac{4 \cdot 26\ \text{m} \cdot 16\ \text{m}}{\pi}} = 23.01\ \text{m} \tag{11.58}$$

$$R_{EM} = \frac{45\ \Omega\text{m}}{2 \cdot 23.01\,\text{m}} = 0.977\ \Omega \tag{11.59}$$

Calculate earth voltage when the single-pole earth short-circuit is 350 A.

$$U_E \le 2 \cdot U_{Tp} \tag{11.60}$$

The condition is

$$I_E \cdot R_E \le 2 \cdot U_{Tp} \tag{11.61}$$

According to EN 50522 Table B.3, we take the fault duration as 0.5 second and $U_{Tp} = 220\,V$; then

$$350\ A \cdot 0.977\ \Omega \leq 2 \cdot 220\ V$$

$$341.95\ V \leq 440\ V$$

If the EPR is below the permissible touch voltage and the requirements of $U_E \leq 2 \cdot U_{Tp}$ are met, the design is complete. Otherwise, additional design evaluation is necessary (see Annex E in EN 50522, conditions for the use of recognized specified measures M to ensure permissible touch voltages).

11.20 Example of Earthing Drawings and Star Point Treatment of Transformers

In Figure 11.24–11.30, some example of earthing drawings and star point treatment of transformers are shown and calculated with Neplan.

1) *Earthing of a transformer station*: In this example, a transformer station is shown. High and low voltage distribution panels are connected to the earthing.

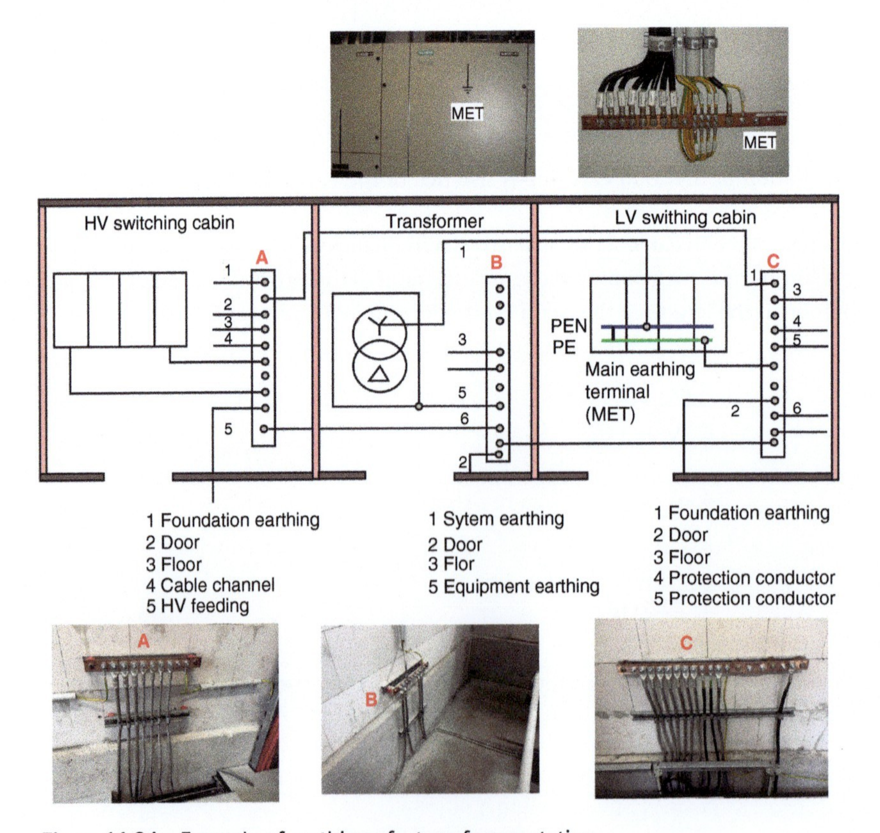

Figure 11.24 Example of earthing of a transformer station.

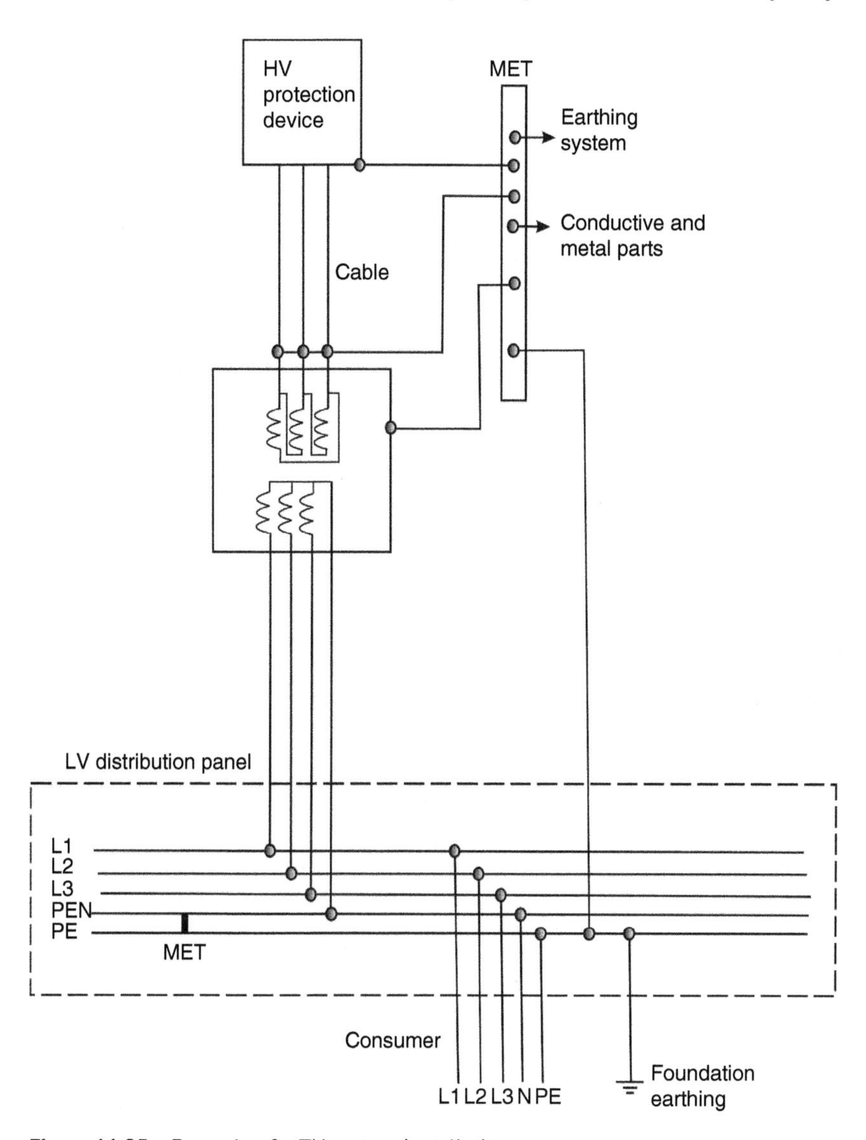

Figure 11.25 Example of a TN system installation.

2) *TN system*: In this example a TN system is shown. A direct connection to earth from either the transformer star point or the generator star point is not allowed. The conductor from either the transformer star points or the generator star points to the PEN busbar in the Low voltage main distribution must be laid isolated. The function of this conductor is similar to that of a PEN conductor, but the conductor must not be connected to electrical equipment. The connection between the interconnected points of the power sources and the PE may only be take place once. This connection must be made in the low voltage main distribution board. Additional earthing of the PE in the plant may be provided.

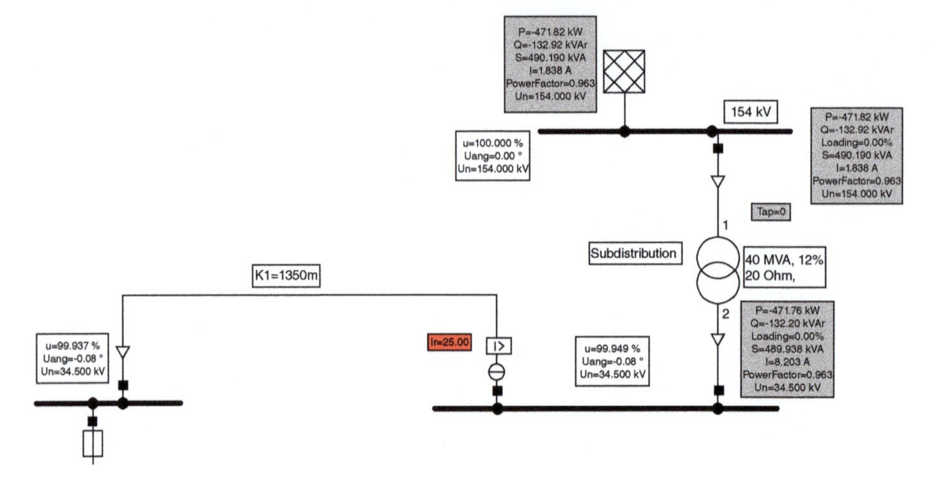

Figure 11.26 Example of a sub-distribution.

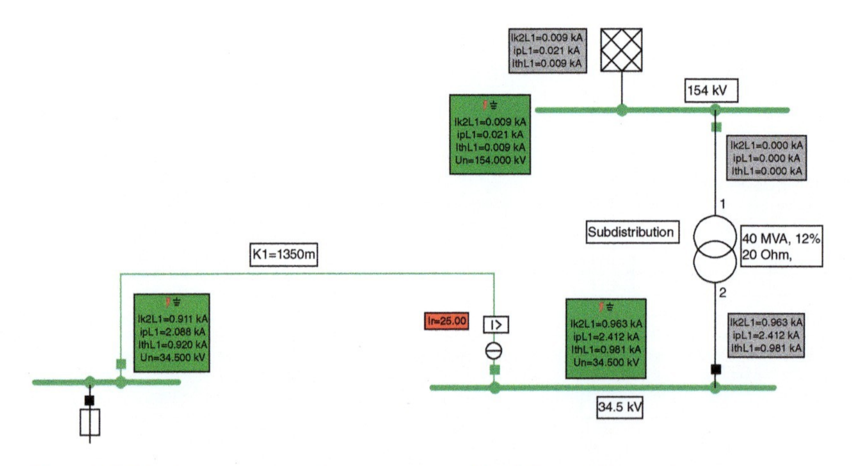

Figure 11.27 Example of a sub-distribution with 20 Ω earthing resistance.

3) *Load calculation*: At each node it is possible to calculate the voltage profile and power flowing into the system.
4) *One-phase short-circuit*: In this example, one-phase short-circuit is calculated by installing a 20 Ω earthing resistance to the star point of the transformer.
   The single-pole short-circuit is important for testing and dimensioning the protective measures "protection by disconnection" of the final circuits. The protective devices must be set to this current. On the other hand, personal protection is of great importance. If the star point is earthed via a resistor a reduced current flows to earth, which is 963 A. For safety reasons, this current for the setup must be 20% less than the calculated current.
5) *Three-phase short-circuit*: In this example, three-phase short-circuit is calculated. The three-pole short-circuit is important for testing and dimensioning the

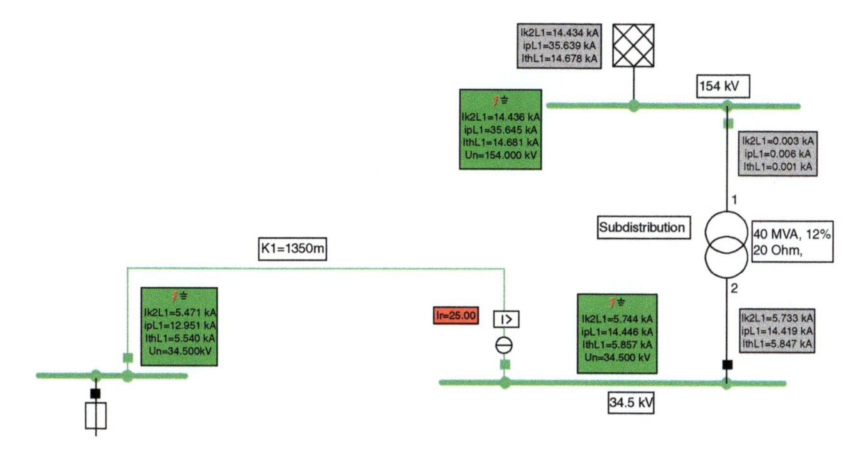

Figure 11.28 Example of an one-phase short-circuit directly earthed.

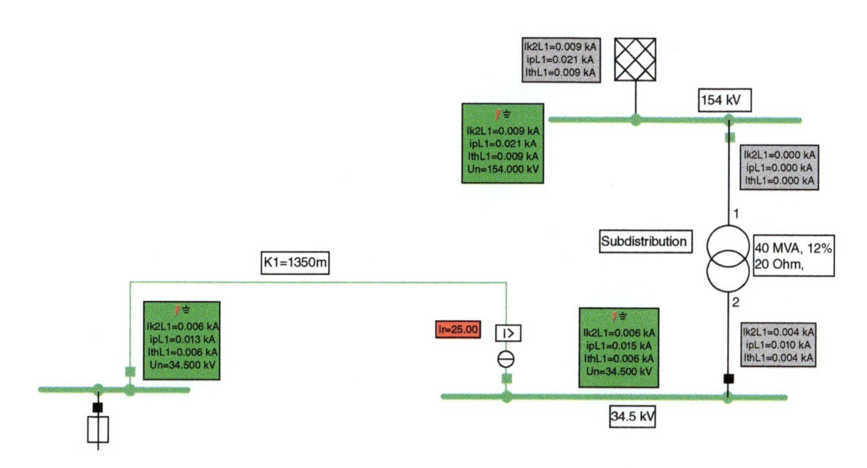

Figure 11.29 Example of a sub-distribution with neutral point isolation.

dynamic short-circuit strength of the electrical power system. The three-phase short-circuit current in this example is 5.744 kA.

6) *Neutral point isolated*: In this example, the star point of the transformer is isolated and earth fault is calculated. Insulated networks are used in medium-voltage systems and in low-voltage systems, for example, in hospitals. The power supply is not interrupted at the first fault. As you can see, the residual current is very small, because the star point is not earthed and almost no current flows.
7) *Protection of the system*: In this example, the Figure shows the time–current diagrams. Electrical installations have to be protected against over-currents. Many protective devices are available. The most important protective devices are used here as examples.

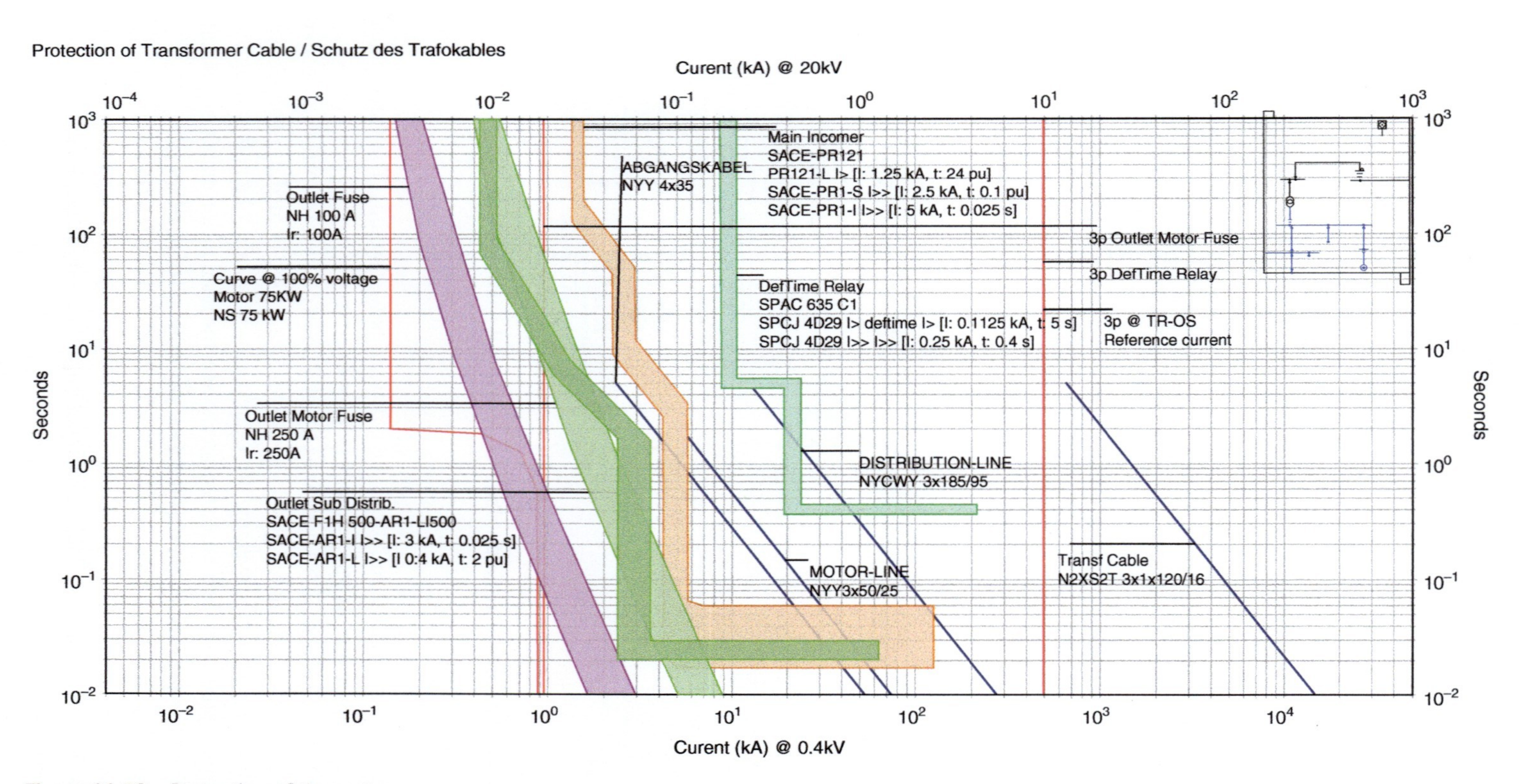

Figure 11.30 Protection of the system.

11.21 Software for Earthing Calculation

This chapter will introduce the software program XGSLab. This program will be used for the calculation of a substation system. XGSLab is one of the most powerful software of electromagnetic simulation for power, grounding, and lightning protection systems and takes into account International (IEC/TS 60479-1:2018), European (EN 50522:2010), and American (IEEE Std. 80-2013) Standards in grounding system analysis.

XGS includes the following modules:

- XGSLab (Grounding System Analysis) for basic application with underground systems
- GSA-FD (Grounding System Analysis in the Frequency Domain) for general applications with underground systems
- XGSA-FD (Over and Underground System Analysis in the Frequency Domain) for general applications with overhead and underground systems
- XGSA-TD (Over and Underground System Analysis in the Time Domain) for general applications with overhead and underground systems
- NETS (Network Solver) solver for multi-conductor and multi-phase full meshed networks

The modules GSA, GSA-FD, XGSA-FD, and XGSA-TD integrate the module SRA (soil resistivity analysis) and SA (seasonal analysis) and XGSA-TD integrates also the module FA (direct/inverse Fourier analysis). The application field of these four modules is wide because they are based on the PEEC (partial element equivalent circuit) method, a numerical method for general applications, powerful and flexible and perfectly suitable for engineering purposes. This method allows the analysis of complex scenarios including external parameters such as voltages, currents, and impedances. The implemented PEEC method solves the Maxwell equations in full wave conditions taking into account the Green functions for propagation, the Sommerfeld integrals for the earth reaction, the Jefimenko equations for electric and magnetic fields, and moving from the frequency to the time domain by means of the Fourier transforms. The module NETS is based on the phase components method and integrates specific routines for the calculation of the parameters of lines, cables, and transformers. All modules are integrated in an "all-in-one" package and provide professional numerical and graphical output useful to investigate any electromagnetic greatness.

11.21.1 Numerical Methods for Grounding System Analysis

The calculation of ground resistances and impedances, ground potential rise (GPR), and safety voltages can be approached with both analytical and numerical methods. The analytical methods in conditions of practical interest can be used under assumptions and simplifications as, for instance, low frequency, uniform soil model, and simple or symmetric layout, which are not always realistic but nevertheless, these methods are still used and are useful for preliminary evaluations.

The numerical methods can overcome limits and constraints of the analytical methods, allow realistic simulations, and are irreplaceable in order to predict performances before construction and for finding optimized solutions.

It is important to highlight that problems involved in electromagnetic simulations of grounding systems are generally complex because most approximations and simplifications used in other fields are often not applicable. This is mainly for the following reasons:

- The grounding system layout can be complex and wide
- The conductors can be either bare or insulated
- The soil is usually heterogeneous
- The soil parameters depend on frequency and climatic conditions

Moreover, in applications at high frequency, the propagation effects cannot be neglected; propagation must be described with Green functions and the earth reaction calculated with Sommerfeld integrals. The application of numerical techniques can consider all these aspects. The most used numerical techniques for grounding simulations are essentially the method of moments (MoM) and the most recent PEEC. The description of these techniques is out of the scope of this book.

Both methods MoM and PEEC are full-wave and then applicable in general conditions. At low frequency (DC and power frequency) the propagation effects can be neglected and quasi-static or static approximations can be used . Moreover, if the system size is small when compared to the wavelength of the electromagnetic field, the equipotential assumption can be adopted.

All above approximations can greatly simplify the numerical model but can be used only when applicable. For instance, the effects of the equipotential assumption can be represented with Figures 11.31 and 11.32 related to a square grid with side 200 m in a uniform soil model. The injection point is in a grid corner and the frequency is 50 Hz.

As evident, the earth surface potential distribution using the equipotential assumption or a rigorous model is completely different and as a consequence, the distribution of touch and step voltages will be different.

The equipotential assumption tends to underestimate the GPR and the impedance to earth. Figure 11.33 summarizes the application limits for models based on the equipotential assumption. The application limits are essentially related to the wavelength. The maximum grid size D (for instance, the diagonal for a rectangular grid) should be lower than 1/10 of the wavelength. The wavelength of an electromagnetic field λ with frequency f in a uniform soil model with resistivity ρ_E can be calculated with the following approximated formula:

$$\lambda \cong 3162 \cdot \sqrt{\frac{\rho_E}{f}} \tag{11.62}$$

The following example is related to the grounding system analysis of a quite small substation. This substation is related to a distribution power system (132/20 kV, for instance). The calculation has been performed using the module GSA of the computer program XGSLab. The example has been performed using the IEEE Std. but

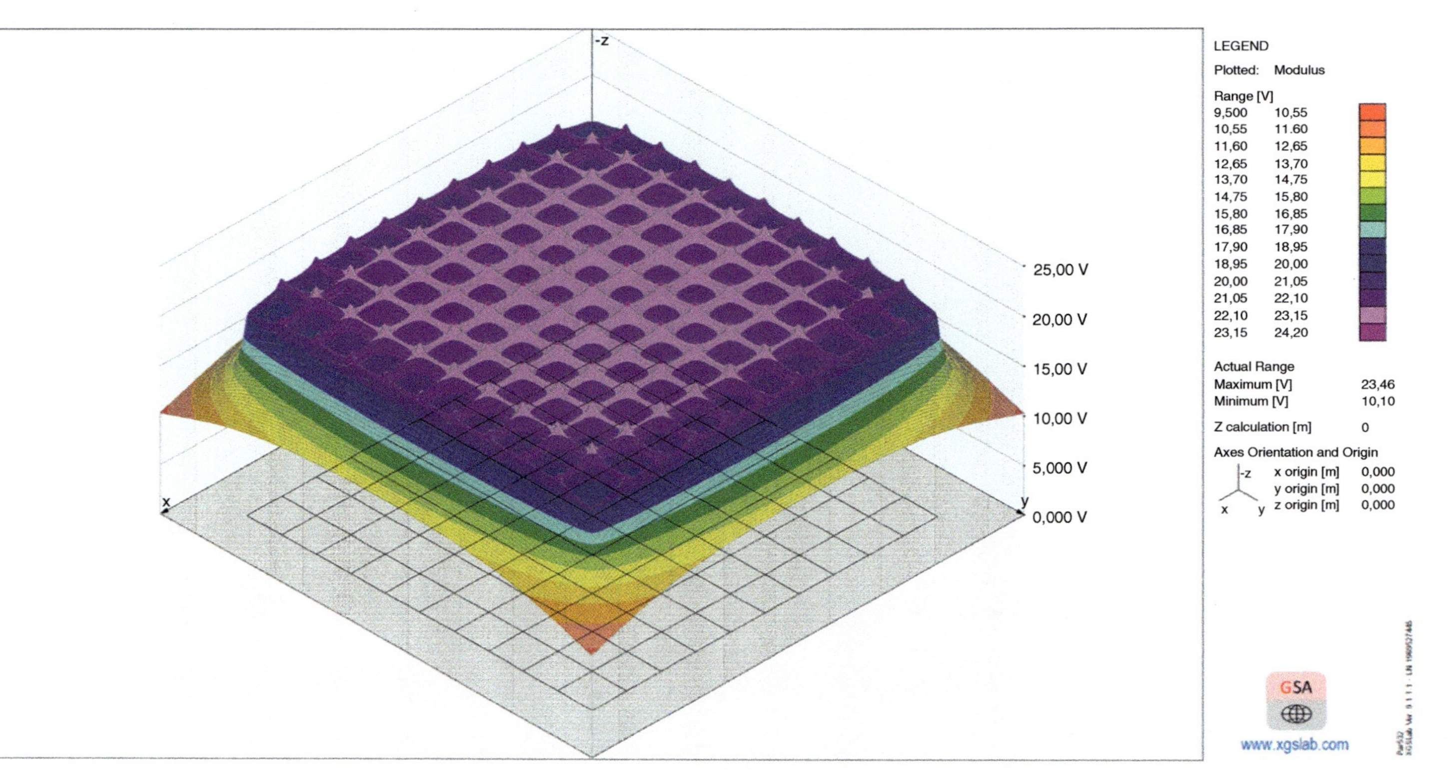

Figure 11.31 Earth surface potential distribution with static model and equipotential assumption (color range 9.5–24.2 V). Source: SINT Srl.

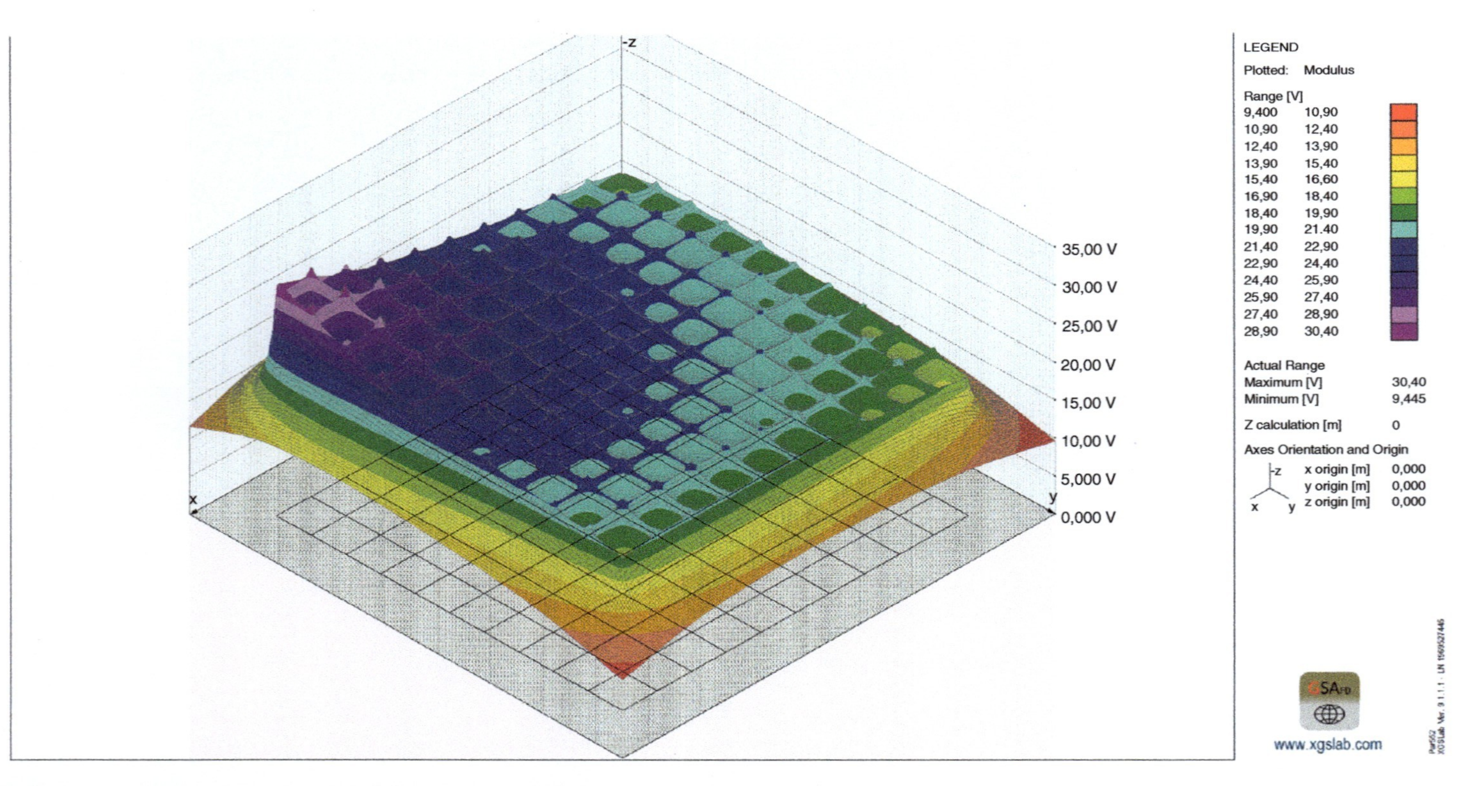

Figure 11.32 Earth surface potential distribution with rigorous model (color range 9.4–30.4 V). Source: SINT Srl.

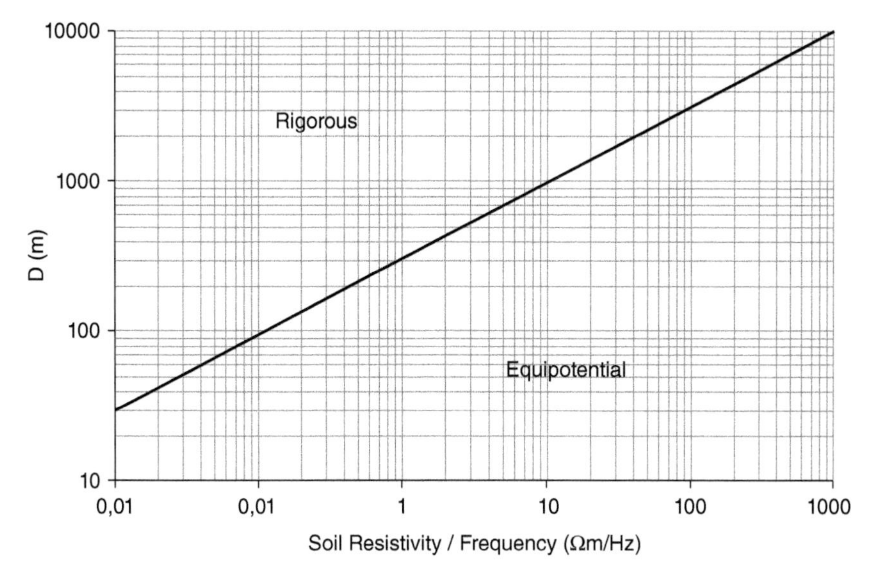

Figure 11.33 Application limits for model based on equipotential assumption.

safety conditions are evaluated using both IEEE and EN Std. This interesting comparison is anyway related to a specific condition and is not general.

11.21.2 IEEE Std. 80 and EN 50522

1) *Input data*: The following data is supposed to be available:
   - Current to earth: $I_e = 15\,\text{kA}$
   - Clearance time: $t_f = 0.5\,\text{second}$
   - Soil resistivity measurements: (Figure 11.34)

Figure 11.34 Measured apparent soil resistivity and resistance as a function of the electrodes spacing.

a (m)	ρ_E (Ωm)	R_W (Ω)
1.000	80.20	12.76
1.500	58.60	6.218
2.000	41.00	3.263
3.000	35.10	1.862
4.500	36.20	1.280
6.000	36.60	0.9708
9.000	44.10	0.7799
13.50	54.30	0.6402
18.00	66.70	0.5898
27.00	79.70	0.4698
36.00	95.00	0.4200
54.00	101.8	0.3000

- System layout: Figure 11.36.
- Conductors: cross section 95 mm^2, outer diameter 12.5 mm

The evaluation of the reduction factor, current to earth, and conductor sizing can be done according to IEEE standard as described above. The same values will be used with the EN standard.

2) *Soil model*: Suppose that the following on-site soil resistivity measured values are available. Measures have been performed with the Wenner method. Measured apparent soil resistivity and resistance are shown as a function of the electrodes spacing (Figure 11.34).

   The multilayer soil model parameters related to previous measurements can be calculated using the soil resistivity analyzer SRA integrated in XGSLab. The module SRA is based on a powerful minimization algorithm called "Trust Region Method." The number of layers can be chosen manually or automatically set by the program. In the specific case, a three-layer soil model allows a good approximation of measured values.

   The three-layers soil model parameters calculated by SRA are as follows:
   - Soil resistivity of the upper layer = 133.0 Ωm
   - Soil resistivity of the central layer = 29.62 Ωm
   - Soil resistivity of the bottom layer = 128.9 Ωm
   - Upper layer thickness = 0.6804 m
   - Central layer thickness = 6.359 m

   The average soil resistivity is 60.77 Ωm (Figure 11.35).

3) *Touch and step permissible step voltages*: Suppose the reference body weight is 70 kg; the touch and step permissible voltages are
   - $U_{STP} = 268.2$ V
   - $U_{SSP} = 406.6$ V

   The IEEE Std. 80-2013 considers the body weight 50 or 70 kg. The value 70 kg is usually used in places not accessible to the public and protected by fences or walls, as for instance, electrical substations.

   The touch and step permissible voltages with body weight 50 kg are
   - $U_{STP} = 198.1$ V
   - $U_{SSP} = 300.4$ V

   The touch permissible voltages with the EN 50522:2010 standard is $U_{TP} = 220.0$ V. The permissible step voltage with the European standard is very high and for this reason usually not considered in safety evaluations.

4) *Layout*: Suppose a grounding system layout as in Figure 11.36.

   With reference to the average soil resistivity and the industrial frequency 60 and 50 Hz, the wavelength of the electromagnetic field is

   $\lambda = 3182$ m with 60 Hz and 3486 m with 50 Hz.

   The equipotential assumption can be used if the maximum system size is lower than (worst case)

$$D = \frac{\lambda}{10} = 318.2\,\text{m} \tag{11.63}$$

   The maximum system size of the substation is about 70 m, well below this value. This justifies the use of a module based on a simplify model such as GSA.

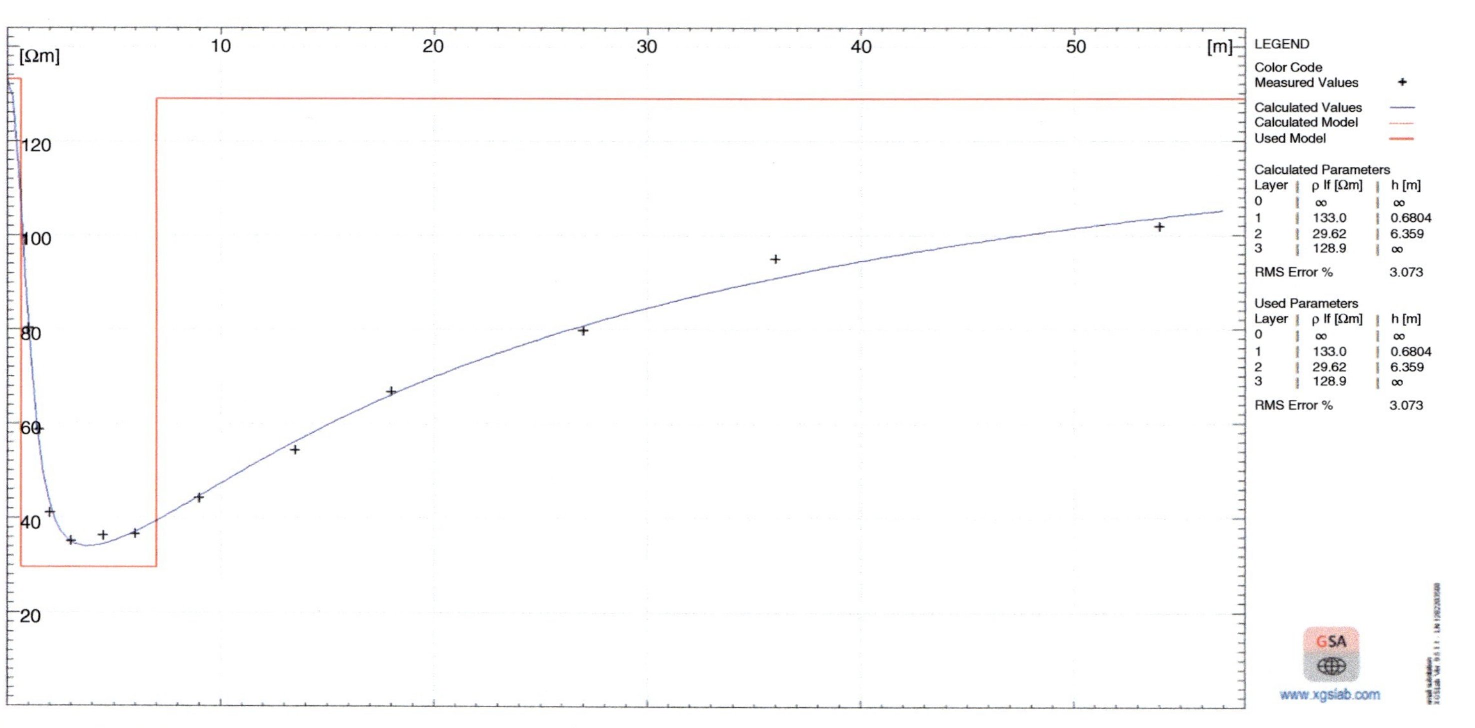

Figure 11.35 Soil resistivity measured values and four-layers soil model. Source: SINT Srl.

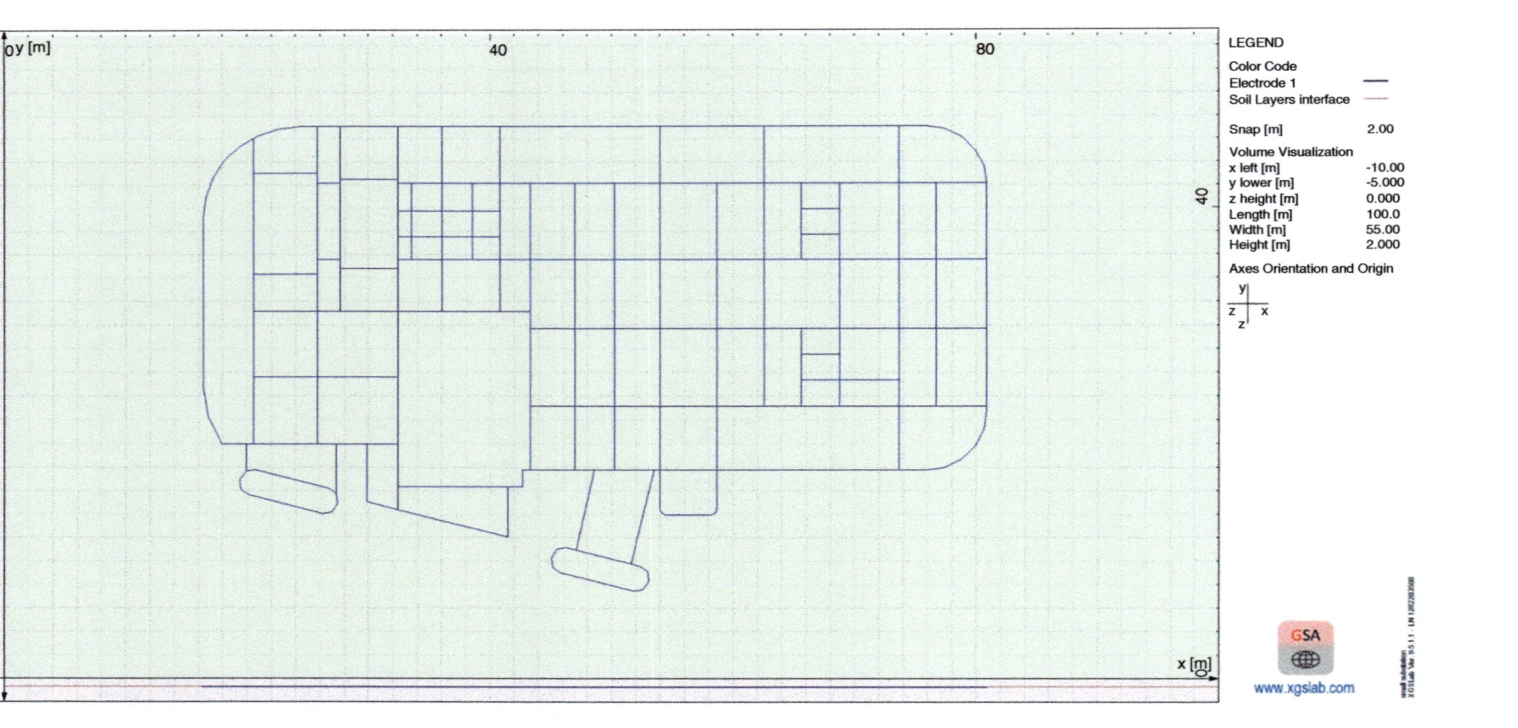

Figure 11.36 Grounding system layout and injection point of the current to earth. Source: SINT SrL.

5) GPR and earthing impedance
   The injection point of the current to earth is irrelevant because of the equipotential assumption. The corresponding values of GPR and resistance to earth are the following:

$$\mathrm{GPR} = 10\,908\ \mathrm{V}$$

   and

$$R_\mathrm{E} = 0.7272\ \Omega$$

   The impedance to earth in this case is a resistance, as expected in small grids at low frequency.
6) *Current and potential distribution on conductors*: The following distributions are related to the modules of leakage currents and potentials on conductors (Figures 11.37 and 11.38).
   The leakage current distribution indicates that the most effective parts of the system are the peripheral conductors, as expected at low frequency.
7) *Potential and touch and step voltages distribution*: The following distributions are related to the modules of earth surface potential, touch voltage, and step voltages on an area including the whole system (Figures 11.39–11.41).
   Touch and step voltages can be compared to the permissible values. The results can be summarized as in Figure 11.42.
   This figure indicates that both step and touch voltages overcome the permissible values diffusely.
   The use of a high resistivity soil covering layer (SCL) usually is effective in reducing touch and step voltages. For instance, a gravel layer with the following data could be used:
   Resistivity: 5000 Ωm, Thickness: 100 mm
   Taking into account this additional SCL, the new touch and step permissible voltages calculated using IEEE Std. 80-2013 Eqs. (29), (30), (32), (33) are

$$U_{\mathrm{STP+SCL}} = 1433\ \mathrm{V}, U_{\mathrm{SSP+SCL}} = 5065\ \mathrm{V}$$

   Again, touch and step voltages with SCL can be compared to the new permissible values. The results can be summarized as in the following figure.
   Safe areas are extended throughout the substation. It is interesting to compare this result with similar results using IEEE Std. 80-2013 body weight 50 kg and EN 50522:2010.
   Further, safe areas distribution with additional soil covering layer are shown according-IEEE Std 80-2013 body weight 70 kg and 50 kg and EN 50522:2010 in Figures 11.43–11.45.
   As expected, considering a body weight of 50 kg safe areas are smaller than areas related to a body weight of 70 kg. The safe areas using the European standard are a compromise between IEEE Std. 80-2013 body weight 50 and 70 kg.

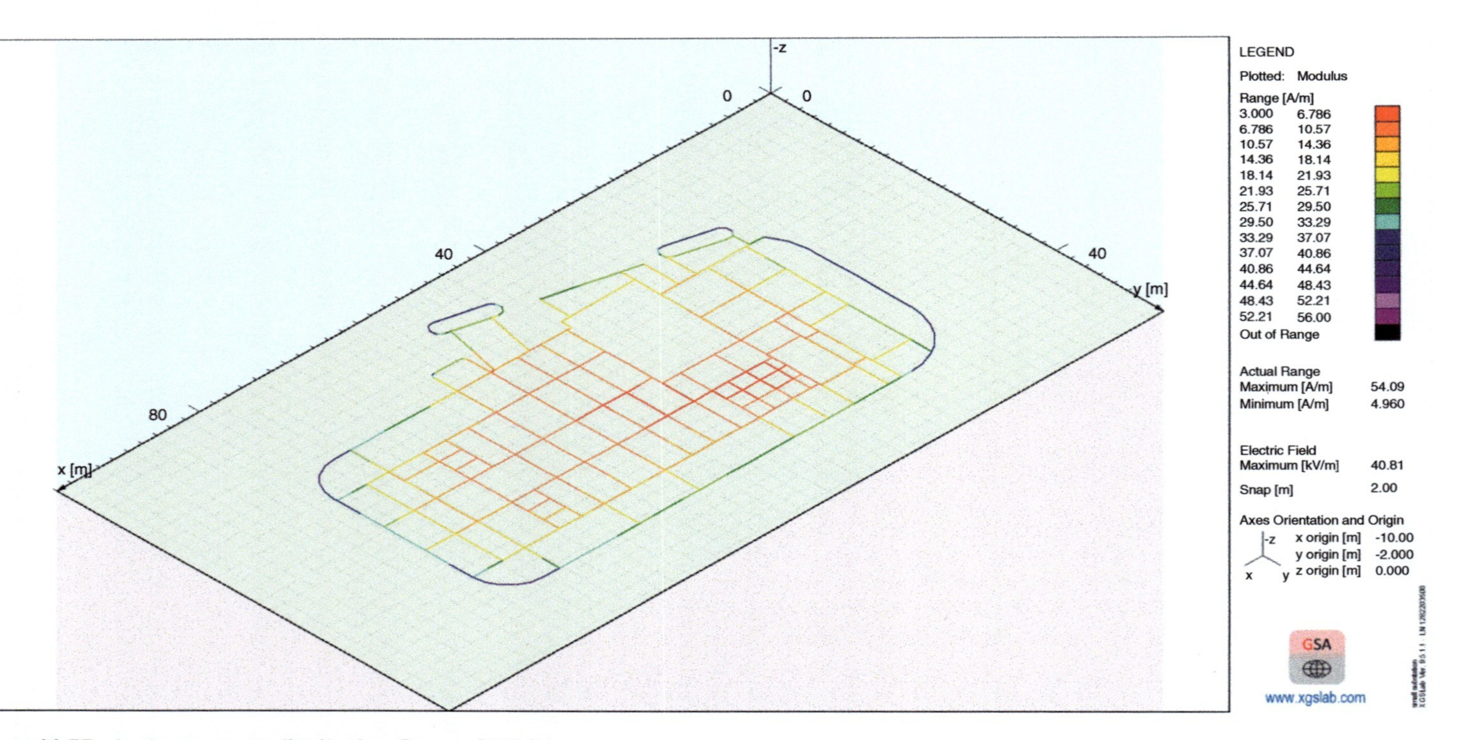

Figure 11.37 Leakage current distribution. Source: SINT Srl.

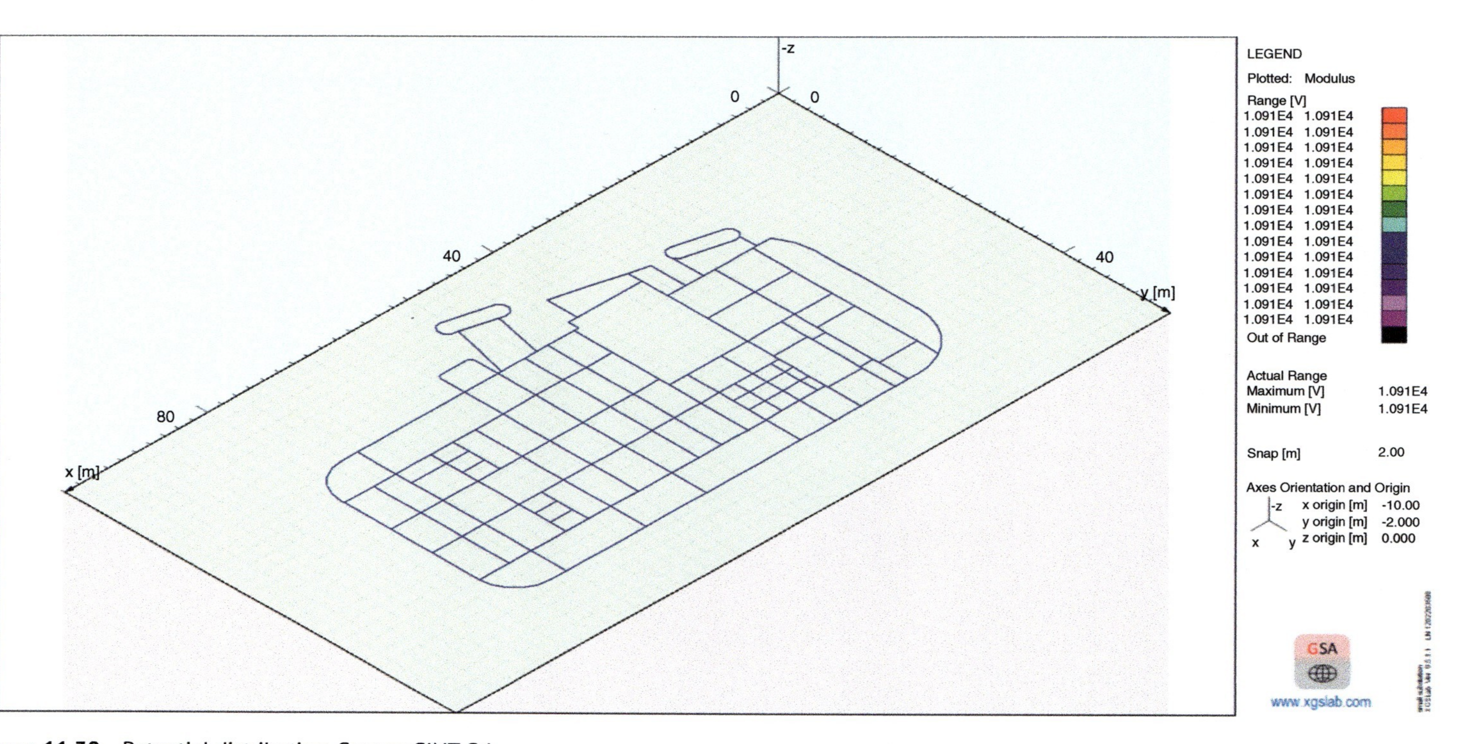

Figure 11.38 Potential distribution. Source: SINT Srl.

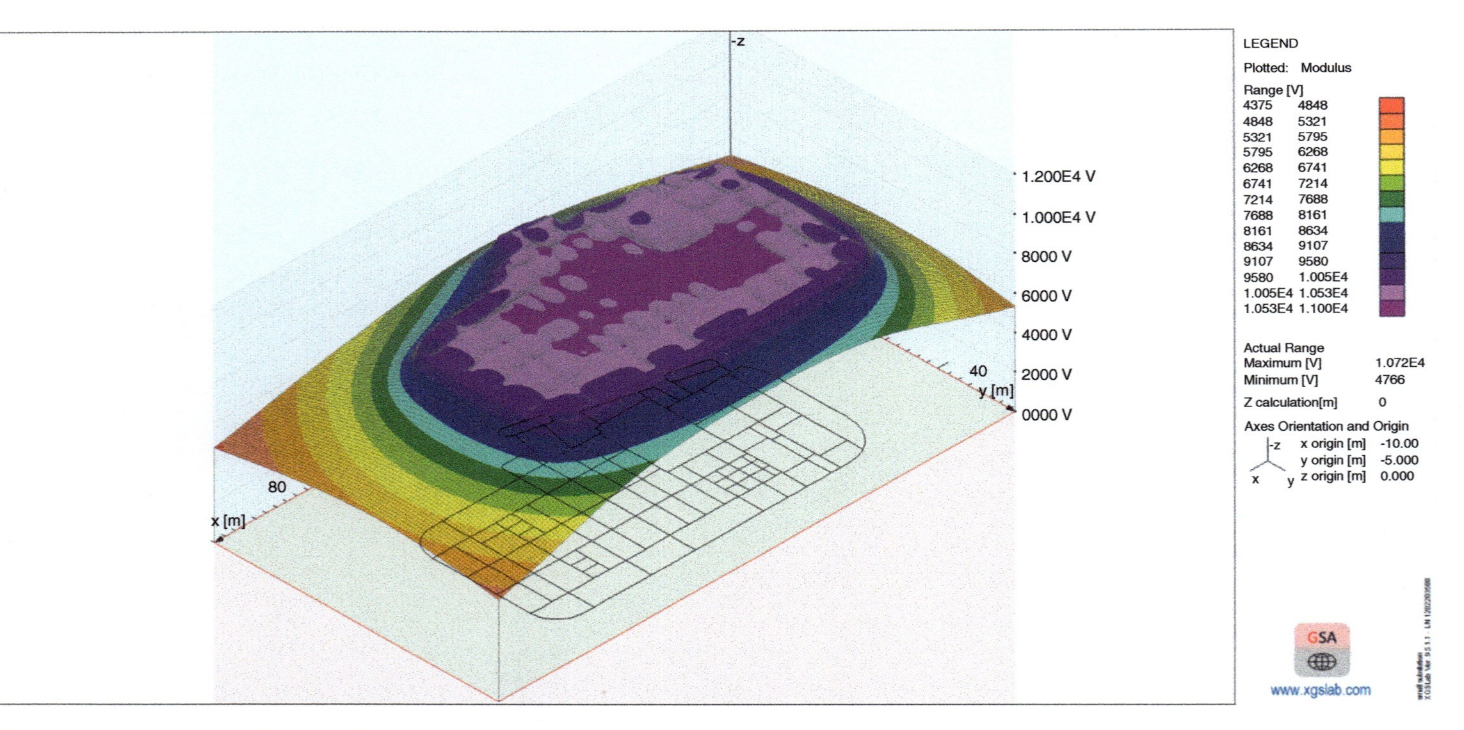

Figure 11.39 Earth surface potential distribution. Source: SINT Srl.

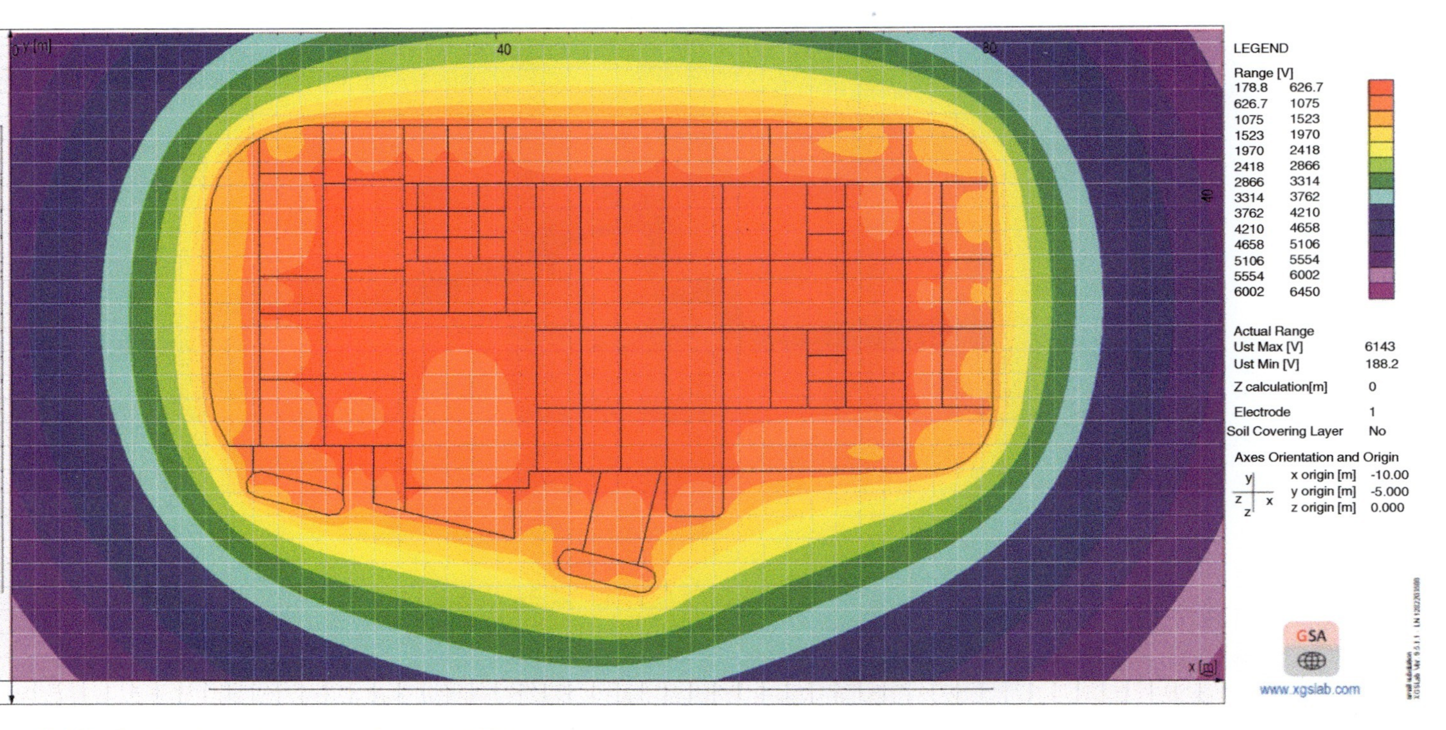

Figure 11.40 Touch voltage distribution. Source: SINT Srl.

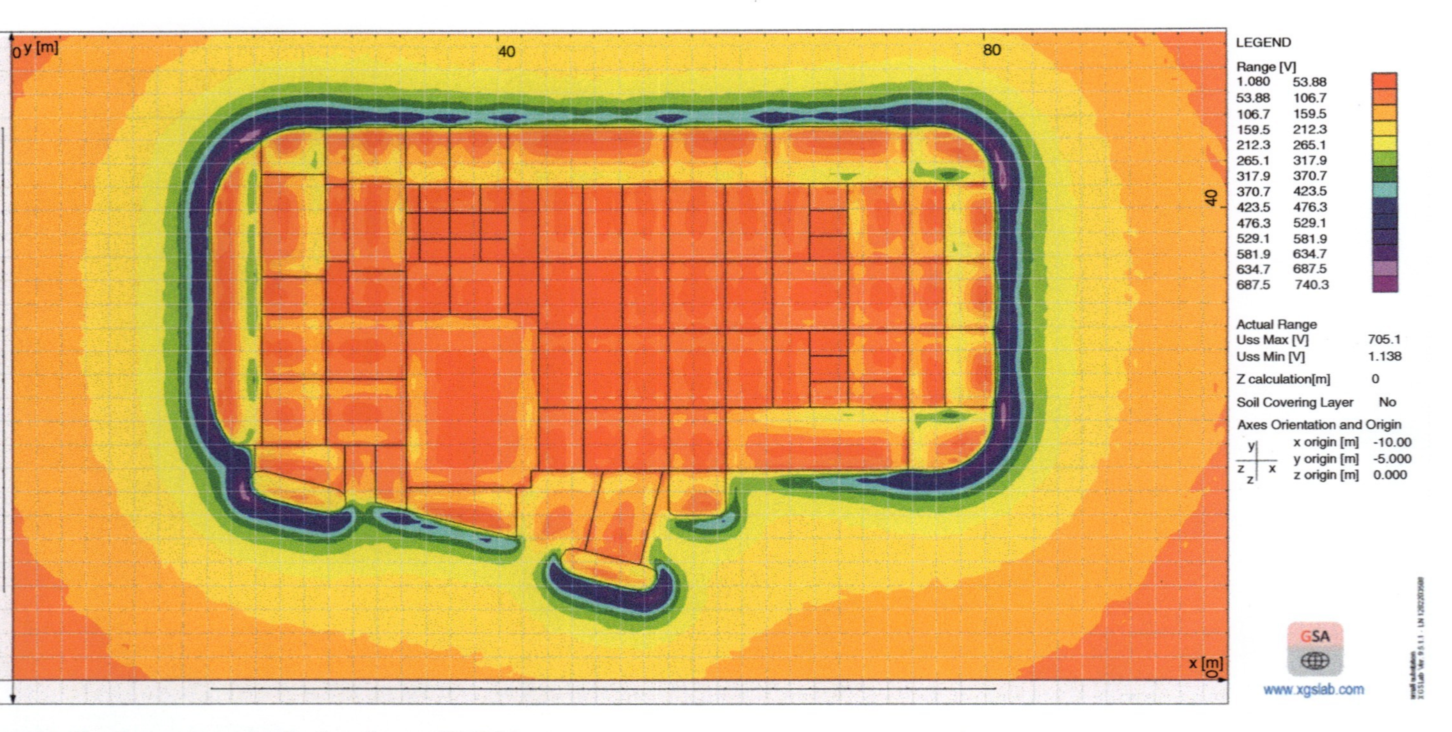

Figure 11.41 Step voltage distribution. Source: SINT Srl.

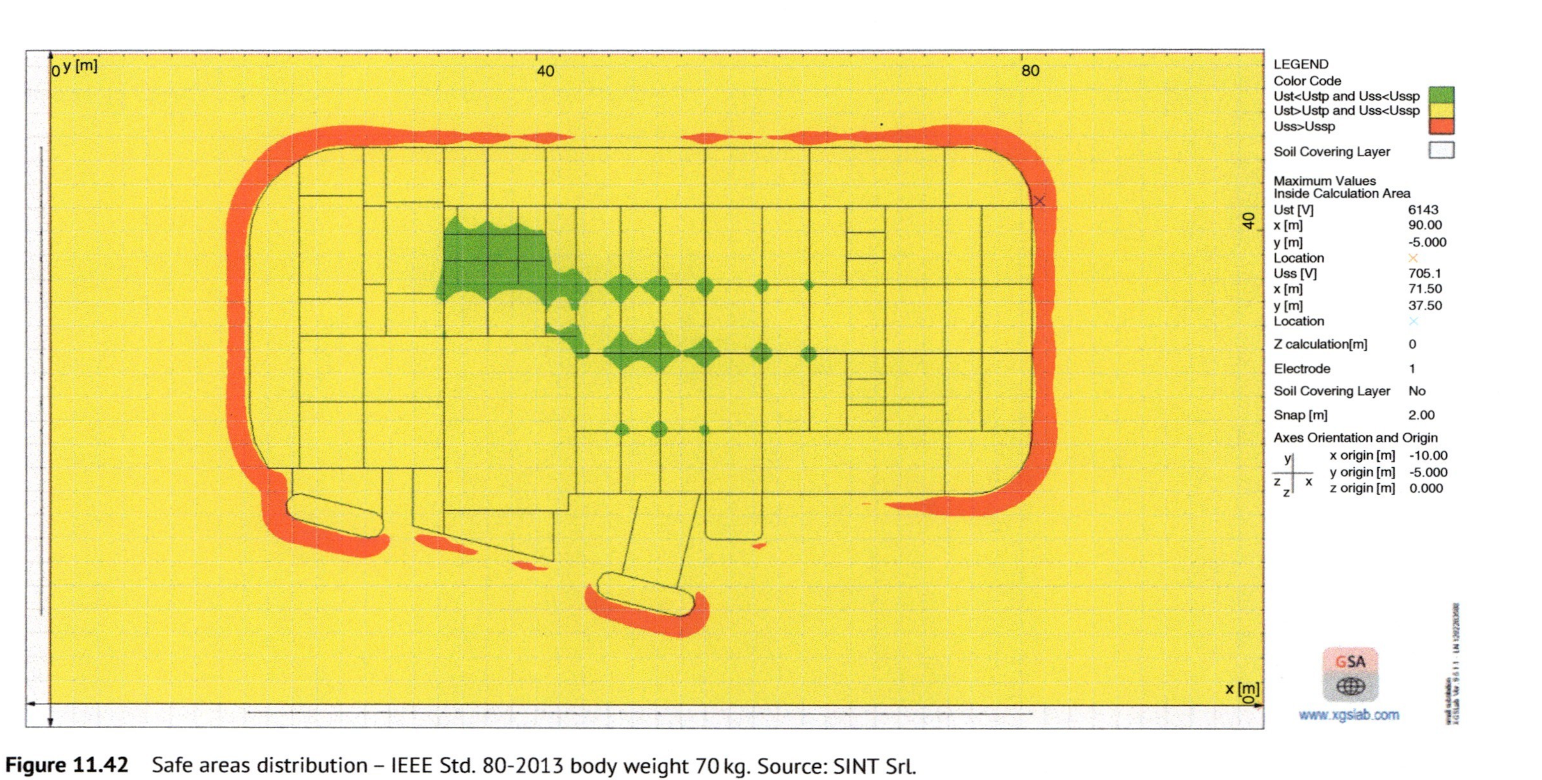

Figure 11.42 Safe areas distribution – IEEE Std. 80-2013 body weight 70 kg. Source: SINT Srl.

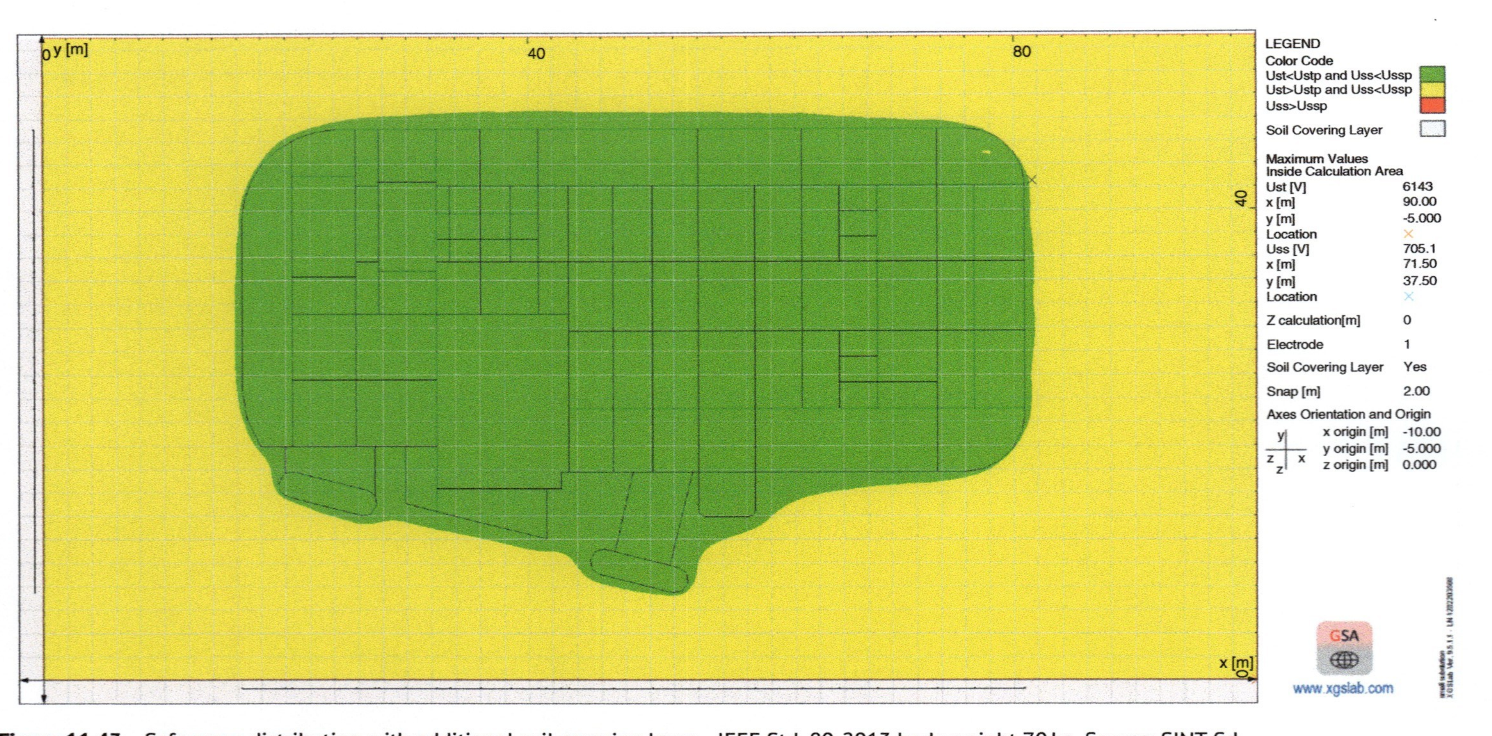

Figure 11.43 Safe areas distribution with additional soil covering layer – IEEE Std. 80-2013 body weight 70 kg. Source: SINT Srl.

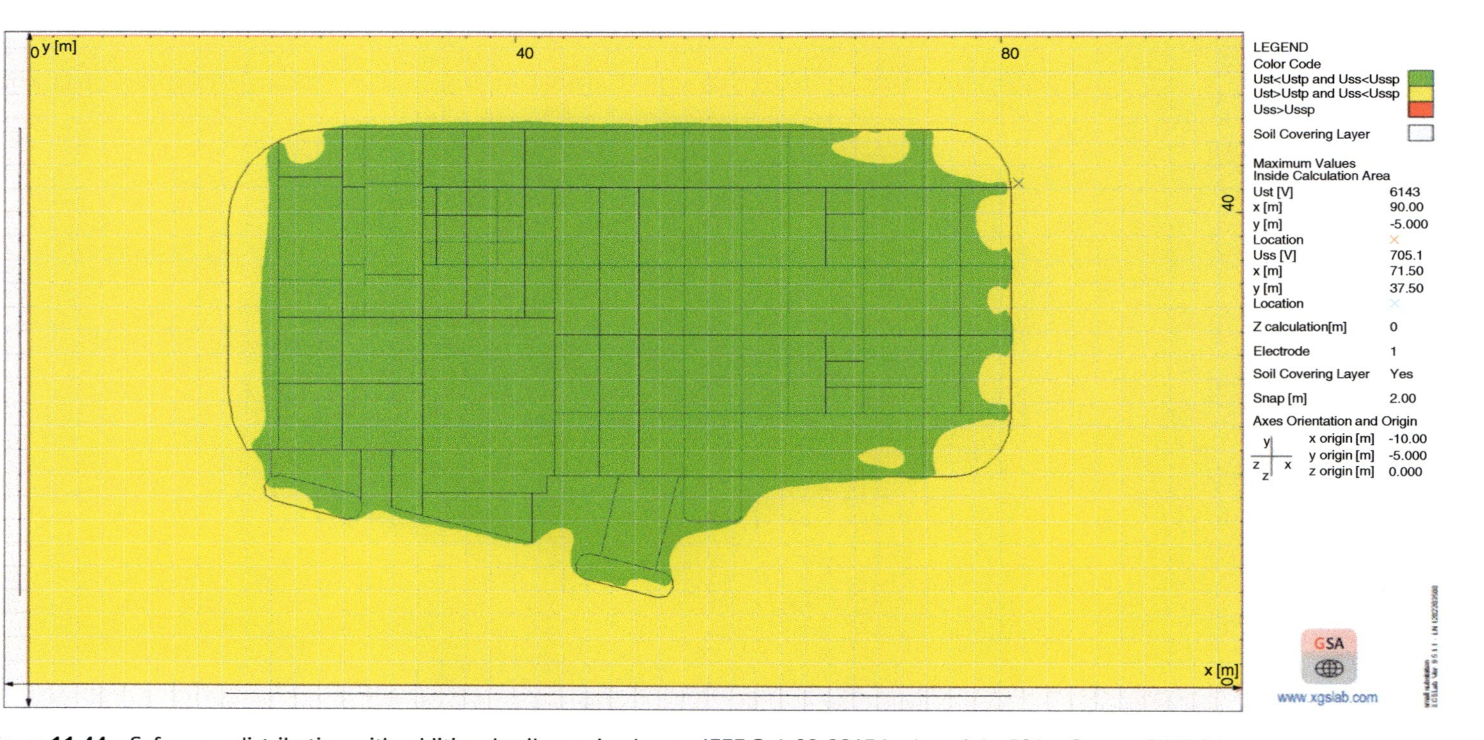

Figure 11.44 Safe areas distribution with additional soil covering layer – IEEE Std. 80-2013 body weight 50 kg. Source: SINT Srl.

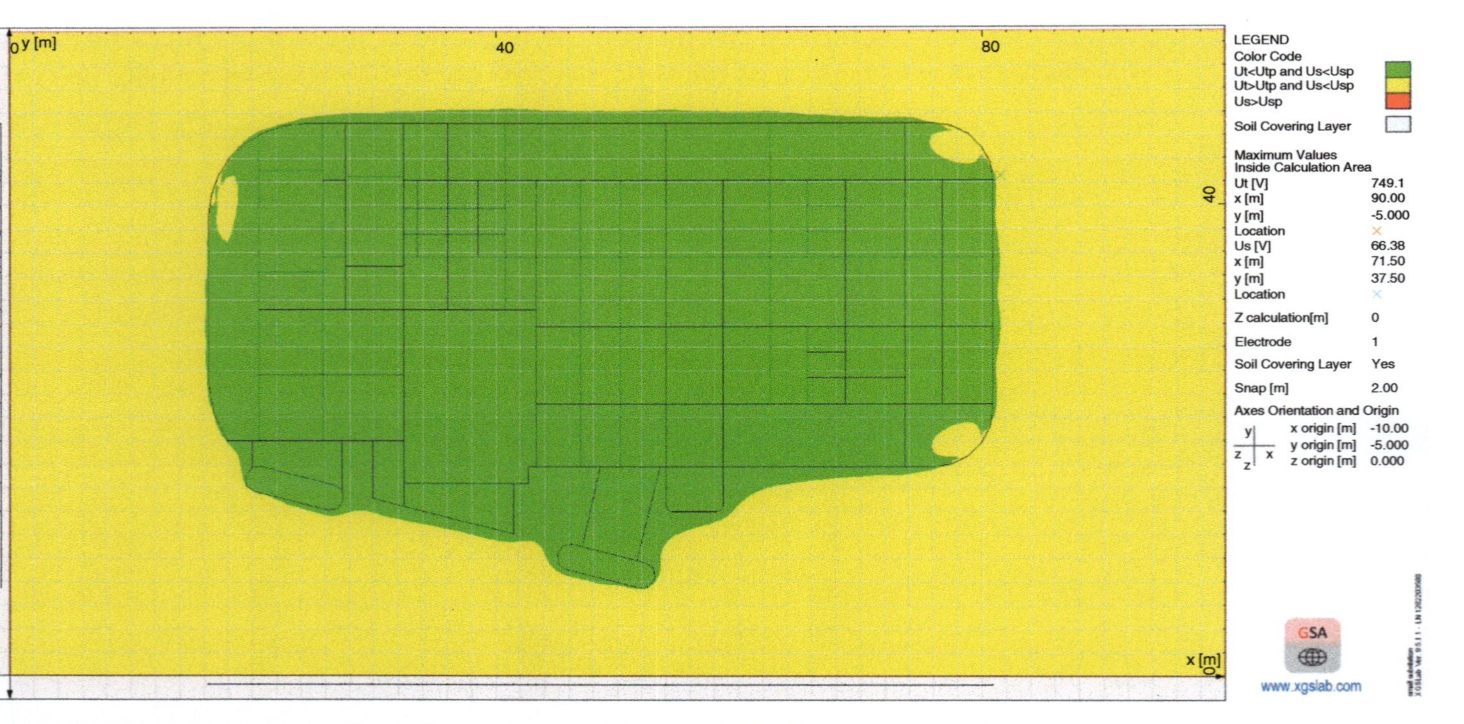

Figure 11.45 Safe areas distribution with additional soil covering layer – EN 50522:2010. Source: SINT Srl.

11.21.3 Summary

Earthing systems serve to protect people, animals, and property in the event of short-circuits, lightning, and switching operations. They must be dimensioned in such a way that the permissible step and touch voltages and certain earthing resistances are not exceeded. IEC 60364-5-54 must be observed in low-voltage systems and IEEE Std. 80 or EN 50522 in high-voltage systems. Apart from lightning protection systems ($< 10\ \Omega$), no value for the permissible earthing resistances is specified in the standard. Nevertheless, earthing resistance below $< 2\ \Omega$ is persistently tried to be maintained. In high voltage, however, a value of ($< 1\ \Omega$) is sought.

12

Protection Against Electric Shock

IEC 60479 describes the effects on the human body when a sinusoidal alternating current in the frequency range above 100 Hz passes through it. In this standard, the effects of current passing through the human body for

- alternating sinusoidal current with DC components,
- alternating sinusoidal current with phase control, and
- alternating sinusoidal current with multicycle control

are given but are only deemed applicable for alternating current frequencies from 15 to 100 Hz. Electric energy in the form of alternating current at frequencies higher than 50/60 Hz is increasingly used in modern electrical equipment. Attention is given to the fact that the impedance of human skin decreases approximately in inverse proportion to the frequency for touch voltages in the order of some tens of volts. IEC 60479-1 contains information about body impedance and body current thresholds for various physiological effects. This information can be combined to derive estimates of AC and DC touch voltage thresholds for certain body current pathways, contact moisture conditions, and skin contact areas. The characteristics of the impedance of the body of livestock and the effects of sinusoidal alternating currents are described in Annex. H of IEC 60479.

Figure 12.1 shows the conventional time/current zones of the effects of AC currents (15–100 Hz) on persons for a current path corresponding to the left hand to feet.

The time/current zones and the current boundaries for the different physiological effects of the pathway hand to feet are given in Table 12.1. For a given current path through the human body, the danger to persons depends mainly on the magnitude and duration of the current flow. However, the time/current zones specified in the following clauses are, in many cases, not directly applicable in practice for designing measures of protection against electrical shock. The necessary criterion is the admissible limit of touch voltage (i.e. the product of the current through the body called touch current and the body impedance) as a function of time. The relationship between current and voltage is not linear because the impedance of the human body varies with the touch voltage, and data on this relationship is therefore required. The different parts of the human body (such as the skin, blood, muscles, other tissues and

Analysis and Design of Electrical Power Systems: A Practical Guide and Commentary on NEC and IEC 60364, First Edition. Ismail Kasikci.

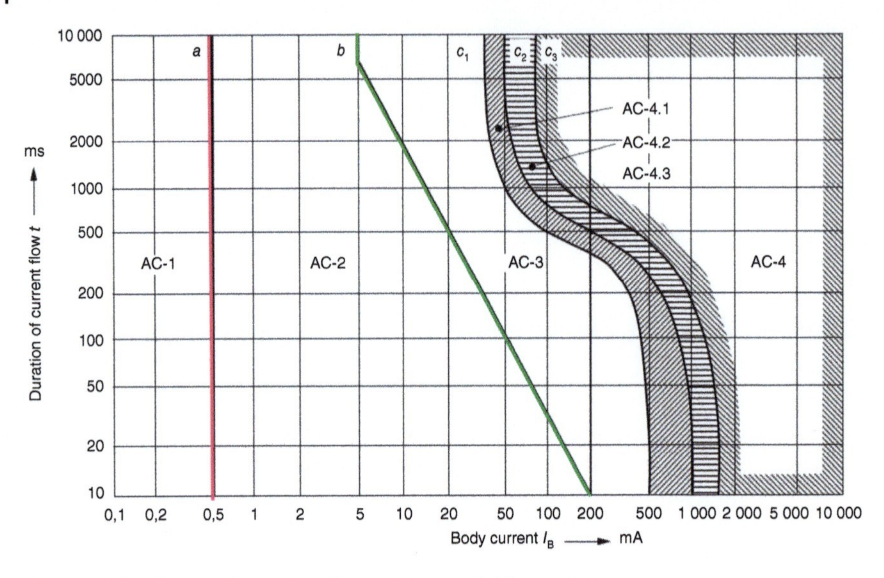

Figure 12.1 Conventional time/current zones of AC currents.

Table 12.1 Description of time/current zones.

AC-1	Up to 0.5 mA curve *a*	Perception possible but usually no "startled" reaction
AC-2	0.5 mA up to curve *b*	Perception and involuntary muscular contractions likely but usually no harmful electrical physiological effects
AC-3	Curve *b* and above	Strong involuntary muscular contractions. Difficulty in breathing. Reversible disturbances of heart function
AC-3	Curve *b* and above	Strong involuntary muscular contractions. Immobilization may occur. Effects increasing with current magnitude. Usually no organic damage to be expected
AC-4	Above curve *c*1	Patho-physiological effects may occur such as cardiac arrest, breathing arrest, and burns or other cellular damage. Probability of ventricular fibrillation increasing with current magnitude and time

joints) present to the electric current certain impedance composed of resistive and capacitive components.

The danger to humans from electrical currents depends on many circumstances. The most important factors of hazard are the following:

- The level of current through the human body
- Duration and effect of the current

- The physical condition, in particular the condition of the skin of the person entering the circuit
- Path of the current through the body
- The resistors in the circuit
- Type of current (direct, alternating, high-frequency)

Protection against electric shock in accordance with IEC 60 364, Part 41, is a very important pilot standard within the harmonization document HD 384.4.41 S2 and IEC 364-4-41. It can be ensured by way of the following measures:

- Through protection under both normal operating conditions and under fault conditions (protection against both direct and indirect contact)
- Through protection under normal operating conditions (protection against direct contact, or basic protection)
- Through protection under fault conditions (protection against indirect contact, or fault protection)

12.1 Voltage Ranges

This standard redefines voltage ranges (Table 12.2) and cut-off times (Tables 12.3 and 12.6).

Definition of voltage ranges according to the rated voltage of the network:

Voltage range I: Protection against electric shock is ensured by the value of the voltage. The voltage is limited for functional reasons.

Voltage range II: These are the voltages in household installations, as well as commercial and industrial systems, public supply systems, etc.

Table 12.2 Voltage ranges.

	For AC currents		
Voltage range	**Grounded networks**		**Isolated or ungrounded networks between external conductors**
	External conductor-ground	**Between external conductors**	
I	$U \leq 50$ V	$U \leq 50$ V	$U \leq 50$ V
II	50 V $< U \leq 600$ V	50 V $< U \leq 1000$ V	50 V $< U \leq 1000$ V
	For DC currents		
I	$U \leq 120$ V	$U \leq 120$ V	$U \leq 120$ V
II	120 V $< U \leq 900$ V	120 V $< U \leq 1500$ V	120 V $< U \leq 1500$ V

12.2 Protection by Cut-Off or Warning Messages

This section deals with protection against electric shock under fault conditions. Persons and useful animals must be protected from the hazards of dangerous currents passing through the body. For this purpose, the coordination of systems with overcurrent protective equipment is of greatest importance. The measures of protection through cut-off in accordance with IEC 60 364, Part 41, for TN and TT systems or through warning for IT systems with equipotential bonding find use here.

12.2.1 TN Systems

In TN systems (Figure 12.2), a large short-circuit current must flow in order for the cut-off to take place within the specified time (0.2, 0.4, and 0.5 second) (Table 12.3). If the tripping condition cannot be met, it is then necessary to lower the rated current of the overcurrent protection device or provide for an RCD or additional equipotential bonding [11]. In TN systems, a disconnection time not exceeding five seconds is permitted for distribution circuits. In Table 12.3, the maximum disconnection time shall be applied to final circuits not exceeding 32 A.

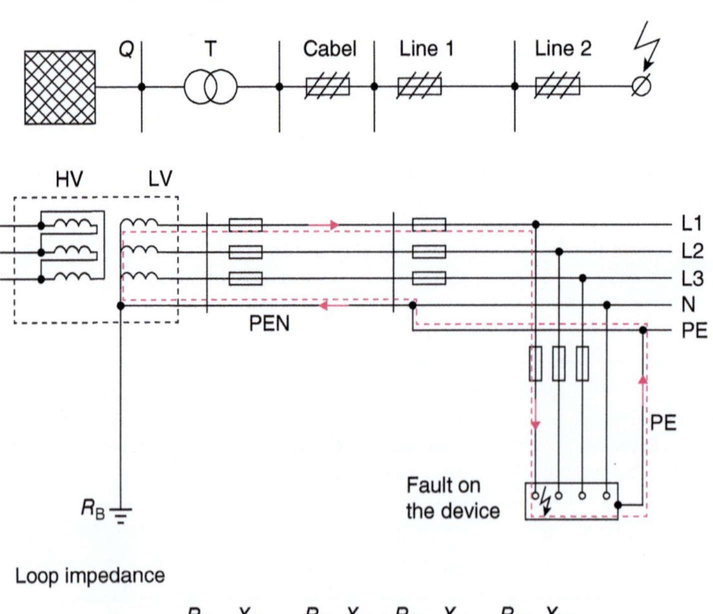

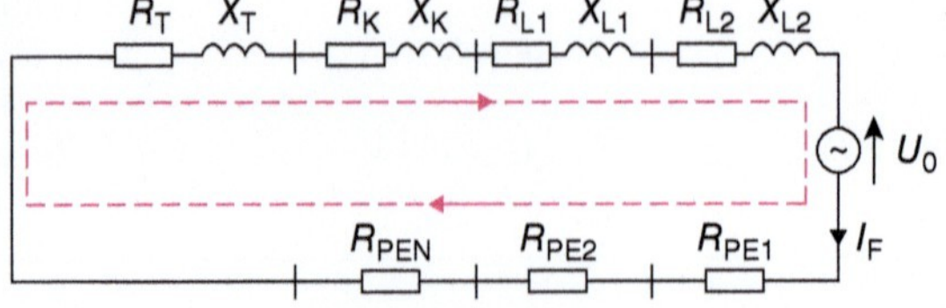

Figure 12.2 Circuitry of a TN system and its equivalent circuit diagram.

Table 12.3 Rated voltages and maximum cut-off times for TN systems.

Voltage U_0 (V)	Cut-off time (s)
230	0.4
400	0.2
>400	0.1

For the TN-S system, from Figure 12.2 we obtain the following cut-off conditions:

- *Loop impedance*:

$$Z_s \leq \frac{U_0}{I_a} \tag{12.1}$$

Since the measurements are carried out with small currents at room temperature, the procedure described below can be used to take into account the increase in conductor resistance with increasing temperature due to faults and to prove that for TN systems the measured value of the fault loop impedance meets the requirements of IEC 60364-6.

The requirements are considered to be fulfilled if the measured value of the fault loop impedance fulfills the following condition:

$$Z_{sm} \leq \frac{2}{3} \cdot \frac{U_0}{I_a} \tag{12.2}$$

- *Single-pole short circuit current*:

$$I''_{k1} = \frac{U_0}{Z_v + \sum Z'\,l} \tag{12.3}$$

Here the meanings of the symbols are

Z_s Loop impedance
U_0 Line-to-ground voltage
I_a Cut-off current of the overcurrent protective device
Z_v Source impedance
Z' Per-unit impedance
l Length of cable or line

The relationship

$$I''_{k1} \geq I_a \tag{12.4}$$

must always hold true. In accordance with IEC 60909-0, for the single-pole short-circuit current

$$I''_{k1} = \frac{\sqrt{3}\, c_{\min}\, U_n}{|\underline{Z}_1 + \underline{Z}_2 + \underline{Z}_0|} \tag{12.5}$$

In accordance with IEC 60364, Supplement 5, we obtain

$$I''_{k1} = \frac{\sqrt{3}\, c_{\min}\, U_n}{3\, \sqrt{(2\, R_L + R_v)^2 + (2\, X_L + X_v)^2}} \tag{12.6}$$

With $R_L = l\, R'_L$ and $X_L = l\, X'_L$, it follows that

$$I''_{k1} = \frac{\sqrt{3}\, c_{\min}\, U_n}{3\, \sqrt{(2\, l\, R'_L + R_v)^2 + (2\, l\, X'_L + X_v)^2}} \tag{12.7}$$

With an impedance phase angle of 28°, we then obtain

$$R_v = Z_v\, \cos 28° \tag{12.8}$$

$$X_v = Z_v\, \sin 28° \tag{12.9}$$

This averaged impedance phase angle is applicable for power supply companies and industrial networks. Here, the symbols have the meanings:

$c_{\min}$ Voltage factor
U_n Rated voltage
I''_{k1} Single-pole short-circuit current
R_v Resistance per unit length
X_v Reactance per unit length
Z_v Loop impedance of supply network from current source to overcurrent protective device

12.2.2 TT Systems

TT systems (Figure 12.3) are characterized by the grounding of the source current and the operational equipment [11]. Cut-off with overcurrent protective equipment is difficult to achieve due to the required low-resistance grounding resistance. As a result, predominantly RCDs are used.

For TT systems, from Figure 12.3 we obtain the following conditions:

- *Grounding resistance with protective equipment*:

$$Z_s \le \frac{U_0}{I_a} \tag{12.10}$$

- *Grounding resistance with RCD*:

$$R_A \le \frac{U_T}{I_{\Delta n}} \tag{12.11}$$

- *Grounding resistance with selective RCD*:

$$R_A \le \frac{U_T}{2\, I_{\Delta n}} \tag{12.12}$$

Structure of the three-phase system

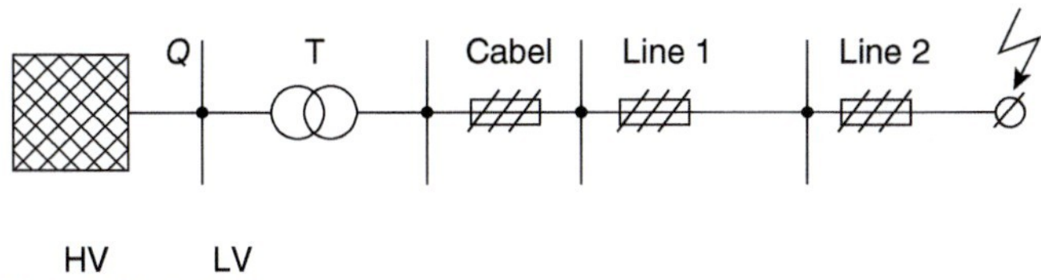

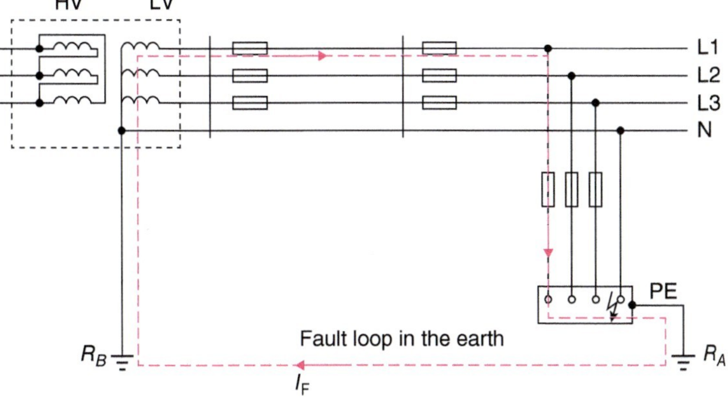

Loop impedance

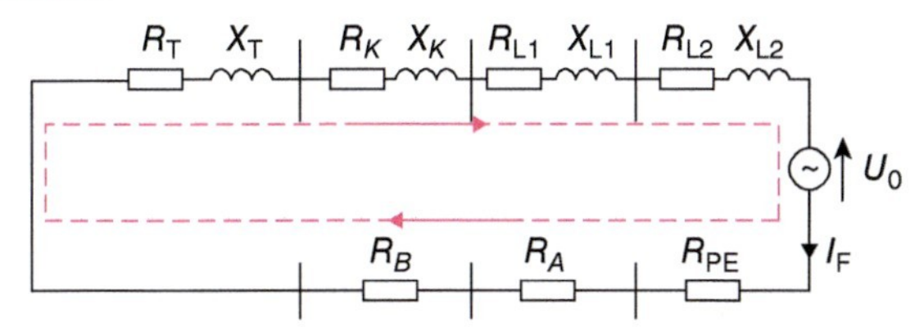

Figure 12.3 Circuitry of a TT system and its equivalent circuit diagram.

- Grounding resistance for several RCDs in parallel
  The total grounding resistance of the system is calculated considering Table 12.5

$$R_A \leq \frac{U_T}{g \sum I_{\Delta n}} \tag{12.13}$$

  The rated voltage and maximum cut-off time of TT Systems is shown in Figure 12.3. In TT systems, a disconnection time not exceeding 1 s is permitted for distribution circuits. In Table 12.4, the maximum disconnection time shall be applied to final circuits not exceeding 32 A.

Table 12.4 Rated voltages and maximum cut-off times for TT systems.

Voltage U_0 (V)	Cut-off time (s)
230	0.2
400	0.07
>400	0.4

Table 12.5 Coincidence factor g.

Number of RCDs	g
2–4	0.5
5–10	0.35
>10	0.25

In distribution circuits, a cut-off time of ≤1 second is permissible to obtain selectivity. Cut-off with protective equipment must take place without delay or, for a short trigger time with rising current, within a maximum time of five seconds (Table 12.5).

- *Single-pole short-circuit current*: For a short-circuit to an exposed conductive part, the fault current becomes a single-pole short circuit current. The following relationships apply:

$$I''_{k1} = \frac{U_n}{\sqrt{3} \cdot Z_s} \tag{12.14}$$

$$I''_{k1} = \frac{\sqrt{3} \cdot U_n}{2 \cdot \underline{Z} + \underline{Z}_0} \tag{12.15}$$

Assuming that the grounding resistances are larger than the network impedances, then (Figure 12.3),

$$I''_{k1} = \frac{U_n}{\sqrt{3} \cdot (R_A + R_B)} \tag{12.16}$$

Here, the symbols have the meanings:

R_A	Sum of resistances of grounding electrode and protective conductor
R_B	Operating impedance
$\underline{Z}_s$	Loop impedance
$\underline{Z}$	Positive-sequence impedance
$\underline{Z}_L$	Line impedance
$\underline{Z}_{0T}$	Zero-sequence impedance of the transformer
U_T	Touch voltage
I_a	Cut-off current of protective device
$I_{\Delta n}$	Rated differential current of RCD

12.2.3 IT Systems

For IT systems the current source is not grounded, i.e. isolated from ground or connected through a high impedance to ground [11]. The operational equipment must be connected to a grounding electrode (Figure 12.4). The leakage current is very small and can be taken from the manufacturer. The first fault is indicated by the isolating equipment and can be localized, i.e. no cut-off may take place here. The second

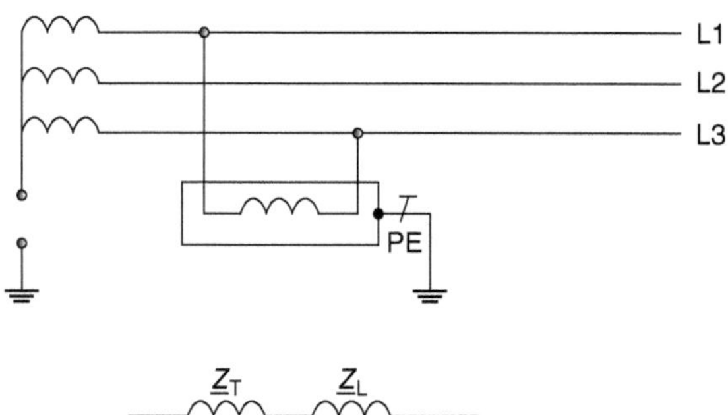

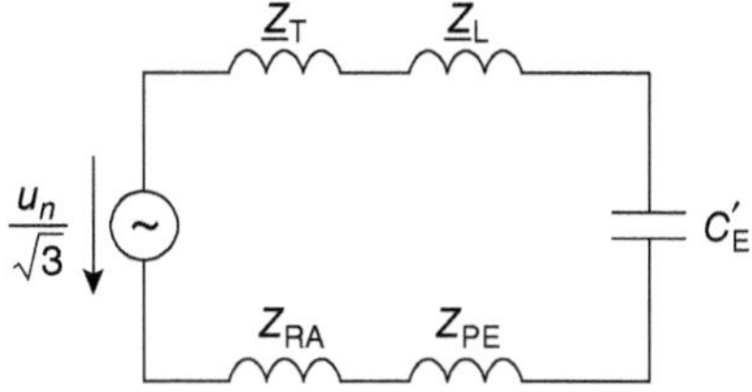

Figure 12.4 Circuitry of an IT system and its Equivalent circuit diagram.

fault must lead to a cut-off. IT systems must be used, for example, in hospitals (rooms according to class of application) and in mining.

For IT systems, from Figure 12.4 we obtain the following cut-off conditions (Table 12.6):

- *Grounding resistance*:

$$R_A \leq \frac{U_\mathrm{T}}{I_d} \tag{12.17}$$

The following condition must be satisfied if the neutral conductor is not distributed:

$$Z_S \leq \frac{U}{2\, I_a} \tag{12.18}$$

The following condition must be satisfied if the neutral conductor is distributed:

$$Z_S' \leq \frac{U_0}{2\, I_a} \tag{12.19}$$

- *Fault current*:

$$I_\mathrm{F} = \frac{\sqrt{3}\, U_n}{3\, Z_\mathrm{F} - \frac{1}{w\, C_E'\, l}} \tag{12.20}$$

$$Z_\mathrm{F} = Z_\mathrm{K} + Z_\mathrm{st} \tag{12.21}$$

Table 12.6 Rated voltages and maximum cut-off times for IT systems (second fault).

Rated voltage of electrical system in *V*	Cut-off time (s) Neutral conductor not distributed	Cut-off time (s) Neutral conductor distributed
230/400	0.4	0.8
400/690	0.2	0.4
580/1000	0.1	0.2

$$Z_0 = -\frac{1}{w\, C'_E\, l} \tag{12.22}$$

This results in

$$I_F = \frac{\sqrt{3}\, U_n}{Z_0 + 3\, Z_F} \tag{12.23}$$

The meanings of the symbols are

R_A Grounding resistance
U_T Touch voltage
I_d Leakage current
C'_E Capacitance to ground
Z_F Fault impedance
Z_0 Zero-sequence impedance
Z_k Impedance of exposed conductive parts
Z_{st} Site impedance
Z_S Impedance of fault loop, consisting of external conductor and protective conductor of the circuit
Z'_S Impedance of fault loop, consisting of neutral conductor and protective conductor of the circuit
U_0 Rated AC voltage between external conductor and neutral conductor
U Rated AC voltage between external conductors

During or following installation every system must, before the user begins operation, be inspected, tested, and measured. The requirements of the following standards must be fulfilled:

- *IEC 60 364, Part 61*: Measurements for the testing of protective measures in heavy current systems before initial startup
- *IEC 60 364 Part 701*: Measurements for the testing of protective measures for electrical equipment
- *IEC 60 364 Part 702*: Measurements for the testing of protective measures following repairs or modifications
- *EN 50110-1*: Measurements for the testing of protective measures for requalification tests

The measurements must take the errors in measurements of the measuring instrument and the method of measurement into account. For measuring instruments it is also necessary to consider the operating error. Along with the operating error, systematic errors (fundamental errors in the method of measurement) must also be considered (Table 12.7).

12.2.4 Summary of Cut-Off Times and Loop Resistances

Table 12.8 summarizes the cut-off currents and loop resistances in a TN system. These values can be used as the basis for the measurement and calculation of electrical systems. They apply at 80 °C line temperature. When measurements are

Table 12.7 Summary of measurement errors and measuring instrument errors.

	Error (%)
Resistance for 80 °C line temperature	
$R_x = R_{20°}[1 + \kappa\Delta\varphi]$	24
Error of measuring instrument	±5
For loop impedance measurement	+30
For single-pole short circuit current measurement	−30

performed at other temperatures it is necessary to correct for the loop resistance. In addition, the operating error of the measuring instrument must be considered (Table 12.7).

12.2.5 Example 1: Checking Protective Measures

On the basis of Figure 12.5, check the protective measures for a series resistance of $R_V = 0.6\ \Omega$.

(a) With an electrical outlet
Loop resistance:

$$R_S = R_V + R_L = 0.6\ \Omega + 1.24\ \frac{2 \cdot 60\ \text{m}}{56\frac{\text{m}}{\Omega\text{mm}^2} \cdot 2.5\ \text{mm}^2} = 1.66\ \Omega$$

According to Table 12.8, the cut-off current for B16A is 80 A. With 0.4 second cut-off time, the fault current is given by

$$I_\text{F} = \frac{230\ \text{V}}{1.66\ \Omega} = 138.5\ \text{A}$$
$$I_\text{F} > I_a \longrightarrow 138.5\ \text{A} > 80\ \text{A}$$

The cut-off condition is therefore fulfilled.

(b) With a heating unit

$$t_a \leq 5\ \text{seconds}$$

$$R_S = R_V + R_L = 0.6\ \Omega + 1.24\ \frac{2 \cdot 50\ \text{m}}{56\frac{\text{m}}{\Omega\text{mm}^2} \cdot 2.5\ \text{mm}^2} = 1.485\ \Omega$$

From Table 12.8, the cut-off current is 70 A for gL16A.

$$I_\text{F} = \frac{230\ \text{V}}{1.485\ \Omega} = 154.88\ \text{A}$$
$$I_\text{F} > I_a \longrightarrow 154.88\ \text{A} > 70\ \text{A}$$

The cut-off current is less than the fault current. This fulfills the cut-off condition.

Table 12.8 Summary of overcurrent protective equipment and loop resistances in TN systems.

	Rated currents of overcurrent protective equipment										
I_n	6	10	16	20	25	35	50	63	80	100	160
Low-voltage fuses of duty class gG in accordance with IEC 269-1											
Cut-off currents (A)											
I_a(0.2 s)	60	100	148	192	250	372	578	750	990	1310	2080
I_a(0.4 s)	48	80	140	180	210	300	450	600	800	1000	1850
I_a(5 s)	28	46	70	85	118	173	260	350	452	573	995
Loop resistances (Ω)											
Z_S(0.2 s)	3.7	2.2	1.5	1.2	0.9	0.6	0.4	0.3	0.22	0.17	0.11
Z_S(5 s)	7.8	4.7	3.2	2.6	1.9	1.3	0.8	0.6	0.5	0.4	0.22
Circuit-breakers in accordance with IEC 898, B characteristic Cut-off current $I_a = 5 \cdot$ rated current of overcurrent protective device											
I_a	30	50	80	100	125	175	250	315	400	500	800
Z_S	7.3	4.4	2.8	2.2	1.8	1.3	0.886	0.55	0.44	0.28	
Circuit-breakers in accordance with IEC 898, C characteristic Power circuit-breakers in accordance with EN 60439-1 for corresponding setting Cut-off current $I_a = 10 \cdot$ rated current of overcurrent protective device											
I_a	60	100	160	200	250	350	500	630	800	1000	1600
Z_S	3.6	2.2	1.4	1.1	0.88	0.63	0.45	0.35	0.27	0.22	0.14
Circuit-breakers in accordance with IEC 898, K characteristic Motor starter in accordance with EN 60439-1, Parts 102 and 104 Cut-off current $I_a = 15 \cdot$ rated current of overcurrent protective device											
I_a	90	150	240	300	375	525	750	945	1200	1500	2400
Z_S	2.4	1.5	0.9	0.7	0.6	0.4	0.29	0.23	0.18	0.15	0.09

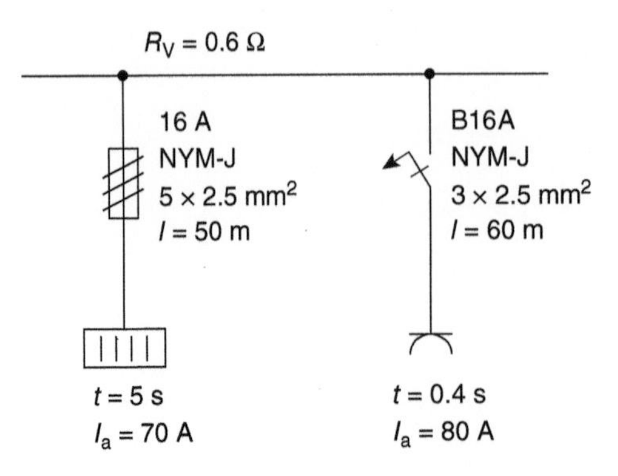

Figure 12.5 Protective measure with heating unit and electrical outlet.

12.2.6 Example 2: Determination of Rated Fuse Current

In a circuit the loop resistance is measured to be $R_S = 1.8\ \Omega$ with $U_0 = 230$ V. Find the rated current of the fuse for 0.4 and 5 seconds.

$$I''_{k1} = \frac{U_0}{R_S} = \frac{230\ \text{V}}{1.8\ \Omega} = 127.7\ \text{A}$$

For cut-off times of 0.4 and 5 seconds, from the time–current characteristic (Figure 13.22) for D02 gL fuse elements we find the rated currents:

$$0.4\ \text{s} \Rightarrow I_n = 20\ \text{A} \quad 5\ \text{s} \Rightarrow I_n = 25\ \text{A}$$

12.2.7 Example 3: Calculation of Maximum Conductor Length

Given:
Rated current of fuse $I_n = 25$ A
Line-to-ground voltage $U_0 = 230$ V
Cross-section of conductor $S = 4\ \text{mm}^2$

The cut-off times can be read from the time–current characteristic (Figure 13.22): for 0.4 seconds: $I_a = 170$ A; for five seconds: $I_a = 80$ A

Calculation of the loop resistance:

$$R_S = \frac{U_0}{I_a} = \frac{230\ \text{V}}{80\ \text{A}} = 2.875\ \Omega$$

$$R_S = \frac{U_0}{I_a} = \frac{230\ \text{V}}{170\ \text{A}} = 1.35\ \Omega$$

Calculation of the length:

$$l_{5\,\text{s}} = \frac{R_S\ \kappa\ S}{2} = \frac{2.875\ \Omega \cdot 56 \frac{\text{m}}{\Omega\text{mm}^2} \cdot 4\ \text{mm}^2}{2} = 322\ \text{m}$$

$$l_{0.4\,\text{s}} = \frac{R_S\ \kappa\ S}{2} = \frac{1.35\ \Omega \cdot 56 \frac{\text{m}}{\Omega\text{mm}^2} \cdot 4\ \text{mm}^2}{2} = 151.2\ \text{m}$$

12.2.8 Example 4: Fault Current Calculation for a TT System

Determination of the rated current for the overcurrent protective device of a TT system:
Given:
Touch voltage $U_0 = 230$ V
Ground resistance $R_A = 5\ \Omega$
Service resistance $R_B = 2\ \Omega$
The cut-off current (fault current) is then

$$I_\text{F} \leq \frac{U_0}{R_A + R_B} = \frac{230\ \text{V}}{7\ \Omega} = 32.85\ \text{A}$$

With this fault current, it is not possible to trip if we take a circuit-breaker MCB 16 A.

An RCD of (0.03 A) results in a resistance of.

$$R_A \leq \frac{U_T}{I_{\Delta n}} = \frac{50\ \text{V}}{0.03\ \text{A}} = 1666\ \Omega$$

By determination of the RCD, there is no need to consider the ground resistance.

12.2.9 Example 5: Cut-Off Condition for an IT System

A warning message is given for the following:

1. Errors reported by the isolation monitoring and cut-off
2. Errors due to tripping of the overcurrent protection device

Calculation of the loop resistance with a cut-off current of 80 A:

$$Z_S \leq \frac{\sqrt{3}\, U_0}{2\, I_a} = \frac{\sqrt{3} \cdot 230\ \text{V}}{2 \cdot 80\ \text{A}} = 2.48\ \Omega \quad \text{without neutral conductor}$$

$$Z_S \leq \frac{U_0}{2\, I_a} = \frac{230\ \text{V}}{2 \cdot 80\ \text{A}} = 1.43\ \Omega \quad \text{with neutral conductor}$$

12.2.10 Example 6: Protective Measure for Connection Line to a House

On a service panel for the connection line to a house the PEN conductor is not properly connected, so that a contact resistance remains. On the house service panel itself, however, the PE is grounded, with $R_{B2} = 10\ \Omega$ (Figure 12.6).

1. What is the touch voltage U_B that occurs for breakage of the PEN conductor when a load with $P = 4\ \text{kW}$ is operated?
2. What is the touch voltage that occurs for a contact resistance of Ω when a load with $P = 4\ \text{kW}$ is operated?
3. What is the short circuit current that flows for a total short -circuit to conductive parts (neglect line impedances)?

Solution 1:

$$P = U\, I$$

$$I = \frac{P}{U_0} = \frac{4\ \text{kW}}{230\ \text{V}} = 17.4\ \text{A}$$

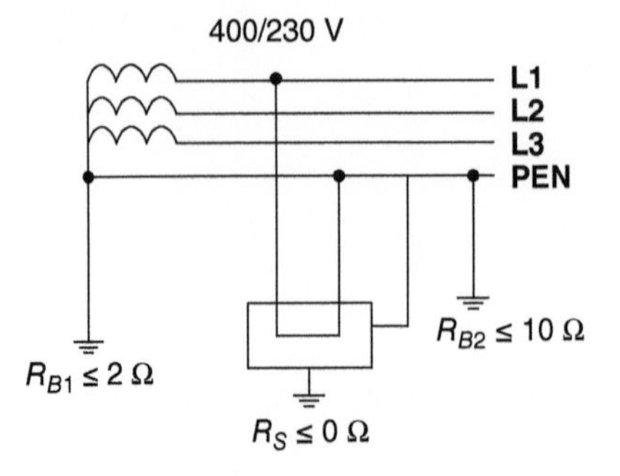

Figure 12.6 Protective measure for connection line to a house.

$$R = \frac{230\ \text{V}}{17.4\ \text{A}} = 13.2\ \Omega$$

$$I_F = \frac{U_0}{R_G} = \frac{230\ \text{V}}{13.2\ \Omega} = 17.4\ \text{A}$$

$$U_T = 10\ \Omega \cdot 17.4 = 174\ \text{V}$$

Solution 2:

$$R_{\text{Total}} = 2\ \Omega + (10\ \Omega + 2\ \Omega)||1\ \Omega = 12.92\ \Omega$$

$$I_1 = \frac{230\ \text{V}}{12.92\ \Omega} = 17.8\ \text{A}$$

$$U = 17.8\ \text{A} \cdot 12.92\ \Omega = 230\ \text{V}$$

$$I_2 = \frac{230\ \text{V}}{12\ \Omega} = 19.16\ \text{A}$$

$$U_T = 19.16\ \text{A} \cdot 10\ \Omega = 191.6\ \text{V}$$

Solution 3:

$$I_k = \frac{U}{R} = \frac{230\ \text{V}}{12\ \Omega} = 19.17\ \text{A}$$

12.2.11 Example 7: Protective Measure for a TT System

With a ground resistance of $R_E = 1\ \Omega$ and a resistivity of $\rho_E = 300\ \Omega\text{m}$ (Figure 12.7) find the following:

(a) The length of the grounding strip

$$R_E \leq \frac{U_T}{I_a} \leq \frac{50\ \text{V}}{50\ \text{A}} \leq 1\ \Omega$$

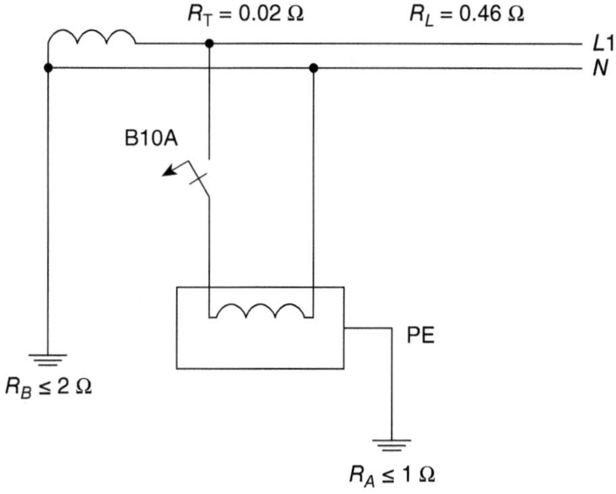

Figure 12.7 Protective measure for a TT system.

$$R_E \approx \frac{3\ \rho_E}{l} \text{ (for } l > 10 \text{ m)(in accordance with EN 50522)}$$

$$l \approx \frac{3 \cdot 300\ \Omega\text{m}}{1\ \Omega} = 900\ \text{m}$$

(b) The fault current

$$I_F = \frac{230\ \text{V}}{(0.02 + 0.46 + 1 + 2)\ \Omega} = 66.1\ \text{A}$$

(c) The voltage on the grounding strip

$$U_E = I_F\ R_A = 66.1\ \text{A} \cdot 1\ \Omega = 66.1\ \text{V}$$

The larger the value of I_a, the more difficult it is to obtain the ground resistance. Here it is necessary to use an RCD.

13
Equipment for Overcurrent Protection

This section will discuss all types of protection equipment, switching Devices, and interconnecting devices in low-voltage systems, including fuses, which in contrast to switching devices function only once and then melt in the event of a fault.

13.1 Electric Arc

When the contacts of a current-conducting switch (Figure 13.1) are opened, the pronounced heating at the base points of the current with a large contact resistance results in an electric arc, which the voltage at the contacts then maintains.

The electrons emitted from the hot cathode are accelerated by the electrical field and thus transported to the anode. At a high enough velocity, collisions occur between atoms and electrons, ionizing the atoms. The ions drift slowly to the cathode and the electrons quickly to the anode. An ion cloud forms in front of the cathode. The same process occurs in front of the anode, only in this case with electrons. Between these is the electric arc, consisting of a neutral plasma of ions and electrons. Positive ions and the negative cathode form the cathode fall, with field strengths up to 10^5 V/cm, which are required to attract the electrons from the cathode and accelerate them.

The cathode spot forms on the cathode. For low-voltage switching devices the cathode fall is the main feature, whereas high-voltage switching devices function with a cooled electric arc. A multiple of the cathode fall is obtained by dividing the arc gap into partial gaps, either by double-break contacts or through plates. The cooling is achieved by extending the arc (thermally or magnetically), blowing with air or gases, heat dissipation, or with oil or water.

13.1.1 Electric Arc Characteristic

If an electric arc is formed in series in a circuit with a resistance (Figure 13.2), a falling characteristic results since with increasing current the ionization process is enhanced. The resistance is necessary to limit the current.

Analysis and Design of Electrical Power Systems: A Practical Guide and Commentary on NEC and IEC 60364, First Edition. Ismail Kasikci.

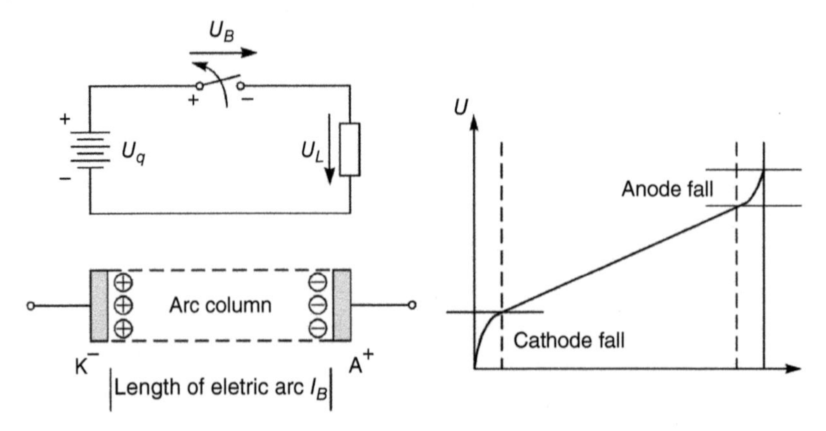

Figure 13.1 Origin of the electric arc.

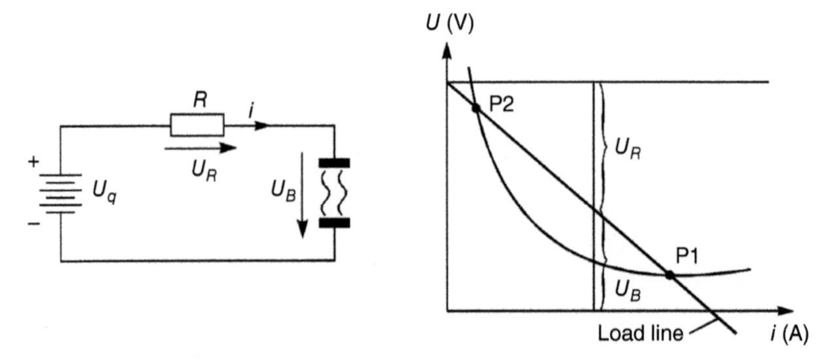

Figure 13.2 Electric arc characteristic.

There are two operating points, P1 and P2, such that P1 is stable and P2 labile. If the current of the arc burning at point P1 increases for any reason (e.g. poorer cooling), a negative induced voltage results, which reduces the current at the point P1.

In other words, the increasing current requires a voltage that exceeds the load line, i.e. the resistance of the arc must increase. If, on the other hand, the current decreases, again starting with point P1, the voltage on the load lines increases and thus lowers the resistance, increasing the current. It is obvious that the conditions for point P2 are exactly the opposite, so that this point is labile.

For systems engineering, there are then the following quenching possibilities:

1. Constant arc and changing resistance
2. Constant resistance in the circuit and changing arc, until the critical point is exceeded

For the DC arc, as expressed by the static characteristic, the energy is balanced and the energy supplied from the current and voltage compensates for the losses. For the AC current, on the other hand, the power supply – as a result of the driving voltage – must then cause a change in temperature and thus in the energy content and the losses. This gives rise to the dynamic characteristic (Figure 13.3). A change in

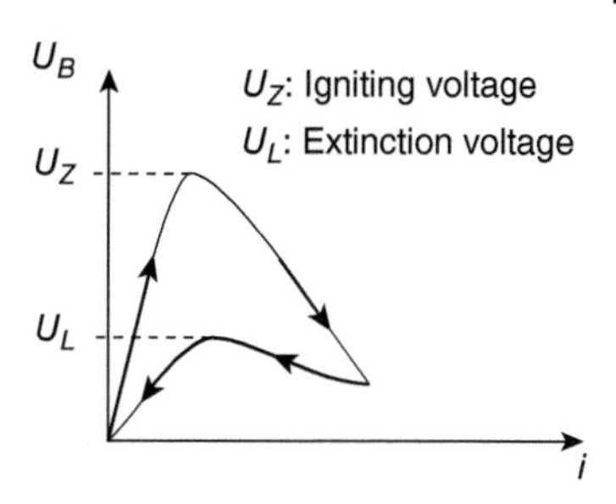

Figure 13.3 Dynamic characteristic.

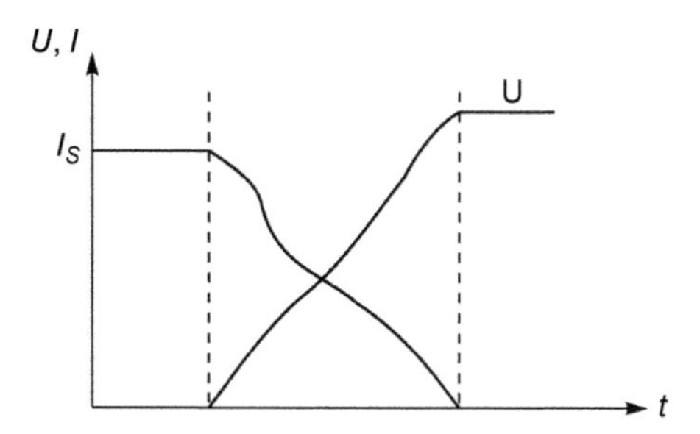

Figure 13.4 DC cut-off.

the energy content results in a change in the power. With increasing current, the arc requires additional power for the temperature rise, whereas with decreasing current the losses can be compensated from the heat content of the arc. This means that for an increasing current the voltage must be greater than for a decreasing current, leading to the development of arc hysteresis. At higher frequencies, a straight line will result because the temperature is not able to follow the fast changes.

13.1.2 DC Cut-Off

The voltage and current of the electric arc will adjust themselves according to the static characteristic (Figure 13.4). If the cooling effect increases over time, the current then falls and the voltage rises. The switch must be able to withstand the temperature sufficiently to execute the cut-off cycle. DC switches present problems. It must also not be forgotten that the arc must be present in order to dissipate the energy stored in the inductivities. In order that no overvoltages arise, the value of di/dt must not be too great. The so-called high-speed circuit-breakers attain cut-off times of about 4 ms. The arc gradient is about 200 V/cm, so that DC high-voltage switches are generally not possible. The switches themselves function with thermal cooling and magnetic blowout.

13.1.3 AC Cut-Off

For AC cut-off we must deal with the dynamic arc, i.e. the arc voltage is greater for increasing current than for falling current (Figure 13.5). After the zero current crossing there must be a short pause without current until the voltage is attained that is required to reignite the arc. For phase-shifted values, however, during the zero

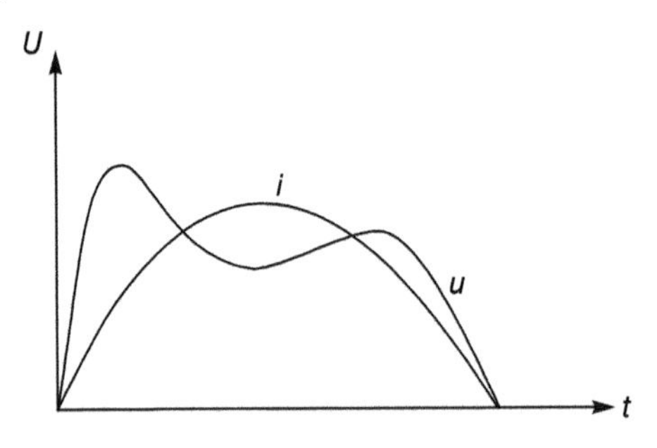

Figure 13.5 AC cut-off.

current crossing the recovery voltage is immediately present with its corresponding instantaneous value. However, it cannot increase to this value in an infinitesimally short time, because as a result of the capacitances and inductances always present a transient process takes place. We will first discuss all AC circuit-breakers without considering the transient process in order not to complicate matters. The voltage present at the contacts leads to a slight, hardly measurable post-arc current, comprised of the residual charge carriers of the electric arc, which are only newly accelerated. This post-arc current is about 1/1000 of the short-circuit current. It flows for about 0.1 ms. Thus, with AC breakers two processes always take place:

1. The arc is cooled and the current therefore reduced. Finally, the charge carriers recombine and the remainder is taken up by the contacts. During the zero current crossing, the current extinguishes.
2. The recovery voltage loads the contact path and also causes a slight post-arc current.

Assuming the switch can withstand both stresses, the arc is then successfully extinguished.

13.1.3.1 Cut-Off for Large Inductances

With increasing distance between contacts or with greater cooling, the arc voltage increases and the current decreases slightly until the igniting voltage is greater than the driving voltage (Figure 13.6). Cut-off is not easy for the unit because during the zero current crossover the peak value of the recovery voltage is present. This

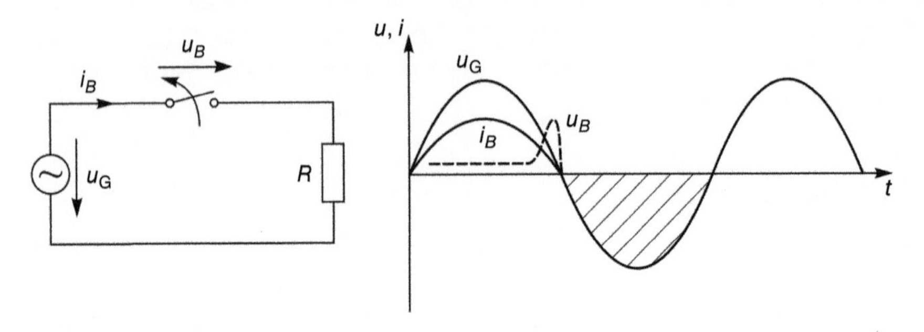

Figure 13.6 Cut-off for large inductances.

means that the arc will then reignite a few times more. We must try to design the cooling conditions for the first zero current crossing adequately to anticipate the first crossover. Inductive cut-off is practically a short-circuit path, since here the circuit is largely inductive due to the generators, transformers, and lines.

13.1.3.2 Cut-Off of Pure Resistances

Here, the current and voltage are in phase, the recovery voltage is practically zero, and the switch is almost without voltage loading (Figure 13.7). The switches can therefore be constructed of a lightweight material.

13.1.3.3 Cut-Off of Capacitances

Let us assume that the switch extinguishes during the first zero current crossing. The contact then lies on the capacitor voltage. Since there is no discharge path, the voltage remains. The recovery voltage follows the generator voltage. As a result, a difference voltage is formed at the switch contacts, which rises to double the peak value. This can easily lead to restriking, in turn causing overvoltages. The capacitive cut-off is therefore critical. It must always be checked whether the capacitive current from the particular switch can be controlled, even at low voltages (Figure 13.8).

13.1.3.4 Cut-Off of Small Inductances

For small currents, that is especially for the magnetization currents of transformers in the no-load state, the cooling can interrupt the current already before the zero current crossover. Owing to the rapidly changing current, a high induced voltage then results, which, as for the capacitances, adds to the generator voltage and can

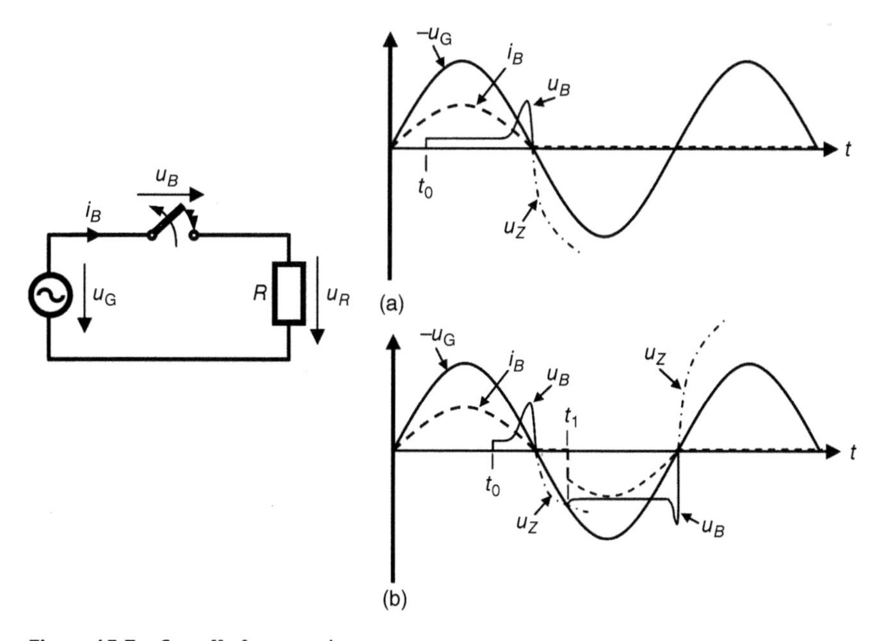

Figure 13.7 Cut-off of pure resistances.

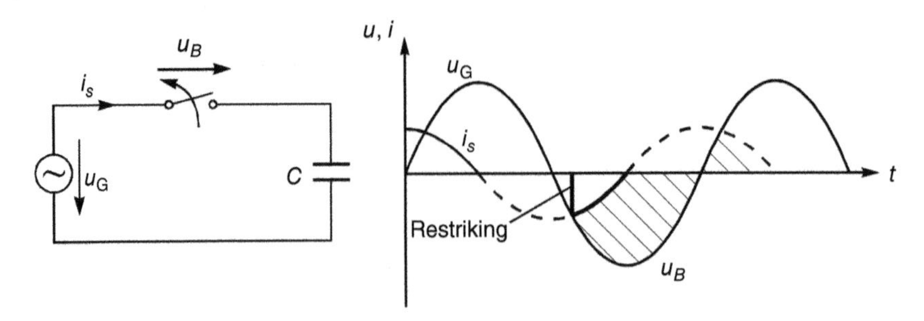

Figure 13.8 Cut-off of capacitances.

cause restriking. Here again, there is a danger of overvoltages. Since the switch is designed for the three-pole short-circuit and not for the small magnetization currents, an overvoltage arrester should be connected to the transformer. The inductive currents for transformers in the no-load state are about 1–5 A.

13.1.4 Transient Voltage

The network consists of inductances and capacitances. The natural frequency can lie between 0.5 and 50 Hz; however, the range from 1 to 5 kHz is rarely exceeded. A damping of the transient voltage is usually achieved with the use of a resistor. There is no difference compared with an AC cut-off, but there are some special considerations since the three currents do not pass through zero simultaneously. After the first zero current crossover, a two-pole short-circuit remains. In detail, the result is as in Figure 13.9:

1. At time t_1 the current flowing in line L1 is the first to pass through zero. It extinguishes during the first crossover.
2. The recovery voltage is shifted by 180° for both currents and in both cases has the value U_{star}.
3. After the first zero current crossover, a two-pole short-circuit current remains at time t_2.

Of the many possible types of load, we will examine only two special cases here (Figure 13.9). One involves the coupling switch between two power stations or networks. If the coupling is opened because a fault has occurred, due to the voltage breakdown the synchronizing moment is no longer present because the active current of the load is missing. The generators then fall out of step. The systems rotate freely and during extinction can come to exactly opposite phases. This results in double the switching power as for normal three-pole cut-off. The short circuit at the end of a long line in networks above 1 kV, the voltage at the switch is not zero, but takes intermediate values depending on the distance. Because of the different inductances on both sides of the switch, different transient values arise on the generator side and the fault side.

The short-line fault is greater than the terminal short-circuit. The terminal short-circuit occurs in the immediate vicinity of the switch. The DC element decays

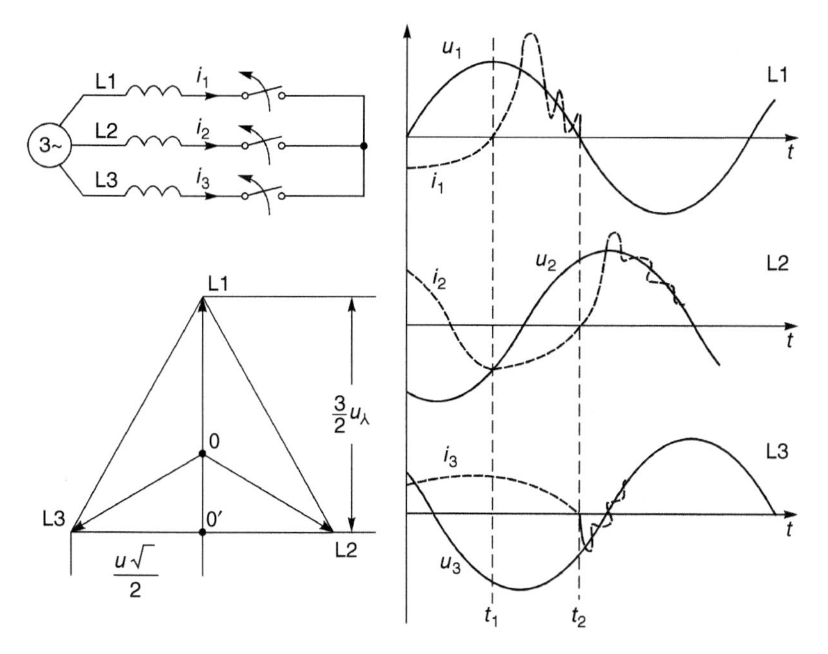

Figure 13.9 Transient voltage.

with the time constant $T = \frac{L}{R}$, for which IEC 45 give 45 ms. In real networks and various operational equipment, this value oscillates between 10 and 450 ms.

13.2 Low-Voltage Switchgear

In switchgear and industrial systems, and also in any electrical systems, a wide range of switchgear is in use today for the reliable control of power, as well as its connection, interruption, and servicing. Since the requirements for low-voltage and high-voltage switchgear are very different, e.g. mechanical problems are foremost with low-voltage units, while there is high electrical loading of high-voltage units, it is advisable to discuss such units separately. A classification of low-voltage switchgear and sizes follows for the connection, interruption, and disconnection of circuits in the following way.

13.2.1 Characteristic Parameters

In accordance with IEC 60947-2 we distinguish between the very different designations for the individual electrical parameters.

- Voltages
  1. Rated voltage
  2. Rated insulation voltage
  3. Operating voltage

   4. Fault voltage
   5. Touch voltage
- Currents
   1. Rated current
   2. Continuous current
   3. Dynamic current
   4. Thermal current
- Capacities
   1. Rated short-circuit breaking capacity I_{cu}
   2. Rated service breaking capacity I_{cs}

Low voltage includes operating voltages up to 1000 V. For various reasons, DC equipment is also included here. The ratings of the individual equipment depend on their classification according to the rated breaking capacity.

13.2.2 Main or Load Switches

Main or load switches are actuated without current. For their electrical dimensioning, in addition to the continuous current the limiting dynamic value is also of interest. The switches must be make-proof and must open by themselves even as a result of short-circuit forces.

The dimensioning parameters are the following:

- Rated voltage and rated insulation voltage
- Rated current
- Continuous current
- Thermal current in kA as RMS value
- Dynamic current in kA as peak value

In the interest of saving space, fuses and disconnectors will be dealt with together. Disconnectors have a defined isolating distance for the protection of the operating personnel (line segment with a definite dielectric strength) and can conduct operating currents and overcurrents, but cannot switch these on or off. Load break switches are load switches with disconnection function. Switches for smaller currents for control circuits up to about 200 A have certain construction features. The dimensioning of the load switch is as for disconnectors; however, in addition the cut-off current or the breaking capacity must be specified. The displacement power factor $\cos \cos \varphi = 0.7$ is defined by the test specifications. However, the motor power is mostly given as the per-unit quantity. Main switches or load switches are used in accordance with IEC 408 for switching operational equipment and system parts on and off under normal operating conditions.

13.2.3 Motor Protective Switches

IEC 292

Three-phase motors are most frequently used in industry, because of their simple design, ruggedness, favorable prices, and operational reliability. Depending on the

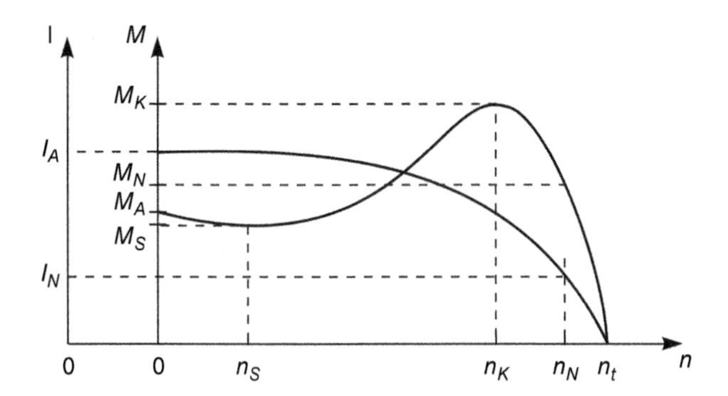

Figure 13.10 Switching of an asynchronous three-phase machine.

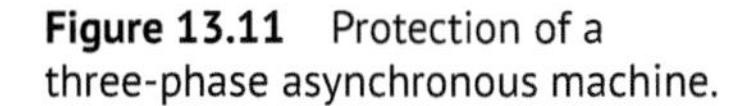
Figure 13.11 Protection of a three-phase asynchronous machine.

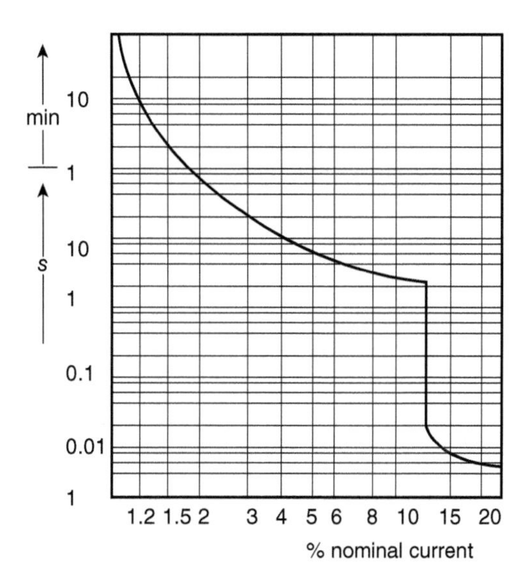

type of operation, they must be continuous switches (Figure 13.10) and protected (Figure 13.11). Motor protective switches are intended to execute a three-pole cut-off in the event of a fault occurrence. They are mostly combined with a bimetallic and short-circuit trip, which can also be replaced by a fuse. For the choice of starters, the rated current of the motor, the type of operation, and the desired contact service life are decisive for the dimensioning. For direct switch-on the motor is loaded briefly with five to six times the rated current. For greater currents, the motor must be switched off in order to protect it from thermal destruction. Dangerous overcurrents arise during operation due to mechanical overloading over a longer time or external conductor breaks.

- *Motor protection with fuses*: Overload protection is ensured by the motor protection relay, while the fuse is responsible for short-circuit protection. The auxiliary switch responds through the overload release only after exceeding a definite time and initiates the all-pole tripping of the motor protection.

- *Motor protection with circuit-breakers*: Circuit-breakers incorporate overload protection through bimetallic elements and short-circuit protection through the instantaneous magnetic release. This switch has the properties of a disconnector and can be used as the main switch.

13.2.4 Contactors and Motor Starters

IEC 60947-4-1
Conditions for fulfilling Classification 1:

- The contactor or starter must not endanger persons or systems in the event of a fault occurrence.
- The contactor or starter does not need to be suitable for continuing operation without repair and replacement.
- Damage to the contactor and overload relay is permissible.

Conditions for fulfilling Classification 2:

- The contactor or starter must not endanger persons or systems in the event of a fault occurrence.
- The contactor or starter must be suitable for continuing operation.
- There must be no damage to the overload relay or other parts.

Table 13.1 lists the utilization categories for typical applications in practice in accordance with IEC 60947-4-1.

13.2.5 Circuit-Breakers

IEC 898
Miniature circuit-breakers (mcb) disconnect the short-circuit current and are used for the protection of lines and cables. They are equipped with thermal and magnetic

Table 13.1 Utilization categories for contactors.

Utilization category	Load	$\frac{I_c}{I_e}$
AC1	Noninductive or weakly inductive load	
	Resistance furnace	1.5
AC2	Starting, switching off	
	Slipring rotor motors	4.0
AC3	Starting, switching off while running	
	Squirrel-cage rotor motors	8.0
AC4	Starting, braking by reversal, jogging	
	Squirrel-cage rotor motors	10.0

I_c: Make and break current; I_e: Operating current.

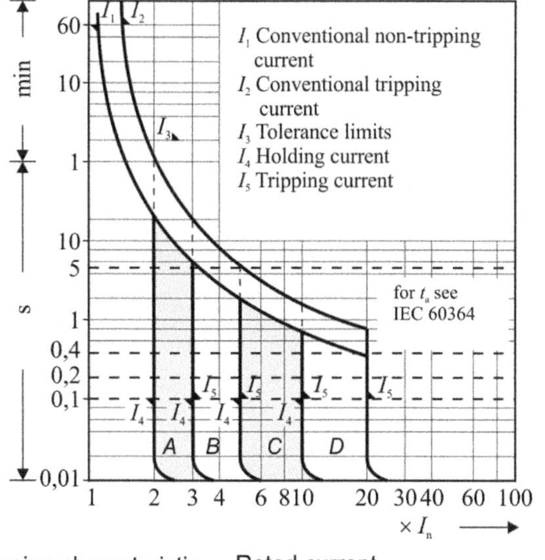

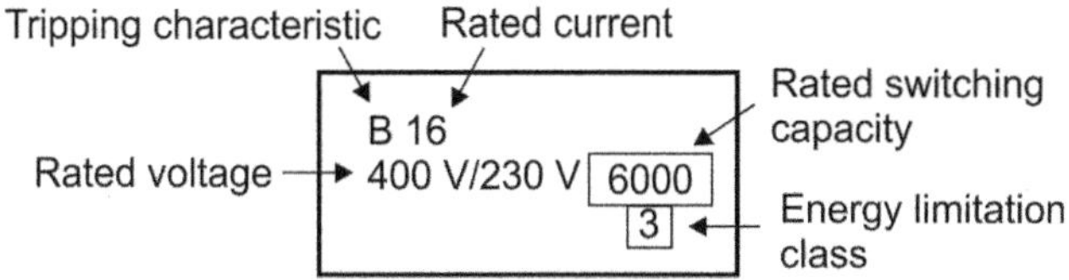

Figure 13.12 Time–current characteristics for circuit-breakers in accordance with IEC 60898. Source: [58].

trip elements and are classified according to cross sections. Different characteristics are necessary in order to optimally match the Current–time behavior of the trip element to the current-time-thermal balance of the objects to be protected. Figure 13.12 shows time–current characteristics for mcb and recommendation for lettering on mcb according to DIN VDE 0641-11. The tripping current is standardized.

These protective devices are manufactured and used in single-pole or three-pole form. If the three-pole short-circuit current at the site of the device is greater than the rated breaking capacity of the switch, it is necessary to install fuses ahead of the device (backup protection). Circuit-breakers switch operating currents and function under fault conditions.

Table 13.2 gives the maximum permissible let-through values for different breaking capacities of the circuit-breakers.

13.2.6 RCDs (Residual Current Protective Devices)

RCDs (residual current operated circuit-breakers, Figure 13.13) consist of a summation current transformer with primary and secondary windings, tripping relay, contact system with latching mechanism, tripping device, and housing. A tripping current circuit serves to control functioning. With the test key, a fault can be simulated and the function of the RCD be checked.

Table 13.2 Current-limiting classes in A^2 s.

Breaking capacity	**Current-limiting class**	**Current-limiting class**
A	**1**	**3**
3 000	31 000	15 000
6 000	100 000	35 000
10 000	240 000	70 000

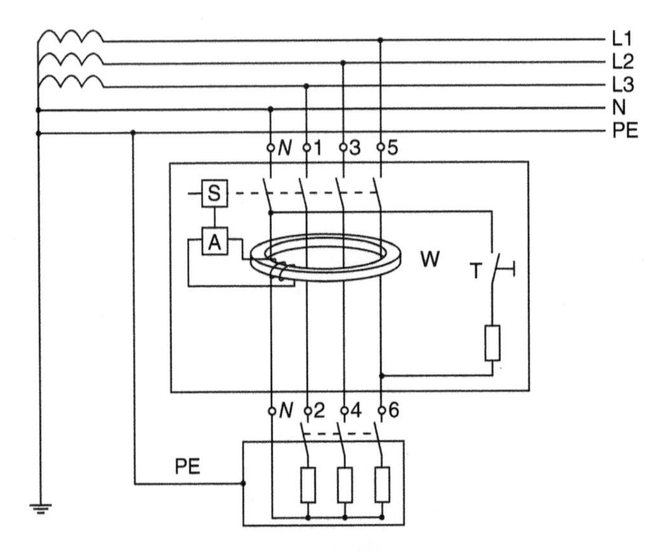

Figure 13.13 Principle of an RCD [57].

RCDs prevent injury to persons and damage to equipment due to electric current and within a small fraction of a second switch to 50–75% of the rated fault current. RCDs enable protection against the remaining excessive fault currents >30 mA, protection against fires igniting for electrical reasons (maximum 300 mA), and protection in the event of direct contact (protection for persons) up to 30 mA, or better 10 mA because 10 mA lies below the human let-go current. In a fault-free electrical installation the operating current flows from the network through the summation current transformer to the consumer, and from there through the summation current transformer again and back to the network. The geometrical sum of the inward and outward currents in the transformer is zero. If in the event of a fault a current is led off to ground, then the current flowing to the consumer is proportionately greater than the outwardly flowing current. The current difference induces a magnetic field in the transformer, causing a current to flow in the secondary winding of the transformer, which in turn causes the magnetic tripping element to respond and thus initiates an all-pole cut-off with the RCD.

RCDs can be used in TN-S, IT, and TT systems. This requires that the neutral point of the network is grounded, the PEN conductor isolated before the RCD, and the protective ground conductor is not led through the RCD. RCDs do not serve the purpose of cutting off overcurrents. It is therefore necessary to protect them using a suitable overload protection equipment, better to combine the two so that the system is also protected against overloading and short-circuits. During planning, it is necessary to take into account that the rated current of the overload protection equipment is not greater than that of the RCD. They must trip reliably when a pulsating DC current, such as occurs especially with such consumers as rectifiers, thyristors, and triacs, flows to the ground. For tripping the fault current must reach at least the value zero within a period.

Unintentional tripping of RCDs can be prevented by installing a delayed RCD of type designation S (selective) in addition to the nonselective RCDs following. This is especially recommended in regions in which there is a danger of lightning striking, as well as for consumers with long lines and for floor heating systems in which capacitive discharge currents arise during switch-on.

RCDs are firmly established in regulations for electrical installations (IEC 60 364) and international regulations. Figure 13.14 shows different examples for planning with RCDs, in combination with circuit-breakers.

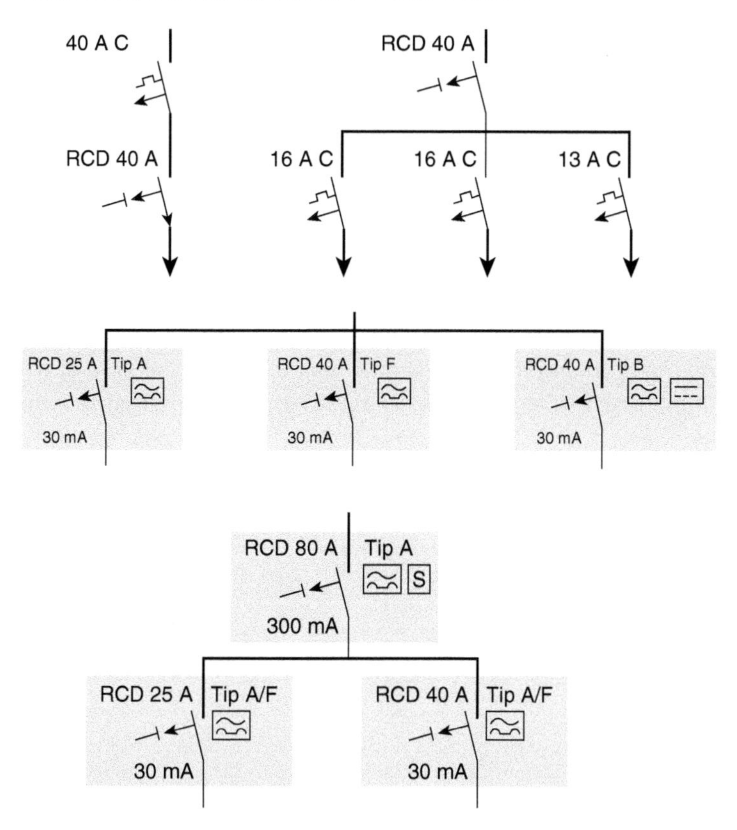

Figure 13.14 Planning with RCDs.

13.2.7 Main Protective Equipment

This section describes primary protection equipment (single-pole or three-pole versions) in the form of SLS (selective circuit-breakers). These overload protection devices may be used according to technical conditions for connection, power supply companies, and utilities as line-side meter fuses in meter mounting boards or in the lower terminal housing instead of D02 and NH00 fuses. Their installation must be carried out according to the instructions of the power supply company or utility.

Properties of these breakers:

- Can be operated by untrained persons (safe operation), but not NH00
- Sealable
- Shock protection in accordance with the accident prevention regulations of the BG
- Backup protection for all overload protection equipment
- Selectivity of breakers among each other
- Disconnector properties with contact display
- Has selectivity to line fuses
- Can be installed without problems in meter mounting boards
- Fulfills the requirements of international and national testing standards for line protection equipment and offers full domain protection
- High making and breaking capacity (min. 25 kA) for meter mounting boards
- Direct installation on top-hat rails or on busbars using a high-quality adapter with special clamps
- Fast return to operation after occurrence of a fault
- The following versions are available: single-pole, double-pole, or three-pole, 25 A, 35 A, 50 A, 63 A, 80 A, 100 A

SLS breakers are comprised of the primary current circuit (i), the secondary current circuit (ii), and the making current circuit (iii) (Figure 13.15).

The primary current circuit includes the input terminal, bimetallic element, electromagnet, a movable double contact, and an output terminal. The secondary current path is comprised of the bimetallic element, the pure resistance, and

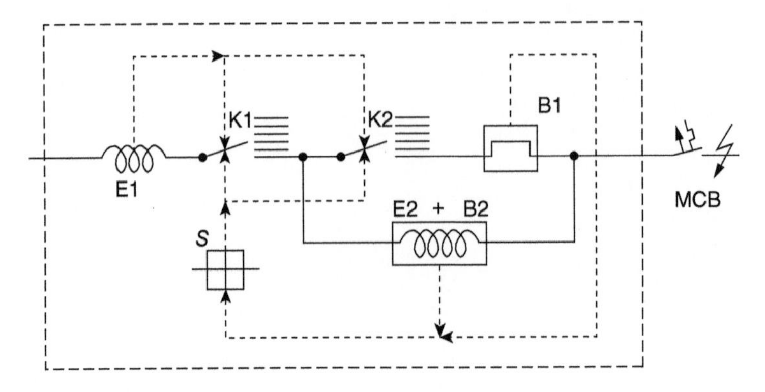

Figure 13.15 Design and function of the SLS breaker [46].

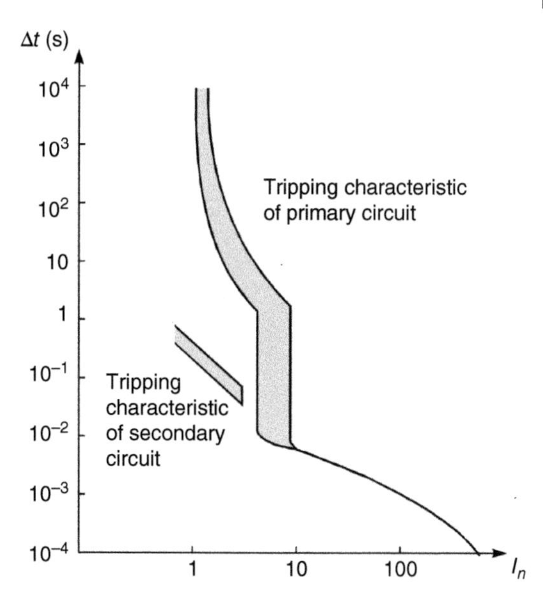

Figure 13.16 Tripping characteristic of SLS breakers [46].

the contact. The making current circuit includes an electromagnet, a movable contact, and a terminal. In the event of overloading the bimetallic element releases latching mechanism S1, which then opens contact K2. With the occurrence of a short-circuit, the electromagnet E1 releases latching mechanism S2, which opens contact K1. The tripping characteristic is similar to that of the circuit-breaker. In addition, there is also the tripping strip of the bimetallic element B2 in the secondary current circuit (Figure 13.16). In the event of a short-circuit, the current commutates following tripping and extinction of the electric arc to the secondary current circuit. Figure 13.17 illustrates the current limiting for SLS breakers for clearing short-circuits. The relative energy content for a 40 kA half-wave is 16×10^6 A^2 s. The short-circuit is initiated at $\alpha = 60°$ (unfavorable point). In 2.2 ms the SLS breaker disconnects the fault location from the network and limits the let-through energy to 112 500 A^2 s. Figure 13.18 shows characteristics of SLS breakers and NH fuses for clearing short-circuits.

13.2.8 Meter Mounting Boards with Main Protective Switch

It is possible to install main automated fuse assemblies and selective Circuit-breakers on meter mounting boards in the following ways:

1. Single-rate connection up to 63 A loading capacity
   Connection takes place in accordance with service panel regulations. A three-pole main switch is provided in the lower terminal housing and a main branch circuit terminal/summation fuse in the upper terminal housing or an overcurrent protective device as a switchable disconnecting device, e.g. a selective main branch circuit-breaker in the lower terminal housing and a main branch circuit terminal in the upper terminal housing.

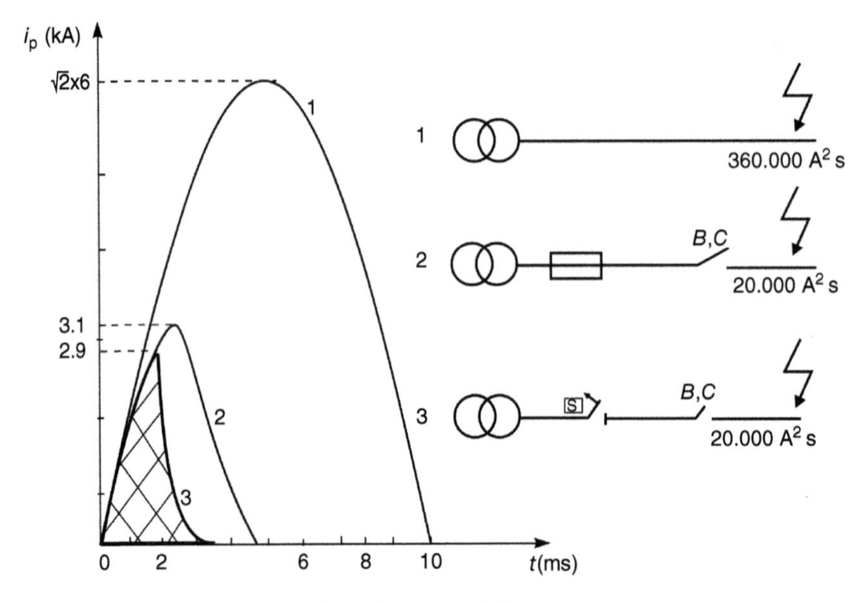

Figure 13.17 Current limiting for SLS breakers [58].

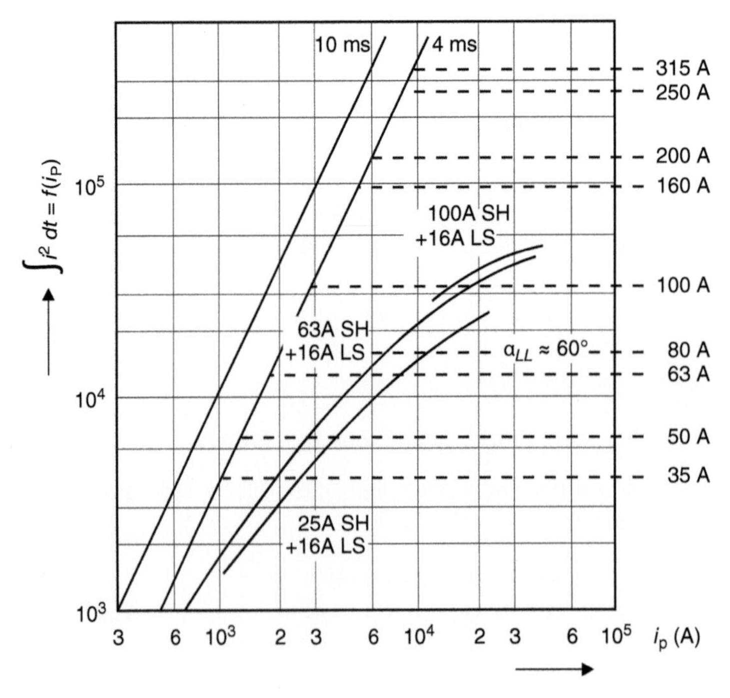

Figure 13.18 Comparison of SLS breakers with downstream circuit -breakers and NH fuses for clearing short-circuits [58].

2. Multi-rate connections for 63–100 A loading capacity
   Connection takes place in accordance with service panel regulations. A three-pole main switch is provided in the lower terminal housing and summation fuses in the upper terminal housing or an overcurrent protective device as a switchable disconnecting device, e.g. a selective main branch circuit-breaker in the lower terminal housing and a main branch circuit terminal in the upper terminal housing.
3. Multi-rate connections for more than 100 A loading capacity
   Group fuses up to 100 A are outside of the meter cabinet, a three-pole main switch in the lower terminal housing, and summation fuses in the upper terminal housing or an overcurrent protective device as a switchable disconnecting device, e.g. a selective main branch circuit-breaker in the lower terminal housing and a main branch circuit terminal in the upper terminal housing.

13.2.9 Fuses

Fine compacted sand serves as the quenching medium for the electric arc. The threshold current melts or vaporizes the conductor. The electric arc that arises maintains the current. The arc in turn melts the sand. For this, a certain Joule heat pulse $\int i^2\,dt$ is necessary.

The fuse always interrupts large short-circuit currents by current limiting before reaching i_p and without waiting for the natural zero current crossing. The electric arc voltage of the fuse must be greater than the overvoltage produced in order to ensure quenching. The type of characteristic determines the function class of the fuse and its external construction its construction type. The current-dependent characteristic of the fuse follows from the simplified equation

$$\int i^2 dt = c\,m\,\Delta\theta \tag{13.1}$$

The assignment of a function class to an object to be protected determines the duty class of a fuse. Fuses are current-limiting switching devices. The peak short-circuit current is limited to the let-through current. Here, it is necessary to distinguish between overloading and short-circuits. In the event of overloading, the left section of the characteristic (Figures 13.19 and 13.20), a fusible element, detects the overload condition, whereas the melting of the bottleneck section interrupts a short-circuit. The complete fuse consists of the base with adapter ring, adapter screw, indicator, screw cap, and plug head.

The rated current of the fuses for motor circuits should be about twice the rated current of the motor under normal startup conditions. In practice, there are graduated tables for the rated currents of fuses that function selectively. At the node points of the meshed networks the fuses are selective, as long as the largest partial current does not exceed approximately 0.8 times the summation current.

Rated parameters for fuses:

1. *Let-through current*: The maximum value of the short-circuit current is not reached, so that it is sufficient here to use the let-through current.

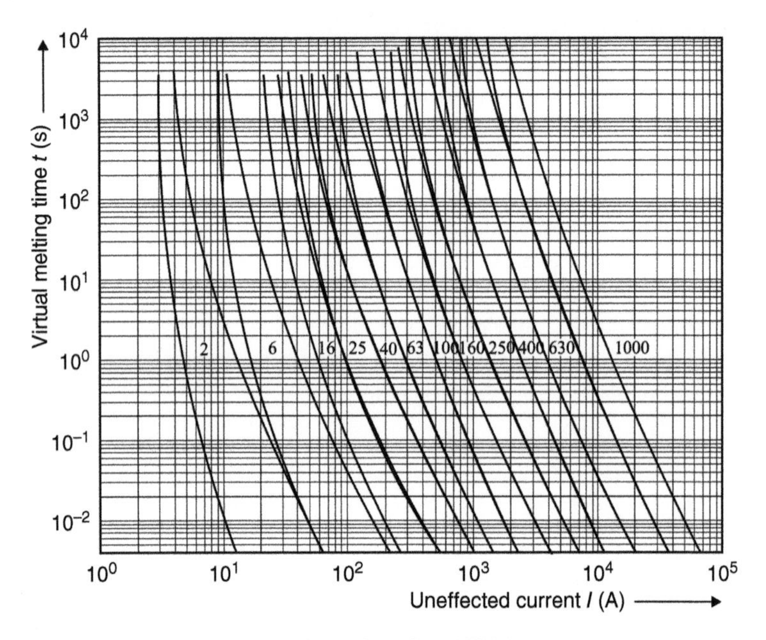

Figure 13.19 Fuse characteristics from 2 to 1000 A.

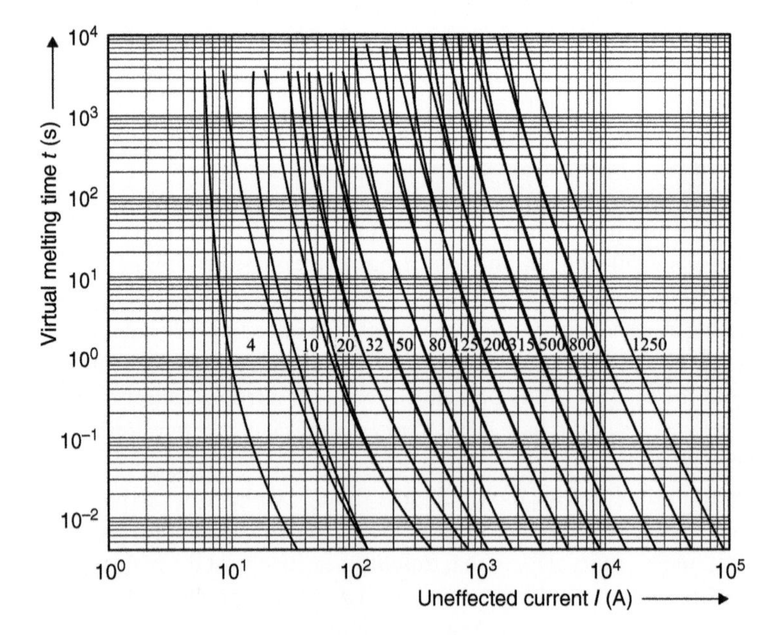

Figure 13.20 Fuse characteristics from 4 to 1250 A.

Table 13.3 Color coding of fuses.

Rated current A	Color
2	Pink
4	Brown
6	Green
10	Red
16	Gray
20	Blue
25	Yellow
35	Black
50	White
63	Copper
80	Silver
100	Red
125	Yellow
160	Copper
200	Blue

2. *Short-circuit strength*: The fuse must not explode or even blow out due to the short-circuit.
3. Color coding of fuses in accordance with Table 13.3
4. The dimensioning follows from the following:
   - Rated voltage
   - Rated current
   - Short-circuit strength
   - Let-through current
   - Time–current characteristic

13.2.9.1 Types of Construction

Here we distinguish between three types of construction:

- Screw-type fuses
  These are characterized by the non-interchangeability of the fuse unit and the protection against electric shock. These fuses can be operated by untrained personnel. The bases of screw-type fuses are constructed to accommodate adapter rings.
- Fuses with blade contact
  These consist of a lower fuse part, the replaceable fuse element, and the operating element for replacing the fuse element. Untrained personnel cannot operate

these fuses. A low-capacity fuse disconnector ensures safe conditions while replacing the fuse elements. It is possible to install and remove fuses with the system under load.

- Operation classes of the fuses
  These duty classes are identified by two letters, of which the first indicates the function class and the second the object to be protected.
- Function classes
  1. Function class
     g Full domain protection (protection against overload and short-circuits)
     a Subdomain fuses (protection against short-circuits)
  2. Defined objects of protection for fuses:

     G Cables and lines (general applications)
     L Cables and lines
     M Switchgear
     T Transformers
     R Semiconductors
     B Mining systems

The rated current ranges for NH fuse elements are listed in Table 13.4. According to the rated current intensities there are different thread sizes for fuse systems (Table 13.5).

The cut-off behavior of fuses in the overload region is determined by the small (no cut-off in the specified checking time) and the large (cut-off during the specified checking time) tripping currents. Table 13.6 summarizes the tripping currents.

The time–current characteristics for gG fuses are shown in Figure 13.19, from 2 to 1250 A. NH fuses of duty class gL protect electrical operational equipment against overloading and short-circuits. With corresponding classification, they can also be used to protect motors against short-circuits.

Table 13.4 Rated current ranges for NH fuse elements.

Size	NH fuse elements				NH fuse bases (A)	Switching strips (A)
	500 V gL (A)	660 V gL (A)	500/660 V aM (kVA)	400 V gT (A)		
00	6–100	6–100	35–100	—	160	160
1	80–250	80–250	80–250	—	250	250
2	125–400	125–400	125–400	50–250	400	400
3	315–630	315–500	315–630	250–400	630	630
4a	500–1250	500–800	630–1250	400–1000	1250	—

Table 13.5 Threads of fuses.

	Thread of D fuses	
E27	2–25 A	D2
E33	36–63 A	D3
R1.24 in.	80–100 A	D4H
	Thread of D0 fuses	
E14	2–16A	D01
E18	20–63A	D02
M30x2	80–100A	D03

Table 13.6 Tripping currents of fuses.

Duty class	Rated current (A)	Small tripping current I_1	Large tripping current I_2	Checking time
		Tripping currents of NH fuses		
gG	<4	1.5 I_n	2.5 I_n	1 h
	4–10	1.5 I_n	1.9 I_n	1 h
	10–25	1.4 I_n	1.75 I_n	1 h
	25–63	1.3 I_n	1.6 I_n	1 h
	63–160	1.3 I_n	1.6 I_n	2 h
	160–400	1.3 I_n	1.6 I_n	3 h
	>400	1.3 I_n	1.6 I_n	4 h
aM	all of the I_n	4 I_n	6.3 I_n	60 s
		Tripping currents of D0 fuses		
	bis 4	1.5 I_n	2.1 I_n	1 h
	4–10	1.5 I_n	1.9 I_n	1 h
gG	10–25	1.4 I_n	1.75 I_n	1 h
	25–63	1.3 I_n	1.6 I_n	1 h
	63–100	1.3 I_n	1.6 I_n	2 h

The current limitation diagrams for gG fuses are drawn in Figure 13.21. Before reaching the peak short-circuit current, the short-circuit current I_k'' is disconnected.

D0 fuses are frequently found in industrial and residential installations.

Figure 13.22 gives the time–current characteristics. The rated breaking capacity is 50 kA for these fuses.

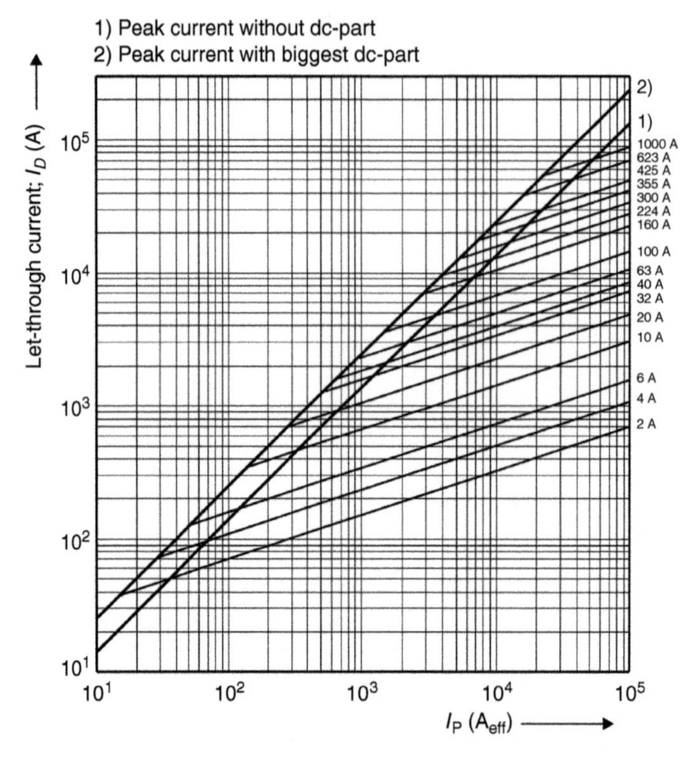

Figure 13.21 Let-through curves for gG fuses.

13.2.10 Power Circuit-Breakers

IEC 60947 In accordance with IEC 60947 a power circuit-breaker is a mechanical switching device, which under normal operating conditions for a given circuit, switches on, directs, and switches off current. In addition, under unusual circumstances such as a short-circuit the device directs and switches off current. Power circuit-breakers are switching devices for the repeated switch-on and switch-off of circuits in normal operation and under fault conditions. Power circuit-breakers always release in three-pole form. Following the occurrence of a fault they are switched on again. Signaling with remote on or off is possible without problems. Power circuit-breakers have a limited breaking capacity and are selective under one another only with certain restrictions. The making and breaking capacity increases linearly with the rated current. The response time of modern power circuit-breakers is between 2 and 10 ms for both making and breaking. The large switching forces required are supplied from compressed air, hydraulic or spring-loaded drives, which must have certain properties such as the following:

- An actuated switch must always be ready for switch-off.
- A switch-on action must be completed once begun.

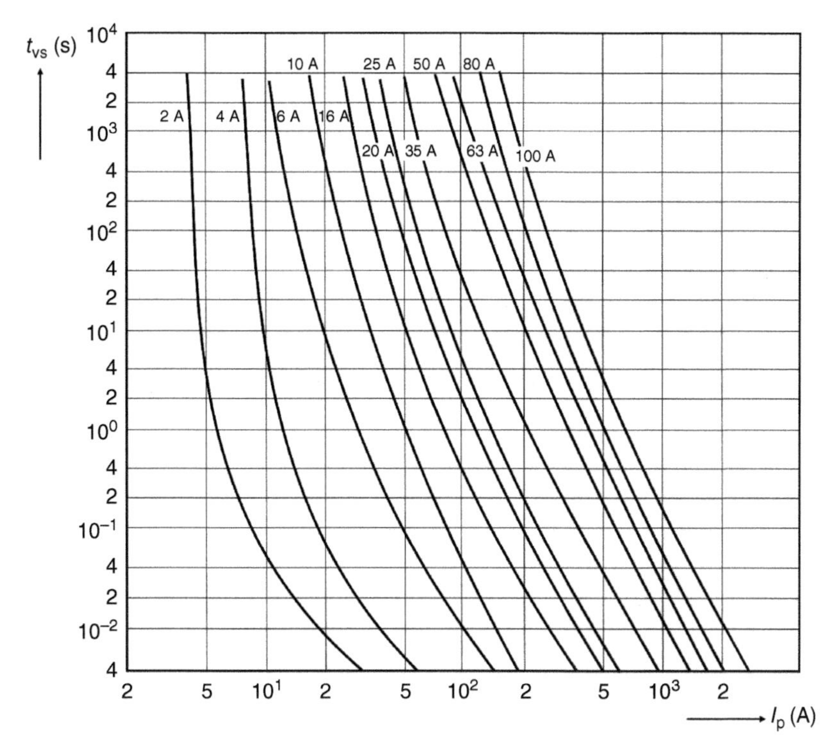

Figure 13.22 Time–current characteristics for DO fuses.

Power circuit-breakers must be able to control equipment under normal operating conditions as well as under fault conditions, i.e. they must be able to cut off reliably at any time. Their breaking capacity is dimensioned according to the effective value of the unaffected prospective short-circuit current. For the making and breaking capacity according to short-circuit category P2, a cut-off and two make–break switches with two interruptions at 1.1 times the rated voltage and the appropriate power factor are required. The power factor is 0.15. For the tests that demonstrate the making and breaking capacity, the test switching sequence must be complied with. Today, all power circuit-breakers are designed according to the building block principle, which enhances the breaking power, divides the applied voltage, and reduces the chamber loading through multiple chambers. The puffer is the most important element of a power circuit-breaker. During cut-off, it is drawn back by a force. This compresses the gas. At the same time, a switching contact coupled to the puffer, which joins the two contact tubes and closes to the make state, is displaced. This results in breaker gap between the upper contact tube and the contact piece. The voltage over the breaker gap drops, causing an electric arc. The line current at first continues to flow. The compressed gas flows into the breaker gap and cools the electric arc (heat dissipation). The current is interrupted, and a transient voltage begins between the line connections, which must be checked for its reliability when dimensioning the switches (recovery voltage). The length of the arc sustained in the contact

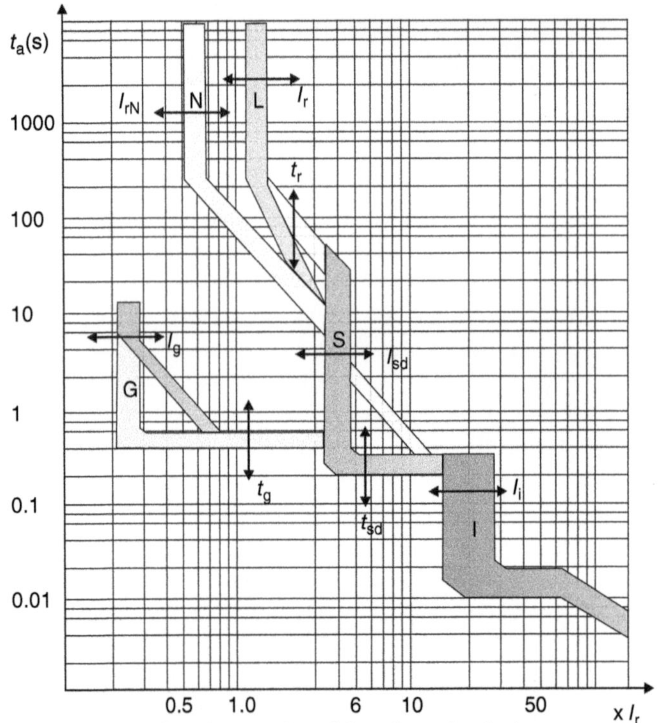

Figure 13.23 Time–current characteristics of power circuit-breakers [6].

tube becomes larger. Since the resistance of the arc also increases, the current and the supplied power are reduced. As we already know, the current extinguishes during the zero current crossing. A current chopping (tolerated only up to 4 A) is not permissible before this quenching, since otherwise large overvoltages would arise according to the law of induction. At switch-on the puffer and the switch contact are displaced upwards within 30 ms. The voltage along the breaker gap drops. Today, power circuit-breakers are manufactured up to 80 kA, and in high-voltage applications power circuit-breakers with several chambers per pole are used. Figure 13.23 gives the tripping characteristics for different power circuit-breakers.

13.2.10.1 Short-Circuit Categories in Accordance with IEC 60947

- Short-circuit category 1

old: P-1, Application for short-circuits

new: I_{cu} in accordance with IEC 60947

I_{cu} rated ultimate short-circuit breaking capacity

Switching sequence for testing the rated ultimate short-circuit breaking capacity:

$O - t - CO$: suitable only for reduced operation,
insignificant displacement of characteristic for rated ultimate short-circuit breaking capacity

- Short-circuit category 2

old: P-2, Application for frequent high short-circuits

new: I_{cs} in accordance with IEC 60947

I_{cs}: service short-circuit breaking capacity

Switching sequence for testing the service short-circuit breaking capacity: $O - t - CO - t - CO$: suitable for normal operation without servicing, no displacement of characteristic

The meanings of the symbols are

O switch-off

CO on-off sequence (onto the short-circuit)

t time (waiting interval of three minutes)

13.2.10.2 Breaker Types

We differentiate between the following breaker types, according to the arc extinguishing medium:

- *Liquid-level circuit-breakers*: The breaking gaps are under oil in a tank. The oil serves as an insulating material and as the arc extinguishing medium. Heat is extracted from the electric arc through vaporization, expansion, following an increase in pressure, and through thermal conduction. This requires large amounts of oil. This type of breaker is no longer manufactured today, since the oil is flammable and represents an environmental hazard.
- *Compressed air or compressed gas circuit-breakers*: The arc extinguishing medium flows independently of the current to be interrupted into the arc space. Following successful quenching, the cold gas under high pressure rapidly produces a surge-proof breaking gap. Today, only type SF6 is still used. The gas is not toxic and is odorless, inflammable, and a good thermal conductor.
- *Vacuum circuit-breakers*: These breakers find use in low-voltage and medium-voltage systems and have very good insulating properties. They require little servicing and have a large number of operating cycles. Owing to the absence of an arc extinguishing medium, flashing and contact wear result. The contacts are slit obliquely in order to prevent these difficulties. Copper-chrome alloys are used in order to produce metal vapor and to keep the chopping current small.
  Properties of vacuum circuit-breakers:
  - Can be switched on again immediately
  - Large number of operating cycles
  - High reliability
  - No open arc
  - Long electrical life
  - High short-circuit breaking capacity
  - Requires little servicing

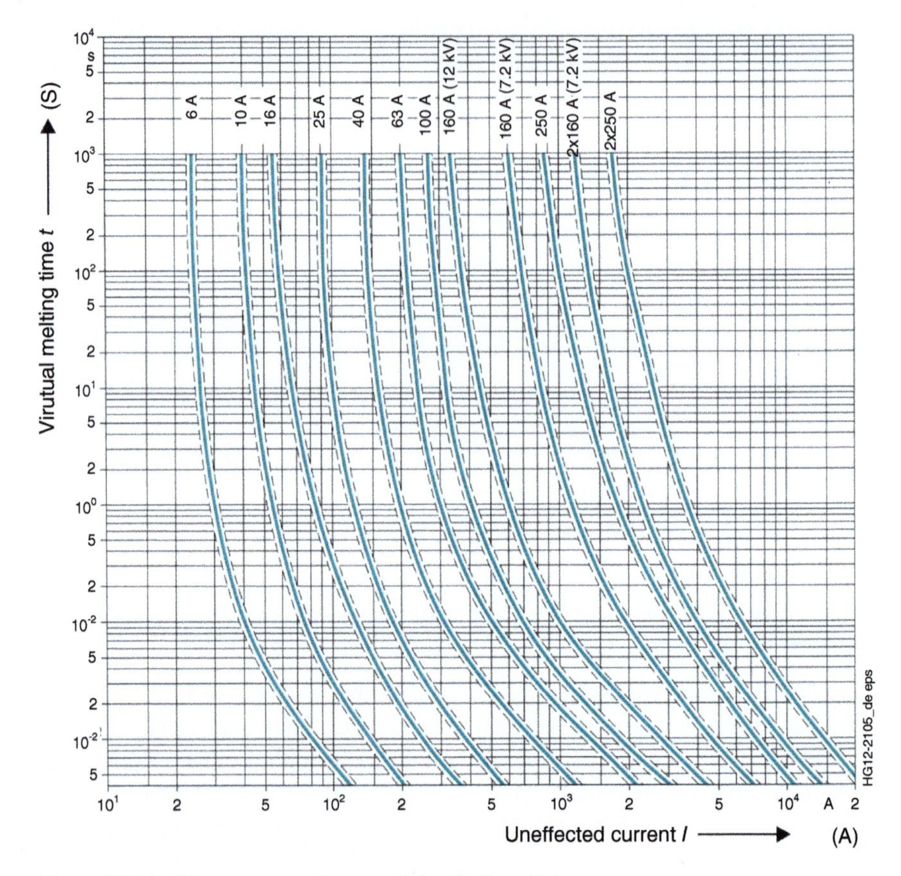

Figure 13.24 Time–current characteristics for fuse links.

13.2.11 Load Interrupter Switches

IEC 408 In medium-voltage networks, load interrupter switches are often used together with fuses and combined with disconnectors. Switch disconnectors are also used in order to save costs. Only the load is switched and a breaking gap is produced. The power factor is $\cos \varphi = 0.7$. Load interrupter switches are switching devices that fulfill the same switching functions as switch disconnectors.

13.2.12 Disconnect Switches

Disconnect switches are necessary in order to de-energize system components, so that these are accessible for checking. The main problems with disconnect switches are the heating of the contact surfaces and disengaging of the contact elements with the occurrence of abrupt short-circuit loading. They have no elements for extinguishing the electric arc, but must still be capable of quenching small currents.

A distinction is made between

- center-break disconnectors
- automatic disconnectors
- pantograph disconnectors.

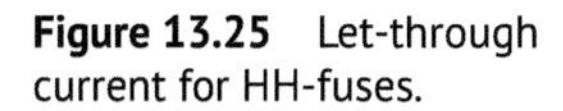

Figure 13.25 Let-through current for HH-fuses.

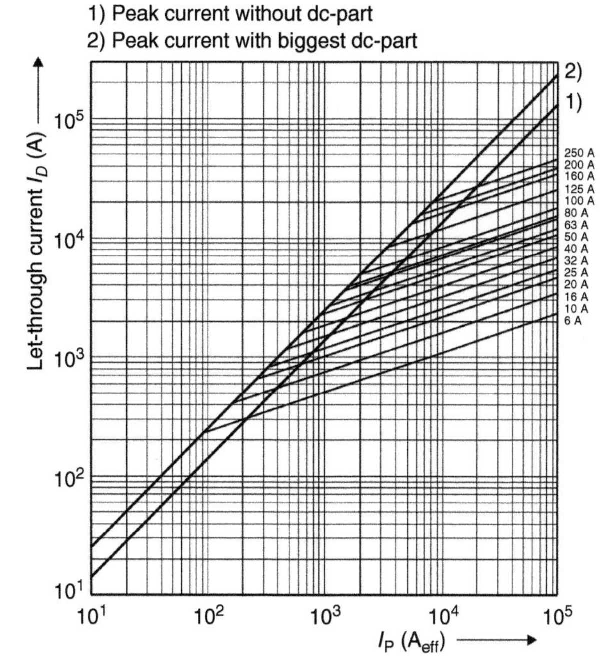

Disconnector drives:

- manual drives
- motor drives
- compressed air drives

Compressed air drives are no longer manufactured. Disconnectors, grounding Electrodes, and circuit-breakers must be provided with interlocks. Disconnectors must be switched only with the circuit-breakers in the OFF position. On the other hand, the grounding electrodes must be actuated only with the disconnectors open.

13.2.13 Fuse Links

This section gives the characteristics for fuse links that provide protection against short-circuits on the medium-voltage side of the transformer. They are used at 6 kV up to 150 A and at 30 kV up to 40 A (Figures 13.24 and 13.25).

13.2.14 List of Components

This section summarizes position units (PU) for the most common built-in units and terminals. From the sum of the position units, with sufficient reserve positions, it is then possible to determine the cabinet types and distributions (Table 13.7). For each field width for the installation of the built-in units 12 position units are available. One position unit has the dimensions 150 mm × 180 mm.

Table 13.7 Determining the position requirements for distribution cabinets.

Type	PU	Type	PU
Circuit-breaker 0.5–63 A		Staircase	1
		Circuit-breaker	
1-pole	1	Time switch analog day	3
2-pole	2	Time switch analog week	4
3-pole	3	Time switch digital 2-channel	2
		Time switch digital 3-channel	4
Fuse element			
63 A	1.5	Air-break contactor 24 A, AC	2
25 A	2.4	Air-break contactor 40–63 A, AC	3
Neozed fuse element			
100 A	2.6	Three-point terminal	0.35
63 A	1.5	Modular terminal 4 mm^2	0.35
16 A	1.5	Modular terminal 16 mm^2	0.6
Breaker 16 A, 1-,2-, and 3-pole	1	Modular terminal 50 mm^2	1.2
Breaker 25 A, 3-pole	1	N isolating terminal, blue, 4 mm^2	0.35
One-way switch 63–80 A, 3-pole	2.5	N isolating terminal, blue, 16 mm^2	0.6
One-way switch 63–100 A, 3-pole	3	PE terminal 4 mm^2	0.45
Master switch 63 A, 3-pole	3	PE terminal 16 mm^2	0.7
Master switch 100 A, 3-pole	4		
Fuse switch disconnector 16 A 1-pole	1	Servo relay	1
Fuse switch disconnector 63 A 1-pole	1.5	Phase monitor	2
Fuse switch disconnector 16 A 3-pole	3	Dimmer	3
Fuse switch disconnector 63 A 1-pole	4.5	Load disconnecting relay	1
Fuse switch disconnector 16 A 4-pole	4	Remote control switch 16 A, 2-pole	10
Fuse switch disconnector 63 A 4-pole	4.5	Remote control switch 16 A, 1-pole	1
Power circuit-breaker 6–6300 A	24	Photoelectric controller switch	4
RCD 16–63 A, 2-pole	2	Network rectifier	6
RCD 25–63 A, 4-pole	4	Miniature transformer 8–24 V	4
Pushbutton and control lamp	1	Incorporated transformer 100 VA	6
Grounding type outlet	2.5	Doorbell transformer 8–12 V, 8 VA	2

14

Current Carrying Capacity of Conductors and Cables

IEC 60364, Part 43

14.1 Terms and Definitions

- *Overcurrent*: Overcurrent refers to every overload current and every short-circuit Current, which is greater than the maximum permissible current carrying capacity I_z.
- *Overload*: The overload is greater than the rated current and arises during the fault-free operating condition. High loading of a motor or the simultaneous use of several loads can lead to overloading of the cable or the line.
- *Short-circuit*: The terms short-circuit and ground fault are used to describe faults in the functional isolation of operational equipment when energized parts are shunted out as a result. In accordance with IEC 60 909, a short-circuit arises through a fault, accidentally or intentionally, between active lines under voltage standing in opposition, through a low resistance or impedance.
- *Short-circuit current*: In accordance with IEC 60 909, a short-circuit current flows as a result of a fault for the duration of the short-circuit. Here it is necessary to distinguish between short-circuit current at the fault location and the transferred short-circuit currents. Lines and cables must be protected against excessive temperature rises as a result of an overcurrent (overload and short-circuit protection) with overcurrent protective equipment.
- *Load carrying capacity*: The load carrying capacity defines, under certain conditions, the highest permissible currents.
- *Loading*: Loading describes the currents to which a cable or line is subjected through a particular mode of operation or a fault condition.
- *Current carrying capacity*: The current carrying capacity of a cable or a line depends on the type of installation, the grouping of lines and cables, the ambient temperature, the operating temperature, the number of core wires and the insulation material, which must be considered in the form of correction factors for the dimensioning of lines and cables.
- *Rated short-time current density*: This value is the effective value of the current density which a cable is able to withstand the rated short-circuit duration.

Analysis and Design of Electrical Power Systems: A Practical Guide and Commentary on NEC and IEC 60364, First Edition. Ismail Kasikci.

- *Rated short time current*: This value is the effective value of the current which electrical operational equipment, under predefined conditions, is able to withstand over the rated short-circuit duration. The rated short-time current and the related rated short-circuit duration are determined by the manufacturer of the operational equipment.

The difference between overload and overcurrent lies in the cause of the fault. For a short-circuit, the overcurrent arises as a result of a defect or a wrong operation. The overload causes a current which exceeds the maximum current carrying capacity.

The choice of circuit-breaker follows on the basis of the tripping behavior and the breaking capacity, which is matched to the tripping behavior. The lines in heavy current systems are made of copper (Cu) and aluminum (Al).

The effective electrical cross section is taken as the cross section of the line. The resistivity for copper at 20 °C is $\rho_{20} = 0.017\,241\,\Omega\,\mathrm{mm}^2/\mathrm{m}$. The temperature coefficient is $3.93 \times 10^{-3}/\mathrm{K}$ and increases or decreases with the conductivity. The resistivity for aluminum at 20 °C is $\rho_{20} = 0.028\,264\,\Omega\,\mathrm{mm}^2/\mathrm{m}$. The temperature coefficient is $4.03 \times 10^{-3}/\mathrm{K}$ and increases or decreases with the conductivity, as with copper.

The temperature dependence of the resistivity is:

$$R_{\varphi} = R_{20\,°\mathrm{C}}\,[1 + \alpha\,(\varphi - 20\,°\mathrm{C})] \tag{14.1}$$

The symbols have the meanings:

α Temperature coefficient
ϑ Temperature

14.2 Overload Protection

The matching of overload protection equipment for cables and lines must be based on the following conditions (IEC 60364, Part 43):

- *Rated current rule*: The overload protection equipment should be chosen so that its rated current I_n or current setting I_e for power circuit-breakers is less than or equal to the current carrying capacity I_z of the cable or line:

$$I_B \le I_n \le I_z \tag{14.2}$$

- *Tripping rule*: The conventional tripping current I_2 must not exceed 1.45 times the current carrying capacity of the cable or line:

$$I_2 \le 1.45\,I_z \tag{14.3}$$

Here, the meanings of the symbols are

I_B Operating current
I_n Nominal current
I_z Permissible current carrying capacity
I_2 Conventional tripping current

The overload protection can be determined:

- from tables
- by calculation
- from the current carrying capacity of the line or cable.

Layout of overload protection equipment:

1. Overload protection devices must be installed at all places for which the current carrying capacity is reduced.
2. Overload protection devices are required at the beginning of the circuit.
3. Overload protection devices can be staggered as long as this does not endanger the overload protection of the circuit.
4. Overload protection can be neglected if the occurrence of an overload is not expected.
5. Overload protection must not be implemented if interruption of the circuit can pose a danger.

14.3 Short-Circuit Protection

Short-circuit currents must be interrupted before they can damage the insulation of lines, connections and terminal connections, or even operational equipment. Here a short-term overtemperature is permissible. The parameters affecting this short-term short-circuit strength and which determine the required cables and lines are:

- Current
- Length of time
- Voltage drop
- Power transferred
- Protective measures
- Electrical resistivity
- Operation temperatures of the system
- Permissible operating temperature of the cable or line
- Cross section of the cable or line
- Specific thermal capacity
- Corrosive stress
- Regulations to be observed.

The short-circuit protection can be determined:

- in accordance with IEC 60909-0
- from tables
- from measurements in the system based on instructions from the power supply company
- from the current carrying capacity of the line or cable.

Layout of overload protection equipment:

1. Overload protection devices must be installed at all places for which the current carrying capacity is reduced.
2. Overload protection equipment against short-circuits can be moved along the line or cable if the part of the line between the reduced cross section and the overload protection device is not more than 3 m long, and there is no reason to expect a fire hazard and injury to persons, or the occurrence of a short-circuit is not expected. Here it is necessary to make sure that the electrical installation is well grounded and secure against short-circuits.

Selection of overload protection equipment:

1. Common overload protection device against short-circuits and overloading
2. Calculation of the permissible cutoff time
3. Limitation of the line length

Short-circuit currents must be interrupted by overload protection devices before the temperature limits for lines and cables are reached. The calculation of the permissible cutoff time up to five seconds can be done as follows:

$$t_2 = \left(\frac{k\,S}{I''_{k1}}\right)^2 \tag{14.4}$$

When the calculated cutoff times are below 0.1 second, the matching of power circuit-breakers and circuit-breakers must then be according to the following equation:

$$k^2\,S^2 > I^2\,t \tag{14.5}$$

This condition is always satisfied when the cross section of the line is at least 1.5 mm^2 and the source fuse is 63 A.

It must always hold true that

$$t_1 < t_2 \tag{14.6}$$

The meanings of the symbols are

t_1 Cutoff time of the overload protection device in seconds
t_2 Permissible cutoff time or short-circuit duration in seconds
I_f Fault current (smallest short-circuit current) in A
k Material factor or specific conductor factor in $\frac{A\sqrt{s}}{mm^2}$ after Table 14.3
S Line cross section in mm^2

The matching of reference values for lines an overload protection equipment is illustrated in Figure 14.1.

Table 14.9, can be used for the current carrying capacity of cables or lines and for the selection of overcurrent protective equipment for protection in the event of overloading in a building wiring installation at 30 °C.

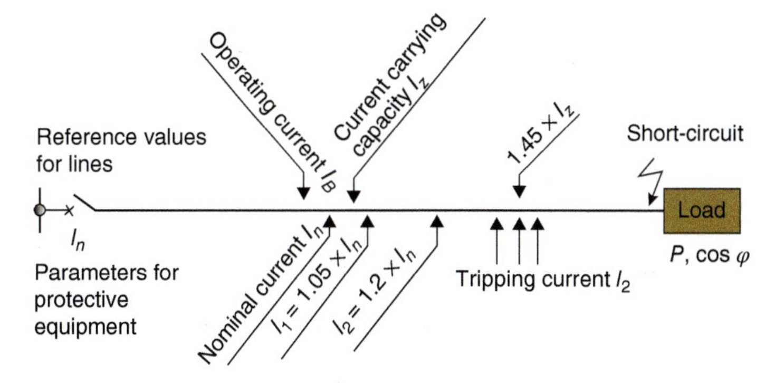

Figure 14.1 Matching of reference values for lines and overload protection equipment.

Table 14.1 Overload protection equipment, conventional tripping current, and cutoff currents.

OPE	Delayed thermal tripping device (overload protection) I_2	Short-time delay magnetic tripping device (short-circuit protection) I_5
A	$1.45\,I_n$	$3\,I_n$
B	$1.45\,I_n$	$5\,I_n$
C	$1.45\,I_n$	$10\,I_n$
D	$1.45\,I_n$	$15\,I_n$
Z	$1.2\,I_n$	$3\,I_n$
K	$1.2\,I_n$	$20\,I_n$
E	$1.2\,I_n$	$6.25\,I_n$

For disconnecting the overload and the short-circuit current, it is necessary to determine the overload and short-circuit tripping devices (Table 14.1).

It is also possible to determine the let-through energy of a circuit-breaker from the energy equation [9]:

$$m\,c\,\Delta\varphi = R \int i^2\,dt \tag{14.7}$$

The temperature rise of the conductor is then given by

$$\Delta\varphi = \frac{\rho \frac{l}{S} \int i^2\,dt}{l\,S\,\rho_m\,c} \tag{14.8}$$

$$\Delta\varphi = \frac{\int i^2\,dt}{S\,\rho_m\,c} \tag{14.9}$$

The meanings of the symbols are

$m = \rho_m \, l \, S$	Mass of conductive material
ρ_m	Density of conductive material
c	Specific thermal capacity of conductive material
$\Delta\varphi$	Temperature rise
$R = \frac{\rho_m \, l}{S}$	Resistance of conductor

For copper:

$$\rho = 0.017\,241 \,\frac{\Omega\,\text{mm}^2}{\text{m}}$$

$$\rho_m = 8.92 \,\frac{\text{kg}}{\text{dm}^3}$$

$$c = 380 \,\frac{\text{Ws}}{\text{kg}\,\text{K}}$$

$$\Delta\varphi = 5.16 \times 10^{-3} \frac{\int i^2 \, dt}{S^2} \cdot \frac{\text{mm}^4}{\text{A}^2\,\text{s}}\text{K}$$

For a known cross section of the conductor and a known maximum temperature rise, we obtain the maximum per-unit energy which a protective device can pass in the event of a short-circuit without overloading the line:

$$i^2 \, dt = 194 \, S^2 \Delta\varphi \,\frac{\text{A}^2\,\text{s}}{\text{mm}^4\text{K}} \tag{14.10}$$

In general a single-pole short-circuit can be assumed, and this must be interrupted within 0.4 second for outlet circuits up to 35 A or within five seconds at 35 A for permanently connected circuits.

For a short-circuit the following must hold true:

- The rated breaking capacity of the overcurrent protective device I_{cn} must be greater than the three-pole short-circuit current at the beginning of the line I''_{k3}.
- The current setting for the short-circuit tripping device I_{rm} must be less than or equal to the single-pole short-circuit current at the end of the line I''_{k1}.

As overcurrent protective devices the following can be used:

1. Devices which protect against overloading
2. Devices which protect against short-circuits
3. Devices which protect against both overloading and short-circuits.

14.3.1 Designation of Conductors

The conductors are described by the

- Construction type symbol
  This always begins with the letter N (standard type). Deviations from the standard are set in parentheses. If N is not given, for this conductor there is no building construction code.
- Number of cores × rated cross section in mm^2

- Number of cores × rated cross section of distributed protective conductor in mm^2
- Rated voltages for conductor in V or kV

Example for a line: PVC-sheathed line NYM-J 3 × 2.5
Here the meanings of the symbols are

N Standard type
Y PVC insulation
M sheathed conductor
J Green-yellow core (PE)
3 Three cores
2.5 Rated cross section of conductor in mm^2

The description of harmonized lines is made up of three parts. The first part describes the range of validity and the rated voltage of the line. The second part characterizes the structural elements. The third part gives the number of cores and the rated cross section, with or without green-yellow core.
Example of a line: Flexible PVC sheathed cable H05 VV-F 3 × G1.5
Here the symbols have the meanings:

H Harmonized type
05 Rated voltage 300/500 V
V PVC insulation
V PVC sheath
F Fine-wire conductor
3 Three cores
G With green-yellow core
1.5 Rated cross section of conductor in mm^2

14.3.2 Designation of Cables

The cables are designated with

- Construction type symbol
  This always begins with the letter *N* (standard type). Deviations from the standard are set in parentheses. If *N* is not given, for this line there is no building construction code.
- Symbol for form and type of conductor
- Rated cross section of shielding (if present)
- Number of cores × rated cross section in mm^2
- Number of cores × rated cross section of distributed protective conductor in mm^2
- Rated voltages of conductor in V or kV, U_0/U

Construction-type symbols for cables and line forms (Figure 14.2):
Example of a cable: PROTODUR cable NAYCWY 3 × 150 SE/150

Solid round conductors

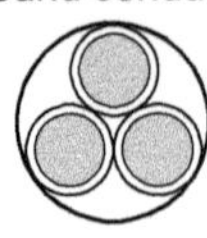

Single-wire sector conductors

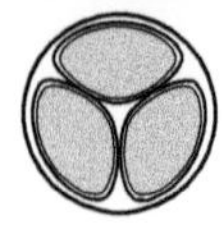

Stranded round conductors

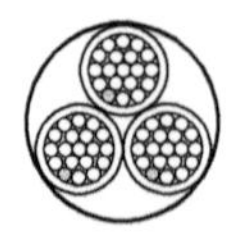

Stranded sector conductors

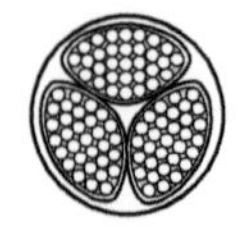

Figure 14.2 Conductor forms.

Here, the symbols have the meanings:

N	Standard type
A	Aluminum conductor
Y	PVC insulation
CW	Wave-shaped concentric conductor
Y	PVC sheath
3	Three cores
150	Rated cross section of conductor in mm^2
SE	Single-core sector-shaped conductor
150	Rated cross section of concentric conductor in mm^2

14.4 Current Carrying Capacity

The selection of the line cross section is made according to the capacity of the cable or line under normal operating conditions and under fault conditions. The current carrying capacity I_z of the line must always be greater than the operating current I_B, and this must be true for all operating modes.

Recommended values for current carrying capacities of cables and lines in buildings are found in IEC 523. The matching of protective equipment has been redefined in IEC 60 364, Part 43. The temperature is assumed to be 30 °C. Protective equipment is standardized for all cross sections.

14.4.1 Loading Capacity Under Normal Operating Conditions

The current load must be limited so that at every point of the cable and line the heat produced can be dissipated under controlled conditions to the environment.

The parameters for the current carrying capacity are

- Rated cross section of cable and line
- Line material
- Number of cores under load
- Insulation material
- Type of installation of cable and line
- Grouping of cables

- Deviating ambient temperatures
- Mode of operation.

For the dimensioning of cables and lines, conventional operating conditions such as operating mode, installation conditions, and ambient temperatures are the basis for the rated currents I_r. For other operating conditions, correction factors are used to determine the current carrying capacity, such as for underground installations:

$$I_z = I_r f_1 f_2 \Pi f \tag{14.11}$$

and for overhead installations:

$$I_z = I_r \Pi f \tag{14.12}$$

The selection of the line cross section under normal operating conditions follows from the relationship:

$$I_z \geq I_B \tag{14.13}$$

Here, the meanings of the symbols are:

I_z Current carrying capacity
I_r Load capability for conventional operating conditions
f_1 Correction factor, e.g. for deviating temperature
f_2 Correction factor, e.g. for grouping
Πf Product of all other correction factors required, e.g. multicore conductors
I_B Load under normal operating conditions

14.4.2 Loading Capacity Under Fault Conditions

With a short-circuit, lines and cables are subjected to both thermal and mechanical stresses. Lines must not heat up to above the permissible short circuit temperature. The initial temperature and the duration of the short circuit must be considered in the calculation. The thermal stressing of the lines depends on the magnitude, the behavior over time and the duration of the short-circuit. In accordance with IEC 865, the thermal equivalent short-circuit current, with the time-dependent thermal effect of the DC and AC current components of the short-circuit current, is used as the basis for testing the short-circuit strength:

$$I_{th} = I_k'' \sqrt{m + n} \tag{14.14}$$

The factor m is the thermal effect resulting from the DC component. It can be read from Figure 8.27 or calculated with the following equation for $f = 50$ Hz:

$$m = \frac{1}{2\pi\, T_k\, \ln(\kappa - 1)} [e^{4 f\, T_k\, ln(\kappa - 1)} - 1] \tag{14.15}$$

For a short-circuit lasting a time $T_k > 1$ second, we can set $m = 0$.

The factor n is the thermal effect resulting from the AC component. For a far from generator short-circuit $n = 1$.

Electrical operational equipment has a sufficient thermal short-circuit strength when:
for $T_k \leq T_{kr}$

$$I_{\text{th}} \leq I_{\text{thr}} \tag{14.16}$$

for $T_k \geq T_{kr}$

$$I_{\text{th}} \leq I_{\text{thr}} \sqrt{\frac{T_{kr}}{T_k}} \tag{14.17}$$

For the calculation of the thermal short-circuit strength for lines, instead of the current the thermal equivalent short-circuit current density can also be used, provided that the following condition is satisfied:

$$S_{\text{th}} \leq S_{\text{thr}} \sqrt{\frac{T_{kr}}{T_k} \frac{1}{\eta}} \tag{14.18}$$

The factor η takes account of the heat dissipated to the insulating material during the time of the short-circuit.

The rated short-circuit current density S_{thr} is given by

$$S_{\text{thr}} = \sqrt{\frac{\kappa_{20}\, c\, \rho}{\alpha_{20}\, T_{kr}}\, ln\, \frac{1 + \alpha_{20}\,(\varphi_e - 20\,°\text{C})}{1 + \alpha_{20}\,(\varphi a_b - 20\,°\text{C})}} \tag{14.19}$$

The meanings of the symbols are

I''_k	Initial symmetrical short-circuit current in kA
I_{th}	Thermal equivalent short-circuit current in kA
I_{thr}	Rated short-time current in kA
T_{kr}	Rated short-circuit duration in seconds
T_k	Short-circuit duration in seconds
S_{thr}	Rated short-circuit current density in A/mm^2
S_{th}	Thermal equivalent short-circuit current density in A/mm^2
φ_e	Temperature of line at end of short-circuit in °C
φ_b	Temperature of line at beginning of short-circuit in °C
ρ	Density of conductive element in g/cm^3
κ	Withstand ratio
α	Temperature coefficient for electrical resistance in 1/°C
η	Factor for heat dissipated to insulating material

The factors for Eq. (14.19) are taken from (Table 14.2).

For the limiting temperature curve, the material coefficient k for the line can be taken from Table 14.3 or calculated. Table 14.4 gives the permissible line temperatures for a Short-circuit, according to the insulating material of the cable. Table 14.5 gives the short-circuit temperatures for overhead lines.

Table 14.2 Parameters for the rated short-circuit current density.

Material	c (J/gK)	ρ (g/cm^3)	κ (m/Ω mm^2)	α_{20} (1/K)
Copper	0.39	8.9	56	0.0039
Aluminum	0.91	2.7	34.8	0.0040
Steel	0.48	7.85	7.25	0.0045

Table 14.3 Factor k.

	PVC	R	XLPE, EPR
Cu	115	135	143
Al	76	87	94

Table 14.4 Short-circuit temperatures for cables in °C.

	PVC	R	XLPE, EPR
Temperature at beginning T_a	70	60	90
Temperature at end T_e	160	200	250

Table 14.5 Short-circuit temperatures T_e for overhead lines in °C.

Single-conductor materials				Composite materials	
Al	Aldrey	Cu	St	Aldrey/Steel	Al/St
130	160	170	200	160	160

The symbols have the meanings:

PVC Polyvinyl chloride insulation
XLPE Cross-linked polyethylene insulation
EPR Ethylene-propylene rubber insulation
R Rubber insulation

14.4.3 Installation Types and Load Values for Lines and Cables

Recommended values for the current carrying capacity of cables and insulated lines for permanent installation in buildings and of flexible lines are described in IEC 523.

Table 14.6 Planning information for cables and lines.

Construction of line	Type of line, number of cores	
	Insulation material	
	Rated cross section in	mm^2
	Form and class of line	
	Rated cross section of shielding in	mm^2
	Number of wires	
	Diameter of line in	mm
	Thickness of insulation in	mm
	Thickness of sheath in	mm
	Outer diameter of line in	mm
	Weight of cable and line in	kg/km
Voltage	Rated voltage of network in	V
	Operating voltage in	V
	Power frequency in	Hz
Protective line	Protection against dangerous shock currents	
	Dimensioning of PEN and PE in	mm^2
Loading capacity under normal operating conditions	Operating mode, e.g.	CO or STO
	Installation conditions in	A
	Minimum time in	s
	Number of cores under load	
	Ambient conditions	
Loading capacity under fault conditions	Initial symmetrical short-circuit current in	A or kA
	Single-pole and three-pole short-circuit currents in	A or kA
	Duration of short-circuit in	s
	Cutoff times in	s
Mechanical properties	Minimum bending radius in	mm
	Permissible tensile force during installation in	N
Electrical properties	DC resistance per unit length at 70 °C in	Ω/km
	DC resistance per unit length at 20 °C in	Ω/km
	Working capacitance per unit length in	μF/km
	Charging current in	A/km
	Ground fault current in	A/km
Voltage drop	Losses in	W
	Power factor	
	Maximum length in	m or km
	Preselected voltage drop in	%

Table 14.6 (Continued)

Protection against overloading and short-circuit	Matching of overload protection equipment	
Maximum length	IEC 60 364, Supplementary Page 5 IEC 60 364, Part 52	
Underground installation	Current carrying capacity in	A
	Resistance per unit length at 20 °C in	Ω/km
	Resistance per unit length at 90 °C in	Ω/km
	Effective resistance per unit length in zero-sequence	Ω/km
	System in Reactance per unit length in zero-sequence	Ω/km
	System in Permissible transfer power	MVA

CO: continuous operation; STO: short time operation.

For the selection of cables and insulated lines for heavy current systems, the following conditions are essential for all tables and calculations:

Loading capacity under normal operating conditions:

1. Operating conditions and loading capacity
   - The mode of operation is continuous operation
   - The loading capacity (rated value) determined in designated I_r in all tables
   - For cables and lines, there are new types of installation
   - The loading capacity values in the tables are specified for 30 °C
   - The maximum operating temperature is 70 °C.
2. Loading capacity under fault conditions
   - IEC 865 is the reference for the calculation of the thermal Short-circuit strength.
   - The permissible short-circuit temperature for PVC lines in 160 °C for a maximum short-circuit duration of five seconds.

Overhead cables and lines:

- Load factor of 1.0
- Continuous operation, power supply company load
- Ambient temperature 30 °C
- Operating temperature 70 °C

Frequently occurring terms for the loading capacity of lines and cables as well as correction factors will be briefly explained here (Table 14.6).

- *Cable tray*: Continuous carrier plate with raised side parts, but with no cover. A cable tray is regarded as perforated when the punched holes cover at least 30% of the total surface.

Table 14.7 Reference installation types A1, A2, B1, B2, C, E, F, and G for cables and lines for permanent installation in buildings at 70 and 30 °C operating temperature [21].

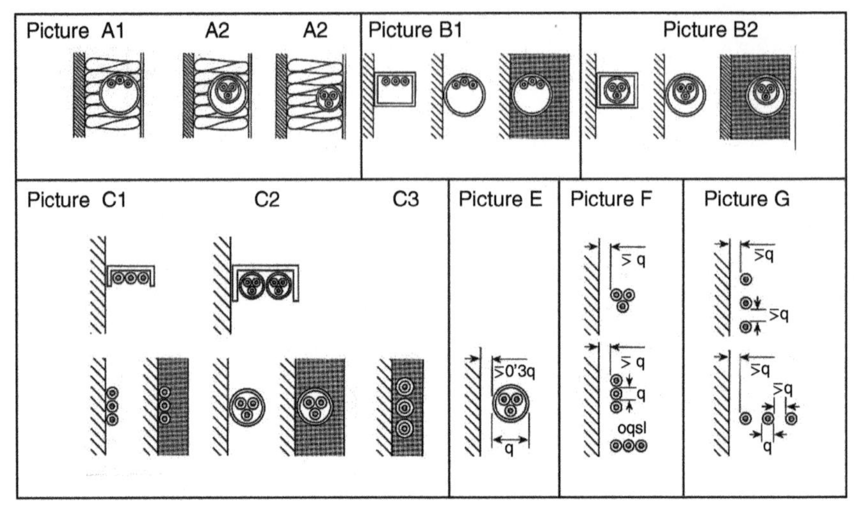

- *Cable rack*: Carrier construction in which the supporting surface is no more than 10% of the total surface.

Tables 14.7–14.17 summarize the values for the current carrying capacity of cables and lines for permanent installation in buildings and of flexible lines Current Carrying Capacity, General, Correction Factors.

14.4.4 Current Carrying Capacity of Heavy Current Cables and Correction Factors for Underground and Overhead Installation

Current carrying capacity in accordance with DIN VDE 0276, Part 603, Correction factors in accordance with DIN VDE 0276, Part 1000

The following conditions are essential for the selection of cables based on all line cross-section tables and calculations for the current carrying capacity of underground cables and overhead cables:

Underground cables

- Power supply company load with a load factor of 0.7
- Installation depth 0.7–1.2 m
- Ground temperature 20 °C
- Thermal resistivity of soil 1.0 K · m/W
- Number of cables/systems 1

Overhead cables and lines

- Load factor of 1.0
- Continuous operation, power supply company load

Table 14.8 Reference installation types A1, A2, B1, B2, C, E, F, and G for cables and lines for permanent installation in buildings at 70 and 30 °C operating temperature [21].

Installation type	A1	A2		B1	B2		C		E	F	G
Illustration	**(A1)**	**(A2)**	**(A2)**	**(B1)**	**(B2)**	**(C1)**	**(C2)**	**(C3)**	**(E)**	**(F)**	**(G)**
Installation conditions	Installation in thermally insulated walls			Installation in electric wiring conduits or closed electric wiring ducting on or in walls or in ducting for underfloor installation		Direct installation or in walls/ceilings or in cable trays		Flat-webbed cable in walls/ ceilings or hollow spaces	Overhead installation, on catenary wires and on cable racks and consoles		
	Non sheathed cables or single-core cables/sheathed cables in electric wiring conduit or ducting	Multicore cables or sheathed cables									
		In electric wiring conduit or ducting	Direct installation	Nonsheathed cables or single- core cables/ sheathed cables	Multicore cables or sheathed cables	Single-core cab- les or sheathed cables	Multicore cables or sheathed cables		Multicore cables or sheathed cables	Single-core cables/ sheathed cables	
										With contact	Without contact, including non-sheathed cables on insulators

Table 14.9 Loading capacity of cables and lines for permanent installation in buildings at 70 and 30 °C operating temperature.

Installation type	A1		A2		B1		B2		C		E		F			G	
Installation	In thermally insulated walls				In electric wiring conduits				Direct				Overhead				
Number of cores simultaneously under load	2	3	2	3	2	3	2	3	2	3	2	3	2	3	3	Horizontal 3	Vertical 3
Nominal cross section in mm²	Current carrying capacity I_z in A																
1.5	15.5	13.5	15.5	13.0	17.5	15.5	16.5	15.0	19.5	17.5	22	18.5	—	—	—	—	—
2.5	19.5	18.0	18.5	17.5	24	21	23	20	27	24	30	25	—	—	—	—	—
4	26	24	25	23	32	28	30	27	36	32	40	34	—	—	—	—	—
6	34	31	32	29	41	36	38	34	46	41	51	43	—	—	—	—	—
10	48	42	43	39	57	50	52	46	63	57	70	60	—	—	—	—	—
16	61	56	57	52	76	68	69	62	85	76	94	80	—	—	—	—	—
25	80	73	75	68	101	89	90	80	112	96	119	101	131	114	110	146	130
35	99	89	92	83	125	110	111	99	138	119	148	126	162	143	137	181	162
50	119	108	110	99	151	134	133	118	168	144	180	153	196	174	167	219	197
70	151	136	139	125	192	171	168	149	213	184	232	196	251	225	216	281	254
95	182	164	167	150	232	207	201	179	258	223	282	238	304	275	264	341	311
120	210	188	192	172	269	239	232	206	299	259	328	276	352	321	308	396	362

Source: ABB STOTZ [21].

Table 14.10 Recommended values for the current carrying capacity of cables and lines with copper conductor and PVC insulation for permanent installation in buildings, continuous operation, operating temperature 70 °C, ambient temperature 25 °C, and matching of rated current of overload protection equipment to conventional tripping current.

Installation type		A1		A2		B1		B2		C		E	
Number of cores simultaneously under load		2	3	2	3	2	3	2	3	2	3	2	3
Nominal cross section in mm²		Current carrying capacity I_z in A Nominal current I_n in A											
1.5	I_z	16.5	14.5	16.5	14.0	18.5	16.5	17.5	16.0	21	18.5	23	19.5
	I_n	16	13	16	13	16	16	16	16	20	16	20	16
2.5	I_z	21	19	19.5	18.5	25	22	24	21	29	25	32	27
	I_n	20	16	16	16	25	20	20	20	25	25	32	25
4	I_z	28	25	27	24	34	30	32	29	38	34	42	36
	I_n	25	25	25	20	32	25	32	25	35	35	40	35
6	I_z	36	33	34	31	43	38	40	38	49	43	54	46
	I_n	35	32	32	25	40	35	40	35	40	40	50	40
10	I_z	49	45	46	41	60	53	55	49	67	60	74	64
	I_n	40	40	40	40	50	50	50	40	63	50	63	63
16	I_z	65	59	60	55	81	72	73	66	90	81	100	85
	I_n	63	50	50	50	80	63	63	63	80	80	100	80
25	I_z	85	77	80	72	107	94	95	85	119	102	126	107
	I_n	80	63	80	63	100	80	80	80	100	100	125	100
35	I_z	105	94	98	88	133	117	118	105	146	126	157	134
	I_n	100	80	80	80	125	100	100	100	125	125	125	125
50	I_z	126	114	117	105	160	142	141	125	178	153	191	162
	I_n	125	100	100	100	160	125	125	125	160	125	160	160

Source: ABB STOTZ [21].

- Ambient temperature 30 °C
- Operating temperature 70 °C

Tables 14.18–14.28 list recommended values for the current carrying capacity of heavy current cables.

The permissible continuous current carrying capacity for overhead lines (Tables 14.26 and 14.27) applies for DC and AC up to 60 Hz, for a wind speed of 0.6 m/s and a temperature of 35 °C

The continuous current carrying capacity of busbars (Table 14.28) applies for an ambient temperature of 35 °C and takes a temperature rise of 30 K, resulting from the current heating losses, into account.

Table 14.11 Recommended values for the current carrying capacity of cables and lines with copper conductor and PVC insulation for permanent installation in buildings, continuous operation, operating temperature 70 °C, ambient temperature 40 °C, and matching of rated current of overload protection equipment to conventional tripping current in accordance with EN 60204, Part 1: 1993-06.

Installation type		B1	B2	C	E
Nominal cross section in mm²		Current carrying capacity I_z in A Nominal current I_n in A			
0.75	I_z	7.6	—	—	—
	I_n	6	—	—	—
1.0	I_z	10.4	9.6	11.7	11.5
	I_n	10	8	10	10
1.5	I_z	13.5	12.2	15.2	16.1
	I_n	13	10	13	16
2.5	I_z	18.3	16.5	21	22
	I_n	16	16	20	20
4	I_z	25	23	28	30
	I_n	25	20	25	25
6	I_z	32	29	36	37
	I_n	32	25	35	35
10	I_z	44	40	50	52
	I_n	40	40	50	50
16	I_z	60	53	66	70
	I_n	50	50	63	63
25	I_z	77	67	84	88
	I_n	63	63	80	80
35	I_z	97	83	104	114
	I_n	80	80	100	100
50	I_z	—	—	123	123
	I_n	—	—	100	100

Source: ABB STOTZ [21].

14.5 Examples of Current Carrying Capacity

14.5.1 Example 1: Checking Current Carrying Capacity

Given Figure 14.3.

- 8 NYM-J 5 × 1.5 mm² Cu-conductor sheathed lines, grouped directly on the wall, in contact with each other (installation type C)
- Applicable standards: IEC 60 364, Part 43.

Table 14.12 Correction factors for grouping of cables and lines with rated load for continuous operation.

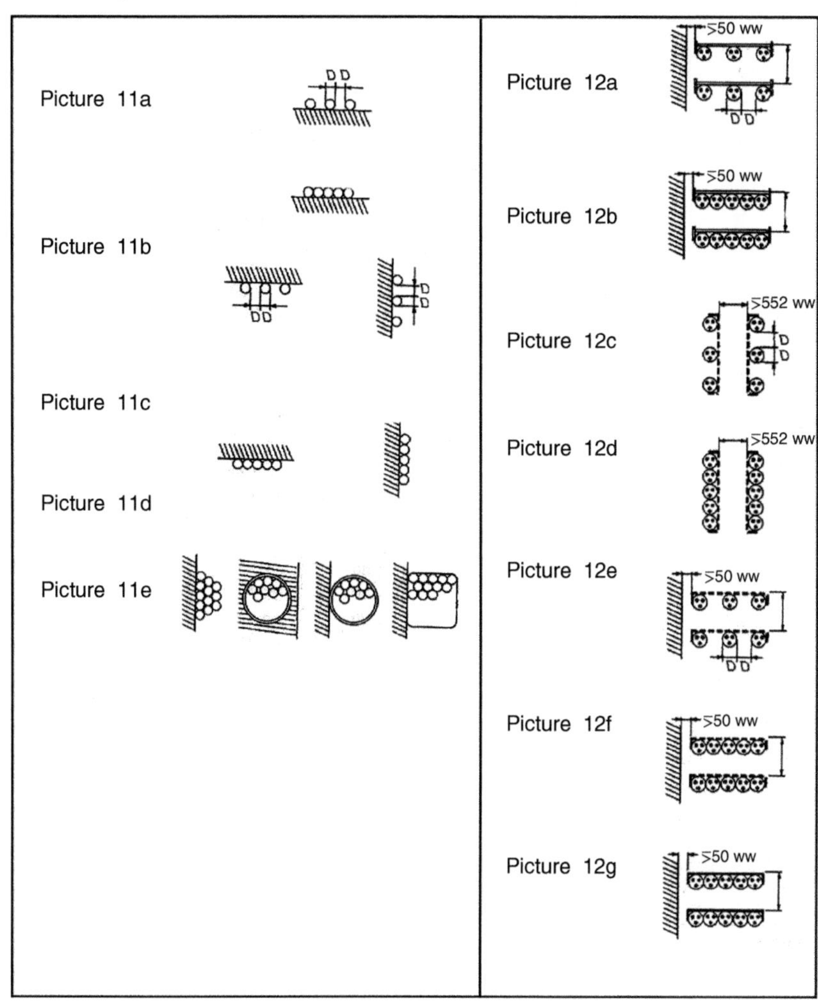

Source: ABB STOTZ [21].

Figure 14.3 Checking the current carrying capacity

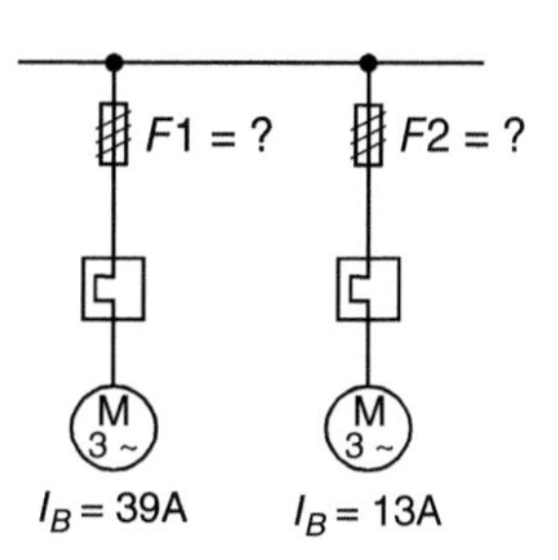

Table 14.13 Correction factors for grouping of cables and lines with rated load for continuous operation.

Layout of installation	Number of multicore cables or lines or number of AC or three-phase circuits with single-conductor cables or lines (two or three current-carrying conductors)														
	1	**2**	**3**	**4**	**5**	**6**	**7**	**8**	**9**	**10**	**12**	**14**	**16**	**18**	**20**
Grouped directly on the wall, on the floor, in electric wiring conduit or ducting, on or in the wall (Picture 11a)	1.00	0.80	0.70	0.65	0.60	0.57	0.54	0.52	0.50	0.48	0.45	0.43	0.41	0.39	0.38
Single layer on the wall or on the floor, with contact (Picture 11b)	1.00	0.85	0.79	0.75	0.73	0.72	0.72	0.71	0.70	0.70	0.70	0.70	0.70	0.70	0.70
Single layer on the wall or on the floor, with spacing equal to outer diameter *d* (Picture 11c)	1.00	0.94	0.90	0.90	0.90	0.90	0.90	0.90	0.90	0.90	0.90	0.90	0.90	0.90	0.90
Single layer under ceiling, with contact (Picture 11d)	0.95	0.81	0.72	0.68	0.66	0.64	0.63	0.62	0.61	0.61	0.61	0.61	0.61	0.61	0.61
Single layer under ceiling, with spacing equal to outer diameter *d* (Picture 11e)	0.95	0.85	0.85	0.85	0.85	0.85	0.85	0.85	0.85	0.85	0.85	0.85	0.85	0.85	0.85

Source: ABB STOTZ [21].

Table 14.14 Correction factors for grouping of multicore cables and lines on cable trays and racks [21].

Layout of installation		Number of trays or racks	Number of multicore cables or lines					
			1	2	3	4	6	9
Unperforated cable trays[a)]	(Picture 12a)	1	0.97	0.84	0.78	0.75	0.71	0.68
		2	0.97	0.83	0.76	0.72	0.68	0.63
		3	0.97	0.82	0.75	0.71	0.66	0.61
		6	0.97	0.81	0.73	0.69	0.63	0.58
Perforated cable trays	(Picture 12b)	1	1.00	0.88	0.82	0.79	0.76	0.73
		2	1.00	0.87	0.80	0.77	0.73	0.68
		3	1.00	0.86	0.79	0.76	0.71	0.66
		6	1.00	0.84	0.77	0.73	0.68	0.64
	(Picture 12c)	1	1.00	1.00	0.98	0.95	0.91	—
		2	1.00	0.99	0.96	0.92	0.87	—
		3	1.00	0.98	0.95	0.91	0.85	—
	(Picture 12d)	1	1.00	0.88	0.82	0.78	0.73	0.72
		2	1.00	0.88	0.81	0.76	0.71	0.70
	(Picture 12e)	1	1.00	0.91	0.89	0.88	0.87	—
		2	1.00	0.91	0.88	0.87	0.85	—
Cable racks	(Picture 12f)	1	1.00	0.87	0.82	0.80	0.79	0.78
		2	1.00	0.86	0.81	0.78	0.76	0.73
		3	1.00	0.85	0.79	0.76	0.73	0.70
		6	1.00	0.83	0.76	0.73	0.69	0.66
	(Picture 12g)	1	1.00	1.00	1.00	1.00	1.00	—
		2	1.00	0.99	0.98	0.97	0.96	—
		3	1.00	0.98	0.97	0.96	0.93	—

a) Perforations cover less than 30% of total surface.
Source: ABB STOTZ [21].

Table 14.15 Correction factors for spooled lines.

Number of layers	Correction factors
1	0.80
2	0.61
3	0.49
4	0.42
5	0.38

For spiral-shaped spooling the correction factor 0.80 is used.

Table 14.16 Correction factors for multicore cables and lines with $S \leq 10\,\text{mm}^2$.

Number of cores under load	Correction factors
5	0.75
7	0.65
10	0.55
14	0.50
19	0.45
24	0.40
40	0.35
61	0.30

Find:

1. The required protective equipment
2. The line cross sections, taking account of all reduction and correction factors.

Protective equipment: Both motors have 16 A fuses. Using the rule of nominal currents

$$I_B \leq I_n \leq I_z = 15\,\text{A} \leq 16\,\text{A} \leq I_z$$

Calculation of the line cross sections:

The reduction factor for a different temperature: from 25 to 30 °C, from Table 14.17, is 1.06

The reduction factor for grouping, from Table 14.12, is 0.52

The required loading of the line is given by

$$I'_z = \frac{I_n}{f_1 f_2} = \frac{16\,\text{A}}{1.06 \cdot 0.52} = 29.03\,\text{A}$$

Table 14.17 Correction factors for deviating ambient temperatures.

Permissible operating temperature (°C)	40	60	70	80	85	90
Ambient temperature (°C)			**Correction factors**			
10	1.73	1.29	1.22	1.18	1.17	1.15
15	1.58	1.22	1.17	1.14	1.13	1.12
20	1.41	1.15	1.12	1.10	1.09	1.08
25	1.22	1.08	1.06	1.05	1.04	1.04
30	1.00	1.00	1.00	1.10	1.00	1.00
35	0.71	0.91	0.94	0.95	0.95	0.96
40	—	0.82	0.87	0.89	0.90	0.91
45	—	0.71	0.79	0.84	0.85	0.87
50	—	0.58	0.71	0.77	—	0.82
55	—	0.41	0.61	0.71	—	0.76
60	—	—	0.50	0.63	—	0.71
65	—	—	0.35	0.55	—	0.65
70	—	—	—	0.45	—	0.58
75	—	—	—	0.32	—	0.50
80	—	—	—	—	—	0.41
85	—	—	—	—	—	0.29

Selection of lines: NYM-J $5 \times 4\,\text{mm}^2$, from Table 14.19 $I_{z\ Tab.} = 32$ A actual current carrying capacity:

$$I_z = I_{z\ Tab.} f_1 f_2 = 32\ \text{A} \cdot 0.52 \cdot 1.06 = 17.3\ \text{A}$$

It then follows that:

$$I_B \leq I_n \leq I_z = 15\ \text{A} \leq 16\ \text{A} \leq 17.3\ \text{A}$$

The requirement is therefore satisfied.

14.5.2 Example 2: Checking Current Carrying Capacity

Check the current carrying capacity of 12 NYM-J $3 \times 1.5\,\text{mm}^2$ lines in a conduit, surface mounted (Figure 14.4). The coincidence factor, according to information from the manufacturer, is 0.8, the ambient temperature is 25 °C, the installation type B2 and $I_B = 6$ A.

Find the current carrying capacity and the fusing.

Correction factors:

Reduction factor for temperature, from Table 14.17: $f_1 = 1.06$

Table 14.18 Loading capacity, underground overhead installation, layout of installation for Tables 11.16 and 11.17.

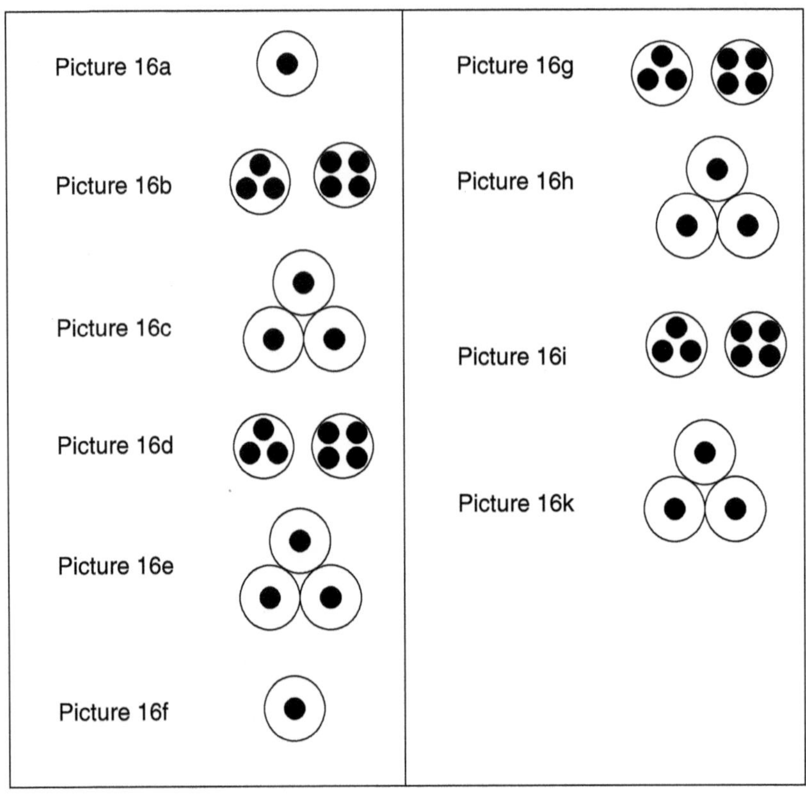

Source: ABB STOTZ [21].

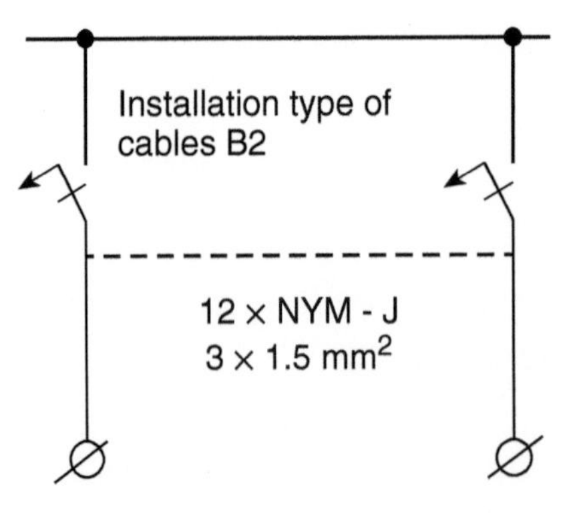

Figure 14.4 Checking the current carrying capacity.

Table 14.19 Loading capacity, underground installation, cable with $U_0/U = 0.6/1\,kV$.

Insulation material	**PVC**									
Permissible operating temperature	**70 °C**									
Construction type symbol	**NYY**			**NYCWY**		**NAYY**			**NAYCWY**	
Layout	**(a)[a]**	**(b)**	**(c)**	**(d)**	**(e)**	**(f)[a]**	**(g)**	**(h)**	**(i)**	**(k)**
Number of cores under load	**1**	**3**	**3**	**3**	**3**	**1**	**3**	**3**	**3**	**3**
Cross section in mm²	**Cu-Conductor Rated current in A**					**Alu-Conductor Rated current in A**				
1.5	41	27	30	27	31	—	—	—	—	—
2.5	55	36	39	36	40	—	—	—	—	—
4	71	47	50	47	51	—	—	—	-	—
6	90	59	62	59	63	—	—	—	—	—
10	124	79	83	79	84	—	—	—	—	—
16	160	102	107	102	108	—	—	—	—	—
25	208	133	138	133	139	160	102	106	103	108
35	250	159	164	160	166	193	123	127	123	129
50	296	188	195	190	196	230	144	151	145	153
70	365	232	238	234	238	283	179	185	180	187
95	438	280	286	280	281	340	215	222	216	223
120	501	318	325	319	315	389	245	253	246	252
150	563	359	365	357	347	436	275	284	276	280
185	639	406	413	402	385	496	313	322	313	314
240	746	473	479	463	432	578	364	375	362	358
300	848	535	541	518	473	656	419	425	415	397
400	975	613	614	579	521	756	484	487	474	441
500	1125	687	693	624	574	873	553	558	528	489
630	1304	—	777	—	636	1011	—	635	—	539
800	1507	—	859	—	—	1166	—	716	—	—
1000	1715	—	936	—	—	1332	—	796	—	—

a) Rated current in DC systems with return line far away.

Table 14.20 Loading capacity, overhead installation, cable with $U_0/U = 0.6/1$ kV.

Insulation material	PVC									
Permissible operating temperature	70 °C									
Construction type symbol	NYY			NYCWY		NAYY			NAYCWY	
Layout	(a)[a]	(b)	(c)	(d)	(e)	(f)[a]	(g)	(h)	(i)	(k)
Number of cores under load	1	3	3	3	3	1	3	3	3	3
Cross section in mm²	Cu-Conductor Rated current in A					Alu-Conductor Rated current in A				
1.5	27	19.5	21	19.5	22	—	—	—	—	—
2.5	35	25	28	26	29	—	—	—	—	—
4	47	34	37	34	39	—	—	—	—	—
6	59	43	47	44	49	—	—	—	—	—
10	81	59	64	60	67	—	—	—	—	—
16	107	79	84	80	89	—	—	—	—	—
25	144	106	114	108	119	110	82	87	83	91
35	176	129	139	132	146	135	100	107	101	112
50	214	157	169	160	177	166	119	131	121	137
70	270	199	213	202	221	210	152	166	155	173
95	334	246	264	249	270	259	186	205	189	212
120	389	285	307	289	310	302	216	239	220	247
150	446	326	352	329	350	345	246	273	249	280
185	516	374	406	377	399	401	285	317	287	321
240	618	445	483	443	462	479	338	378	339	374
300	717	511	557	504	519	555	400	437	401	426
400	843	597	646	577	583	653	472	513	468	488
500	994	669	747	626	657	772	539	600	524	556
630	1180	—	858	—	744	915	—	701	—	628
800	1396	—	971	—	—	1080	—	809	—	—
1000	1620	—	1078	—	—	1258	—	916	—	—

a) Rated current in DC systems with return line far away.

Table 14.21 Correction factors for grouping of overhead lines, single-core cables in three-phase systems.

Layout of installation **Level installation** **Spacing = cable diameter *d***		**Number of trays/racks over one another**	**Number of systems[b] next to one another**		
			1	**2**	**3**
On the ground	(Picture 19a)	1	0.92	0.89	0.88
Unperforated cable trays[c]	(Picture 19b)	1	0.92	0.89	0.88
		2	0.87	0.84	0.83
		3	0.84	0.82	0.81
		6	0.82	0.80	0.79
Perforated cable trays[c]	(Picture 19c)	1	1.0	0.93	0.90
		2	0.97	0.89	0.85
		3	0.96	0.88	0.82
		6	0.94	0.85	0.80
Cable racks[d] (cable gratings)	(Picture 19d)	1	1.00	0.97	0.96
		2	0.97	0.94	0.93
		3	0.96	0.93	0.92
		6	0.94	0.91	0.90
On supporting frames or on the wall or laid out vertically on perforated cable trays	(Picture 19f)	Number of trays/racks over one another	Number of systems over one another		
			1	2	3
		1	0.94	0.91	0.89
		2	0.94	0.90	0.86
Layout of installation **Grouped installation** **Spacing = 2*d***		**Number of trays/racks over one another**	**Number of systems[b] next to one another**		
			1	**2**	**3**
On the ground	(Picture 19g)	1	0.98	0.96	0.94
Unperforated cable trays[c]	(Picture 19h)	1	0.98	0.96	0.94
		2	0.95	0.91	0.87
		3	0.94	0.90	0.85
		6	0.93	0.88	0.82

Table 14.21 (Continued)

Layout of installation Grouped installation Spacing = 2*d*		Number of trays/racks over one another	Number of systems[b)] next to one another		
			1	2	3
Perforated cable trays[c)]	(Picture 19i)	1	1.0	0.98	0.96
		2	0.97	0.93	0.89
		3	0.96	0.92	0.85
		6	0.95	0.90	0.83
Cable racks[d)] (cable gratings)	(Picture 19j)	1	1.00	1.00	1.00
		2	0.97	0.95	0.93
		3	0.96	0.94	0.90
		6	0.95	0.93	0.87
On supporting frames or on the wall or laid out vertically on perforated cable trays	(Picture 19k)	Number of trays/racks over one another	Number of systems over one another		
			1	2	3
		1	1.0	0.91	0.89
		2	1.0	0.90	0.86

a) In confined spaces or for the grouping of large number of cables if the heat loss from the cables causes a temperature rise in the air, the correction factors for deviating air temperatures in Table 12 must also be used (see 5.3.2.3).
b) Factors in accordance with CENELEX report R064.001 for HD 384.5.523:1991.
c) A cable tray is s continuous carrier plate with raised side parts, but with no cover. A cable tray is regarded as perforated when the punched holes cover at least 30% of the total surface.
d) A cable rack is a carrier construction in which the supporting surface is no more than 10% of the total surface.

For grouped installation, no load reduction is necessary if the spacing between neighboring systems is at least four times the cable diameter, provided that there is no rise in the ambient temperature as a result of heat losses.

Reduction factor for grouping, from Table 14.12: $f_2 = 0.45$
Coincidence: 12 lines · 0.8 = 9.6 lines
$I_{z\,Tab.} = 16.5$ A, so that:

$$I'_z = I_{z\,Tab.} f_1 f_2 = 16.5\text{ A} \cdot 1.06 \cdot 0.45 = 7.87\text{ A}$$

With a 6 A fuse:

$$I_B \leq I_n \leq I_z = 6\text{ A} \leq 6\text{ A} \leq 7.87\text{ A}$$

14.5.3 Example 3: Protection of Cables in Parallel

An additional NYY-J 4 × 50 mm² cable (Figure 14.5) was laid in parallel with the existing NYCWY 4 × 70 mm² cable. Find the fuse required for the cables in parallel.

Table 14.22 Correction factors for grouping of overhead lines, multiple-core cables and single-core DC cables, layout of installation.

Layout of installation Spacing = cable diameter *d*		Number of trays/racks over one another	Number of cables[d] next to one another				
			1	**2**	**3**	**4**	**6**
On the ground	(Picture 20a)	1	0.97	0.96	0.94	0.93	0.90
Unperforated cable trays[b]	(Picture 20b)	1	0.97	0.96	0.94	0.93	0.90
		2	0.97	0.95	0.92	0.90	0.86
		3	0.97	0.94	0.91	0.89	0.84
		6	0.97	0.93	0.90	0.88	0.83
Perforated cable trays[b]	(Picture 20c)	1	1.0	1.0	0.98	0.95	0.91
		2	1.0	0.99	0.96	0.92	0.87
		3	1.0	0.98	0.95	0.91	0.85
		6	1.0	0.97	0.94	0.90	0.84
Cable racks[c] (cable gratings)	(Picture 20d)	1	1.0	1.0	1.0	1.0	1.0
		2	1.0	0.99	0.98	0.97	0.96
		3	1.0	0.98	0.97	0.96	0.93
		6	1.0	0.97	0.96	0.94	0.91
On supporting frames or on the wall or laid out vertically on perforated trays	(Picture 20e)		Number of cables over one another				
			1	2	3	4	6
		1	1.0	0.91	0.89	0.88	0.87
		2	1.0	0.91	0.88	0.87	0.85

Layout of installation In contact with one another		Number of trays/racks over one another	Number of cables[d] next to one another					
			1	**2**	**3**	**4**	**6**	**9**
On the ground	(Picture 20f)	1	0.97	0.96	0.94	0.93	0.90	0.90
Unperforated cable trays[b]	(Picture 20g)	1	0.97	0.96	0.94	0.93	0.90	0.90
		2	0.97	0.95	0.92	0.90	0.86	0.90
		3	0.97	0.94	0.91	0.89	0.84	0.90
		6	0.97	0.93	0.90	0.88	0.83	0.90
Perforated cable trays[b]	(Picture 20h)	1	1.0	1.0	0.98	0.95	0.91	0.90
		2	1.0	0.99	0.96	0.92	0.87	0.90
		3	1.0	0.98	0.95	0.91	0.85	0.90
		6	1.0	0.97	0.94	0.90	0.84	0.90

Table 14.22 (Continued)

Layout of installation In contact with one another		Number of trays/racks over one another	Number of cables[d)] next to one another					
			1	2	3	4	6	9
Cable racks[c)] (cable gratings)	(Picture 20i)	1	1.0	1.0	1.0	1.0	1.0	0.90
		2	1.0	0.99	0.98	0.97	0.96	0.90
		3	1.0	0.98	0.97	0.96	0.93	0.90
		6	1.0	0.97	0.96	0.94	0.91	0.90
Laid out vertically on perforated cable trays	(Picture 20j)		Number of cables over one another					
			1	2	3	4	6	9
		1	1.0	0.91	0.89	0.88	0.87	0.87
		2	1.0	0.91	0.88	0.87	0.85	0.87
Laid out on supporting frames or on the wall	(Picture 20k)		1.0	0.91	0.89	0.88	0.87	0.87

a) In confined spaces or for the grouping of large number of cables if the heat loss from the cables causes a temperature rise in the air, the correction factors for deviating air temperatures in Table 12 must also be used (see 5.3.2.3).
b) A cable tray is s continuous carrier plate with raised side parts, but with no cover. A cable tray is regarded as perforated when the punched holes cover at least 30% of the total surface.
c) A cable rack is a carrier construction in which the supporting surface is no more than 10% of the total surface.
d) Factors in accordance with CENELEC report R064.001 for HD 384.5.523:1991.
No load reduction is necessary if the horizontal or vertical spacing between neighboring systems is at least twice the cable diameter, provided that there is no rise in the ambient temperature as a result of heat losses.

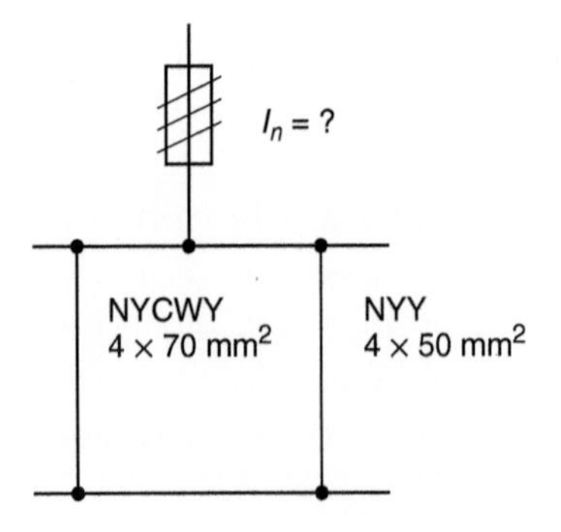

Figure 14.5 Protection of cables in parallel.

Table 14.23 Correction factors for underground grouping of cables, ground temperature 20 °C, thermal resistivity of soil 1.0 K·m/W, spacing between cables/Systems 7 cm, delta installation for single-core cables.

Load factor 1							
Insulation	**Cable type**	**Number of cables/systems**					
		1	**2**	**4**	**6**	**8**	**10**
Paper	Belted cable 1–6 kV	0.82	0.66	0.52	0.47	0.43	0.40
PVC	Multicore cable 1–6 kV	0.81	0.66	0.52	0.46	0.43	0.40
XLPE	Multicore cable 1–30 kV	0.81	0.66	0.52	0.47	0.43	0.41
Load factor 0.7							
			2	**4**	**6**	**8**	**10**
Paper	Belted cable 1–6 kV		0.86	0.72	0.65	0.61	0.58
PVC	Multicore cable 1–6 kV		0.86	0.71	0.64	0.60	0.57
XLPE	Multicore cable 1–30 kV		0.85	0.70	0.63	0.59	0.56

Conditions: Ambient temperature 25 °C, overhead installation, two trays each with one cable. The current carrying capacity I_z for NYCWY $4 \times 70\,\text{mm}^2$, from Table 14.20, is 202 A.

Correction factor for the ambient temperature, from Table 14.24: $f_1 = 1.06$, correction factor for grouping, from Table 14.22: $f_2 = 0.97$

$$I_z = 202 \times 1.06 \times 0.97 = 207.7\,\text{A}$$

$$I_{zges} = I_z \left(1 + \frac{S_1}{S_{\max}}\right)$$

$$I_{zges} = 207.7\,\text{A} \left(1 + \frac{50\,\text{mm}^2}{70\,\text{mm}^2}\right) = 356.06\,\text{A}$$

$$I_n(gL) = 315\,\text{A (NH2)} \qquad \text{selected}$$

14.5.4 Example 4: Connection of a Three-Phase Cable

Given the connection of a three-phase NYY-J $4 \times 10\,\text{mm}^2$ cable 20 m in length (Figure 14.6) to a boiler with 30 kW, protection by circuit-breaker, C characteristic. The loop resistance is $Z_S = 200\,\text{m}\Omega$, $U = 400\,\text{V}$. Find the permissible length in accordance with IEC 60 364, Supplementary Page 5.

Table 14.24 Correction factors for deviating ambient temperatures.

Permissible operating temperature (°C)	Air temperature (°C)								
	10	**15**	**20**	**25**	**30**	**35**	**40**	**45**	**50**
90	1.15	1.12	1.08	1.04	1.00	0.96	0.91	0.87	0.82
80	1.18	1.14	1.10	1.05	1.00	0.95	0.89	0.84	0.77
70	1.22	1.17	1.12	1.06	1.00	0.94	0.87	0.79	0.71
65	1.12	1.20	1.13	1.07	1.00	0.93	0.85	0.76	0.65
60	1.29	1.22	1.15	1.08	1.00	0.91	0.82	0.71	0.58

Table 14.25 Correction factors for multicore cables with conductor cross sections from 1.5 to 10 mm^2; underground or overhead installation.

Number of cores under load	Underground installation	Overhead installation
5	0.70	0.75
7	0.60	0.65
10	0.50	0.55
14	0.45	0.50
19	0.40	0.45
24	0.35	0.40
40	0.30	0.35
61	0.25	0.30

Solution:

- Operating current $I_B = \frac{P}{\sqrt{3}\,U} = \frac{30\ \text{kW}}{\sqrt{3}\cdot 400\ \text{V}} = 43.3\ \text{A}$
- Rated current of overload protection device $I_n = 50\ \text{A}$

From Supplementary Page 5 of IEC 60 364, Table 6, the following values are found: Maximum line length = 55 m, minimum short-circuit current = 500 A.

Line length for the given voltage drop $\Delta U = 3\%$.

$$l_{\text{max}} = 9.56\ \text{m}\ \frac{400\ \text{V}}{43.3\ \text{A}} = 88.3\ \text{m}$$

14.5.5 Example 5: Apartment Building Without Electrical Water Heating

An apartment building without electrical water heating, with a central meter mounting board is to be planned according to Figure 14.7. Answer the following questions:

Table 14.26 Continuous current carrying capacity of overhead lines.

S_r	S_s	Number × d		Continuous current carrying capacity (A)		
(mm²)	(mm²)	(mm)	d_s	Copper	Aluminum	AAAC
16	15.89	7 × 1.7	5.1	125	110	105
25	24.25	7 × 2.1	6.3	160	145	135
35	34.36	7 × 2.5	7.5	200	180	170
50	48.35	19 × 1.8	9.0	250	225	210
70	65.81	19 × 2.1	10.5	310	270	255
95	93.27	19 × 2.5	12.5	380	340	320
120	117.00	19 × 2.8	14.0	440	390	365

The meanings of the symbols are

S_r Rated cross section
S_s Specified cross section
d Diameter
d_s Line diameter

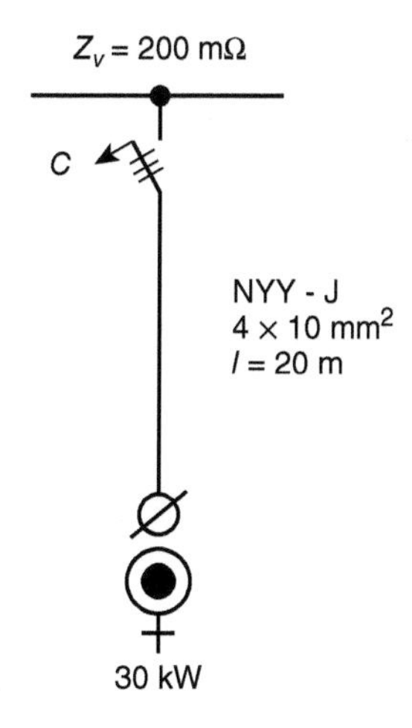

Figure 14.6 Determination of line length for a boiler.

A Determination of power requirement

1. Central meter mounting board, 15 apartment units on a main line, 100 A fuse, $P_1 \approx 64$ kVA, NH00 100 A fuse
2. Total requirement $P_2 = 11\text{ kVA} \cdot 0.7 = 7.7\text{ kVA}$
3. Office $P_3 = 80\text{ kVA} \cdot 0.7 = 56\text{ kVA}$
4. $\Sigma P_G = P_1 + P_2 + P_3 = 127.7\text{ kVA}$

Table 14.27 Continuous current carrying capacity of overhead lines.

S_r (mm^2)	S_s (mm^2)	Al/St ratio	Continuous current carrying capacity (A)
16/2.5	17.8	5.4	105
25/4	27.8	6.8	140
35/6	40.1	8.1	170
50/8	56.3	9.6	210
70/12	81.3	11.7	290
95/15	109.7	13.6	350
120/20	141.4	15.5	410

Table 14.28 Continuous current carrying capacity of busbars.

Width × Thickness	Cross section	Loading capacity of busbars (A)	
(mm)	(mm^2)	Enamelled	Without enamel
12 × 2	23.5	202	182
15 × 2	29.5	240	212
20 × 2	39.5	302	264
20 × 3	59.5	394	348
25 × 3	74.5	470	412
30 × 3	89.5	544	476
40 × 3	119	692	600
50 × 5	249	1140	994
60 × 5	299	1330	1150
80 × 5	399	1680	1450
100 ×5	499	2010	1730
120 ×10	1200	3280	2860
200 ×10	2000	4970	4310

B Calculation of the rated currents, selection of lines and overcurrent protection devices:

a) *Operating current with* $\cos\varphi = 0.95$:

$$I_B = \frac{P}{\sqrt{3}\,U\,\cos\varphi} = \frac{127.7\ \text{kVA}}{\sqrt{3}\cdot 400\ \text{V}\cdot 0.95} = 194\ \text{A}$$

b) *Protective equipment in service panel: NH1-200* A

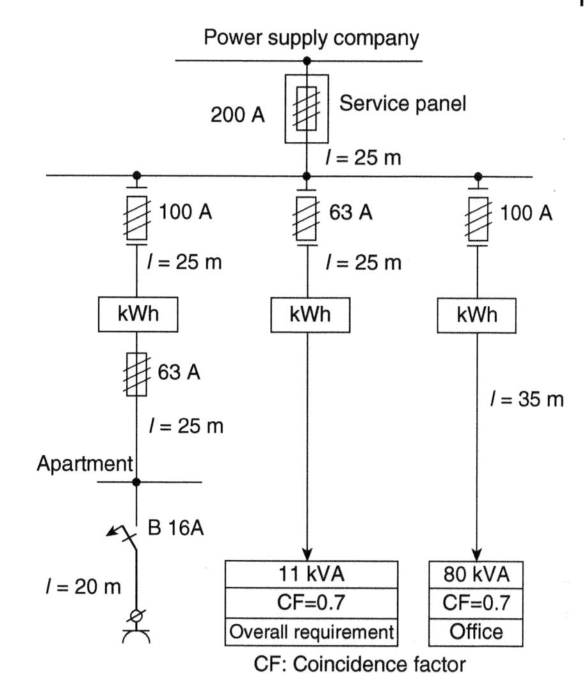

Figure 14.7 Apartment building with a central meter mounting board.

c) *Selection of line*:
   Wall-mounted installation, 25 °C, no grouping. Correction factor for temperature: $f_1 = 1.06$, correction factor for wall-mounted installation: $f_2 = 0.95$.

$$I_z = \frac{I_n}{f_1 f_2} = \frac{200\,\text{A}}{1.06 \cdot 0.95} = 198.6\,\text{A}$$

   From Table 14.20, I_z' is 202 A with NYCWY 4 × 70 mm^2.

1. *Feeder to meter mounting board for overall requirement*:
   Given: $I_n = 63$ A with NYM-J 5 × 16 m^2 (TN-S system):
   Type C installation, no grouping. From Table 14.10, the current carrying capacity I_z is then 81 A and

$$I_n \leq I_z = 63\,\text{A} \leq 81\,\text{A}$$

2. *Feeder to office*:
   Type B2 installation, with 25 °C

$$I_B = \frac{P}{\sqrt{3}\,U \cos\varphi} = \frac{56\,\text{kW}}{\sqrt{3} \cdot 400\,\text{V} \cdot 0.95} = 85.08\,\text{A}$$

   Rated current of overcurrent protection device selected: 100A
Current carrying capacity from Table 14.10: 101 A

$$I_n \leq I_z = 100\,\text{A} \leq 101\,\text{A}$$

   Cross section of line: NYM-J 5 × 35 mm^2

3. *Feeder to meter distribution system*:
   Type B2 installation, with 25 °C, no grouping

$$I_B = \frac{P}{\sqrt{3}\,U\cos\varphi} = \frac{64\,\text{kW}}{\sqrt{3}\cdot 400\,\text{V}\cdot 0.95} = 97.2\,\text{A}$$

   Rated current of overcurrent protection device selected: 100 A
   Current carrying capacity from Table 14.10: 101 A

$$I_n \leq I_z = 100\,\text{A} \leq 101\,\text{A}$$

   Cross section of line: NYM-J 5 × 35 m^2
4. *Feeders to apartment units*:
   Type B1 installation, $I_n = 63$ A, current carrying capacity I_z from Table 14.10 = 72 A

$$I_n \leq I_z = 63\,\text{A} \leq 72\,\text{A}$$

   Cross section of line: NYM-J 5 × 16 mm^2
5. *Electric circuits for apartment installation*:
   Type B1 installation, $I_n = 16$ A, current carrying capacity I_z from Table 14.10 = 18.5 A

$$I_n \leq I_z = 16\,\text{A} \leq 18.5\,\text{A}$$

   Cross section of line:: NYM-J 3 × 1.5 m^2
   C *Voltage drops*:
   The calculation is for an operating temperature of 50 °C, with the correction factor 1.12.
1. *Feeder for main line*:
   $I_n = 200$ A, $S = 4 \times 70\,\text{mm}^2$, $l = 25$ m, $\cos\varphi = 0.95$

$$\Delta u = \frac{1.12\,\sqrt{3}\,l\,100\%\,I_n\,\cos\varphi}{\kappa\,S\,U_n}$$

$$\Delta u = \frac{1.12\cdot\sqrt{3}\cdot 25\,\text{m}\cdot 100\%\cdot 200\,\text{A}\cdot 0.95}{56\frac{\text{m}}{\Omega\,\text{mm}^2}\cdot 70\,\text{mm}^2\cdot 400\,\text{V}} = 0.59\%$$

   According to the technical conditions for connections the voltage drop for 100–250 kVA systems is 1% maximum. The requirement is therefore satisfied.
2. *Main meter – individual meter feeder*:
   $I_n = 63$ A, $S = 5 \times 16\,\text{m}^2$, $l = 25$ m, $\cos\varphi = 1$

$$\Delta u = \frac{1.12\cdot\sqrt{3}\cdot 25\,\text{m}\cdot 100\%\cdot 63\,\text{A}\cdot 1}{56\frac{\text{m}}{\Omega\,\text{mm}^2}\cdot 16\,\text{mm}^2\cdot 400\,\text{V}} = 0.85\%$$

   Remaining voltage drops for end circuits (maximum 4%)

$$\Delta u \text{ for the main line} + \Delta u \text{ for the meters} = 0.59\% + 0.85\% = 1.44\%$$

$$4\% - 1.44\% = 2.56\%$$

3. *Feeder to office*:
$I_n = 100\,\text{A}, S = 5 \times 35\,\text{mm}^2, l = 35\,\text{m},\ \cos\varphi = 0.95$

$$\Delta u = \frac{1.12 \cdot \sqrt{3} \cdot 35\,\text{m} \cdot 100\% \cdot 100\,\text{A} \cdot 0.95}{56\frac{\text{m}}{\Omega\,\text{mm}^2} \cdot 35\,\text{mm}^2 \cdot 400\,\text{V}} = 0.823\%$$

Remaining voltage drops for the end circuits (maximum 4%)

$$\Delta u \text{ for the main line} + \Delta u \text{ for the meters} = 0.59\% + 0.823\% = 1.413\%$$

$$4\% - 1.413\% = 2.58\%$$

4. *Feeder to meter distribution system*:
$I_n = 100\,\text{A}, S = 5 \times 35\text{m}^2,\ l = 25\,\text{m},\ \cos\varphi = 0.95$

$$\Delta u = \frac{1.12 \cdot \sqrt{3} \cdot 25\,\text{m} \cdot 100\% \cdot 100\,\text{A} \cdot 1}{56\frac{\text{m}}{\Omega\,\text{mm}^2} \cdot 35\,\text{mm}^2 \cdot 400\,\text{V}} = 0.619\%$$

5. *Apartment areas*:
$I_n = 63\,\text{A}, S = 5 \times 16\,\text{mm}^2, l = 25\,\text{m},\ \cos\varphi = 1$

$$\Delta u = 0.85\%$$

3% is permitted from the meter mounting board to the end circuits, so that 2.15% remain for the end circuits.

6. *Electrical outlet circuits*:
$I_n = 16\,\text{A}, S = 3 \times 1.5\,\text{mm}^2, l = 20\,\text{m},\ \cos\varphi = 1, U_n = 230\,\text{V}, T_U = 20\,°\text{C}$

$$\Delta u = \frac{2 l \cdot I_n \cdot 100\%}{\kappa\, S\, U_n} = \frac{2 \cdot 20\,\text{m} \cdot 16\,\text{A} \cdot 100\%}{56\frac{\text{m}}{\Omega\,\text{mm}^2} \cdot 1.5\,\text{mm}^2 \cdot 230\,\text{V}} = 3.31\%$$

The voltage drops are too high. A maximum of 2.15% is permissible.
Measures:

1. Increase the cross section by one step.
2. Reduce the circuit-breaker to 10 A.

D Selectivity control:
For the selectivity condition the rule applies:
The rated current of the upstream fuse must be at least 1.6 times the rated current of the downstream fuse. This drop is ensured except for the circuit between the apartment and the meter distribution system.
E Cutoff conditions:
Main feeder:
The factor **1.24** is the temperature increase from 20 to 80 °C, and the reactance for the cable is $x' = 0.08\frac{\text{m}\Omega}{\text{m}}$, see Table 16.5.

$$R_L = 1.24\,\frac{2\,l\,1000}{\kappa\,S} = 1.24 \cdot \frac{2 \cdot 25\,\text{m} \cdot 1000}{56\frac{\text{m}}{\Omega\,\text{mm}^2} \cdot 70\,\text{mm}^2} = 15.8\,\text{m}\Omega$$

$$X_L = 2\,l\,x'_L = 2 \cdot 25\text{m} \cdot 0.08\frac{\text{m}\Omega}{\text{m}}\ \text{m} = 4\,\Omega$$

$$Z_S = \sqrt{R_L^2 + X_L^2} = \sqrt{15.8^2 + 4^2}\ \text{m}\Omega = 16.3\ \text{m}\Omega$$

Meter distribution – apartment:

$$Z_S = 1.24 \cdot \frac{2 \cdot 25\,\text{m} \cdot 1000}{56 \frac{\text{m}}{\Omega\,\text{mm}^2} \cdot 35\,\text{mm}^2} = 31.6\ \text{m}\Omega$$

Sub-distribution for apartments:

$$Z_S = 1.24 \cdot \frac{2 \cdot 25\,\text{m} \cdot 1000}{56 \frac{\text{m}}{\Omega\,\text{mm}^2} \cdot 16\,\text{mm}^2} = 69.2\ \text{m}\Omega$$

Electrical outlet:

$$Z_S = 1.24 \cdot \frac{2 \cdot 20\,\text{m} \cdot 1000}{56 \frac{\text{m}}{\Omega\,\text{mm}^2} \cdot 1.5\,\text{mm}^2} = 590.5\ \text{m}\Omega$$

The total short-circuit impedance is:

$$Z_K = Z_V + Z_{\text{system}} = 300\ \text{m}\Omega + 707.6\ \text{m}\Omega \approx 1\Omega$$

Single-pole short-circuit current at the electrical outlet:

$$I''_{k1} = \frac{c\,U_n}{\sqrt{3}\,Z_K} = \frac{0.9 \cdot 400\ \text{V}}{\sqrt{3} \cdot 1\ \Omega} = 207.84\ \text{A}$$

Breaking current of circuit- breaker $I_a = 5 \cdot 16\ \text{A} = 80\ \text{A}$,

$$I''_{k1} \geq I_a$$

207.84 A > 80 A, so that the requirement is satisfied.

14.6 Examples for the Calculation of Overcurrents

14.6.1 Example 1: Determination of Overcurrents and Short-Circuit Currents

In a heating system, a motor-driven pump with a power of 200 kW, $\cos\varphi = 0.88$, $\eta = 0.93$ is connected through an overhead cable with $l = 45$ m (Figure 14.8). Calculate the:

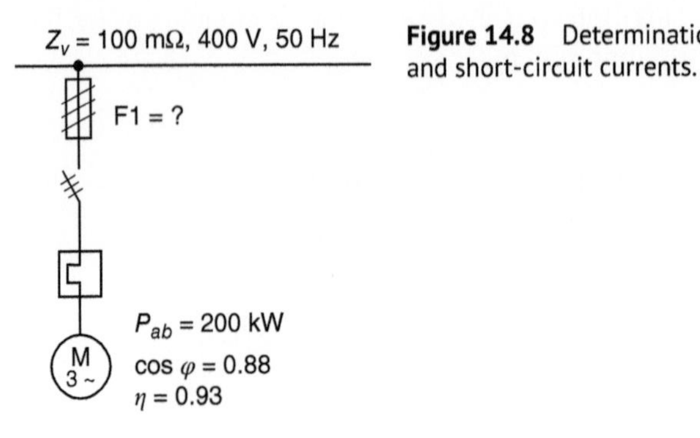

Figure 14.8 Determination of overload and short-circuit currents.

1. Operating current

$$I_B = \frac{P}{\sqrt{3}\,U_n\,\cos\varphi\,\eta} = \frac{200\ \text{kW}}{\sqrt{3}\cdot 400\ \text{V}\cdot 0.88\cdot 0.93}$$
$$I_B = 353\ \text{A}$$

2. Overload protection, $I_n = 400$ A selected.
From Table 14.20, we find a current carrying capacity of 445 A for an NYY cable with a cross section of 240 mm².

$$I_B \le I_n \le I_z$$
$$353\ \text{A} \le 400\ \text{A} \le 445\ \text{A}$$

3. 2% voltage drop is permissible (20 °C)

$$\Delta U = \frac{\sqrt{3}\,1I\,\cos\varphi}{\kappa\,S}$$
$$\Delta U = \frac{\sqrt{3}\cdot 45\ \text{m}\cdot 353\ \text{A}\cdot 0.88}{56\frac{\text{m}}{\Omega\ \text{mm}^2}\cdot 240\ \text{mm}^2} = 1.8\ \text{V}$$
$$\Delta u = \frac{100\%\ \Delta U}{U_n} = \frac{100\cdot 1.8\ \text{V}}{400\ \text{V}} = 0.45\%$$

4. Protection by cutoff
Requirement: $I''_{k1} > I_a$ must always hold true:
The cutoff current of the 400 A gL fuse in five seconds is (Figure 13.19):

$$I_a = 2800\ \text{A}$$
$$I''_{k1} = \frac{c\,U_n}{\sqrt{3}Z_k}$$
$$R_l = 1.24\,\frac{2\,l\cdot 1000}{\kappa\,S} = 1.24\cdot\frac{2\cdot 45\ \text{m}\cdot 1000}{56\frac{\text{m}}{\Omega\ \text{mm}^2}\cdot 240\ \text{mm}^2} = 8.30\ \text{m}\Omega$$
$$X_l = 2\,l\,x'_1 = 2\cdot 45\ \text{m}\cdot 0.08\frac{\text{m}\Omega}{\text{m}}\ \text{m} = 7.2\ \Omega$$
$$Z_k = \sqrt{R_l^2 + X_l^2} = \sqrt{8.30^2 + 7.2^2}\ \text{m}\Omega = 10.99\ \text{m}\Omega$$
$$Z_k = Z_v + Z_l = 100\ \text{m}\Omega + 10.99\ \text{m}\Omega \approx 111\ \text{m}\Omega$$
$$I''_{k1} = \frac{0.9\cdot 400\ \text{V}}{\sqrt{3}\cdot 0.111\ \Omega} = 1872.48\ \text{A}$$

The condition:

$$I''_{k1} > I_a$$

must always hold true.
Since in this example $I''_{k1} < I_a$, i.e. 1872.48 A < 2800 A, the condition is not satisfied. Measure:

a) Increase the cross section of the cable, if possible
b) Use a smaller value fuse.

5. Calculational proof of short-circuit protection

$$t_{\text{permissible}} = \left(\frac{k\,S}{I''_{k1}}\right)^2 = \left(\frac{115\,\frac{\text{A}\,\sqrt{\text{s}}}{\text{mm}^2}\cdot 240\,\text{mm}^2}{1976.5\,\text{A}}\right)^2$$

$$t_{\text{permissible}} = 217.26\ \text{seconds}$$

6. Overcurrent relay
   Current setting for a power circuit-breaker of 240–400 A to $I_{\text{rM}} = 353$ A:
   Protection: 400 A-AC3
7. Direct starting
   With $I_{an}/I_{\text{rM}} = 7$ and a start time of maximum 10 seconds: $I_n = 500$ A. An NH fuse is required.

14.6.2 Example 2: Overload Protection

1. Overload protection
   In accordance with IEC 60364, Part 43 determine the overload and short circuit protection with an operating current of 170 A (Figure 14.9).
   For a load factor of 0.7 and one cable, we obtain a current carrying capacity of:
   $I_z = 280$ A (Table 14.18)
   $I_n = 200$ A, $I_B = 170$ A
   From the rule of nominal currents (Figure 14.12):

$$\begin{aligned} I_B &\le I_n \le I_z \\ 170\text{A} &\le 200\ \text{A} \le 280\ \text{A} \end{aligned}$$

With the tripping rule:
Fuse

$$\begin{aligned} I_2 &\le 1.45 \cdot I_z \\ 1.6 \cdot I_n &\le 1.45 \cdot I_z \\ 1.6 \cdot 200\ \text{A} &\le 1.45 \cdot 280\ \text{A} \\ 320\ \text{A} &\le 406\ \text{A} \end{aligned}$$

Overload protection is therefore ensured.

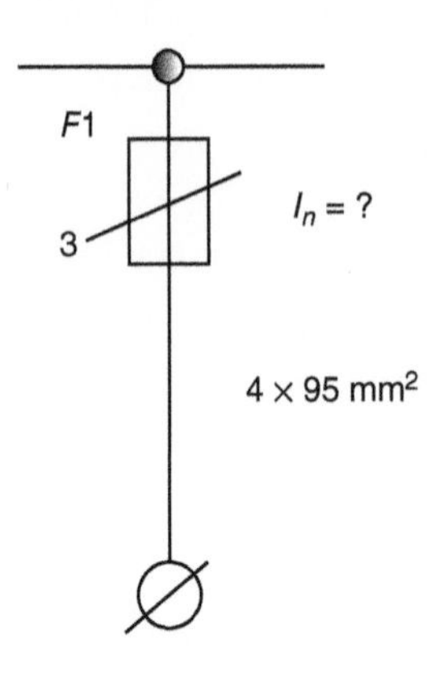

Figure 14.9 Example of overload protection.

Figure 14.10 Short-circuit strength of a line.

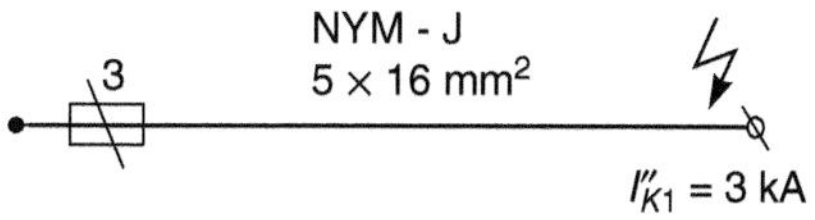

2. Short-circuit protection
   The let-through energy of the protective device and the cable is found for $t_a < 0.1$ second from:

$$I^2\, t \leq k^2\, S^2$$

where t is read from the characteristic

at five seconds: $I''_{k1} = 1.3\ \text{kA}$

$$(1.3\ \text{kA})^2 \cdot 5\ \text{seconds} \leq (115\ \text{A}\sqrt{\text{s}}/\text{mm}^2)^2 \cdot (95\ \text{mm}^2)^2$$

$$8450000\ \text{A}^2\,\text{s} \leq 119355625\ \text{A}^2\,\text{s}$$

Short-circuit protection is therefore ensured as well.

14.6.3 Example 3: Short-Circuit Strength of a Conductor

At the end of a 16 mm² Cu line (Figure 14.10), a short circuit current of 3 kA was measured. Prove that this line has sufficient short-circuit strength.

$$I''_{k1} = 3\ \text{kA} \Rightarrow S = 16\ \text{mm}^2$$

$$t_{zul} = \left(\frac{k\,S}{I''_{k1}}\right) = \left(\frac{115\ \text{A}\ \sqrt{s}}{\text{mm}^2}\frac{16\ \text{mm}^2}{3\ \text{kA}}\right)^2 = 0.3765\ \text{s}$$

A comparison of the let through energies at $t_a < 0.1\text{s}$ is necessary:

$$I^2\, t \leq k^2\, S^2$$

$$(3\ \text{kA})^2 \cdot 0.1\ \text{s} \leq (115\ \text{A}\sqrt{\text{s}}/\text{mm}^2)^2 \cdot (16\ \text{mm}^2)^2$$

$$900000\ \text{A}^2\ \text{s} \leq 3385600\ \text{A}^2\ \text{s}$$

How large is the short-time current density for a short-circuit lasting one second with $\vartheta_e = 160\ °C$ and $\vartheta_a = 30\ °\text{C}$?

Using the constants from Table 14.2, we obtain the rated short-time current density J_{thr}:

$$J_{\text{thr}} = \sqrt{\frac{\kappa_{20}\, c\, \rho}{\alpha_{20}\, T_{kr}} \ln \frac{1 + \alpha_{20}\,(\varphi_e - 20\ °C)}{1 + \alpha_{20}\,(\vartheta_b - 20\ °C)}}$$

$$I_{\text{thr}} = \sqrt{\frac{56 \cdot 10^3\ \text{mm} \cdot 8.9 \cdot 10^{-3}\,\text{g} \cdot \text{K} \cdot 0.39\ \text{Ws}}{\Omega \cdot \text{mm}^2 \cdot \text{mm}^3 \cdot \text{g} \cdot \text{K} \cdot 0.0039 \cdot 1\ \text{s}} \ln \frac{1 + 0.0039 \cdot 1/\text{K} \cdot (160 - 20)\text{K}}{1 + 0.0039 \cdot 1/\text{K} \cdot (30 - 20)\text{K}}} \tag{14.20}$$

$$J_{\text{thr}} = 140 \approx 143 \frac{\text{A}}{\text{mm}^2}$$

rated short-time current:

$$I_{\text{thr}} = J_{\text{thr}}\, S = 143 \frac{\text{A}}{\text{mm}^2} \cdot 16\ \text{mm}^2 = 2.288\ \text{kA}$$

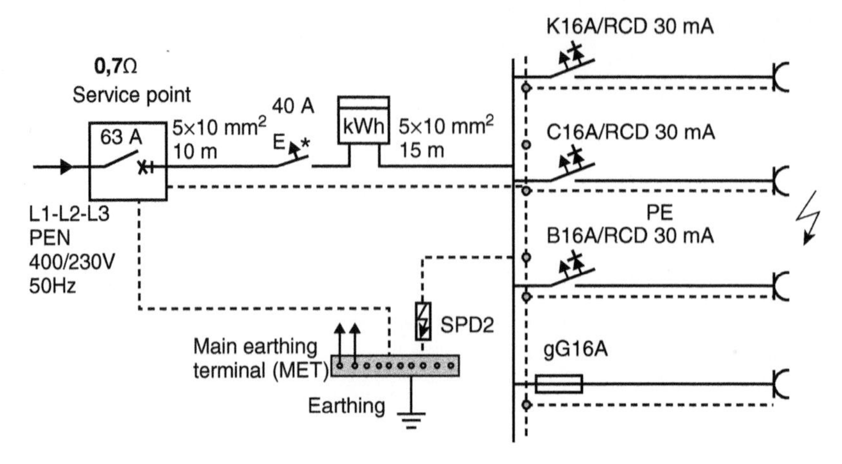

Figure 14.11 Terminal diagram for protective equipment.

Permissible short-time current for another short-time period:

$$I_{thz} = I_{thr}\sqrt{\frac{t_{kr}}{t_k}} = 2.288\ \text{kA} \cdot \sqrt{\frac{1\ \text{s}}{2.3\ \text{s}}} = 1.508\ \text{kA}$$

14.6.4 Example 4: Checking Protective Measures for Circuit-Breakers

Given the system (Figure 14.11) with different circuit breakers, which are to be compared with each other.

All exit cables: NYM-J $3 \times 1.5\ \text{mm}^2$, $P = 3\ \text{kW}T_{zu} = 70^0$, $l = 20\ \text{m}$, type B2 installation, $\cos\varphi = 1$

Operating current:

$$I_B = \frac{3\ \text{kW}}{230\ \text{V}} = 13\ \text{A}$$

Check the:

1. *Single-pole short-circuit current*:
   For NYM-J $4 \times 10\ \text{mm}^2$ the impedance is $Z'_L = 0.00449\ \text{m}\Omega/\text{m}$ (outward and return line):

$$\begin{aligned}
Z_G &= \sum Z_L = \sum (Z'_L\, l) \\
Z_G &= Z_{EVU} + Z'_{L1}\, l + Z'_{L2}\, l + Z'_{L3}\, \text{l} \\
Z_G &= 0.7\ \Omega + 0.00449\frac{\Omega}{\text{m}} \cdot 10\ \text{m} + 0.00449\frac{\Omega}{\text{m}} \cdot 15\ \text{m} \\
&\quad + 0.03001\frac{\Omega}{\text{m}} \cdot 20\ \text{m} \\
Z_G &= 1.41245\ \Omega \\
I''_{k1} &= \frac{c\,U_n}{\sqrt{3}\,Z_G} = \frac{0.9 \cdot 400\ \text{V}}{\sqrt{3} \cdot 1.41245\ \Omega} = 147.15\ \text{A}
\end{aligned}$$

2. *Current carrying capacity*:
NYM-J 3 × 1.5 mm², for type B2 installation Table 14.10 gives a current carrying capacity of $I_z = 16.5$ A at 25°.
3. *Overload protection*:
From the rule of nominal currents:

$$I_B \leq I_n \leq I_z$$
$$13\ \text{A} < 16\ \text{A} < 16.5\ \text{A}$$

Conventional tripping current: $I_2 = k \cdot I_n$
Tripping rule: $I_2 \leq 1.45 \cdot I_z$

$$\begin{aligned} I_2 &\leq 1.45 \cdot I_z \\ k \cdot I_n &\leq 1.45 \cdot I_z \\ &\leq 1.45 \cdot 16.5\ \text{A} \\ &\leq 23.9\ \text{A} \end{aligned}$$

$$\begin{aligned} \text{gG} &: I_2 = 1.75\, I_n = 1.75 \cdot 16\ \text{A} = 28\ \text{A} \\ \text{B} &: I_2 = 1.45\, I_n = 1.45 \cdot 16\ \text{A} = 23.2\ \text{A} \\ \text{C} &: I_2 = 1.45\, I_n = 1.45 \cdot 16\ \text{A} = 23.2\ \text{A} \\ \text{K} &: I_2 = 1.2\, I_n = 1.2 \cdot 16\ \text{A} = 19.2\ \text{A} \end{aligned}$$

For gL and L overload protection is not ensured.
4. *Cutoff currents for the overcurrent protection equipment*:

$$\begin{aligned} \text{B} &: I_a = 5\, I_n = 5 \cdot 16\ \text{A} = 80\ \text{A} \\ \text{C} &: I_a = 10\, I_n = 10 \cdot 16\ \text{A} = 160\ \text{A} \\ \text{gG} &: I_a = \text{at 0.2 second mit } I_n = 16\ \text{A} = 148\ \text{A} \\ \text{K} &: I_a = 14\, I_n = 14 \cdot 16\ \text{A} = 224\ \text{A} \end{aligned}$$

For the cutoff condition:

$$I''_{k1} > I_a$$
$$155.3\ \text{A} > I_a$$

For the *C* and *K* characteristics, the cutoff conditions are not satisfied (Figure 14.13).
5. *Utilization factor*:
The utilization factor *N* should be as large as possible, so that the 100% capability of the line is used.

$$N = \frac{I_n}{I_z}\ 100\%$$
$$N = \frac{16\ \text{A}}{15\ \text{A}} \cdot 100\% = 106\%\ \text{with}\ 1.5\ \text{mm}^2$$
$$N = \frac{16\ \text{A}}{20\ \text{A}} \cdot 100\% = 80\%\ \text{with}\ 2.5\ \text{mm}^2$$

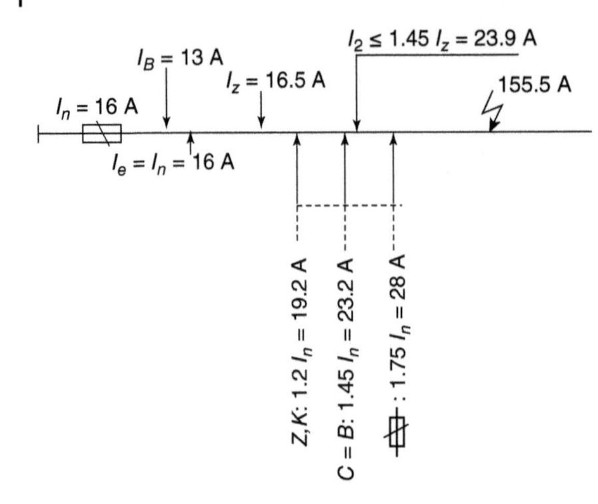

Figure 14.12 Coordination of overcurrent protection equipment characteristics.

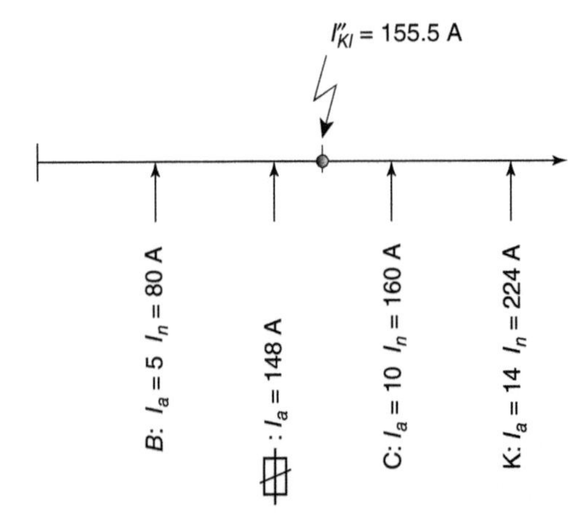

Figure 14.13 Cut-off currents of protective equipment.

6. *Degree of protection:*
   The degree of protection S should be as small as possible. The loading capacity of the line and the overtemperature are then kept low and the life of the line increases.

$$S = \frac{I_2}{I_z}$$

$$S = \frac{k\,I_n}{I_z}$$

$$S = \frac{1.75 \cdot 16\text{ A}}{16.5\text{ A}} = 1.69 \text{ with the fuse}$$

$$S = \frac{1.9 \cdot 16\text{ A}}{16.5\text{ A}} = 1.84 \text{ with the circuit-breaker}$$

The factor above is the ratio of the tripping current to the nominal current for the overcurrent protection device and can serve for the calculation of the tripping current, e.g. $I_2 = x \cdot I_n$.

Both examples show that the degree of protection of $S \leq 1.45$ is not satisfied. It is therefore necessary to choose a smaller I_n.

7. *Selection of fuse or circuit-breaker*:

$$I_n \leq \frac{1.45}{k} I_z$$
$$\text{gG}: I_n \leq \frac{1.45}{1.75} \cdot 16.5\,\text{A} = 13.6\,\text{A}$$
$$\text{B}: I_n \leq \frac{1.45}{1.45} \cdot 16.5\,\text{A} = 16.5\,\text{A}$$
$$\text{C}: I_n \leq \frac{1.45}{1.45} \cdot 16.5\,\text{A} = 16.5\,\text{A}$$
$$\text{K}: I_n \leq \frac{1.45}{1.2} \cdot 16.5\,\text{A} = 19.94\,\text{A}$$

8. *Factors for grouping and temperature*:
   Lines in conduit

$$f_1 = 0.8$$
$$T = 25^\circ$$
$$f_2 = 1.06$$
$$I_{z\text{Tab}} = \frac{I_n}{f_1 f_2} = \frac{16\,\text{A}}{0.8 \cdot 1.06} = 18.86\,\text{A}$$

   NYM-J $3 \times 2.5\,\text{mm}^2$ selected.

9. *Voltage drop calculation at* $20\,^\circ C$ *and* $\cos\varphi = 1$:
   a) Service panel meter distribution system (Figure 14.11):

$$\Delta u = \frac{\sqrt{3} \cdot 10\,\text{m} \cdot 100\% \cdot 63\,\text{A}}{56 \frac{\text{m}}{\Omega\,\text{mm}^2} \cdot 10\,\text{mm}^2 \cdot 400\,\text{V}} = 0.49\%$$

   According to the technical conditions for connections only 0.5% is permissible.

   b) Service panel meter distribution – sub-distribution:

$$\Delta u = \frac{\sqrt{3} \cdot 15\,\text{m} \cdot 100\% \cdot 63\,\text{A}}{56 \frac{\text{m}}{\Omega\,\text{mm}^2} \cdot 10\,\text{mm}^2 \cdot 400\,\text{V}} = 0.73\%$$

   c) Sub-distribution – end circuits:

$$\Delta u = \frac{2 \cdot 10\,\text{m} \cdot 100\% \cdot 16\,\text{A}}{56 \frac{\text{m}}{\Omega\,\text{mm}^2} \cdot 2.5\,\text{mm}^2 \cdot 230\,\text{V}} = 1\%$$

   The sum of all voltage drops is

$$0.49\% + 0.73\% + 1\% = 2.22\%$$

   (maximum permissible is 4%).

15

Selectivity and Backup Protection

15.1 Selectivity

Electrical operational and consumer equipment must be protected against stresses resulting from short-circuit circuits by the selective disconnection of the faulty systems. IEC 60364, Part 53 requires selectivity in low-voltage networks. The most important selectivity conditions for protective devices are described. An electrical system is equipped with several overcurrent protective devices connected in series such as fuses and circuit-breakers, which are selective because in the event of a fault only the overcurrent protective device located directly before the fault location in the direction of current flow responds. Investigating the selectivity conditions requires a comparison of the time-current characteristics for the overcurrent protective devices or the fuse melt integrals with one another. In the following, the most important selectivity cases are described.

In electrical systems, different overcurrent protective devices are used for overload and short-circuit protection. The effects of these faults can be significantly reduced when the protective devices are correctly chosen.

Figure 15.1 illustrates time-current characteristics for different possible devices in a low-voltage network.

Current selectivity is obtained through the use of protective devices with different tripping currents. Time selectivity is obtained by delaying the release of the upstream protective devices.

Conditions for the selectivity of fuses:

- The characteristics must not coincide at any point.
- This is achieved when the upstream fuse has at least 1.6 times the rated current of the downstream fuse.
- With less than 1.6 times the rated current, no protection exists.
- The deviations of the scatter vary by ±7%.
- The cutoff time t_A is the sum of the melting time t_S and the arc extinction time t_L.
- For short-circuit currents ≥20 times the rated current of the fuse and melting times ≤10 ms, characteristics cannot give reliable information. In such cases, it is necessary to check the $i^2\ dt$ values.

Analysis and Design of Electrical Power Systems: A Practical Guide and Commentary on NEC and IEC 60364, First Edition. Ismail Kasikci.

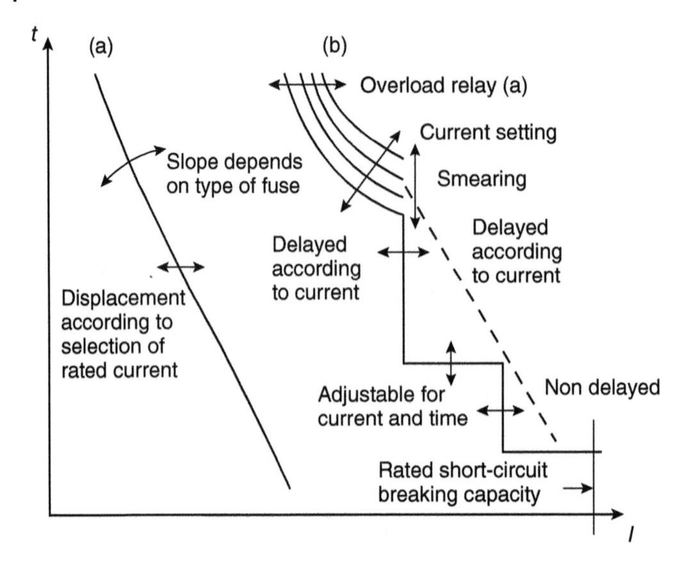

Figure 15.1 Tripping characteristics of low voltage fuses (NH) (a) and power circuit breakers with overcurrent relays and (b) in a low-voltage network. Source: Siemens [57, 58].

Advantages and disadvantages of fuses:

- The short-circuit breaking capacity of fuses is sufficient for nearly all network conditions.
- Fuses cannot be adjusted.
- After cutoff fuses cannot be used again.
- Planning with fuses is simple.
- Systems in which the power levels are not yet known can be adapted to later requirements by simply changing the fuses.

A system with a transformer has a power of 1200 kVA, 6% short-circuit voltage and a 50 kVA short-circuit current at the feed-in (Figure 15.2).

Conditions for the selectivity of power circuit-breakers:

- Time or current grading is possible.
- Power circuit-breakers are selective under one another only with certain restrictions.
- The total cutoff time t_A of the downstream switching device must be less than the minimum command time t_m of the upstream switching device of the upstream switching device.
- The total cutoff time t_A is the sum of the contact parting time t_{OV} and the arcing time t_L.
- The delay time is about 50 ms.
- The interrupting current is 1.2 times the current setting I_e.
- The short-circuit current is 12 times the rated current I_n.
- The power circuit breaker must be able to control the maximum short circuit current at the location of installation.

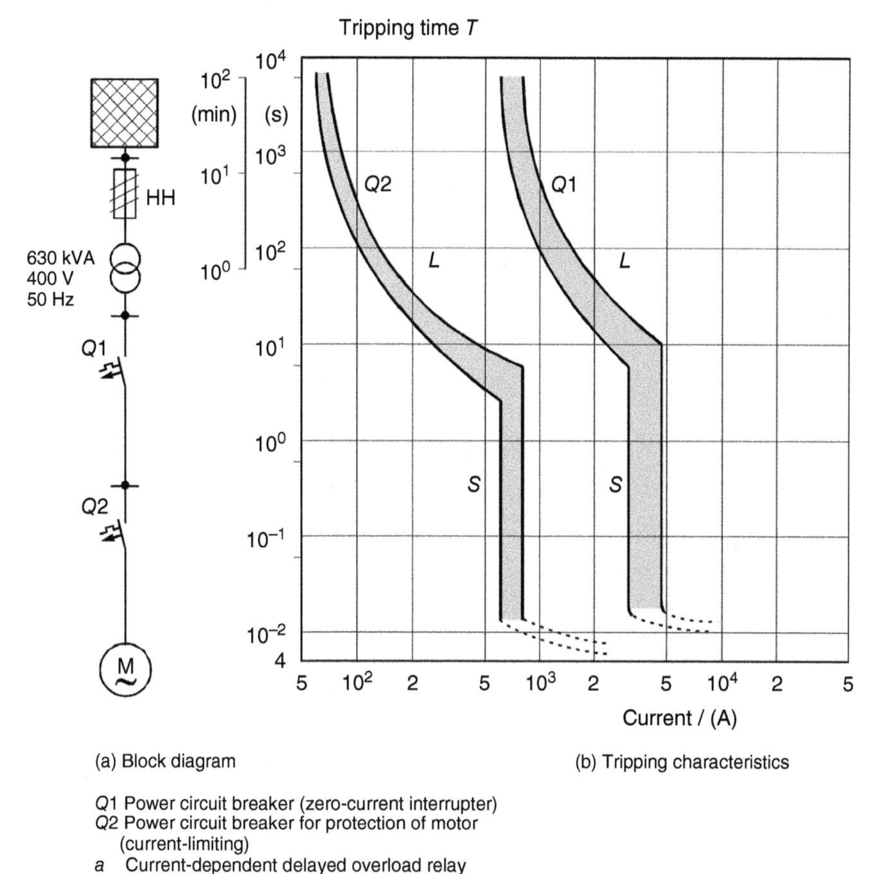

Figure 15.2 Time selectivity of two power circuit breakers in series [47].

Advantages and disadvantages of power circuit breakers:

- Power circuit breakers have a limited short-circuit breaking capacity.
- For a small rated current the breaking capacity is also smaller.
- For correct planning of a system with power circuit breakers, it is necessary to calculate the short-circuit currents.

Important remarks about power circuit breakers and fuses:

- The grading time between a power circuit breaker and the fuse must be at least 100 ms.
- The let-through capacity of the fuse must be greater than that of the power circuit breaker.
- The breaking capacity of the power circuit breaker must always be taken into account.
- With fuses it is necessary to consider the type, state, aging, Manufacture, and characteristics.

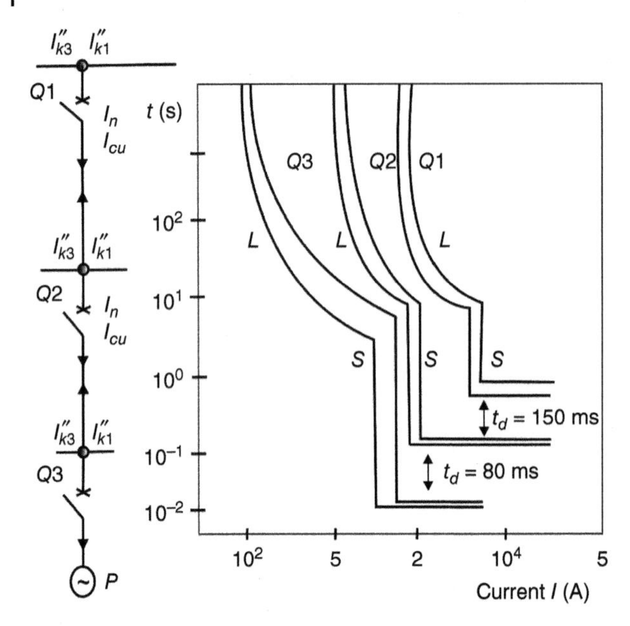

Figure 15.3 Selectivity in the overload region: Fuse upstream from power Circuit-breaker [47].

Current selectivity is obtained through grading of the response currents and non-delayed short-circuit tripping when the short-circuit currents at the installation location differ greatly (Figure 15.2). The nondelayed short-circuit tripping device can be set to the calculated short-circuit values. To test the selectivity, another possibility is to compare the tripping characteristics with each other.

When the short-circuit currents of the upstream and downstream power circuit breakers in a system are approximately equal, it is not possible to make use of time selectivity (Figure 15.3). The delay time must then be chosen so that a downstream power circuit breaker has the required time to cutoff by itself.

Selectivity in the overload region is provided only when the characteristic of the fuse does not touch the characteristic of the power circuit breaker (Figures 15.4–15.6). For larger short circuit currents, the fuse responds so quickly that the power circuit breaker with 100 ms delay time never cuts off. Downstream fuse links switch selectively to a power circuit-breaker when the operating current of the overload relay in the breaker has four to five times the value of the rated current for the fuse (Figure 15.7).

The primary side of the transformers is mostly protected with high voltage fuses (HH) and combined load disconnector switches. On the low-voltage side, power circuit-breakers are used. The HH fuses are standardized, depending on the power of the transformer, and provide protection against short circuits only on the medium-voltage side. For selectivity, it is necessary to take the manufacturers' data and scatter ranges of both characteristics into account. Power circuit-breakers with a protective relay are used in place of HH fuses and provide the best protection. Figure 15.8 illustrates a motor exit cable with a short circuit current of 20 kVA

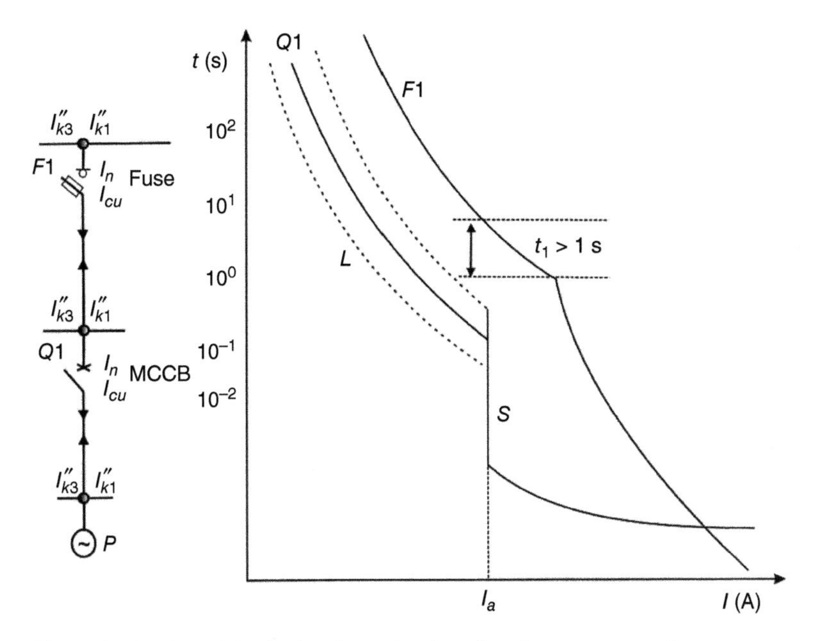

Figure 15.4 Selectivity in the short-circuit region: Fuse upstream from power circuit-breaker [47].

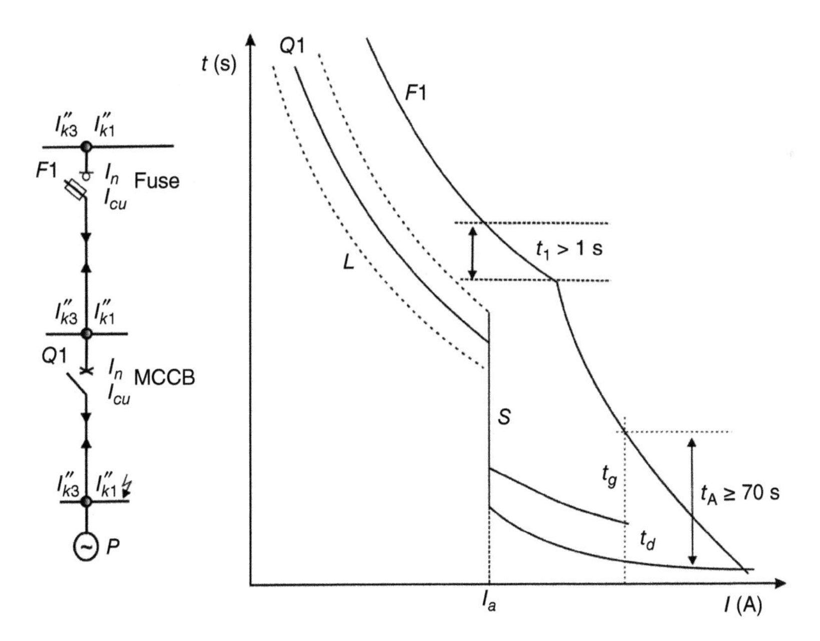

Figure 15.5 Selectivity in the overload region: Power circuit breaker upstream from fuse [47].

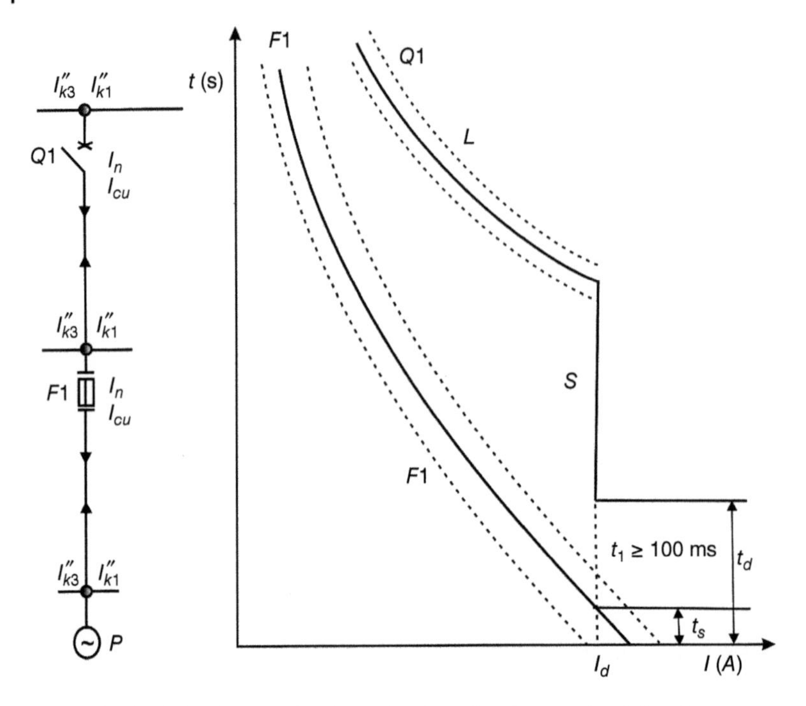

Figure 15.6 Selectivity in the short-circuit region: Power circuit breaker upstream from fuse [47].

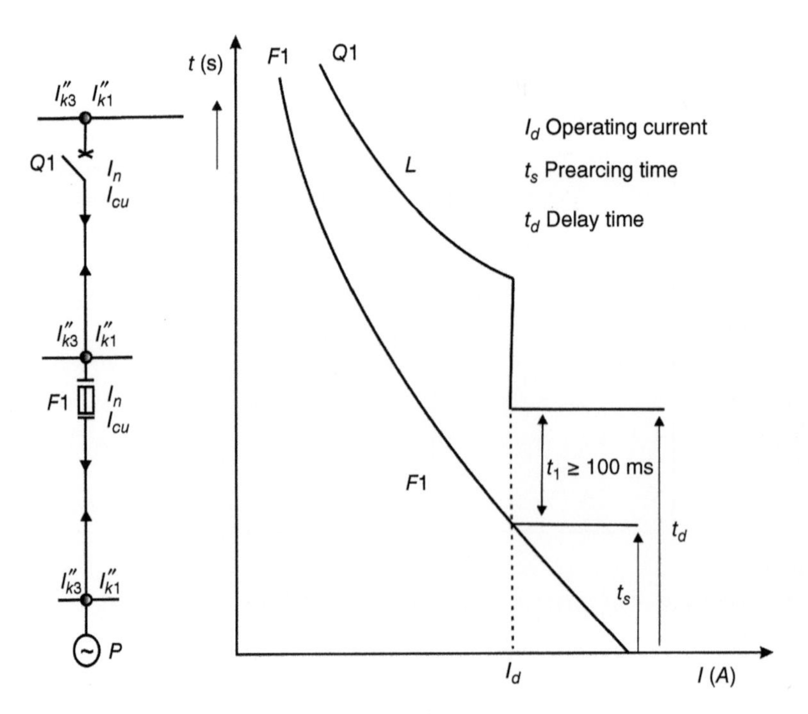

Figure 15.7 Current selectivity for two power circuit-breakers in series [47].

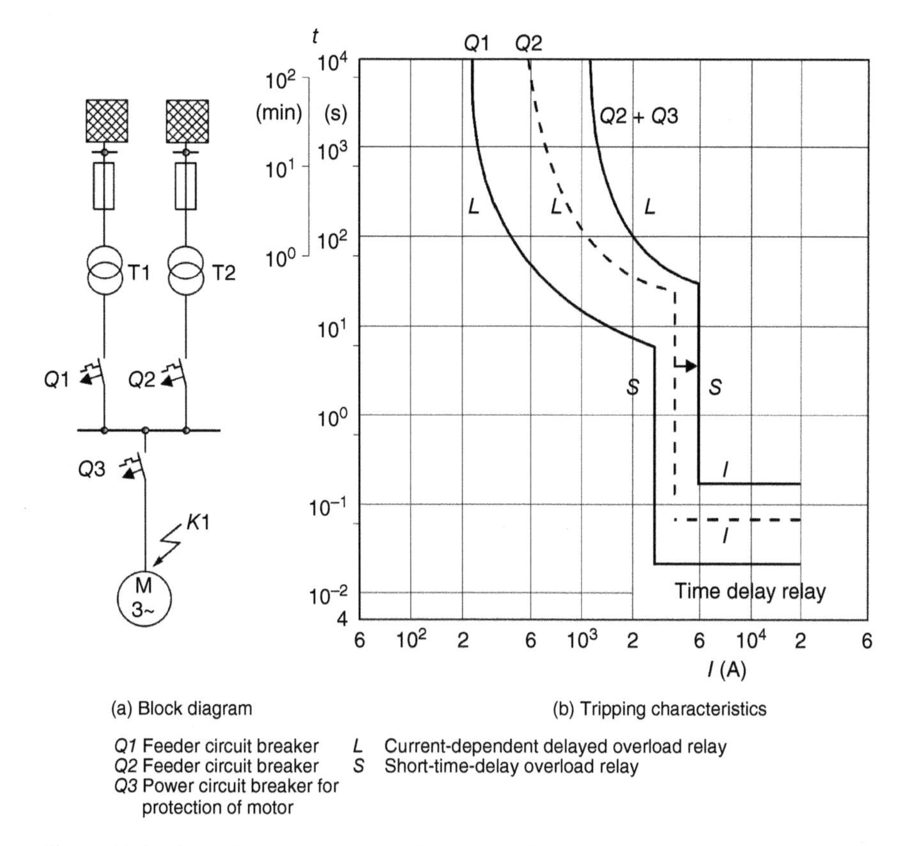

(a) Block diagram (b) Tripping characteristics

Q1 Feeder circuit breaker
Q2 Feeder circuit breaker
Q3 Power circuit breaker for protection of motor
L Current-dependent delayed overload relay
S Short-time-delay overload relay

Figure 15.8 Selectivity in a system with two transformers [47].

connected directly to the busbar of a transformer station and its short-circuit current, which is supplied from two feeder circuit-breakers with 10 kVA each. Here, the trigger characteristic is shifted by a factor of two to the right on the current scale.

The breaking capacity of the fuses is over 100 kVA. The calculation of the short-circuit currents at the installation location is not necessary. The selectivity is provided since the upstream fuse element has ≥1.6 times the rated current of the downstream fuse element. Figure 15.9 shows a system with fuses. On the basis of the melting time characteristics for the fuse elements, we have proof of selectivity (Figure 15.10).

At the installation location of the power circuit-breaker, a calculation of the short-circuit current is always necessary in order to be able to choose the correct power circuit-breaker and current setting. It is always possible to use either a compact power zero-current interrupter (Figure 15.11). The zero-current interrupter quenches the electric arc only during the natural passage of the current through zero. It must be able to withstand the dynamic and thermal short-circuit stresses of its full breaking capacity for a short time. If the short circuit current is interrupted by a current-limiting power circuit breaker the contacts then open very quickly,

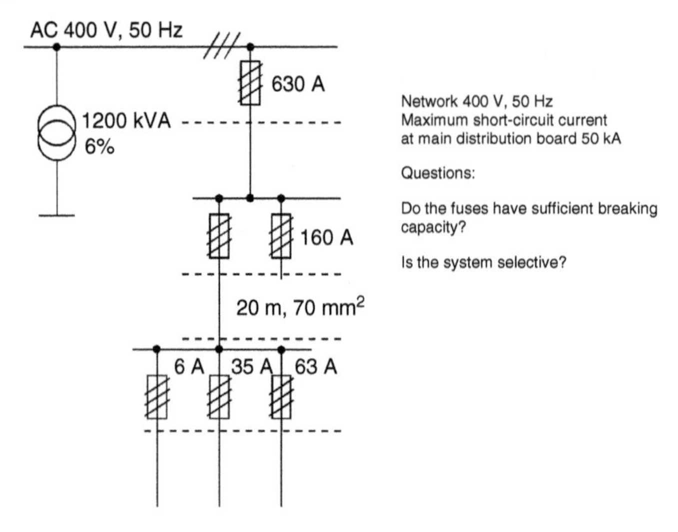

Figure 15.9 Evaluation of a system with fuses.

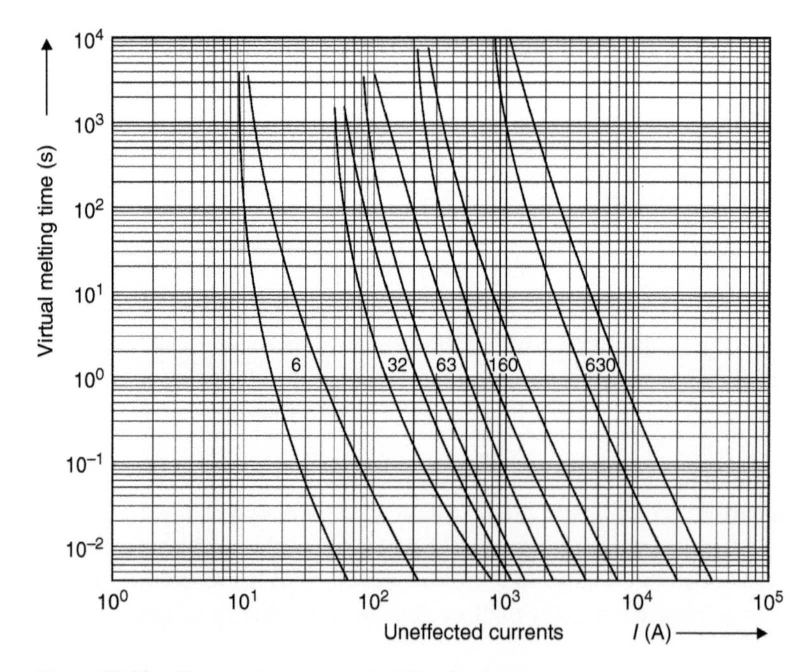

Figure 15.10 Time–current characteristics for fuses.

so that already after only a few milliseconds an arc is ignited which represents an additional resistance that rapidly increases and limits the short-circuit current to the curve form shown with the continuous line. This is a considerable relief for the system and breaker.

In the following example (Figure 15.12), power circuit-breakers with the same rated values are used in place of fuses. Referring to the characteristic in Figure 15.13,

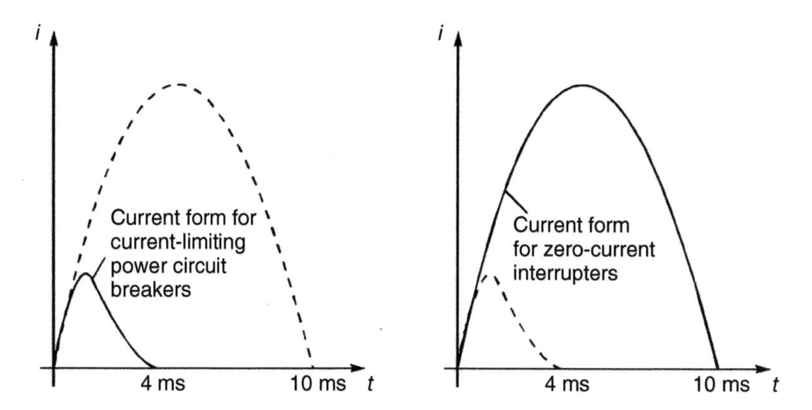

Figure 15.11 Current form for breaking a short-circuit.

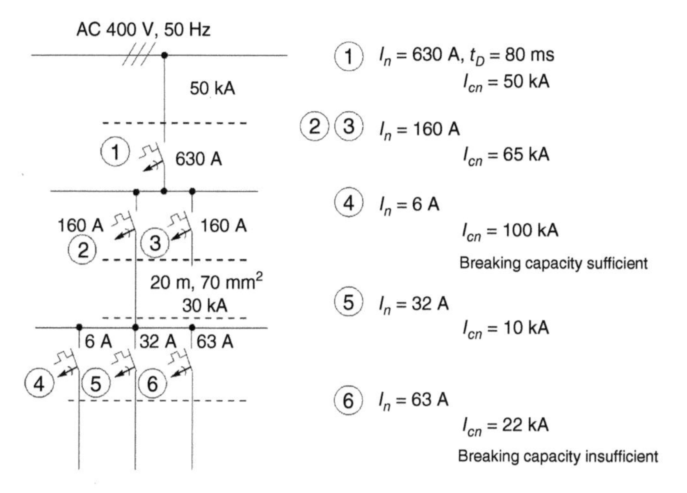

Figure 15.12 Evaluation of a system with power circuit-breakers.

we see that the power circuit breakers are too similar in their characteristics to be selective. Then, we see that the power circuit-breakers are too similar in their characteristics to be selective. The 630 A breaker is chosen with *a z n* release. The *z* release is delayed by 80 ms, and the *n* release sets to infinity. This makes the breaker selective with respect to the others. Since in the sub-distributions the short-circuit currents are >6 kA, a backup protective device must be installed there.

15.2 Backup Protection

Back-up protection is ensured when in the event of a short-circuit a downstream overcurrent protective device is protected by an upstream overcurrent protective device (Figure 15.14). This means that the rated breaking capacity of the downstream breaker can be smaller than the short-circuit current at the installation location. The three-pole short-circuit current must be determined at the installation location and

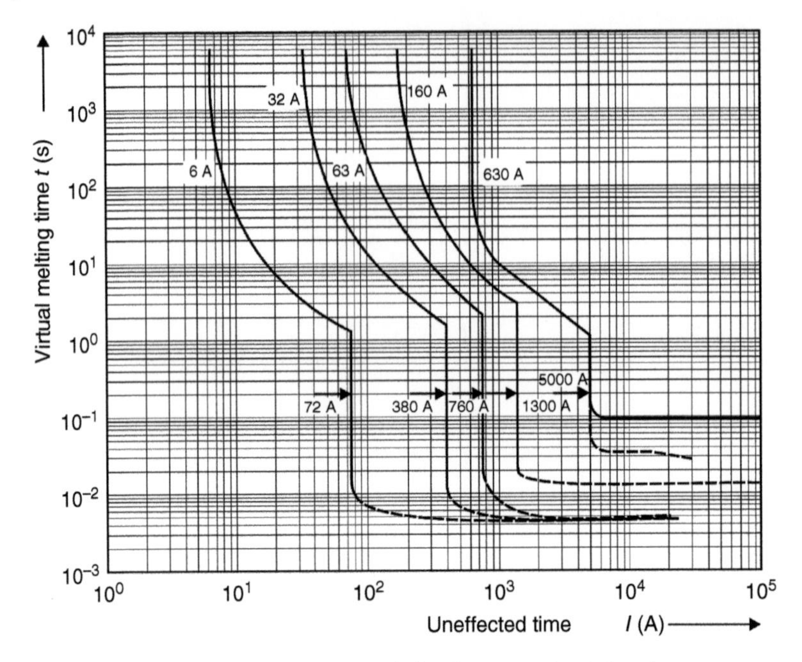

Figure 15.13 Time–current characteristics for power circuit breakers.

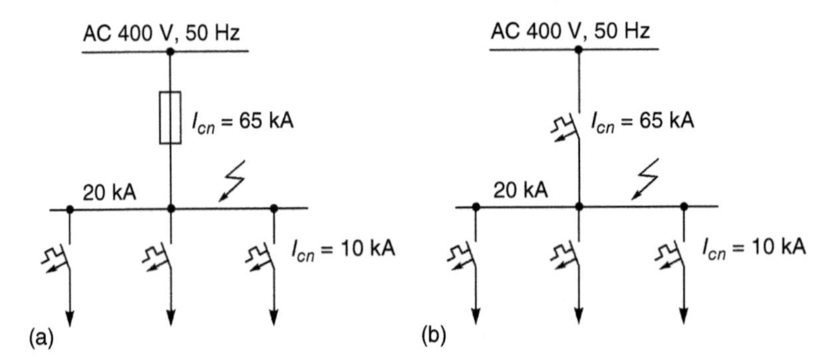

Figure 15.14 Back-up protection in a system.

the rated breaking capacity of the protective devices chosen checked with respect to whether backup protection is required or not. The selectivity limits and the backup protection can also be taken from the manufacturers' information.

Figure 15.15 shows a complete system with different branches. The following power circuit-breakers are chosen:

For the 6 kVA branch, we choose a power circuit-breaker with a breaking capacity of 100 kVA. Back-up protection is not necessary. For the 32 kVA branch, with a breaking capacity of 10 kVA, backup protection of 65 kVA is required.

For the 63 kVA branch, with a breaking capacity of 22 kVA, 65 kVA backup protection is already provided. The backup protection into account is sufficient to enable the power circuit-breakers to keep the required 30 kA short-circuit current under control.

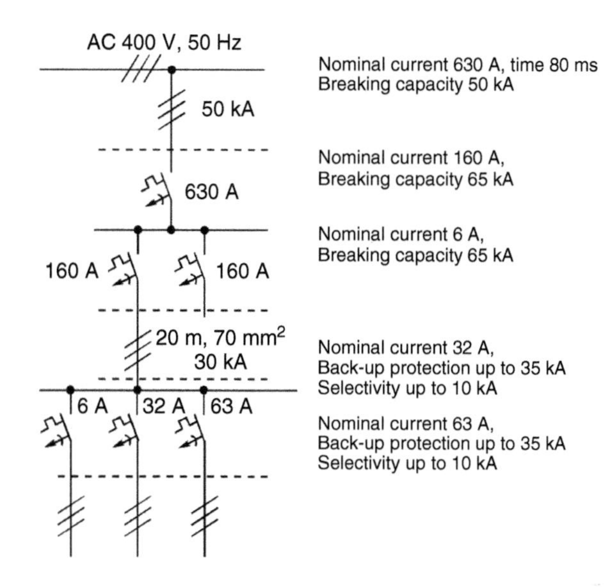

Figure 15.15 Selectivity and backup protection for a system with power circuit-breakers.

16

Voltage Drop Calculations

IEC 60364, Part 52

Cables and lines are the most important part of a system during planning and configuring. They must be able to withstand the mechanical and thermal stresses and transfer the power from connected equipment with as little loss as possible. This requires the careful dimensioning of the electrical networks. Present-day systems and household appliances are matched to the nominal voltage of the network. This may fluctuate only within established limits, since otherwise a normal power output cannot be ensured and the equipment can be destroyed. This section discusses the voltage drop and the maximum line length for AC and three-phase networks, on the basis of the existing regulations and standards [50–52].

16.1 Consideration of the Voltage Drop of a Line

The voltage drop of a line depends on the magnitude and direction the operating current and the impedance of the line. The amount ΔU along a line is not identical to the difference of the amounts of the voltages at the beginning and end of the line. This can be seen when one draws an arc with the radius U_2 around the jump of the pointer diagram and an arc with the radius ΔU around the arrowhead of U_1 (Figure 16.1). The calculation of three-phase networks is carried out with the help of the complex calculation. For this purpose, the initial circuit diagram with the pointer diagram of currents and voltages is used below to determine the voltage drops [53]. The rated current of an electrical circuit (called load current or operating current) causes a voltage drop across the impedance, which is derived from the voltage drop $\underline{I}_w \cdot R$ in phase with the current and an inductive voltage drop $\underline{I}_b \cdot X$, which hurries 90° the current, as shown in the diagram in Figure 16.1. The voltage at the beginning of the line:

$$\underline{U}_A = \underline{U}_E + \Delta\underline{U} = \underline{U}_E + \underline{I} \cdot \underline{Z} \tag{16.1}$$

The pointer diagram shows that the voltage drop $\Delta\underline{U}$ can be divided into a real and imaginary part.

$$\Delta\underline{U} = \Delta\underline{U}_l + j\,\Delta\underline{U}_q \tag{16.2}$$

Analysis and Design of Electrical Power Systems: A Practical Guide and Commentary on NEC and IEC 60364, First Edition. Ismail Kasikci.

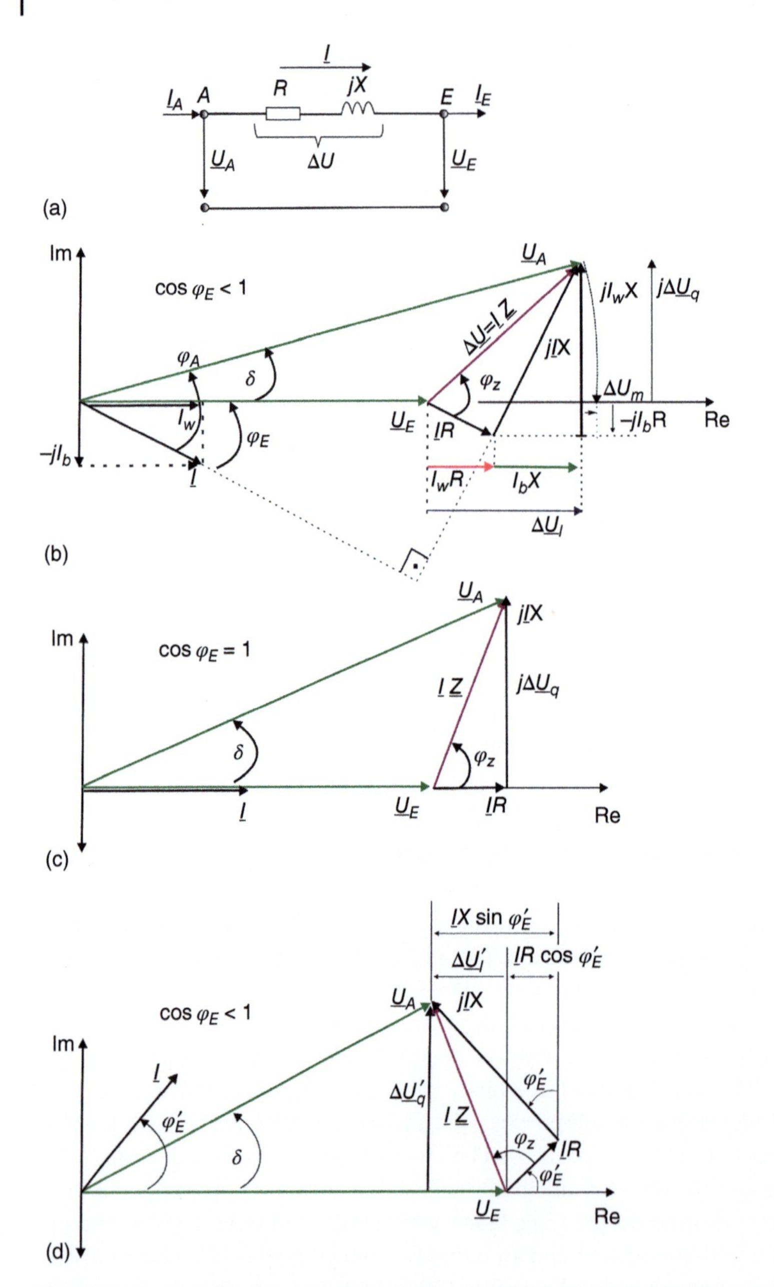

Figure 16.1 Voltage drop calculation with (a) equivalent circuit diagram, (b) inductive load, (c) active load, and (d) capacitive load.

The amount ΔU of the complex voltage difference $\Delta \underline{U}$ along a line, for example, in the pointer diagram of Figure 16.1, is not is identical to the difference of the amounts of the node voltages at the beginning and end of the line, the so-called effective measurable voltage difference $\Delta \underline{U}_m$ [52]. To simplify the calculation of $\Delta \underline{U}_m$, the terms longitudinal and transverse voltage drop were introduced. The real part is called longitudinal voltage drop $\Delta \underline{U}_l$ because it lies in the direction of voltage $\underline{U}_E$, while the imaginary part is called transverse voltage drop $\Delta \underline{U}_q$ because it lies across 90° leading to voltage $\underline{U}_E$. For a load with inductive cosφ, the current is

$$\underline{I} = \underline{I}_w - j\,\underline{I}_b \tag{16.3}$$

inserted in Eq. (16.4) results:

$$\Delta U = \underline{I} \cdot \underline{Z} \Rightarrow \underline{Z} = R + jX \tag{16.4}$$

$$\Delta \underline{U} = (I_w - jI_b) \cdot (R + jX) \tag{16.5}$$

$$(I_w \cdot R + I_b \cdot X) + j(I_w \cdot X - I_b \cdot R)$$

From this, the longitudinal and transverse voltage drop, relative to the string voltage, can be calculated as follows.

$$\Delta \underline{U}_1 = (I_w \cdot R + I_b \cdot X) = I \cdot (R \cdot \cos\varphi + X \cdot \sin\varphi) \tag{16.6}$$

$$\Delta \underline{U}_q = (I_w \cdot X - I_b \cdot R) = I \cdot (X \cdot \cos\varphi - R \cdot \sin\varphi) \tag{16.7}$$

The transverse voltage drop is a measure for the line angle ϑ or for the transmitted voltage and is particularly effective in ring lines and in high-voltage networks. with $X \gg R$. If $\cos\varphi_E = 1$, the transverse voltage drop is related to the string voltage

$$\Delta U_q = I \cdot X \tag{16.8}$$

The voltage, related to the chained voltage, can be set at the beginning of the line using the Pythagoras

$$U_A = \sqrt{(U_E + \Delta U_l)^2 + \Delta U_q^2} \tag{16.9}$$

can be determined.

For medium- and low-voltage networks, set $U_q \ll U_2 + U_l$, so that the voltage difference becomes $\Delta U_m \approx U_l$. For $\cos\varphi = 0.7 \cdots 0.9$, this difference is negligible.

And the tension at the end:

$$U_E = \sqrt{(U_A^2 - \Delta U_q)^2} - \Delta U_l \tag{16.10}$$

The cable angle (generally also called stability angle) results from the difference of the phase angles of the voltages at the beginning and end of the line to

$$\delta = \varphi_A - \varphi_E = Arctan\left(\frac{\Delta U_q}{U_A}\right) = Arctan\left(\frac{\Delta U_q}{U_E + \Delta U_l}\right) \tag{16.11}$$

To be able to make a statement, one refers to the absolute voltage drop to the nominal voltage at the beginning of the line.

$$\Delta u = \frac{\Delta U_l}{U_A} \cdot 100\% \tag{16.12}$$

For the low-voltage networks ($R \gg X$) applies

$$\Delta U_l = \underline{I} \cdot R \cdot \cos\varphi \tag{16.13}$$

As is the case in Figure 16.1d, the voltage triangle with the capacitive load current rotates in its direction. This changes the sign in the equations of longitudinal and transverse voltage drop. This occurs when overcompensation is caused by capacitors or PV systems. Then the voltage at the end of the line is higher than at the beginning of the line.

The transverse voltage drop can be reduced to zero at a certain capacitive power $\Delta U_l' = 0$, if the following condition is fulfilled:

$$I \cdot R \cdot \cos\varphi_E' - I \cdot X \cdot \sin\varphi_E' = 0 \tag{16.14}$$

$$R \cdot \cos\varphi_E' = X \cdot \sin\varphi_E' = 0 \tag{16.15}$$

Furthermore

$$\frac{R}{X} = \frac{\sin\varphi_E'}{\cos\varphi_E'} \tag{16.16}$$

It follows:

$$\cot\varphi_z = \tan\varphi_E' \tag{16.17}$$

The transverse voltage drop reaches its maximum value when the tangent in the capacitive range of the connected load is equal to the cotangent of the impedance angle.

In low-voltage systems, the inductive component X of cables and wires can be neglected. The current causes the longitudinal and transverse voltage drop with its active and reactive component.

$$\Delta\underline{U}_l = I \cdot R \cdot \cos\varphi_E \tag{16.18}$$

$$\Delta\underline{U}_q = I \cdot X \cdot \sin\varphi_E \tag{16.19}$$

In low-voltage systems, only the magnitude of the voltage drop is important. Therefore, the calculation of the longitudinal voltage drop is important (Figure 16.1). It should be mentioned that the longitudinal voltage drop differs from the measurable voltage drop. The error is small in the range of $\cos\varphi_E = 1$ to 0.7.

For one string this means the voltage drop:

$$\Delta\underline{U}_l = I \cdot R \cdot \cos\varphi_E \tag{16.20}$$

where $\cos\varphi$ is the power factor of the consumer.

Table 16.1 contains three boundary cases of the pointer diagrams (Figure 16.1).

Table 16.1 Cases.

Optimum ratios $\cos\varphi = 1$	$\Delta U_l = I \cdot R \Rightarrow \min$ $\Delta U_q = I \cdot X \Rightarrow \max$
High voltage drop $\cos\varphi = 0$, ind.	$\Delta U_l = I \cdot X \Rightarrow \max$ $\Delta U_q = -I \cdot R$
Voltage increase at line end $\cos\varphi = 0$, chapter $U_E > U_A$	$\Delta U_l = -I \cdot X$ $\Delta U_q = I \cdot R$

16.2 Example: Voltage Drop on a 10 kV Line

Data of the network are given on the picture.
Calculate:

1. the consumer current

$$\underline{I}_2 = I_{2w} - I_{2b} = \frac{S_2 \cdot \cos\varphi_2}{\sqrt{3} \cdot U_n} - j\frac{S_2 \cdot \sin\varphi_2}{\sqrt{3} \cdot U_n} \tag{16.21}$$

$$\underline{I}_2 = \frac{1\,\text{MVA} \cdot 0.8}{\sqrt{3} \cdot 10\,\text{kV}} - j\frac{1\,\text{MVA} \cdot 0.6}{\sqrt{3} \cdot 10\,\text{kV}} = (46.19 - j34.64)\,\text{A}$$

2. the voltage drop on the line

$$\Delta U = I_{2w} \cdot R_L - I_{2b} \cdot X_L = 46.19\,\text{A} \cdot 3.19\,\Omega - 34.64\,\text{A} \cdot 3.3\,\Omega = 33.03\,\text{V}$$

$$\Delta u = \frac{\Delta U}{U_n/\sqrt{3}} = \frac{33.03\,\text{V}}{10\,\text{kV}/\sqrt{3}} = 0.572\%$$

3. the voltage angle

$$\delta = \arcsin\frac{\Delta U_q}{U_2 + \Delta U_l} = \arcsin\frac{X_L \cdot I_{Lw} - R_L \cdot I_{Lb}}{U_2 + \Delta U_l} \tag{16.22}$$

$$\delta = \arcsin\frac{3.3\,\Omega \cdot 46.19\,\text{A} - 3.19\,\Omega \cdot 34.64\,\text{A}}{10\,\text{kV}/\sqrt{3} + 0.03303\,\text{kV}} = 0.416°$$

16.3 Example: Line Parameters of a Line

Given: 20 kV overhead line, $l = 8\,\text{km}$, $A = 70\,\text{mm}^2$ Al/St, $R' = 0.435\,\Omega/\text{km}$, $X' = 0.36\,\Omega/\text{km}$, $C' = 10.04\,\text{nF/km}$, load at end of line: $S_L = 2.5\,\text{MVA}$, $P_L = 1.8\,\text{MW}$, voltage at end of line 20.2 kV

We are looking for the sizes: $R, X, C, I, I_b, I_w, U_z, \Delta U, \varphi$.
Solution:

$$R = R' \cdot l = 8\,\text{km} \cdot 0.435\,\Omega/\text{km} = 3.48\,\Omega$$

$$X = X' \cdot l = 8\,\text{km} \cdot 0.36\,\Omega/\text{km} = 2.88\,\Omega$$

$$C = C' \cdot l = 8\,\text{km} \cdot 10.04\,\text{nF}/\text{km} = 80.32\,\text{nF}$$

$$X_C = \frac{1}{\Omega C} = \frac{1}{\Omega \cdot 80.32\,\text{nF}} = 39.63\,\text{k}\Omega$$

$$I_L = \frac{S_L}{\sqrt{3} \cdot U_2} = \frac{2.5\,\text{MVA}}{\sqrt{3} \cdot 20\,\text{kV}} = 72.17\,\text{A}$$

$$\cos\varphi = \frac{P_L}{S_L} = \frac{1.8\,\text{MW}}{2.5\,\text{MVA}} = 0.72 = 43.95°$$

$$I_w = \frac{P_L}{\sqrt{3} \cdot U_2} = \frac{1.8\,\text{MW}}{\sqrt{3} \cdot 20\,\text{kV}} = 71.96\,\text{A}$$

$$I_b = \sqrt{I_L^2 - I_w^2} = \sqrt{72.17^2\,\text{A} - 71.96^2\,\text{A}} = -50.09\,\text{A}$$

Complex voltage drop:

$$U_z = I_L \cdot \sqrt{R^2 - X^2} = 72.17\,\text{A} \cdot \sqrt{3.48^2\Omega - 2.88^2\Omega} = 326\,\text{V} \cdot e^{j43.95°}$$

$$U_z = 326\,\text{V} \cdot e^{j43.95°} = 234.7\,\text{V} + j226\,\text{V}$$

$$\underline{U}_z = \underline{I} \cdot \underline{Z} = 72.17\,\text{A} \cdot e^{43.95°} \cdot (3.48\,\Omega - 2.88\,\Omega)$$
$$= 326.21 \cdot e^{j43.34°} = (325.27 - j24.56)\,\text{V}$$

Longitudinal voltage drop:

$$\underline{U}_{zl} = I_w \cdot R - I_b \cdot X = 51.96\,\text{A} \cdot 3.48\,\Omega - (50.09\,\text{A}) \cdot 2.88\,\Omega = 325.08\,\text{V}$$

$$\Delta u_l = \frac{\Delta U_l}{U_N/\sqrt{3}} = \frac{325.08\,\text{V}}{20\,\text{kV}/\sqrt{3}} = 2.82\%$$

Outer conductor voltage:

$$U_l = U_2 + \sqrt{3} \cdot \Delta U_l = 20.2\,\text{kV} + \sqrt{3} \cdot 325.08\,\text{kV} = 20.76\,\text{kV}$$

Exact calculation of stress:

$$U_l = \sqrt{(U_2 + U_{zl} \cdot \sqrt{3})^2 + (U_{zg}\sqrt{3})^2} = 20.763\,\text{kV}$$

Lateral voltage drop:

$$\underline{U}_{zq} = I_w \cdot X + I_b \cdot R = -24.69\,\text{V}$$

Voltage angle:

$$\varphi = \arctan\frac{\Delta U_g}{U_2 + U_l} = \frac{-24.56\,\text{V}}{20.2\,\text{kV} + 325.08\,\text{V}} = -0.071°$$

16.4 Example: Line Parameters of a Line

Given: 110 kV single-wire oil cable, $l = 30\,\text{km}$, $A = 240\,\text{mm}^2$ Al/St, $R' = 0.138\,\Omega/\text{km}$, $X' = 0.148\,\Omega/\text{km}$, $C'_b = 0.31\,\mu\text{F/km}$, load at end of line: $S_L = 100\,\text{MVA}$, $P_L = 100\,\text{MW}$, voltage at end of line 112 kV

We are looking for the sizes: $R, X, C, I, I_b, I_w, U_z, \Delta U, \varphi$.

Solution:

$$R = R' \cdot 1 = 30\,\text{km} \cdot 0.138\,\Omega/\text{km} = 4.14\,\Omega$$

$$X = X' \cdot 1 = 30\,\text{km} \cdot 0.148\,\Omega/\text{km} = 4.44\,\Omega$$

$$C = C' \cdot 1 = 30\,\text{km} \cdot 0.31\,\mu\text{F/km} = 9.3\,\mu\text{F}$$

$$X_C = \frac{1}{\Omega C} = \frac{1}{\Omega \cdot 9.3\,\mu\text{F}} = 342.27\,\Omega$$

$$X_{Cb/2} = 648.54\,\Omega$$

$$I_L = \frac{S_L}{\sqrt{3} \cdot U_2} = \frac{100\,\text{MVA}}{\sqrt{3} \cdot 110\,\text{kV}} = 524.86\,\text{A}$$

$$cos\varphi = \frac{P_L}{S_L} = \frac{100\,\text{MW}}{100\,\text{MVA}} = 1$$

Complex voltage drop:

$$\underline{U}_z = \underline{I} \cdot \underline{Z} = 524.86\,\text{A} \cdot (4.14\,\Omega + j4.44\,\Omega) = 3186.26\,\text{V}$$

Longitudinal voltage drop:

$$\underline{U}_{zl} = R \cdot I_w - X \cdot I_b = 4.14\,\Omega \cdot 524.86\,\text{A} - 4.44\,\text{A}) \cdot 0\,\text{A} = 2172.92\,\text{V}$$

$$\underline{U}_{zl} = R \cdot I_b + X \cdot I_w = 4.14\,\Omega \cdot 0\,\text{A} + 4.44\,\text{A} \cdot 524.86\,\text{A} = 2330.38\,\text{V}$$

$$U_l = \sqrt{(U_2 + U_{zl} \cdot \sqrt{3})^2 + (U_{zg}\sqrt{3})^2}$$
$$= \sqrt{(112\,\text{kV} + 2172.92\,\text{V} \cdot \sqrt{3})^2 + (2330.38\,V\sqrt{3})^2}$$

$$= 115.834\,\text{kV} \Rightarrow U_{1Y} = \frac{U_1}{\sqrt{3}} = 66.88\,\text{kV}$$

$$\Delta U_l = \frac{\Delta U_{zl}}{U_n/\sqrt{3} + \underline{U}_{zl}} = \frac{2172.92\,\text{V}}{110\,\text{kV} \cdot /\sqrt{3} + 2172.92\,\text{V}} = 1.997\%$$

Voltage angle:

$$\varphi = \arctan \frac{\Delta U_{zg}}{U_2 + U_l} = \frac{2330.38\,\text{V}}{20.2\,\text{kV} + 325.08\,\text{V}} = -0.071°$$

X_C considered:

$$I_{XCb/2} = \frac{U_n/\sqrt{3}}{X_{Cb/2}} = \frac{110\,\text{kV}/\sqrt{3}}{-j684.54\,\Omega} = +j92.776\,\text{A}$$

$$\underline{I} = 524.86\,\text{A} - j92.776\,\text{A} = 533\,e^{+j10°}$$

$$\underline{U}_z = \underline{I} \cdot \underline{Z} = 533\,e^{+j10°} \cdot 6.07 \cdot e^{+j57.02°} = 3235.29 \cdot e^{+j57.02°}\,\text{V}$$
$$= (1761.12 + j2713.96)\,\text{V}$$

$$\underline{U}_{zl} = 1761.12 \quad \underline{U}_{zq} = 2713.96\,\text{V}$$

$$U_l = \sqrt{(U_2 + U_{zl} \cdot \sqrt{3})^2 + (U_{zg}\sqrt{3})^2}$$
$$= \sqrt{(112\,\text{kV} + 1761.12\,\text{V}\sqrt{3})^2 + (2713\,\text{V}\sqrt{3})^2}$$
$$= 115.15\,\text{kV} \Rightarrow U_{1Y} = \frac{U_1}{\sqrt{3}} = 66.88\,\text{kV}$$

$$U_l = U_2 + \sqrt{3} \cdot U_l = 112\,\text{kV} + 1761.12\,\text{V} = 115.05\,\text{kV} \Rightarrow U_{lY} = 66.42\,\text{kV}$$

$$\Delta U_l = \frac{\Delta U_{zl}}{U_n/\sqrt{3} + \underline{U}_{zl}} = \frac{1761.12\,\text{V}}{110\,\text{kV}/\sqrt{3}} = 2.773\%$$

$$\varphi = \arctan\frac{\Delta U_{zg}}{U_2 + U_l} = \frac{2713.96\,\text{V}}{110\,\text{kV}/\sqrt{3} + 1761.12\,\text{V}} = 2.340°$$

16.5 Voltage Regulation

Cables and lines are the means of transfer that enables the transport and distribution of electrical energy, predominantly by way of AC and three-phase current. When an electrical current flows through a cable or a line, heat is generated and is defined according to $P = I^2R$. The amount of heat generated over a defined period of time is given by $Q = I^2\,R\,t$. This heat loss must be dissipated before the insulation is destroyed.

The line resistance in a network is always assumed to be at 20 °C. For the calculation of the voltage drop, the permissible operating temperature is 70 °C. In low-voltage networks, the capacitive current is very small, so that in the equivalent circuit all operating capacities can be neglected. Thus, the cable or line can be mapped with an effective resistance and a reactance. The loads are ohmic-inductive.

For short lines, the Joule heat is decisive, and for long lines the voltage drop is decisive. A phase shift arises between the current and the voltage.

16.5.1 Permissible Voltage Drop in Accordance With the Technical Conditions for Connection

For a power requirement of more than 100 kVA between the point of supply from the power supply company and the measuring instruments, in accordance with Table 16.2 a greater voltage drop than 0.5% is permissible.

Table 16.2 Maximum permissible voltage drop according to technical conditions for connection.

Power requirement (kVA)	Maximum permissible voltage drop (%)
<100	0.50
100–250	1.00
250–400	1.25
Over 400	1.50

16.5.2 Permissible Voltage Drop in Accordance With Electrical Installations in Buildings

The permissible voltage drop in the electrical system before the measuring equipment can be taken from the technical conditions for connection of the power supply company. The voltage drop in the electrical system behind the measuring equipment may not exceed 3%, in consideration of IEC 60 364 Part 52. The basis for the calculation of the voltage drop is the rated current of the upstream overcurrent protection equipment.

16.5.3 Voltage Drops in Load Systems

IEC 60 364 Part 52

The voltage drop from the intersection between the distribution network and the load system to the point of connection of the consumer equipment (electrical outlet or equipment terminals) may not be greater than 4% of the nominal network voltage. In accordance with the technical conditions for connection, lines from the power supply company in medium-voltage and high-voltage networks have a voltage drop of +10% and −10%.

The calculation of the voltage drop in electrical networks can be done with the following equations:

- For DC currents
  1. Voltage drop in V

$$\Delta U = \frac{2 l I}{\kappa S} \tag{16.23}$$

  2. Power in W

$$P = U I \tag{16.24}$$

- For single-phase AC currents
  1. Voltage drop in V

$$\Delta U = \frac{2 l I \cos\varphi}{\kappa S} \tag{16.25}$$

2. Power in W

$$P = U\,I\,\cos\varphi \tag{16.26}$$

- For three-phase currents
  1. Voltage drop in V

$$\Delta U = \frac{\sqrt{3}\,l\,I\,\cos\varphi}{\kappa\,S} \tag{16.27}$$

  2. Power in W

$$P = \sqrt{3}\,U\,I\,\cos\varphi \tag{16.28}$$

- Percent voltage drop

$$\Delta u = \frac{\Delta U}{U_n}\,100\% \tag{16.29}$$

For a symmetrically loaded three-phase network:

$$\Delta U = \sqrt{3}I\,l\,(R'_L\,\cos\varphi + X'_L\,\sin\varphi) \tag{16.30}$$

with

$$R'_L = \frac{1}{\kappa\,S} \tag{16.31}$$

for the voltage drop in %:

$$\Delta u = \frac{\sqrt{3}\,I_n\,l\,(\frac{1}{\kappa\,S}\,\cos\varphi + X'_L\,\sin\varphi)}{U_n}\,100\,\% \tag{16.32}$$

16.5.4 Voltage Drops in Accordance With IEC 60364

For the installation of heavy current systems up to 1 kV, it is necessary to take the maximum permissible lengths for cables and lines into account Above all, protection against indirect shock, protection in the event of a short circuit, and limitation of the voltage drop must be considered during the planning and configuring of electrical systems. Supplement 5 describes the following parameters for the maximum permissible line lengths, leading in each case to different results.

16.5.5 Parameters for the Maximum Line Length

1. Requirements for basic protection
2. Requirements for fault protection
3. Protection against short circuits and overloading
4. Cut-off with the use of an overcurrent protection device, such as
   - Fuses
   - Circuit breakers
   - Power circuit breakers
   - Fault current circuit breakers (RCDs)

5. Compliance with shock hazard protection for the cut-off time t_a:
   - 0.4 second: In circuits up to 35 A rated current with electrical outlets
   - 0.4 second: In circuits of protection class I, for manually operated equipment
   - Five seconds: In all other circuits for permanently installed equipment
6. I_a: Cut-off current (breaking current) for:
   - Fuses: upper limit of time-current characteristic
   - Circuit breakers: I_a = tripping current $(I_5)\, I_n$, within 0.1 second
   - Power circuit breakers: $I_a \geq 1.2\, I_e$
   - RCDs: $I_a = I_{\Delta r}$ (rated residual current)
7. Single-pole short circuit current: I''_{k1}, at 80 °C
8. Coordination of length limiting
9. Breaking capacity of protective device I''_{k3}

For the calculation of the maximum line length, it is necessary to first calculate the smallest single-pole short circuit current in accordance with IEC 60 364, Supplement 5 (Figure 16.2).

$$I''_{k1\,\min} = \frac{\sqrt{3}\, c_{\min}\, U_n}{3 \cdot \sqrt{(2\, l\, R'_L + R_v)^2 + (2\, l\, X'_L + X_v)^2}} \tag{16.33}$$

The source impedance Z_v of the source network (power supply feeder network, transformer and feeder cable) is given by:

$$\underline{Z}_v = R_v + jX_v = \frac{2R_N + R_{0N}}{3} + j\frac{2X_N + X_{0N}}{3} \tag{16.34}$$

The resistance in the positive-sequence system of the source network is:

$$R_N = R_Q + R_T + R_{LN}$$

The resistance in the zero-sequence system of the source network is:

$$R_{0N} = R_{0T} + R_{0LN}$$

The reactance in the positive-sequence system of the source network is:

$$X_N = X_Q + X_T + X_{LN}$$

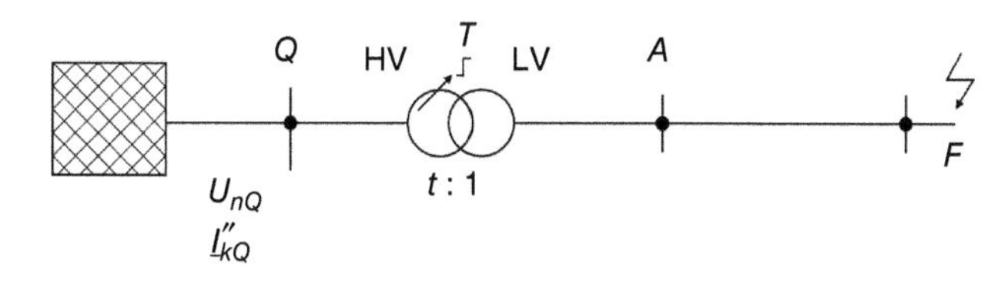

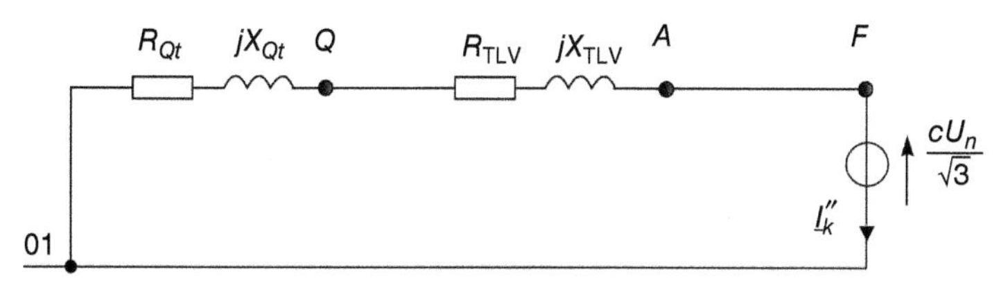

Figure 16.2 Schematic of the network.

The reactance in the zero-sequence system of the source network is:

$$X_{0N} = X_{0T} + X_{0LN}$$

The resistance of the transformer in the zero-sequence system is:

$$R_{0T} = \text{Table value} \cdot R_T$$

The reactance of the transformer in the zero-sequence system is:

$$X_{0T} = \text{Table value} \cdot X_T$$

The resistance in the zero-sequence system of the feeder cable is:

$$R_{0LN} = \text{Table value} \cdot R_L$$

The reactance in the zero-sequence system of the feeder cable is:

$$X_{0LN} = \text{Table value} \cdot X_L$$

For cables and lines with reduced PEN or protective conductor cross sections:

$$R'_L = \frac{R'_{L1} + R'_{L2}}{2} \tag{16.35}$$

The following conditions must be taken into account here:

- The factor $c_{\min} = 0.9$ is taken from Table 8.2.
- The motors are neglected.
- The effective resistances of the lines at 80 °C are used.
- The network layout is chosen so that the smallest short circuit current flows.

In practice, the smallest single-pole short circuit current is calculated with a simplified method. The error in calculation can then be as much as 20%:

$$I''_{k1\,\min} = \frac{c\,U_n}{\sqrt{3}\,Z_k} \tag{16.36}$$

The meanings of the symbols are:

$c_{\min}$	Voltage factor
U_n	Nominal network voltage
R_Q, X_Q	Resistance, inductance of source network
R_T, X_T	Resistance, inductance of transformer
R_L, X_L	Resistance, inductance of line network
R_{0T}, X_{0T}	Zero-sequence resistance, inductance of transformer
R_{0L}, X_{0L}	Zero-sequence resistance, inductance of line network
R_v	Loop resistance of source network
X_v	Loop reactance of source network
R'_L	Resistance per unit length of cable
R'_{L1}	Resistance per unit length of external line
R'_{L2}	Resistance per unit length of PEN or protective conductor
X'_L	Reactance per unit length of cable
l	Line length

16.5.6 Summary of Characteristic Parameters

1. Cross section of line
2. Insulation and type of cable or line
3. Overcurrent protection device
4. Rated current or current setting
5. Cutoff time for shock protection
6. Source impedance
7. Voltage drop limiting
8. Protection against indirect shock in TN, TT, and IT systems

In AC and three-phase networks, the inductive reactance can be neglected for cables and lines under 16 mm^2. Here it is sufficient to calculate with the DC resistance. At this point, it is worth mentioning other calculations for the maximum line lengths:

$$l = \frac{\frac{c\,U_n}{\sqrt{3}\,I''_{k1}} - Z_v}{2\,z_L} \tag{16.37}$$

$$l = \frac{\kappa\,S \times 10^{-3}\left(\frac{U_0 \times 10^3}{I''_{k1}} - Z_v\right)}{n} \tag{16.38}$$

For n we can use:

$n = 2$ for identical cross sections
$n = 3$ for a return line with half cross section.

Calculation of the maximum transmission length, considering the voltage drop:

$$l_{\max} = \frac{U_n\,\Delta u \cos\varphi}{P\,(R'_L \cos\varphi + X'_L \sin\varphi) \cdot 10^{-3}} \tag{16.39}$$

Here the meanings of the symbols are:

I	Current in A
U_0	Line-to-ground voltage in V
U_n	External line voltage in V
P	Power in kW
R'_L	Effective per unit line resistance in Ω/m
X'_L	Effective per unit line reactance in Ω/m
Z_v	Source impedance in mΩ/m
S	Line cross section in mm^2
κ	Conductivity in $\frac{\text{A}\sqrt{\text{s}}}{\text{mm}^2}$
Δu	Percent voltage drop in %
ΔU	Voltage drop in V
l	Line length in m
$\cos\varphi$	Power factor
$\sin\varphi$	Reactive factor

Table 16.3 Maximum line lengths with source impedances up to the meter mounting board for a cutoff time of t_a= 0.4 seconds.

Cross section mm²	OPE A	l_{max} in m for		
		Z_v = 200 mΩ	Z_v = 400 mΩ	Z_v = 600 mΩ
1.5	B/C10	145/69	138/62	131/55
1.5	B/C13	108/50	101/44	94/37
1.5	B/C16	88/40	81/33	74/26
1.5	B/C20	69/31	62/24	55/17
2.5	B/C16	146/67	135/56	123/44
2.5	B/C25	115/51	103/40	91/28
2.5	B/C32	89/39	78/27	65/16
4	B/C20	182/82	165/64	146/45
4	B/C25	143/62	124/44	106/25
4	B/C32	108/45	89/26	71/3
6	B/C25	215/94	187/66	160/38
6	B/C32	162/67	134/39	107/12
10	B/C32	272/112	225/66	138/20

16.5.7 Lengths of Conductors With a Source Impedance

Tables 16.3 and 16.4 give the line lengths for the cut-off conditions 0.4 and 5 seconds, arranged according to line protection devices. The source impedances in the overhead lines are $Z_v = 300 \cdots 600$ mΩ, according to the distance from the feeder, and in cable networks $Z_v = 100 \cdots 300$ mΩ by comparison.

Table 16.5 gives the effective resistances per unit length.

16.6 Examples for the Calculation of Voltage Drops

16.6.1 Example 1: Calculation of Voltage Drop for a DC System

Given the DC system shown in Figure 16.3, calculate the voltage drop in %.

$$\Delta U = \frac{2 l I_n}{\kappa S} = \frac{2 \cdot 30\,\text{m} \cdot 4\text{A}}{56 \frac{\text{m}}{\Omega\text{mm}^2} \cdot 1.5\,\text{mm}^2}$$
$$= 2.85\,\text{V}$$
$$\Delta u = \frac{\Delta U}{U_n} \cdot 100\% = \frac{2.85\,\text{V}}{24\,\text{V}} \cdot 100\% = 11.875\%$$

In accordance with IEC 60 364, Part 52 only 4% is permissible.

Table 16.4 Maximum line lengths with source impedances up to the meter mounting board for a cutoff time of $t_a = 5$ seconds

Cross section	OPE	l_{max} in m for		
mm²	A	$Z_v = 200\ m\Omega$	$Z_v = 400\ m\Omega$	$Z_v = 600\ m\Omega$
1.5	B/C10	145	138	131
1.5	B/C13	108	101	94
1.5	B/C16	88	81	74
1.5	B/C20	69	62	55
2.5	B/C16	146	135	123
2.5	B/C25	115	103	91
2.5	B/C32	89	78	65
4	B/C20	182	165	146
4	B/C25	143	124	106
4	B/C32	108	89	71
6	B/C25	215	187	160
6	B/C32	162	134	107
10	B/C32	272	225	138

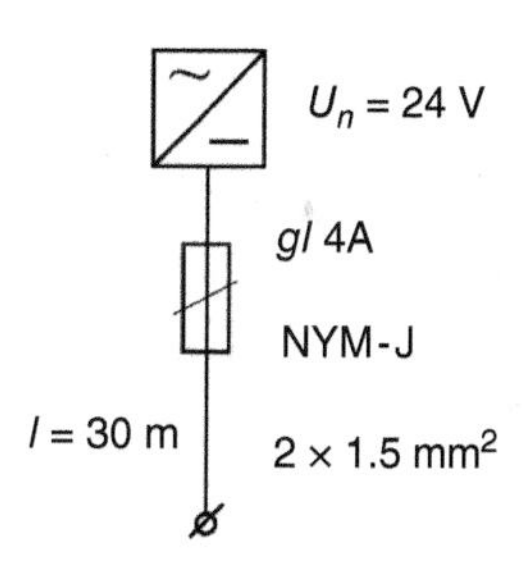

Figure 16.3 Voltage drop for a DC system.

16.6.2 Example 2: Calculation of Voltage Drop for an AC System

Given the motor shown in Figure 16.4 with $l = 30\ \text{m},\ \cos\varphi = 0.8$ and $\eta = 0.74$, calculate the voltage drop in %.

$$I_b = \frac{P}{U_n\ \cos\varphi\ \eta} = \frac{0.75\ \text{kW}}{230\ \text{V} \cdot 0.8 \cdot 0.74} = 5.5\ \text{A}$$

$$\Delta U = \frac{2lI\ \cos\varphi}{\kappa\ S} = \frac{2 \cdot 30\ \text{m} \cdot 5.5\ \text{A} \cdot 0.8}{56\frac{\text{m}}{\Omega\text{mm}^2} \cdot 2.5\ \text{mm}^2} = 1.89\ \text{V}$$

$$\Delta u = \frac{\Delta U}{U_n} \cdot 100\% = \frac{1.89\ \text{V}}{230\ \text{V}} \cdot 100\% = 0.82\%$$

Table 16.5 Resistance per unit length for (Cu) cable with plastic insulation.

Cross section S (mm^2)	R'_L at 70 °C (Ω/km)	X'_L (Ω/km)	Z'_L at $\cos\varphi$ 0.95 (Ω/km)	0.9 (Ω/km)	0.8 (Ω/km)
4 × 1.5	14.47	0.115	13.8	13.1	11.65
4 × 2.5	8.71	0.110	8.31	7.89	7.03
4 × 4	5.45	0.107	5.21	4.95	4.42
4 × 6	3.62	0.100	3.47	3.30	2.96
4 × 10	2.16	0.094	2.08	1.99	1.78
4 × 16	1.36	0.090	1.32	1.26	1.14
4 × 25	0.863	0.086	0.847	0.814	0.742
4 × 35	0.627	0.083	0.622	0.6	0.55
4 × 50	0.463	0.083	0.466	0.453	0.42
4 × 70	0.321	0.082	0.331	0.326	0.306
4 × 95	0.232	0.082	0.246	0.245	0.306
4 × 120	0.184	0.080	0.2	0.2	0.195
4 × 150	0.150	0.080	0.168	0.17	0.168
4 × 185	0.1202	0.080	0.139	0.143	0.144
4 × 240	0.0922	0.079	0.112	0.117	0.121
4 × 300	0.0745	0.079	0.0954	0.101	0.107

16.6.3 Voltage Drop for a Three-Phase System

In accordance with IEC 60364, Part 52 the voltage drop must not exceed 4%. Given the system with a main distribution system and two sub-distribution systems shown in Figure 16.5, check whether this requirement is satisfied.

If no other resistances per unit length exist, we can use the following values [58]:

$X'_L \approx 0.08$ mΩ/m for cables and lines

$X'_L \approx 0.33$ mΩ/m for overhead lines

$X'_L \approx 0.12$ mΩ/m for busbars

Here:

$$\Delta U = \sqrt{3} I\, l\, (R'_L \cos\varphi + X'_L \sin\varphi)$$

With

$$R'_L = \frac{1}{\kappa\, S} \quad \text{and} \quad \Delta u = \frac{\Delta U}{U_n} \cdot 100\%$$

this results in:

$$\Delta u = \frac{\sqrt{3} I_n\, l\, (\frac{1}{\kappa\, S} \cos\varphi + X'_L \sin\varphi)}{U_n} 100\%$$

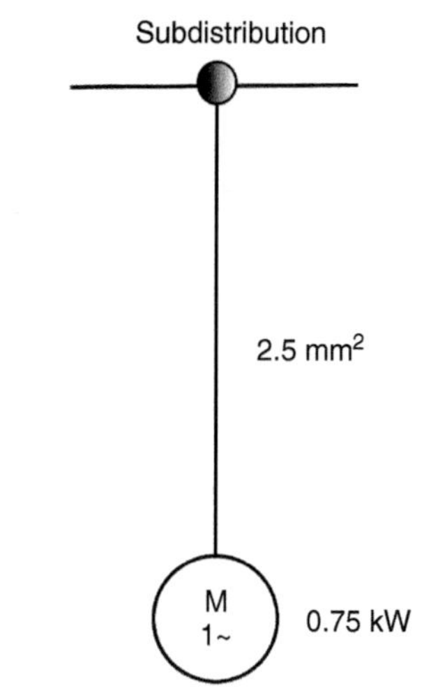

Figure 16.4 Voltage drop for an AC system.

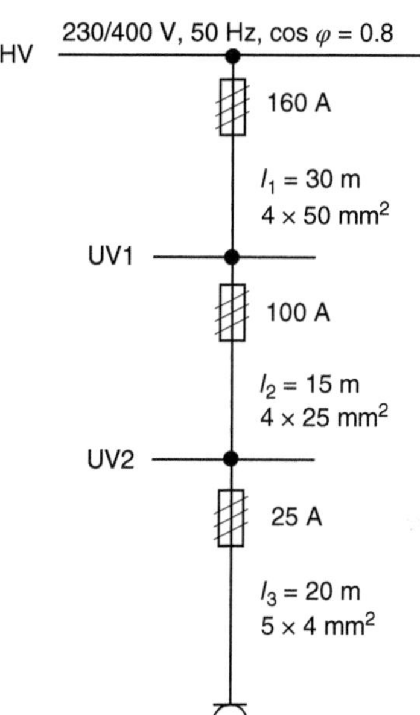

Figure 16.5 Voltage drop for a three-phase system.

a) Overhead line between main distribution and sub-distribution

$$\Delta u = \frac{\sqrt{3} \cdot 160\ \text{A} \cdot 30\ \text{m} \cdot \left(\frac{1}{56 \frac{\text{m}}{\Omega\text{mm}^2} \cdot 50\ \text{mm}^2} \cdot 0.8 + 0.33\ \text{m}\Omega/\text{m} \cdot 0.6 \right)}{400\ \text{V}} \cdot 100\%$$
$$= 1\%$$

b) Cable between sub-distribution 1 and sub-distribution 2

$$\Delta u = \frac{\sqrt{3} \cdot 100\ \text{A} \cdot 15\ \text{m} \cdot \left(\frac{1}{56 \frac{\text{m}}{\Omega\text{mm}^2} \cdot 25\ \text{mm}^2} \cdot 0.8 + 0.33\ \text{m}\Omega/\text{m} \cdot 0.6 \right)}{400\ \text{V}} \cdot 100\%$$
$$= 0.5\%$$

c) Line between sub-distribution 2 and electrical outlet

$$\Delta u = \frac{\sqrt{3} \cdot 25\ \text{A} \cdot 20\ \text{m} \cdot \left(\frac{1}{56 \frac{\text{m}}{\Omega\text{mm}^2} \cdot 4\ \text{mm}^2} \cdot 0.8 + 0.33\ \text{m}\Omega \cdot 0.6 \right)}{400\ \text{V}} \cdot 100\%$$
$$= 0.82\%$$

The total voltage drop at the electrical outlet is $\Delta u = 2.32\%$.

16.6.4 Example 4: Calculation of Voltage Drop for a Distributor

Given a distribution panel with the following data: $P = 200\ \text{kW}$, $Un = 400\ \text{V}$, $l = 200\ \text{m}$, $\Delta u = 4\%$, $\cos\varphi = 0.9$.

How large is the voltage drop at the distributor?

$$I_B = \frac{P}{\sqrt{3}U_n \cos\varphi} = \frac{200\ \text{kW}}{\sqrt{3} \cdot 400\ \text{V} \cdot 0.9} = 321\ \text{A}$$

$$\Delta U = \frac{\Delta u}{100\%} U_n = \frac{4\%}{100\%} \cdot 400\ \text{V} = 16\ \text{V}$$

$$Z = \frac{\Delta U}{\sqrt{3} l I}$$
$$= \frac{16\ \text{V}}{\sqrt{3} \cdot 200\ \text{m} \cdot 321\ \text{A}} = 0.144 \frac{\Omega}{\text{km}}$$

From Table 16.5, we obtain a cross section of $4 \times 185\ \text{mm}^2$ with $Z = 0.143 \frac{\Omega}{\text{km}}$. For this cross section, the voltage drop is then:

$$\Delta U = \sqrt{3} l I (R'_L \cos\varphi + X'_L \sin\varphi)$$
$$= \sqrt{3} \cdot 200\ \text{m} \cdot 321\ \text{A} \cdot (0.143 \frac{\Omega}{\text{km}}) = 15.88\ \text{V}$$

$$\Delta u = \frac{\Delta U}{U_n} \cdot 100\% = \frac{15.88\ \text{V}}{400\ \text{V}} \cdot 100\% = 3.97\%$$

16.6.5 Calculation of Cross Section According to Voltage Drop

Given a three-phase motor with the following data: $Un = 400\ \text{V}$, $P = 5.5\ \text{kW}$, $\eta = 0.83$, $\cos\varphi = 0.85$, with a permissible voltage drop of 5%.

The rated current of the fuse is $I_n = 63\ \text{A}$, and the threshold current of the nondelayed electromagnetic tripping device is $I = 160\ \text{A}$.

The motor connection is to be made through a line to a transformer. The internal resistance of the network and transformer can be neglected. The length of the line (simple) is $l = 75\ \text{m}$. Find the line cross section S.

$$I = \frac{P_{zu}}{\sqrt{3}\, U_n \cos\varphi\, \eta} = \frac{5.5\ \text{kW}}{\sqrt{3} \cdot 400\ \text{V} \cdot 0.85 \cdot 0.83} = 11.26\ \text{A}$$

$$\Delta U = \sqrt{3} Z l I$$

$$Z = \frac{\Delta U}{\sqrt{3} l I} = \frac{20\ \text{V}}{\sqrt{3} \cdot 0.075\ \text{km} \cdot 11.26\ \text{A}} = 13.68 \frac{\Omega}{\text{km}}$$

The cable is selected from Table 16.5: $4 \times 2.5\ \text{mm}^2$.

16.6.6 Example 6: Calculation of Voltage Drop for an Industrial Plant

In an industrial plant the main distribution system is fused with 200 A at a distance of 200 m. How large is the voltage drop in the line? The basis for the calculation is the upstream overcurrent protection device.

$$Z_L = \frac{\Delta U}{\sqrt{3}\, l\, I_n} = \frac{16\,\text{V}}{\sqrt{3} \cdot 0.2\,\text{km} \cdot 200\,\text{A}} = 0.231\,\Omega/\text{km}$$

$$\Delta u = \frac{\sqrt{3}\, l\, I_n\, Z_L}{U_n} \cdot 100\%$$

$$= \frac{\sqrt{3} \cdot 0.2\,\text{km} \cdot 200\,\text{A} \cdot 0.231\,\Omega/\text{km}}{400\,\text{V}} \cdot 100\% = 4\%$$

16.6.7 Example 7: Calculation of Voltage Drop for an Electrical Outlet

An electrical outlet is to be installed at a distance of 35 m from a sub-distribution system. A B16 A power circuit breaker is to be used as the protective device. The line cross section is 2.5 mm^2. Calculate the voltage drop at 20 °C.

$$\Delta u = \frac{2\, l\, I \cdot 100\%\, \cos\varphi}{\kappa\, S\, U_n}$$

$$= \frac{2 \cdot 35\,\text{m} \cdot 16\,\text{A} \cdot 100\% \cdot 1}{56 \frac{\text{m}}{\Omega\text{mm}^2} \cdot 2.5\,\text{mm}^2 \cdot 230\,\text{V}} = 3.47\%$$

16.6.8 Example 8: Calculation of Voltage Drop for a Hot Water Storage Unit

A 4 kW hot water storage unit is connected at a distance of 40 m from a line. How large is the voltage drop of the selected line?

Current consumption of unit:

$$P = \sqrt{3}\, U\, I$$

$$I = \frac{P}{\sqrt{3}\, U} = \frac{4\,\text{kW}}{\sqrt{3} \cdot 400\,\text{V}} = 5.77\,\text{A}$$

From IEC 60 364, Part 43, the cross section is $S = 1.5\,\text{mm}^2$

Voltage drop:

$$\Delta u = \frac{\sqrt{3}\, l\, P \cdot 100\%}{\kappa\, S\, U_n^2} = \frac{\sqrt{3} \cdot 40\,\text{m} \cdot 4\,\text{kW} \cdot 100\%}{56 \frac{\text{m}}{\Omega\text{mm}^2} \cdot 2.5\,\text{mm}^2 \cdot (230\,\text{V})^2} = 2.06\%$$

16.6.9 Example 9: Calculation of Voltage Drop for a Pump Facility

The following values were determined for a pump facility:

$P = 90\,\text{kW},\ U = 210\,\text{V},\ \cos\varphi = 0.88,\ X_L' = 0.33\,\text{m}\Omega/\text{m}.$

Calculate the voltage drop at the motor:

$$P = \sqrt{3}\, U\, I\, \cos\varphi$$

$$I = \frac{P}{\sqrt{3}\,U\,\cos\varphi} = \frac{90\,\text{kW}}{\sqrt{3}\cdot 210\,\text{V}\cdot 0.88} = 281.18\,\text{A}$$

$$I_w = I\,\cos\varphi = 281.18\,\text{A}\cdot 0.88 = 247.43\,\text{A}$$

$$I_B = I\,\sin\varphi = 281.18\,\text{A}\cdot 0.474 = 133.28\,\text{A}$$

$$R = \frac{1}{\kappa A} = \frac{400\,\text{m}}{56\frac{\text{m}}{\Omega\,\text{mm}^2}\cdot 70\,\text{mm}^2} = 0.102\,\Omega$$

$$X = 0.33\,\text{m}\Omega/\text{m}\cdot 400\,\text{m} = 0.132\,\Omega$$

$$\Delta U = I\,R\,\cos\varphi + I\,X\,\sin\varphi = I_w\,R + I_b\,X$$

$$= 247.43\,\text{A}\cdot 0.102\,\Omega + 133.28\,\text{A}\cdot 0.132\,\Omega = 42.8\,\text{V}$$

$$U_1 = 210\,\text{V} + 22.5\,\text{V} = 252.8\,\text{V}$$

$$\Delta u = \frac{\Delta U\cdot 100\%}{U_1} = \frac{42.8\,\text{V}\cdot 100\%}{252.8\,\text{V}} = 16.9\%$$

16.6.10 Example: Calculation of Line Parameters

Following data are given: 110 kV single conductor oil cable, $l = 30\,\text{km}, A = 240\,\text{mm}^2$ Al/St, $R' = 0.138\,\Omega/\text{km}$, $X' = 0.148\,\Omega/\text{km}$, $C_b' = 0.31\,\mu\text{F/km}$, load at the end of the line: $S_L = 100\,\text{MVA}$, $P_L = 100\,\text{MW}$, voltage at the end of the line 112 kV.

Calculate following sizes: $R, X, C, I, I_b, I_w, U_z, \Delta U, \varphi$.

Solution:

$$R = R'\cdot l = 30\,\text{km}\cdot 0.138\,\Omega/\text{km} = 4.14\,\Omega$$

$$X = X'\cdot l = 30\,\text{km}\cdot 0.148\,\Omega/\text{km} = 4.44\,\Omega$$

$$C = C'\cdot l = 30\,\text{km}\cdot 0.31\,\mu\text{F/km} = 9.3\,\mu\text{F}$$

$$X_C = \frac{1}{\omega C} = \frac{1}{\omega\cdot 9.3\,\mu\text{F}} = 342.27\,\Omega$$

$$X_{Cb/2} = 648.54\,\Omega$$

$$I_L = \frac{S_L}{\sqrt{3}\cdot U_2} = \frac{100\,\text{MVA}}{\sqrt{3}\cdot 110\,\text{kV}} = 524.86\,\text{A}$$

$$\cos\varphi = \frac{P_L}{S_L} = \frac{100\,\text{MW}}{100\,\text{MVA}} = 1$$

Complex voltage drop:

$$\underline{U}_z = \underline{I}\cdot\underline{Z} = 524.86\,\text{A}\cdot(4.14\,\Omega + j4.44\,\Omega) = 3186.26\,\text{V}$$

Longitudinal voltage drop:

$$\underline{U}_{zl} = R\cdot I_w - X\cdot I_b = 4.14\,\Omega\cdot 524.86\,\text{A} - 4.44\,\text{A}\cdot 0\,\text{A} = 2172.92\,\text{V}$$

$$\underline{U}_{zl} = R\cdot I_b + X\cdot I_w = 4.14\,\Omega\cdot 0\,\text{A} + 4.44\,\text{A}\cdot 524.86\,\text{A} = 2330.38\,\text{V}$$

$$U_l = \sqrt{(U_2 + U_{zl}\cdot\sqrt{3})^2 + (U_{zg}\sqrt{3})^2}$$

$$= \sqrt{(112\,\text{kV} + 2172.92\,\text{V}\cdot\sqrt{3})^2 + (2330.38\,\text{V}\sqrt{3})^2}$$

$$= 115.834\,\text{kV} \Rightarrow U_{1Y} = \frac{U_1}{\sqrt{3}} = 66.88\,\text{kV}$$

$$\Delta U_l = \frac{\Delta U_{zl}}{U_n/\sqrt{3} + \underline{U}_{zl}} = \frac{2172.92\,\text{V}}{110\,\text{kV}\cdot/\sqrt{3} + 2172.92\,\text{V}} = 1.997\%$$

Voltage angle:

$$\varphi = \arctan\frac{\Delta U_{zg}}{U_2 + U_l} = \frac{2330.38\,\text{V}}{20.2\,\text{kV} + 325.08\,\text{V}} = -0.071^\circ$$

X_C taken into consideration:

$$I_{XCb/2} = \frac{U_n/\sqrt{3}}{X_{Cb/2}} = \frac{110\,\text{kV}/\sqrt{3}}{-j684.54\,\Omega} = +j92.776\,\text{A}$$

$$\underline{I} = 524.86\,\text{A} - j92.776\,\text{A} = 533\,e^{+j10^\circ}$$

$$\underline{U}_z = \underline{I}\cdot\underline{Z} = 533\,e^{+j10^\circ}\cdot 6.07\cdot e^{+j57.02^\circ} == 3235.29\cdot e^{+j57.02^\circ}\,V$$

$$= (1761.12 + j2713.96)\,\text{V}$$

$$\underline{U}_{zl} = 1761.12\,\text{V} \quad \underline{U}_{zq} = 2713.96\,\text{V}$$

$$U_l = \sqrt{(U_2 + U_{zl}\cdot\sqrt{3})^2 + (U_{zg}\sqrt{3})^2}$$

$$= \sqrt{(112\,\text{kV} + 1761.12\,\text{V}\sqrt{3})^2 + (2713\,\text{V}\cdot\sqrt{3})^2}$$

$$= 115.15\,\text{kV} \Rightarrow U_{1Y} = \frac{U_1}{\sqrt{3}} = 66.88\,\text{kV}$$

$$U_l = U_2 + \sqrt{3}\cdot U_l = 112\,\text{kV} + 1761.12\,\text{V} = 115.05\,\text{kV} \Rightarrow U_{lY} = 66.42\,\text{kV}$$

$$\Delta U_l = \frac{\Delta U_{zl}}{U_n/\sqrt{3} + \underline{U}_{zl}} = \frac{1761.12\,\text{V}}{110\,\text{kV}/\sqrt{3}} = 2.773\,\%$$

$$\varphi = arctan\frac{\Delta U_{zg}}{U_2 + U_l} = \frac{2713.96\,\text{V}}{110\,\text{kV}/\sqrt{3} + 1761.12\,\text{V}} = 2.340^\circ$$

17

Switchgear Combinations

EN 60439-1-2-3-4-5

Until 1986, switchgear could be erected on the construction site as switchgear assemblies not ready for manufacture. From 1967, it was possible to produce "fabricated switchgear assemblies" in production plants. Switchgear assemblies have not been manufactured on site since 1984. When planning distributors, a distinction must be made between installation regulations according to DIN VDE 0100 and construction regulations according to DIN VDE 0660.

Switchgear assembly means that electrical and electromechanical equipment, in addition control, measuring, signaling, and protective devices and electronic components or power semiconductors are installed. Switchgear Combinations (SG) may be assembled in transport units or parts at the place of use or in a workshop outside the manufacturer's works. The assembly must be carried out according to the manufacturer's instructions and in accordance with the standard. Switchgear assembly consists of modules such as busbars, NH switchgear, circuit-breakers, and in-line devices.

17.1 Terms and Definitions

For the application of the standard DIN EN 60439, the most important terms and definitions are summarized (Figures 17.1 and 17.2). At this point, it should be mentioned that this standard uses different voltage and current terms and symbols contrary to IEC 60909-0 and IEC 60038.

- *Short circuit current I_c (short-circuit current)*: Overcurrent that occurs in the event of a short circuit due to a fault or incorrect connection in an electrical circuit.
- *Unaffected short circuit current I_{cp} (prospective short-circuit current)*: Effective value of the current that would flow if the supply line of the circuit were to pass through a conductor with negligible impedance.
- *Rated voltage U_n (rated voltage)*: The maximum rated mains voltage, AC voltage (rms value), or DC voltage specified by the switchgear assembly manufacturer for which the main circuits of the switchgear assembly are designed.

Analysis and Design of Electrical Power Systems: A Practical Guide and Commentary on NEC and IEC 60364, First Edition. Ismail Kasikci.

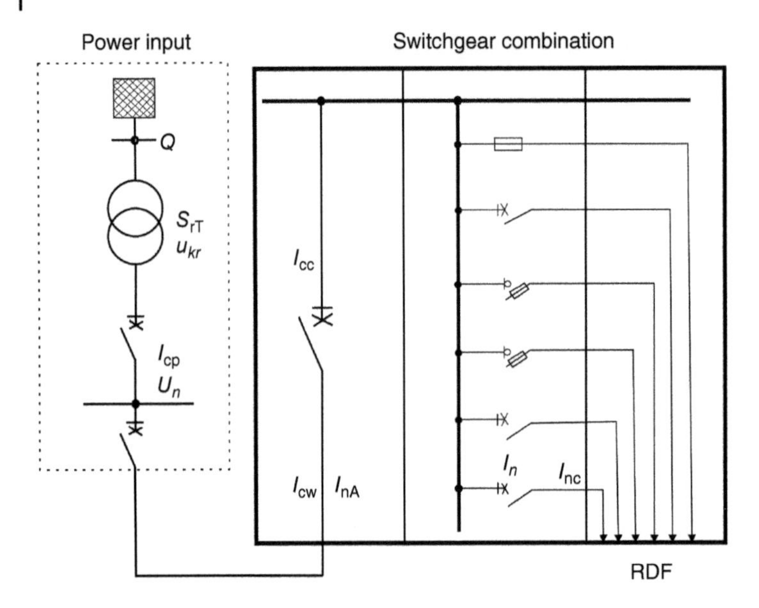

Figure 17.1 Construction of a switchgear.

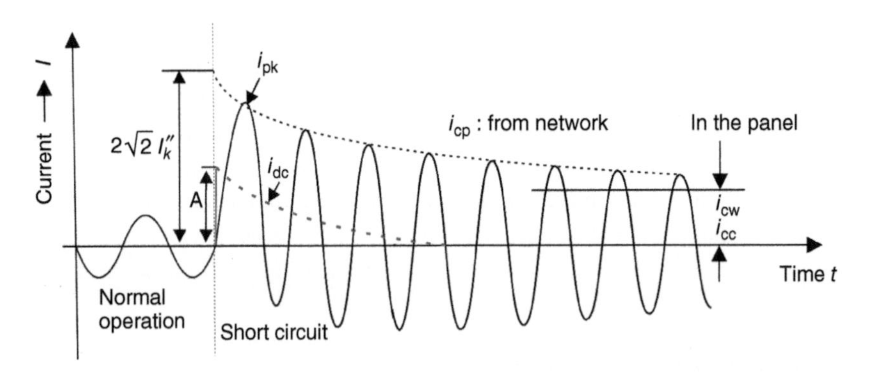

Figure 17.2 Definitions of short-circuit currents.

- *Rated current*: The value of the current specified by the manufacturer of the switchgear assembly which can be carried under specified conditions without exceeding the specified limit overtemperatures of the various parts of the switchgear assembly. For the rated current of the switchgear assembly (I_{nA}) and for the rated current of a circuit (I_{nc}).
- *Rated peak withstand current I_{pk} (rated peak withstand current)*: The maximum instantaneous value of the short-circuit current specified by the switchgear assembly manufacturer and withstood under the specified conditions.
- *Rated short-time withstand current I_{cw} (rated short-time withstand current)*: RMS value of the short-time current specified by the switchgear assembly manufacturer, expressed as current and time that can be resisted without damage under specified conditions.

- *conditional rated short-circuit current I_{cc} (rated conditional short-circuit current)*: The value of the unaffected short-circuit current specified by the manufacturer of the switchgear assembly, which the circuit protected by a short-circuit protection device (SCPD) can withstand under-defined conditions during the total switch-off time (current flow duration) of the device.
- *Rated diversity factor (RDF)*: Percentage value of the rated current specified by the manufacturer of the switchgear assembly with which the outputs of a switchgear assembly can be loaded continuously and simultaneously taking mutual thermal influences into account.
- *Air gap*: Distance between two conductive parts along a yarn which is tensioned along the shortest path between these parts.
- *Creepage distance*: Shortest distance between two conductive parts along a solid insulating surface.

For the new standards of the series DIN EN 61439 (VDE 0660-600):2010-06, IEC 61439 the following parts are provided [49]:

- DIN EN 61439-1 (VDE 0660-600-1):2010-06 Part 1: General requirements
  General specifications list the terms used and describe operating conditions, Construction requirements, technical characteristics, and requirements as well as verification possibilities.
- DIN EN 61439-2 (VDE 0660-600-2):2010-06 Part 2: Energy switchgear and controlgear assemblies
  Energy switchgear and control gear assemblies deal with switchgear systems that generate electrical energy for distribute all types of loads in industrial, commercial, and similar applications and control. It is not intended for laymen. Operating currents or the pending short circuit power at the input terminals is not limited.
- DIN EN 60439-3 (VDE 0660-504):2002-05 Part 3: Distribution boards for layman operation (DBO)
  Distribution boards are used to distribute electrical energy to places where operation is not possible and can be used in networks with up to AC 300 V to earth. The rated current of the distribution board is limited to 250 A. The maximum rated current of an outgoing circuit is 125 A. In addition to the switchgear, the SCPDs for operation contained in the input and output circuits by laymen.
- DIN EN 60439-4 (VDE 0660-501):2005-06 Part 4: Construction power distribution boards (BV)
  Distribution boards are generally transportable and suitable for use on construction sites, in indoors and outdoors. Because of the special conditions on construction sites, particular importance is attached to the mechanical strength of the enclosures and the IP protection class.
- DIN EN 60439-5 (VDE 0660-503):2007-05 Part 5: Cable distribution cabinets
  Switchgear assemblies in public power distribution networks.
  Cable distribution cabinets can be used in public power distribution networks as well as in both indoors and outdoors. Only authorized persons have access. Protection against vandalism must be taken into account.

- DIN EN 60439-2 (VDE 0660-502):2006-07 Part 6: Busbar trunking systems
  Busbar trunking systems are busbars that control the electrical energy in horizontal or vertical direction. Vertical form in the building – from the transformer station to the last sub-distribution.
  Outlet boxes with SCPDs enable variable tapping electrical energy to machines and sub-distribution boards. The distributors must, conditionally due to the design, special requirements with regard to the suspension distances and the mechanical strength and – due to the linear expansion – special have electrical properties.
- Draft Part 7: Switchgear assembly for sites, premises, and installations of a special kind, such as marinas, camping sites, marketplaces, and charging stations for electric vehicles.
- DIN EN 61439-1 Supplement 1:2014-06; VDE 0660-600-1 Supplement 1:2014-06: Low-voltage switchgear and controlgear assemblies – Part 1: General requirements; Supplement 1: Guideline for the specification of switchgear and controlgear assemblies (IEC/TR 61439-0:2013).
  From the user's point of view, this guide specifies the functions and characteristic values that should be specified when specifying switchgear assemblies.
- DIN EN 61439-1 VDE 0660-600-1 Supplement 2:2016-02 Low-voltage switchgear and controlgear assemblies Part 1: General requirements; Supplement 2: Method for verifying the heating of low-voltage switchgear and control gear assemblies by calculation.
  This supplementary sheet describes a method for verifying the heating of low-voltage switchgear and control gear assemblies by calculation. The method can be used for switchgear assemblies in closed design or for panels of switchgear assemblies divided by partitions without forced ventilation. It is used to determine the heating of the air inside the housing. However, it is not applicable if proof of heating has been provided in accordance with the respective product standard of the IEC 61439 series.

The structure and type of verification have been redefined in the DIN EN 60439 series as follows:

1. Structuring into a general part and corresponding product parts in accordance with the IEC 60947 switchgear standard.
2. Classification of the product parts in such a way that they can be clearly assigned to the application.
3. Open for the current and future product range of switchgear and distributors
4. The switchgear assembly must be designed as a "black box" in its interfaces.
5. The rated values must be verified depending on the area of application of the switchgear assembly and with regard to the protective goals to be met.
6. Differentiation between original manufacturer and manufacturer.

DIN VDE 0603 (VDE 0603-1):2017-06:

This standard applies to meter stations used to accommodate equipment for power transmission, distribution and measurement in electrical installations for commercial and domestic installations (e.g. residential buildings, schools, administrative, and office buildings, etc.), indoors and outdoors, up to a rated voltage of 400 V.

- *Part 1*: Meter locations General requirements
- *Part 2-1*: Meter positions for direct measurement up to 63 A
- *Part 2-2*: Meter positions for semi-direct measurement up to 1000 A
- *Part 3-1*: Main branch terminal (HLAK)
- *Part 3-2*: Mounting and contacting device (BKE) for electronic domestic meters (eHZ)
- *Part 3-3*: Meter plug-in terminal (ZSK)
- *Part 100*: Integration of the required measuring systems

The following basic standards must still be taken into account when planning and projecting low-voltage switchgear and controlgear assemblies:

DIN VDE 0100-410: Protection against electric shock, specifications for personal protection, DIN VDE 0100-540: Installation of low-voltage systems Part 5-54: Selection and installation of electrical equipment – earthing systems and protective conductors, DIN VDE 0110: Calculation and test methods, insulation coordination, DIN VDE 0470-1: Rated bases, test equipment, and test methods, DIN EN 60909-0 (VDE 0102):2016-12: Short-circuit currents in three-phase networks Part 0: Calculation of currents.

17.2 Design of the Switchgear

The parameters of the interfaces must first be clarified. Then the switchgear can be dimensioned:

17.2.1 Data for Design

1. *Connection to the electrical network*:
   Information from planners or customers and from manufacturers must be provided. The user should provide a single-pole overview circuit diagram.
   - nominal voltage of the supply
   - network system, system according to type of earth connection
   - rated current
   - short-circuit resistance
   - Transient Overvoltages
   - Rated frequency
   - connection cable
   - protection against electric shock (basic protection, fault protection)
   - short-circuit protection device
2. *Circuits and consumers*:
   - distribution circuits for downstream sub-distribution boards
   - final circuits

For proof of heating by calculation up to 1600 A, the I_{nc} must not exceed 80% of the I_n of the equipment.

$$I_{nc} = 0.8 \cdot I_n \tag{17.1}$$

3. *Installation and ambient conditions*:
   - indoor installation
   - outdoor installation
   - dimensions for transport and installation

   Operation by:
   - Device operation regulation of access

17.2.2 Design of the Distributor and Proof of Construction

The design verification must be provided by the manufacturer of the low-voltage switchgear and controlgear assembly.

17.2.3 Short-Circuit Resistance Proofing

The proof of short-circuit resistance to mechanical and thermal stress of the electrical systems is in accordance with DIN VDE 0102 and DIN VDE 0103. The switchgear assembly must be dimensioned in such a way that the possible short-circuit currents that can occur at any level of the distribution system in the event of a fault can be safely controlled or switched off. The shock short-circuit current I_p is used to assess the mechanical strength. The thermal effects of the short-circuit current can be assessed by the effective value I_{cp}.

Proof may be waived

- in auxiliary circuits with transformer,
- for current limiting ÜSE ($I_{cu} \leq 15kA$),
- with an uninfluenced rated short-circuit current ($I_{cu} \leq 15$ kA).

For the mechanical short-circuit load of the switchgear assemblies, the surge short-circuit current i_p must be determined with the initial short-circuit AC current I''_k and the surge factor n according to DIN EN 61439-1 (VDE 0660-600-1) Table 7 (Table 17.1).

A switchgear combinations comprise several circuits or a part of them. In DIN EN 61439-1 (VDE 0660-600-1) the term "rated load factor" has been introduced, which takes into account alternating and not simultaneous load of the individual main circuits and is used for the dimensioning of feeds and busbars. The factor n applies to all equipment whose power dissipation is proportional to the current, such as switches, pushbuttons, indicator lamps. The factor n' applies to all equipment whose power dissipation is quadratically dependent on the current, such as CB switches, contactors, switch-disconnectors, circuit-breakers, fuse links. Table 17.2 shows values for the assumed load of the main circuits.

Table 17.1 Impact factor depending on i_p and cos φ.

Effective value of surge short-circuit current (kA)	cos φ	Impact factor n
$I \leq 5$	0.7	1.5
$5 < I \leq 10$	0.5	1.7
$10 < I \leq 20$	0.3	2
$20 < I \leq 50$	0.25	2.1
$I > 50$	0.2	2.2

Table 17.2 Design load factor as a function of the number of main circuits.

Type of load	Assumed
	Design load factor n
Power supply	1
Power distribution – 2 and 3 circuits	0.9
Power distribution – 4 and 5 circuits	0.8
Power distribution – 6–9 circuits	0.7
Power distribution – 10 and more circuits	0.6
Actuator	0.2
Motors < 100 kW	0.8
Motors > 100 kW	1

17.2.4 Proof of Heating

All equipment must be operated at an ambient temperature which is less than or equal to the maximum permissible ambient temperature.

17.2.5 Determination of an Operating Current

The operating current I_B is necessary to verify the permissible heating (power loss). In addition to the already determined rated current of the circuit (I_{nc}), the number of circuits is also taken into account. Depending on the number of circuits, an assumed load factor can be used to calculate the operating current (I_B). The operating current is calculated:

$$I_B = I_{nc} \cdot \text{ load factor} \tag{17.2}$$

17.2.6 Determination of Power Losses

The allowable power loss P_v for the entire manifold is calculated from the difference between installed power loss through built-in devices, bus bars, and wiring and radiated power loss of the enclosures in the form of heat.

$$I_{max} = I_{30°} \cdot k_1 \cdot k_2 \tag{17.3}$$

$$P_v = I_{max}^2 \cdot R_{20} \cdot [1 + \alpha \cdot (T_C - 20°)] \tag{17.4}$$

17.2.7 Determination of a Design Loading Factor RDF

Rated diversity factor (RDF) is the percentage of the rated current specified by the manufacturer of the switchgear assembly with which the outputs of a switchgear assembly are continuously and simultaneously loaded taking mutual thermal influences into account.

When the operating current is known:

$$\text{RDF} = \frac{I_B}{I_{nc}} \quad \text{or} \quad \text{RDF} = \sqrt{\frac{\text{radiated power loss}}{\text{installed power loss}}} \tag{17.5}$$

For the entire switchgear assembly:

$$\text{RDF} = \frac{I_{nA}}{\Sigma I_{nc}} \tag{17.6}$$

1. *Proof of heating up to 630 A*: The comparison of the installable power dissipation with the dissipable power dissipation in the current range up to 630 A (only applicable with a cabinet if no horizontal bulkheads are installed). The standard assumes a ambient temperature of 35 °C. The components may only be loaded with maximum 80% of their rated current. All conductors must have a minimum cross section corresponding to 125% of the allowable rated current of the associated circuit.
2. *Proof of heating up to 1600 A*: Proof that limit overtemperatures are not exceeded in the distribution, this applies to the current range up to 1600 A (according to DIN EN 60890). The components may only be loaded with maximum 80% of their rated current. All conductors must have a minimum cross section corresponding to 125% of the allowable rated current of the associated circuit.
3. *Proof of heating above 1600 A*: Proof of heating above 1600 A shall be provided by testing or derivation of design values for similar variants.

17.2.8 Determination of an Operating Current

The operating current I_B is necessary to verify the permissible heating (power loss). In addition to the already determined rated current of the circuit (I_{nc}), the number of circuits is also taken into account. Depending on the number of circuits, an assumed load factor can be used to calculate the operating current (I_B). The operating current is calculated:

$$I_B = I_{nc} \cdot \text{load factor} \tag{17.7}$$

17.2.9 Check of Short-Circuit Variables

Short circuits can occur during the planning, commissioning, and operation of electrical systems. The most common case is the single-pole short circuit at the consumer. Three-pole short circuits can also occur within the SG combination. The most important variables are

I_{pk}: Characteristic value for the mechanical design and strength of the SG combination, peak value of the current.
I_{cw} / I_{cc}: Parameters for thermal design, rms value of the current for one second.

I_{cp} is the value of the short-circuit current which the network can deliver. I_{cc} is the effective value of the current which may queue before the protective device. This can be used to specify:

$$I_{cp} < I_{cw}, I_{cc}$$

Rated peak withstand current:

$$I_{pk} = I_{cp} \cdot n \tag{17.8}$$

No evidence is required if

$$\frac{I_{cp}}{I_{pk}} < \frac{10\ \text{kA}}{17\ \text{kA}}$$

If the short-circuit parameters $I_{cw} = 10\ \text{kA}$, $I_{cc} = 17\ \text{kA}$, $I_{cp} = 17\ \text{kA}$ are exceeded, new checks are generally required when using other makes (than during the test).

17.2.10 Construction and Manufacturing of the Distribution

The specifications in the catalogues, technical manuals, operating, and maintenance manuals must be observed during the construction and manufacture of the distributor. Observe installation instructions.

- assembly of individual parts/assemblies in housings/cabinets
- installation of the devices
- wiring within the switchgear
- inlet and outlet terminals for conductors inserted from outside
- assembly of doors, covers and claddings
- inscriptions/documentation

The arrangements and configurations of switchgear assemblies are different. DIN EN 60439-1 Supplement 1 contains various forms of subdivision in Annex. B.

17.2.11 CE Conformity

The low-voltage directive (2006/95/EC) provides that any product made available to the market, insofar as it is listed under this directive applies, bears the CE marking and the corresponding declaration of conformity is issued. The EU directive is implemented in the German legal system in the Product Safety Act. Declarations of conformity can be issued under the responsibility of the manufacturers themselves, so there is no need to involve a further institution. This corresponds to the product liability borne by the manufacturer

17.3 Proof of Observance of Boundary Overtemperatures

The calculation procedure for the assessment of "overtemperature limits" is very time consuming and difficult. Software programs and tables [49] can be used to save time. This section describes the basics for the verification of compliance with the limit overtemperatures according to DIN VDE 0600 Part 500. The calculated assessment (without fan) determines the course of the excess temperature of the air in the housing/casing. This is composed of the ambient temperature of the switch devices combination (outside the housing) and the excess temperature of the air inside the housing. As a result of the calculation, it shall be determined whether the equipment can operate perfectly at the rated currents used to determine the power dissipation and at the calculated excess temperature of the enclosure internal air. This statement applies to both the built-in switchgear and the electrical connections, such as rails and insulated cables. Unless otherwise specified, the ambient temperature (24-hour average) is 35 °C, the cabinetin-temperature 55 °C and the lower limit −5 °C. In addition, the following conditions must be met.

- The power losses are almost evenly distributed in the housing.
  The air circulation must not be obstructed.
- There must not be more than three horizontal partitions in a switchgear assembly or in a panel divided by partitions.
- For air openings, the cross section of the air exit openings must be at least 10% larger than the cross section of the air entry openings.
- The built-in equipment is suitable for DC and AC voltages up to 60 Hz and for a maximum supply current of 3150 A.

The following points must be known when calculating the overtemperature:

- IP protection class and protection class
- Installation type of housing/casing
- Housing/sheathing dimensions

- Design (wall-mounted or free-standing distribution board)
- Assembly, heat loss, space determination
- Place of installation (height, width, and depth)
- Structure of the lines the number of internal partitions
- Type selection
- Parts list, drawings, space reserve.

17.4 Power Losses

The power losses of the individual equipment must be taken from the manufacturer's specifications and added together.

If the equipment is operated with an operating current deviating from its rated current, the power loss is reduced, which can be divided into four groups.

1. Power losses change quadratically with the current, e.g. main current paths of the devices, busbars, lines:

$$P_v = P_{Vr}\left(\frac{I_B}{I_r}\right)^2 \tag{17.9}$$

2. Power losses change almost proportionally with the current, e.g. rectifiers, thyristors:

$$P_v \sim P_{Vr}\frac{I_B}{I_r} \tag{17.10}$$

3. Power losses show uneven behavior:

$$P_v = P_{Fe} + P_{Cu}\left(\frac{I_B}{I_r}\right)^2 \tag{17.11}$$

4. Power losses remain constant, e.g. solenoid coils of contactors, incandescent lamps:

$$P_v = P_{Vr} \tag{17.12}$$

It means:

P_v Power dissipation

P_{Vr} Loss power of the devices

I_B Operating current

I_r Rated current

P_{Fe} Iron losses

P_{Cu} Copper losses.

18

Compensation for Reactive Power

18.1 Terms and Definitions

- *Reactive power*: Power required to build up the electromagnetic fields. It cannot be converted into useful energy.
- *VAr controller*: Measures the reactive power which arises and gives switching commands to the contactors, which switch capacitors on or off as required.
- *Power factor*: Ratio of effective power to apparent power.
- *Compensation*: Through compensation the economy of operation of electrical systems, i.e. the power factor, is improved. This relieves the electrical operational equipment so that a greater effective power can be transmitted.
- *Individual compensation*: A capacitor is connected directly to the terminals of the load and switched together with this (e.g. a motor).
- *Group compensation*: The compensation equipment (controller unit) is connected to the load (e.g. a motor group or fluorescent lamp group).
- *Central compensation*: The compensation equipment (controller unit) is installed centrally for all loads.
- *Permanent compensation*: One or more capacitors are connected to a load.
- *Harmonic*: Oscillation at a whole-number multiple of the basic frequency of oscillation.
- *Filter circuits*: Series resonance circuits consisting of chokes and capacitors tuned to the frequencies of the harmonic currents.

Reactive current arises in every electrical system. Not only large loads but also smaller loads as well require reactive power. Generators and motors produce reactive power, which causes unnecessary burdens to and power losses in the lines. Figure 18.1 shows the block diagram for the network loading.

Reactive power is necessary to generate magnetic fields, e.g. in motors, Transformers, and generators. This power oscillates between the source and the load and represents an additional loading. Power supply companies and the consumers of this electrical energy are interested in reducing these disadvantages as well as possible. On the other hand, nonlinear loads and phase-controlled inverters cause harmonics, which lead to voltage changes and a decrease in the power factor. In order to reduce these harmonics, series resonant (filter) circuits are used.

Analysis and Design of Electrical Power Systems: A Practical Guide and Commentary on NEC and IEC 60364, First Edition. Ismail Kasikci.

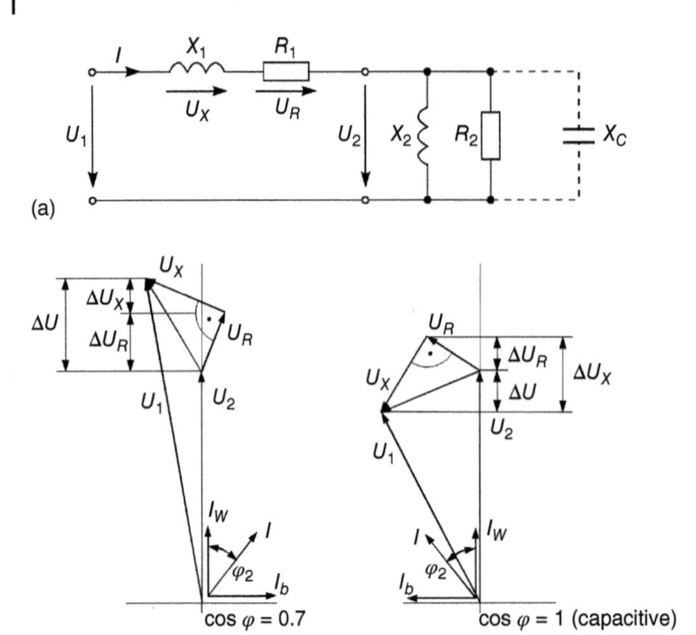

Figure 18.1 Equivalent circuit diagram of a network with different loading. (a) Equivalent circuit. (b) Phasor diagram.

For resistive, inductive, and capacitive loads, the power factor $\cos\varphi$ is different because the relationships between the voltage and current are different. Figure 18.2 shows the different phase relations for resistive, inductive, and capacitive loads.

Figure 18.3 illustrates current and power triangles and Figure 18.4 power triangles with compensation.

The apparent power absorbed in an electrical network is

$$S = U\,I \qquad \text{or} \qquad S = \sqrt{3}\,U\,I \quad \text{(VA)} \tag{18.1}$$

The effective power is

$$P = U\,I\,\cos\varphi \qquad \text{or} \qquad P = \sqrt{3}\,U\,I\cos\varphi \quad \text{(W)} \tag{18.2}$$

The reactive power is

$$Q = \sqrt{S^2 - P^2} \quad \text{(kvar)} \tag{18.3}$$

The reactances are

$$X_L = \omega\,L \qquad \text{and} \qquad X_C = \frac{1}{\omega\,C} \tag{18.4}$$

the power factor is

$$\cos\varphi = \frac{P}{S} \tag{18.5}$$

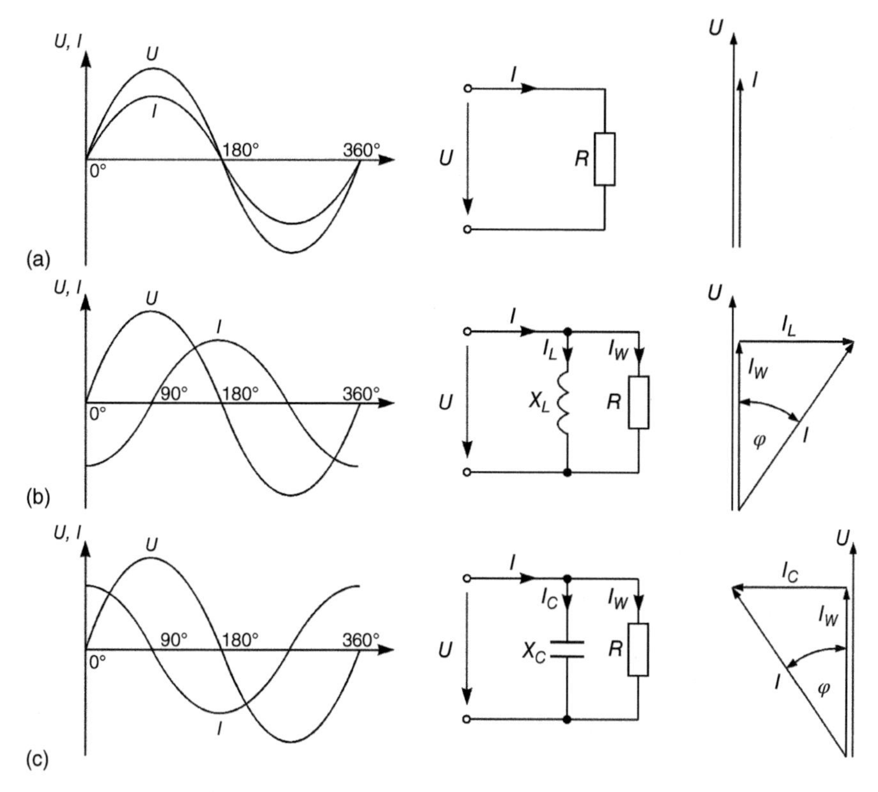

Figure 18.2 Relation between current, voltage, and power for (a) a resistive load, (b) an inductive load, and (c) a capacitive load.

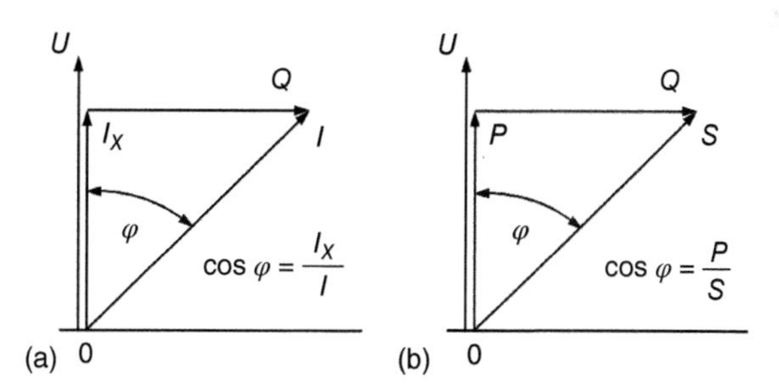

Figure 18.3 Current and power phasors in networks. (a) Current triangle. (b) Power triangle.

The effective current is:

$$I_w = I_n \cos\varphi \tag{18.6}$$

The reactive current is

$$I_b = I_n \sin\varphi \tag{18.7}$$

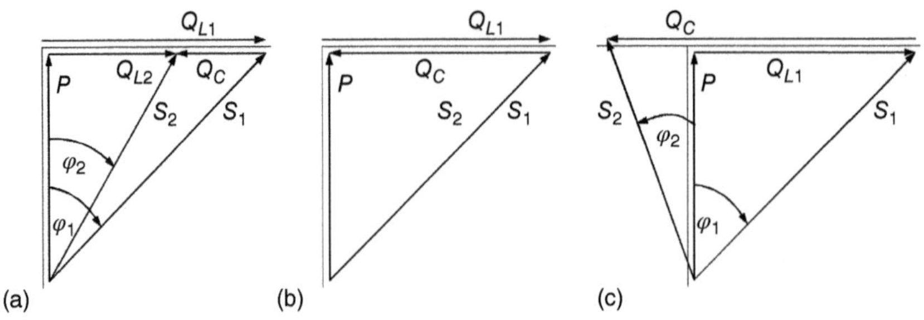

Figure 18.4 Power triangle with compensation. (a) Partial compensation. (b) Full compensation. (c) Overcompensation.

18.2 Effect of Reactive Power

In electrical networks, voltage drops occur due to loading of the internal resistances of the network.

$$\Delta U = \sqrt{3} I_b \, l \, (R_L' \cos \varphi + X_L' \sin \varphi) \tag{18.8}$$

$$\Delta u = \frac{\Delta U}{U_n} \cdot 100\% \tag{18.9}$$

The voltage drop can be calculated from the power as follows:

$$\Delta u = \frac{S}{S_k'' \sqrt{1 + (\frac{R}{X})^2}} \left(\frac{R}{X} \cos \varphi + \sin \varphi \right) \cdot 100\% \tag{18.10}$$

Neglecting the resistive part gives:

$$Q = S \, \sin \varphi \tag{18.11}$$

then

$$\Delta u = \frac{Q}{S_k''} \cdot 100\% \tag{18.12}$$

18.3 Compensation for Transformers

Transformers are designed according to the reactive power absorption and not the maximum required reactive power. Usually standard values can be taken from tables, irrespective of the power and the rated voltage. It is also possible to calculate the reactive power of transformers from the following relationship:

$$Q_{\mathrm{T}} \approx S_0 = \frac{I_0}{100} \, S_{\mathrm{rT}} \tag{18.13}$$

The meanings of the symbols are

Q_T Transformer no-load reactive power in kVAr
I_0 No-load current in A
S_{rT} Rated power of transformer
S_0 Transformer no-load apparent power in kVA

18.4 Compensation for Asynchronous Motors

In practice, the capacitor power is at maximum 90% of the motor power with no load, so that no dangerous self-excitation can occur. The maximum permissible capacitor power is then given by

$$Q_C = 0.9\ \sqrt{3}\ I_0\ U_n\ \sin\varphi_0 \cdot 10^{-3}\ \text{kvar} \tag{18.14}$$

Under no-load conditions, with $\sin\varphi_0 \approx 1$, we obtain:

$$Q_C = 0.9\ \sqrt{3}\ I_0\ U_n \cdot 10^{-3}\ \text{kvar} \tag{18.15}$$

Usually, the motor power and $\cos\varphi$ are known, and the no-load current cannot be measured. In this case, the calculation is based on the motor power P_{rM}:

- up to 40 kW, 40% of the rated motor power
- above 40 kW, 35% of the rated motor power

$$Q_C = \frac{P_{rM}}{\eta}(\tan\varphi_1 - \tan\varphi_2) \tag{18.16}$$

The meanings of the symbols are

η Efficiency of motor
I_0 No-load current in A
0.9 Factor for required no-load reactive power

18.5 Compensation for Discharge Lamps

Modern lighting systems are normally operated with inductive control gear, for which the power factor is between 0.3 and 0.6. By comparison, with electronic control gear the power factor is 0.95. Single-phase or three-phase group compensation for discharge lamps is also possible. For the uniform distribution of light fixture groups over the individual phases, only one-third of the capacitor power is required, which is calculated from:

$$Q_C = (P_L + P_v)\ (\tan\varphi_1 - \tan\varphi_2) \tag{18.17}$$

The required capacity of the condenser is then:

$$C = \frac{Q_C}{2\ \pi\ f_n\ U^2} \tag{18.18}$$

$$C_{\text{str}} = \frac{Q_C}{3\ U_n^2\ 2\ \pi\ f_n} \tag{18.19}$$

Here, the meanings of the symbols are

P_{L} Lamp power
P_v Power loss of control gear
C_{str} Capacity of winding phase
f_{n} Network frequency

18.6 c/k Value

The c/k value defines the magnitude of the required reactive power which must be present before a connection or shut-off takes place. In practice, this response threshold is set to about 60% of the smallest step. The c/k value is calculated from:

$$c/k = (0.6 \cdots 0.8)\frac{C}{\sqrt{3}\ U_n\ k} \tag{18.20}$$

Here, the meanings of the symbols are

c Smallest power step
C Capacitor power in var
U_n Voltage of external line in V
k Transformation ratio of transformer (e.g. 100 A/5 A)

18.7 Resonant Circuits

Resonant circuits are undesirable phenomena in networks. A series resonant circuit (Figure 18.5) consists of an inductance L and a capacitance C. Since the effective resistance is small relative to L and C, it can be neglected. At low frequencies, the reactance of the capacitor predominates, i.e. the series resonant circuit is capacitive. Above the resonance frequency f_{r}, the series resonant circuit behaves inductively. Figure 18.6 shows the behavior of the impedance, current, and voltage.

A parallel resonant circuit consists of an inductance, on the one hand, and a capacitance and resistance, on the other hand, in parallel (Figure 18.7). Here, the behavior is the opposite of that for the series resonant circuit (Figure 18.8).

18.8 Harmonics and Voltage Quality

The supply system is 230/400 V, 50 Hz, and has sine-wave form current and voltage curves. Nonlinear loads produce undesirable harmonics (Figure 18.9 illustrates the principle), which flow directly into the network and thus give rise to the following problems:

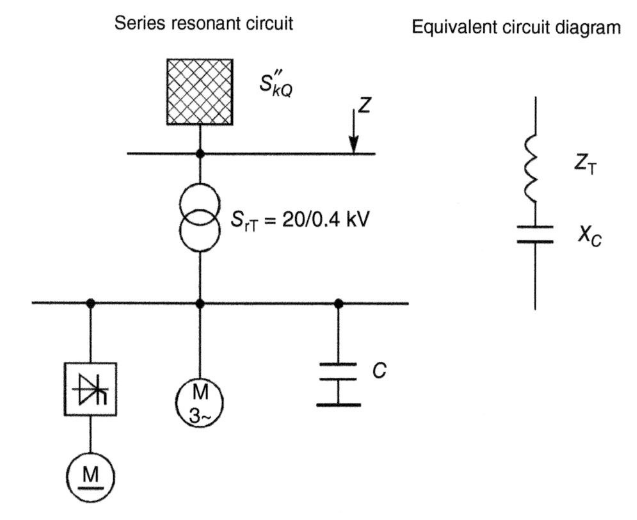

Figure 18.5 Series resonant circuit and equivalent circuit diagram.

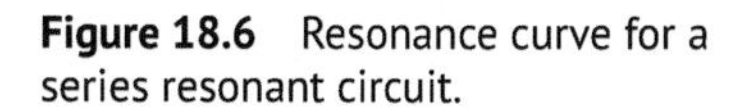

Figure 18.6 Resonance curve for a series resonant circuit.

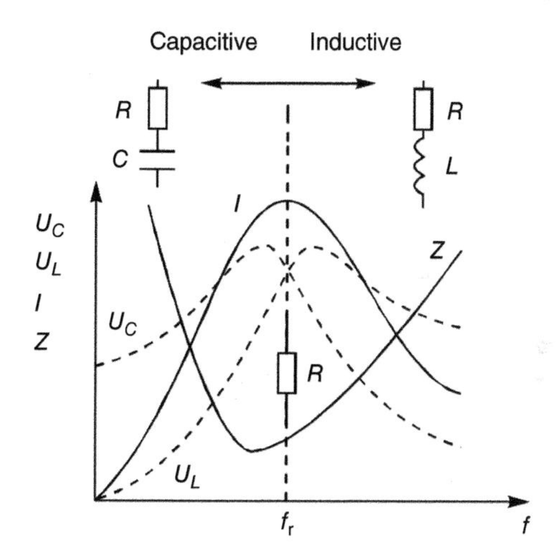

1) Asymmetry
2) Meter errors
3) Power cable errors
4) Voltage distortions
5) Series and parallel resonances
6) Decrease in the power factor
7) Temperature rise and overloading of transformers
8) Communication errors
9) Reduction of motor powers

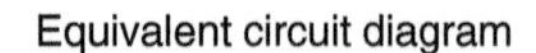

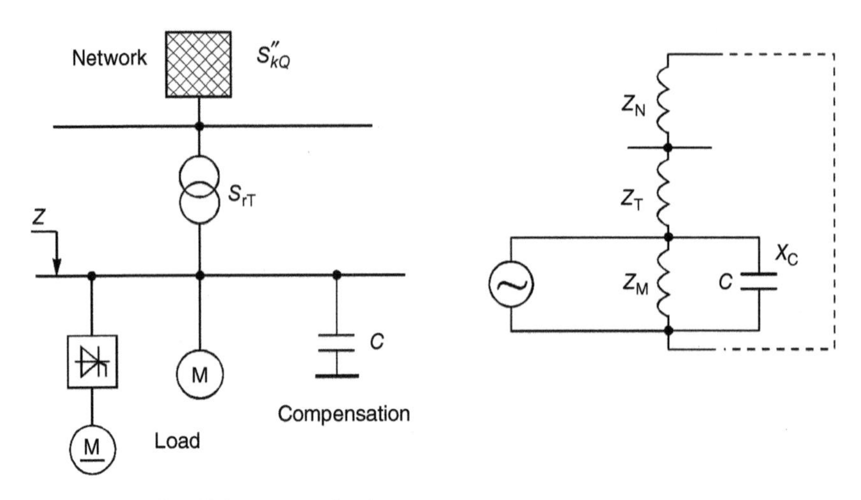

Figure 18.7 Parallel resonant circuit.

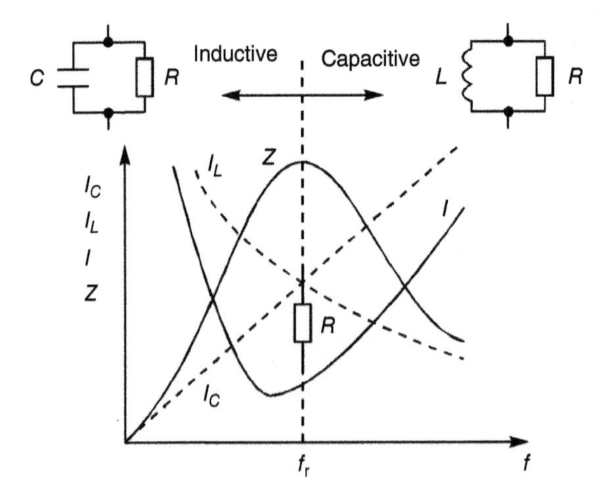

Figure 18.8 Resonance curve for a parallel resonant circuit.

18.8.1 Compensation With Nonchoked Capacitors

Capacitors combine with the reactance of the transformer to form a resonant circuit (Figure 18.10). Figure 18.11 shows the impedance curve for the network, as seen by the consumer generating the harmonics.

The capacitor power for a given harmonic order ν is found from the relationship:

$$Q_C \leq \frac{S_{\mathrm{rT}} 100}{\nu^2 \, u_k} \tag{18.21}$$

This corresponds to a frequency:

$$f_r = f_n \sqrt{\frac{S_{\mathrm{rT}} \, 100}{Q_C \, u_k}} \tag{18.22}$$

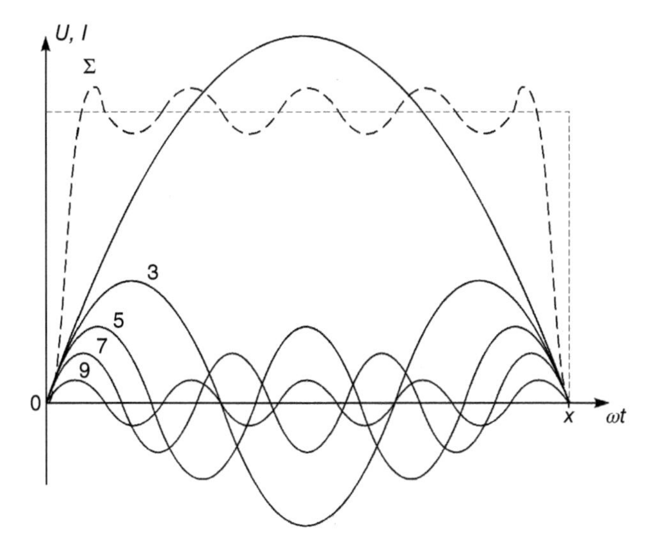

Figure 18.9 Harmonics.

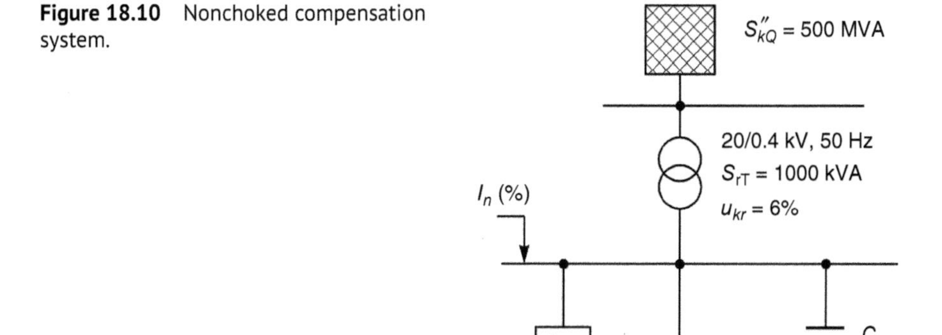

Figure 18.10 Nonchoked compensation system.

In order to prevent resonances, the installed reactive power must be less than the calculated (critical) reactive power. In general, no overcurrents occur in networks with nonlinear loads amounting to less than 20% of the total load.

18.8.2 Inductor–Capacitor Units

With a choked system, the capacitor and the choke coil are in series (Figure 18.12). Figure 18.13 shows the impedance curve for such a network.

When the nonlinear components constitute more than 20% of the total load, critical overcurrents arise. The choking factor can be calculated by summing the powers of the harmonic components and dividing by the nominal power of the transformer,

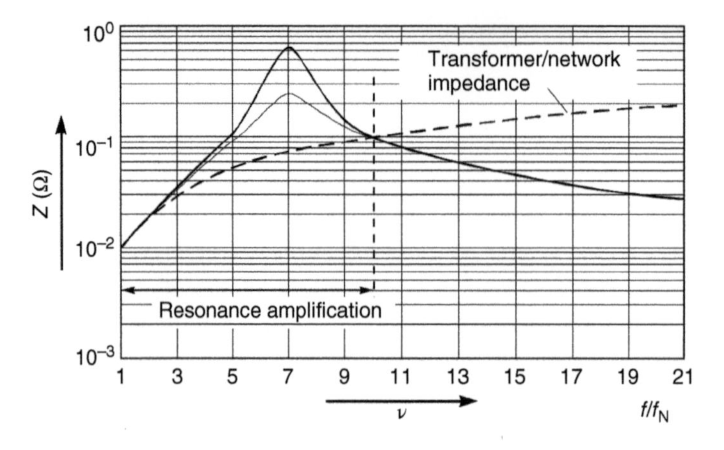

Figure 18.11 Impedance characteristic of a nonchoked network.

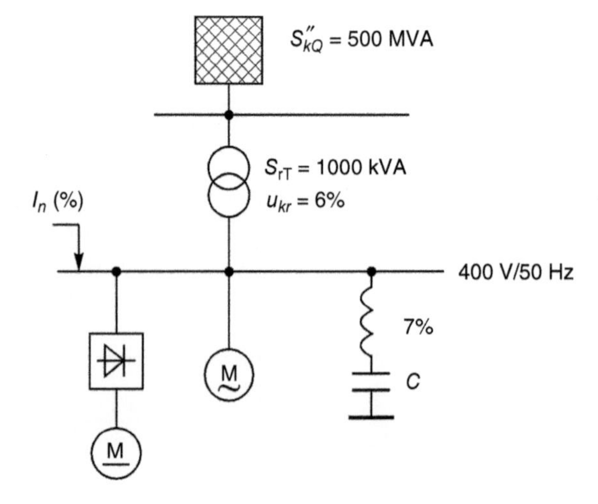

Figure 18.12 Compensation system with 7% choking.

i.e.:

$$V_{PV} = \frac{\sum S_H}{S_{rT}} \tag{18.23}$$

When the capacitor and choke coil are properly matched, the resonance frequency lies below the frequencies of critical harmonic currents.

With the use of choke coils, it is necessary to take the voltage increase of the capacitor into account. Increasing the dielectric strength prolongs the useful life. The increase in the voltage and the capacitor power can be calculated as follows:

$$U_C = \frac{U_n}{1 - p} \tag{18.24}$$

$$Q_C = \frac{Q_{Cn}}{1 - p} \tag{18.25}$$

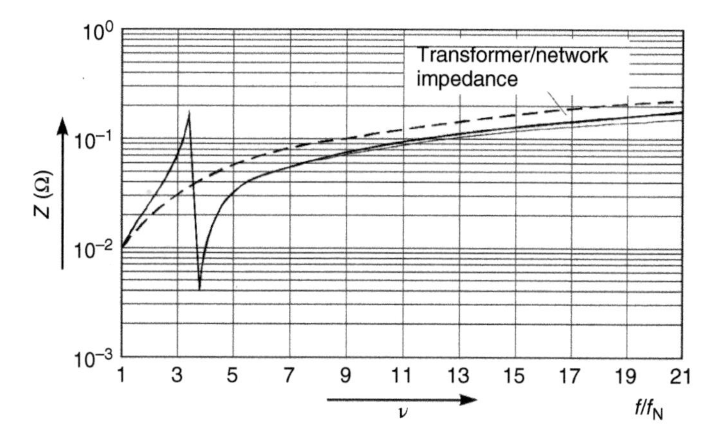

Figure 18.13 Impedance characteristic of a choked network.

The choking factor p in % gives the ratio of the choked reactance to the capacitive reactance at the network frequency:

$$f_\mathrm{r} = f_\mathrm{n}\frac{1}{\sqrt{p}} \tag{18.26}$$

The relative resonance frequency is

$$n_\mathrm{r} = \frac{f_\mathrm{r}}{f_\mathrm{n}} \tag{18.27}$$

$$p = \frac{1}{n_\mathrm{r}^2} \tag{18.28}$$

The choking factor is also a measure for the suppression of harmonics, i.e. the smaller the value of p the more effective is the filtering.

18.8.3 Series Resonant Filter Circuits

In order to prevent resonance phenomena, choke coils are placed upstream of the capacitors (choking). Filter circuits are tuned exactly to the frequencies of the harmonic currents and therefore suppress up to 90% of these currents. In general, it can be said that the filter circuits serve not only for the compensation of reactive power loads but also for the suppression of harmonic currents. Figure 18.14 schematically illustrates the principle of filter circuits.

Figure 18.15 gives the impedance characteristic of a network with load attenuation.

18.9 Static Compensation for Reactive Power

Reactive power can be controlled quickly and precisely with power electronics components. Continuous control and the improved stability of networks are

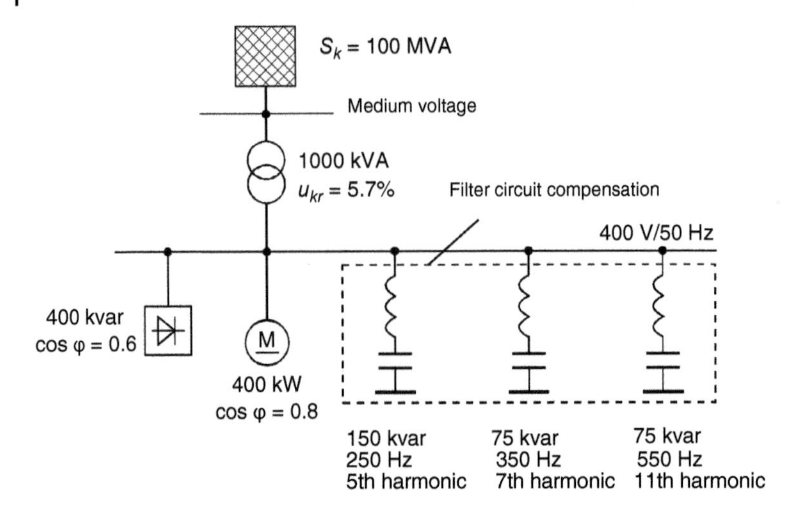

Figure 18.14 Series resonant filter circuits.

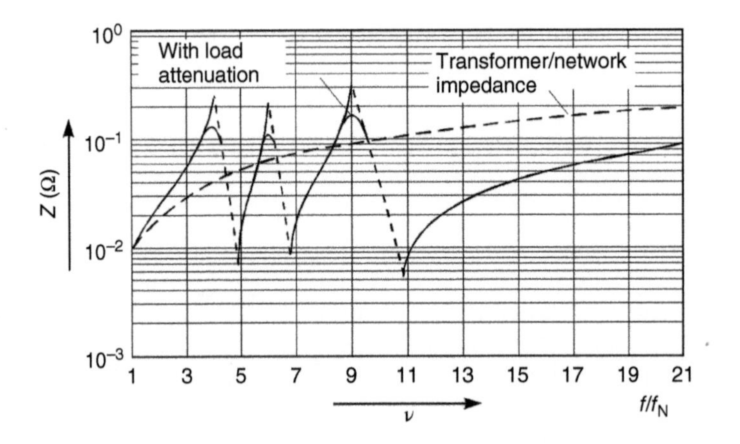

Figure 18.15 Frequency-impedance characteristic.

possible [30]. This section provides a brief description of the equipment used in electrical systems for reactive power compensation.

1) *Inductances*: Inductances are in parallel with the networks in order to absorb the reactive power. They are connected to the tertiary winding of a transformer, e.g. to 12 kV, or directly to the busbar of a generator.
2) *Capacitances*: Capacitances are connected either as a permanent capacitor or as a capacitor bank to the network in order to reduce the reactive power. The economic solution is voltage-controlled reactive power compensation with improved power factor.
3) *Thyristor-controlled static compensation*: These static compensators are reactive power units connected in parallel, in which the thyristor control generates or absorbs reactive power.

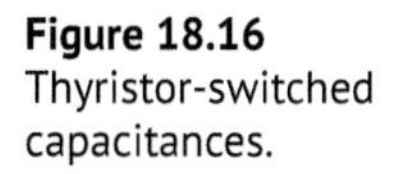
Figure 18.16 Thyristor-switched capacitances.

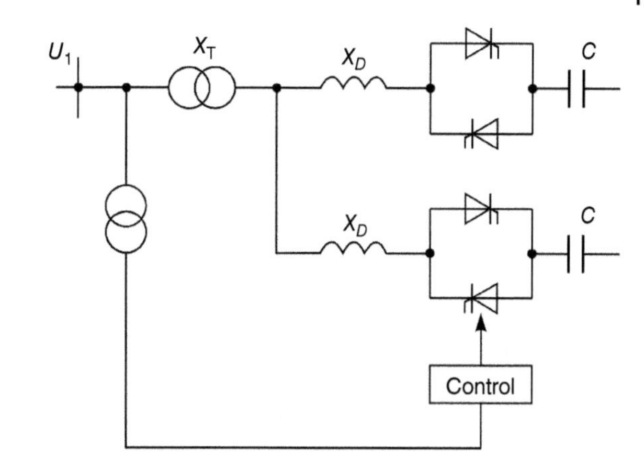

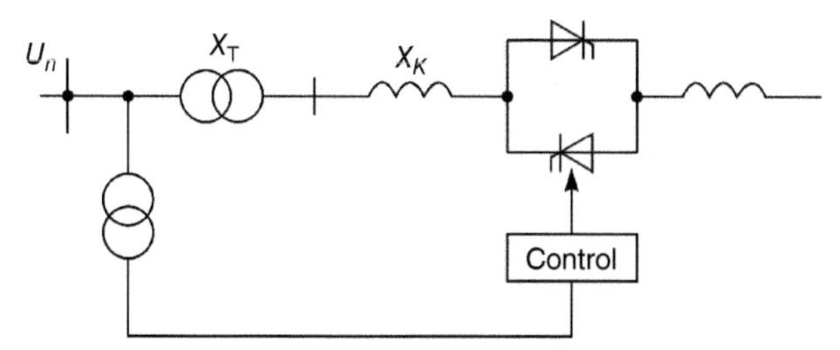

Figure 18.17 Thyristor-controlled inductances.

4) *Thyristor-switched capacitances*: Figure 18.16 schematically illustrates the principle of these capacitances. The capacitance C is switched on and off by the thyristor. The inductance L limits the current rise through the thyristors and causes resonances within the system.
   Properties of these systems are:
   - High costs
   - No transients
   - Stepped control
   - Avoidance of or generation of compensating harmonics
   - Very low losses at the output of the compensator.
5) *Thyristor-controlled inductances*: Figure 18.17 schematically shows the principle of this method of compensation.
   Properties of these systems are:
   - High costs
   - Generation of compensating harmonics
   - Continuous control
   - Continuously applied voltage
   - Losses at the output of the compensator.
6) *Capacitances in series*: Capacitances connected in series are well suited to reactance compensation, but not to reactive power compensation. However, they of course also affect reactive power compensation for transmitted power. Figure 18.18 illustrates this principle. The reactive power generated in these

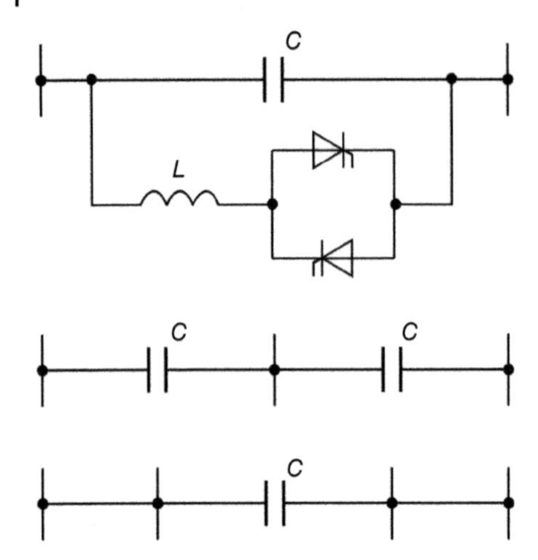

Figure 18.18 Capacitances in series.

capacitances increases with the power transmitted. The reactive power is between 100 and 800 Mvar. The greatest problem is protecting the capacitances from fault currents.

18.9.1 Planning of Compensation Systems

For the planning and configuration of compensation systems, it is necessary to distinguish between existing and new electrical systems. The actual state of existing networks must be determined by measurements at the substation and on the primary loads. Possible methods are, e.g. bills from the power supply company, meter readings, effective and reactive power recorders and harmonic current analysis. Following the analysis of the measured values and the ambient influences, the type of compensation can be decided.

For new systems, it is necessary to work partly with estimated data. Concurrence, utilization factor and extensions of electrical operational equipment must all be considered. It is recommended that the system be dimensioned with a relatively low compensation power and that the correct power level of the capacitors be determined following installation of the system. The calculation and arrangement of a reactive power compensation system can be done best with the use of computer simulations [59, 60].

18.10 Examples of Compensation for Reactive Power

18.10.1 Example 1: Determination of Capacitive Power

A load has an effective power of $P = 50$ kW at 400 V and the power factor is to be compensated from $\cos\varphi_1 = 0.75$ to $\cos\varphi_2 = 0.95$. Determine the required capacitive power. The power and current before compensation are:

$$S_1 = \frac{P_1}{\cos\varphi_1} = \frac{50\ \text{kW}}{0.75} = 66.66\ \text{kVA}$$

$$I_1 = \frac{S_1}{\sqrt{3}\,U} = \frac{66.66\text{ kVA}}{\sqrt{3}\cdot 400\text{ V}} = 96.22\text{ A}$$

The power and current after compensation are

$$S_2 = \frac{P_1}{\cos\varphi_2} = \frac{50\text{ kW}}{0.95} = 52.63\text{ kVA}$$

$$I_2 = \frac{S_2}{\sqrt{3}\,U} = \frac{52.63\text{ kVA}}{\sqrt{3}\cdot 400\text{ V}} = 76\text{ A}$$

The required capacitive power is

$$Q_C = P\,(\tan\varphi_1 - \tan\varphi_2)$$
$$= 50\text{ kW }(0.88 - 0.32) = 28\text{ kVAr}$$

18.10.2 Example 2: Capacitive Power With *k* Factor

The capacitive power can be determined with the factor k for a given effective power. The k factor is read from a table and multiplied by the effective power. The result is the required capacitive power.

For an increase in the power factor from $\cos\varphi = 0.75$ to $\cos\varphi = 0.95$, from the table [60] we find a factor $k = 0.55$:

$$Q_C = P\,k = 50\text{ kW}\cdot 0.55 = 27.5\text{ kVAr}$$

18.10.3 Example 3: Determination of Cable Cross Section

A three-phase power of 250 kW, with $U_n = 400$ V, at 50 Hz is to be transmitted over a cable 80 m in length. The voltage drop must not exceed 4%≙16 V. The power factor is to be increased from $\cos\phi = 0.7$ to $\cos\phi = 0.95$.

What is the required cable cross-section?

$$P = \sqrt{3}\,U\,I\,\cos\varphi$$

The current consumption before compensation is

$$I = \frac{P}{\sqrt{3}\,U\,\cos\varphi} = \frac{250\text{ kW}}{\sqrt{3}\cdot 400\text{ V}\cdot 0.7} = 515.5\text{ A}$$

The current consumption after compensation is

$$I = \frac{P}{\sqrt{3}\,U\,\cos\varphi} = \frac{250\text{ kW}}{\sqrt{3}\cdot 400\text{ V}\cdot 0.95} = 379.8\text{ A}$$

The effective resistance per unit length for 516 A is:

$$(R_L'\cos\varphi + X_L'\sin\varphi) = \frac{\Delta U}{\sqrt{3}\,I\,l}$$
$$= \frac{16\text{ V}}{\sqrt{3}\cdot 515.5\text{ A}\cdot 0.08\text{ km}}$$
$$= 0.224\ \Omega/\text{km}$$

According to Table 16.5 we must choose a cable with a cross section of $4 \times 95\ \text{mm}^2$.

The effective resistance per unit length for 380 A is

$$(R'_L \cos\varphi + X'_L \sin\varphi) = \frac{\Delta U}{\sqrt{3}\, I\, l} = \frac{16\ \text{V}}{\sqrt{3} \cdot 380\ \text{A} \cdot 0.08\ \text{km}} = 0.304\ \Omega/\text{km}$$

Here, a cable cross section of $4 \times 70\ \text{mm}^2$ is required.

As this example illustrates, the improved power factor leads to lower costs because of the reduced cross section.

18.10.4 Example 4: Calculation of the c/k Value

Given a 150 kVAr condenser battery, i.e. five stages of 30 kVAr each, a supply voltage of 400 V, and an instrument transformer with a k of 500 A/5 A, how large is the c/k value?

The ratio c/k is given by:

$$c/k = 0.65 \frac{C}{\sqrt{3}\, U_n\, k} = 0.65 \frac{30\ \text{kVAr}}{\sqrt{3} \cdot 400\ \text{V} \cdot \frac{500\ \text{A}}{5\ \text{A}}} = 0.281\ \text{kvar}$$

The controller is set to 0.3.

19

Lightning Protection Systems

IEC 62305 considers new technical knowledge and is based on the current state of technology. Its use is for the planning and installation of lightning protection systems up to 60 m high and offers safe protection of buildings.

This draft standard provides information about the following areas:

- Lightning protection classes
- Specifications for exterior lightning protection
- Specifications for ground electrodes
- Exposure distance determination
- Equipotential lightning protection bonding

For the installation, planning, extension, and modification of lightning protection systems IEC 62305 apply. In the applicable building ordinances, there is information about the need for lightning protection in buildings. Lightning protection is necessary for buildings which lightning can easily enter and cause severe damage. Such buildings include, e.g. schools, churches, apartment buildings, administrative buildings, hospitals, railroad stations, telecommunications towers, banks, airports, sport arenas, museums, explosion hazard areas, kindergartens, commercial buildings, and high-rise buildings.

Lightning protection systems include the following:

1. Exterior lightning protection
2. Interior lightning protection
3. Overvoltage protection of electronic equipment.

EN 50522 Grounding Systems in AC Systems above 1 kV and IEC 60 364, Parts 20 and 54 Heavy Current Systems up to 1 kV provide comprehensive explanations and terms for grounding systems. Figure 19.1 explains the terms necessary for understanding.

Here, the meanings of the symbols are:

U_S Step voltage
φ Ground-to-electrode potential
FE Concrete-footing ground electrode
SE Potential grading ground electrode (ring ground electrode)

Analysis and Design of Electrical Power Systems: A Practical Guide and Commentary on NEC and IEC 60364, First Edition. Ismail Kasikci.

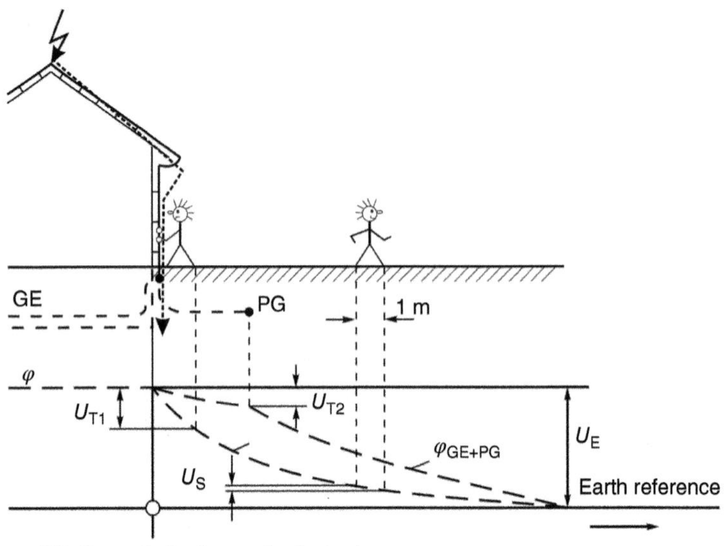

Figure 19.1 Ground-to-electrode potential and voltages for grounding electrodes with current flow. Source: Dehn+Söhne [63].

x	Distance to concrete-footing ground electrode
U_E	Ground potential rise
U_T	Touch voltage
U_{B1}	Touch voltage without potential grading ground electrode (on concrete-footing ground electrode)
U_{B2}	Touch voltage without potential grading ground electrode (on concrete-footing ground electrode and potential grading ground electrode)

- *Ground resistance*: In accordance with EN 50522, the ground resistance is the resistance of the grounding system between the grounding electrode and the ground reference plane.
- *Touch voltage*: The touch voltage is a part of the ground potential rise which can be shunted out through the human body.
- *Ground reference plane*: The ground reference plane is the part of ground, in particular the area outside the range of influence of a ground electrode or a grounding system, in which no measurable voltages occur between any two points as a result of the grounding current.
- *Ground termination network*: Grounding for leading the lightning discharge current to ground.
- *Ground*: Ground is the conductive soil.
- *Ground potential rise*: The ground potential rise is voltage occurring between a grounding system and the ground reference plane.
- *Soil resistivity*: The resistivity of the ground in Ωm.

- *Potential grading ground electrode*: The potential grading ground electrode is a grounding electrode which, due to its shape and arrangement, serves for controlling the potential.
- *Step voltage*: The step voltage is the part of the ground potential rise which can be shunted out in a one meter long section, such that the current flows through the human body from foot to foot.

19.1 Lightning Protection Class

In order to be able to plan a lightning protection system, it is first necessary to determine the protection class for the building in planning. Lightning protection classes I–IV are assigned according to different values of the efficiency of the lightning protection system, the mesh size and the lightning sphere size (Table 19.1). For each protection class, we can determine the shielding angle of a building according to the height of the air terminal above the ground (Figure 19.2). For the determination of the position of the air terminal, in general three methods are used:

1. The shielding angle method (α) for simple shapes
2. The lightning sphere method (radius r) for complicated cases
3. The mesh method (w) for flat surfaces.

Examples for protection classes (Table 19.2):

Table 19.1 Characteristic of the lightning protection classes.

Lightning protection class	Radius of sphere (m)	Mesh size (m)	Efficiency (%)
I	20	5 × 5	98
II	30	10 × 10	95
III	45	15 × 15	90
IV	60	20 × 20	80

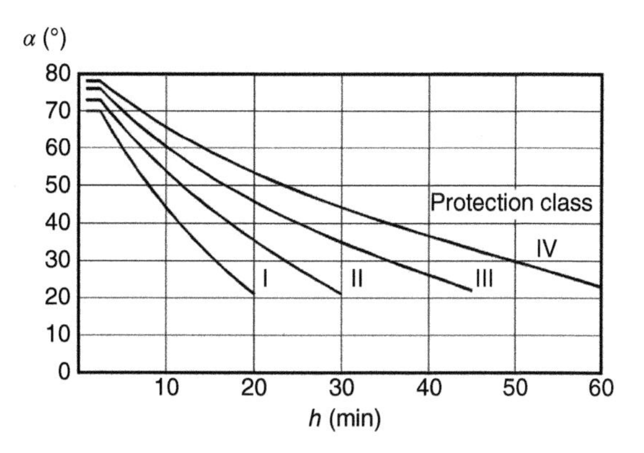

Figure 19.2 Shielding angle. Source: Dehn+Söhne [61].

Table 19.2 Examples for protection classes.

Class	Installations
Class I	Biological and nuclear installations
Class II	Telecommunications towers, cathedrals, industrial facilities
Class III	Apartment buildings, yards, schools, theaters, banks
Class IV	Weather-protected structures, refuge shelters

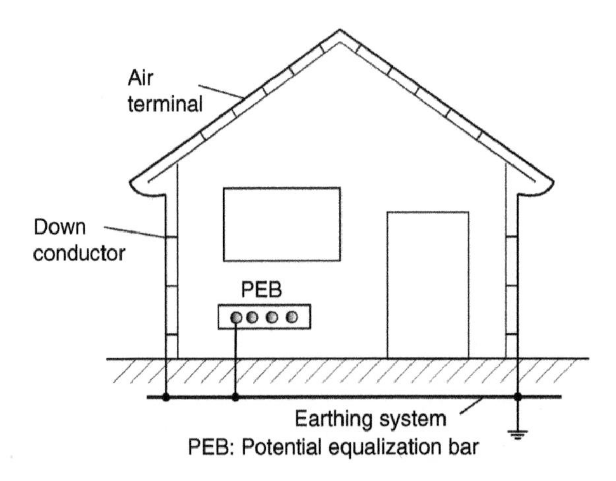

Figure 19.3 Exterior lightning protection.

19.2 Exterior Lightning Protection

Exterior lightning protection includes all equipment for collecting and leading off lightning discharge current to the grounding system (Figure 19.3).

19.2.1 Air Terminal

The air terminal (Figure 19.4) serves as the striking point for lightning. The mesh size of the air terminals is 10 m × 20 m maximum for normal buildings and 10 m × 10 m for hospitals. The mesh must be installed so that no point on the surface of the roof is more than five meters from an air terminal conductor (Figure 19.5). The height of the lightning rod may not be more than 20 m, and the distance from the building must be at least 2 m (Figure 19.6). The building is regarded as protected when the angle is 45°. The area protected by the lightning rods is shown for $H < 30$ m in Figure 19.7 and for $H > 30$ m in Figure 19.8. The roof superstructures, of electrically nonconductive material, may not be longer than 0.3 m (Figure 19.9). A roof ventilation system may not be attached without being first included in the lightning protection concept. If the roof ventilation system is protected by a lightning rod, then the distance between the ventilation system and the lightning rod must be determined in accordance with Figure 19.10.

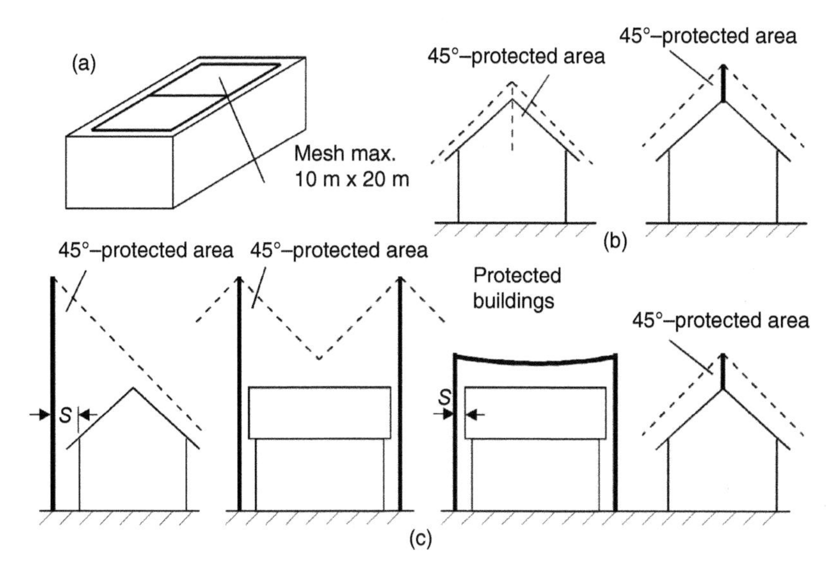

Figure 19.4 Air terminals. Source: Dehn+Söhne [63].

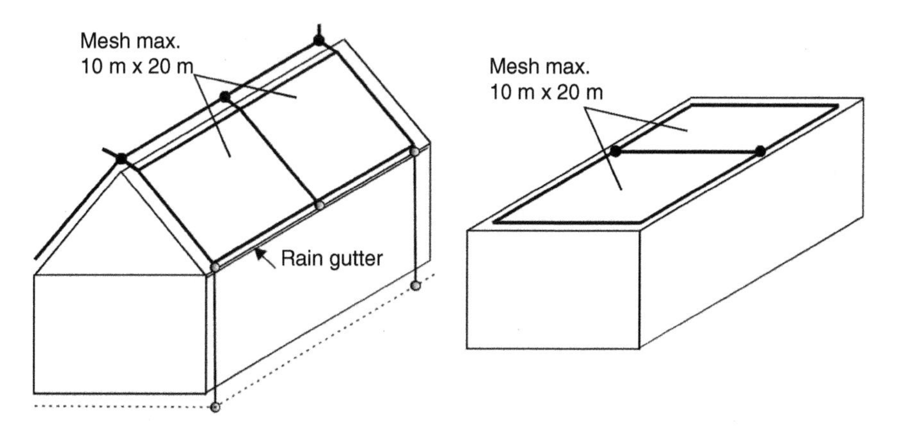

Figure 19.5 Air terminal – mesh. Source: Dehn+Söhne [63].

19.2.2 Down Conductors

The down conductor (Table 19.3) connects the air terminal to the grounding system. For every 20 m, measured along the edges of the roof, a down conductor must be installed, outward from the corners of the building. Through the air terminal, a zone of protection with a shielding angle of 45° is then formed.

Down conductors must have test joints in order to enable later measurements on the system. The required number of down conductors for buildings can be calculated, in any case, at least two down conductors must be arranged.

Figures (19.11–19.13) show different arrangements of down conductors. Down conductors must be installed so that

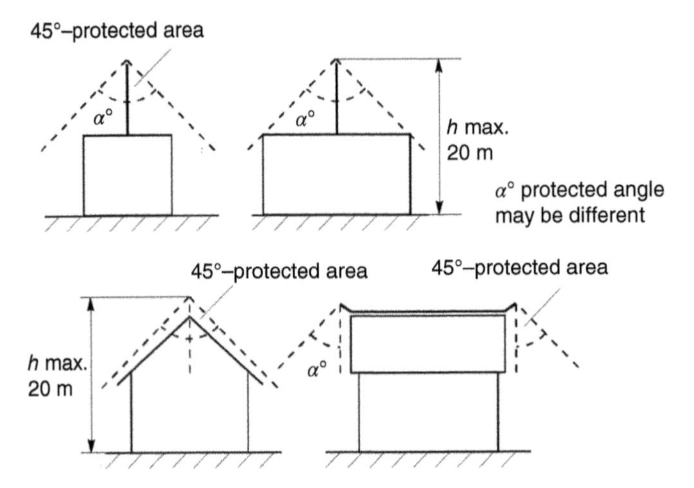

Figure 19.6 Air terminal with protected area. Source: Dehn+Söhne [63].

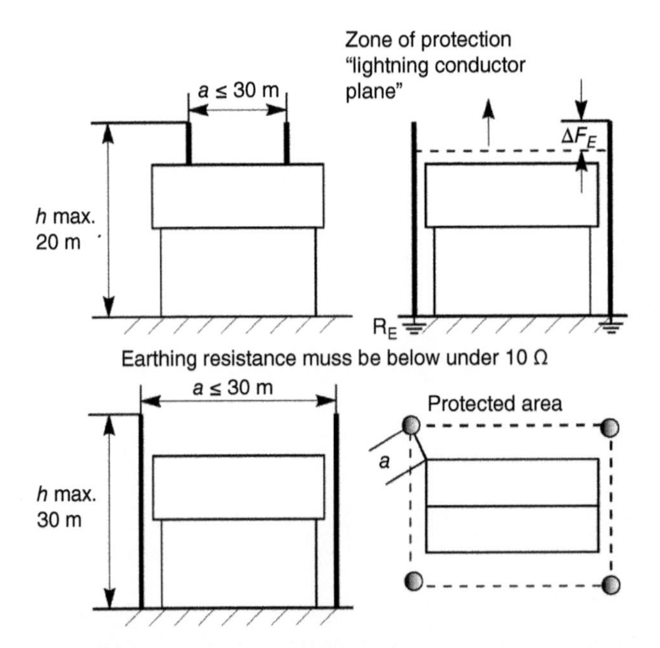

Figure 19.7 Air terminal – lightning rod with protected area up to maximum 30 m. Source: Dehn+Söhne [63].

1. Several parallel current paths exist
2. The length of the current paths must be kept as short as possible
3. Connections to the equipotential bonding are made
4. The down conductors have no effect on safety areas
5. They can be connected to each other near the ground.

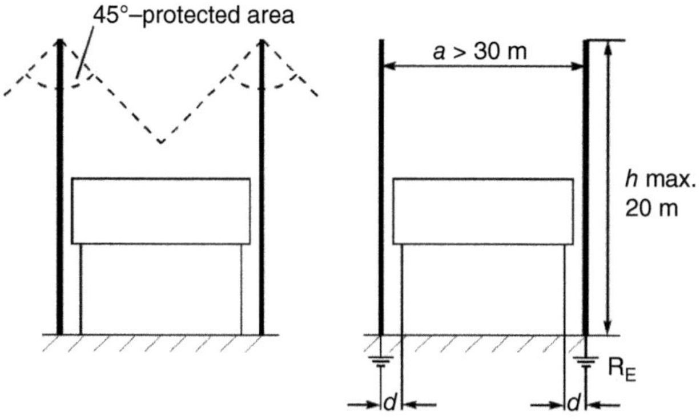

Figure 19.8 Air terminal – lightning rod with protected area greater than 30 m. Source: Dehn+Söhne [63].

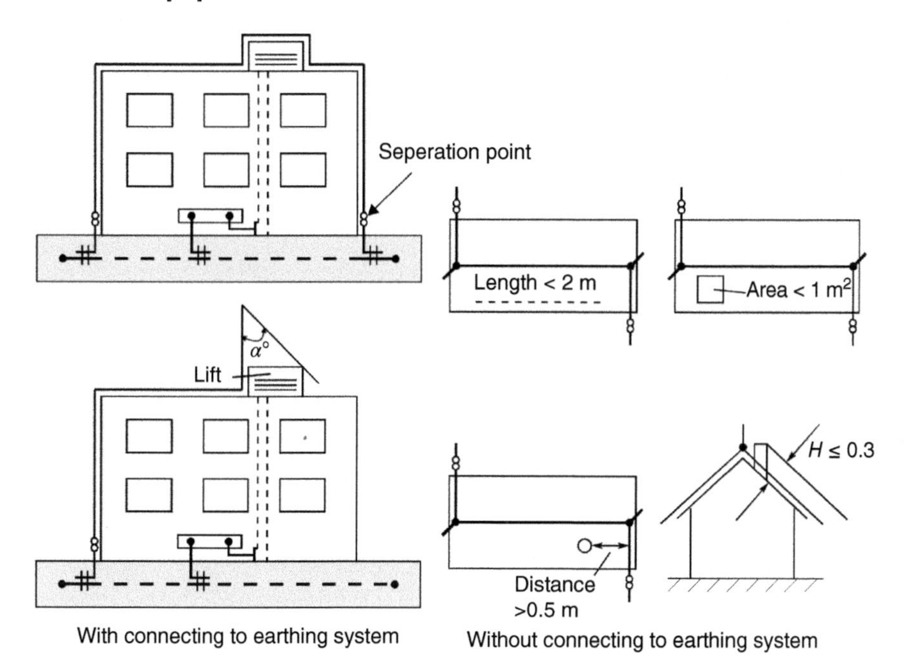

Figure 19.9 Air terminal – roof superstructures of electrically conductive material. Source: Dehn+Söhne [63].

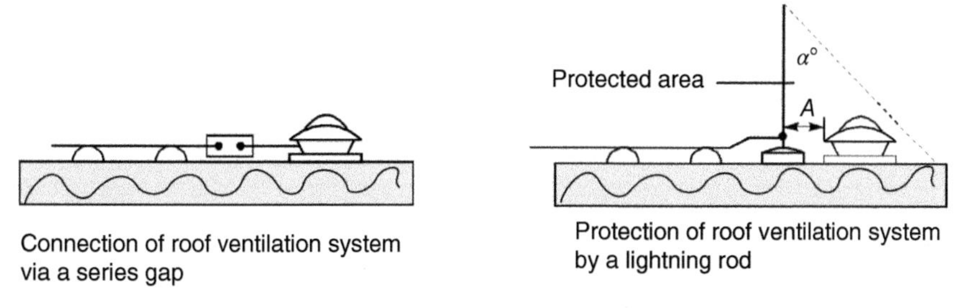

Figure 19.10 Air terminal for smaller roof superstructures. Source: Dehn+Söhne [63].

Table 19.3 Down conductors.

Length of outer roof edges	Number of down conductors		
	Symmetrical building	Unsymmetrical building	Ridged roof up to max. 12 m width or length
			max. 12 m
... 20 m	1	1	1
21...49 m	2	2	2
50...69 m	4	3	2
70...89 m	4	4	4
90...109 m	6	5	4
110...129 m	6	6	6
130...149 m	8	7	6

Source: Dehn+Söhne [63].

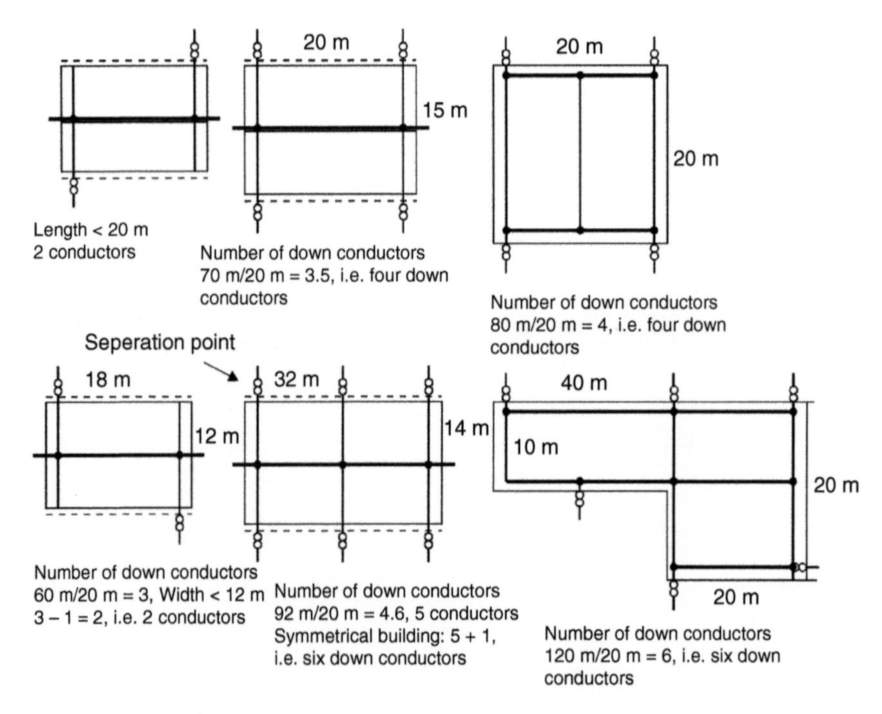

Figure 19.11 Arrangement of down conductors. Source: Dehn+Söhne [63].

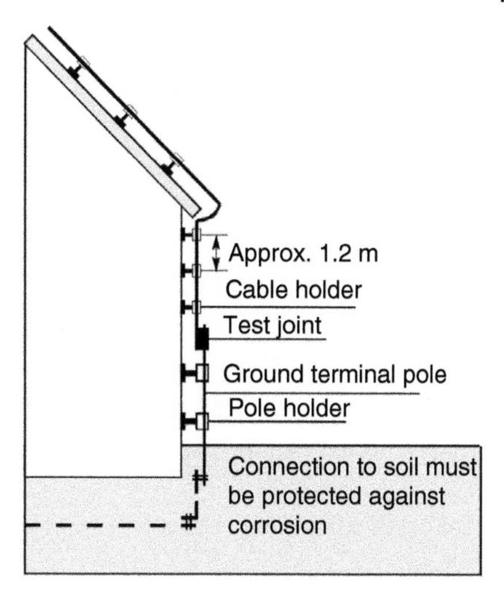

Figure 19.12 Down conductor in accordance with IEC 1024-1, Part 1, Section 5.2. Source: Dehn+Söhne [63].

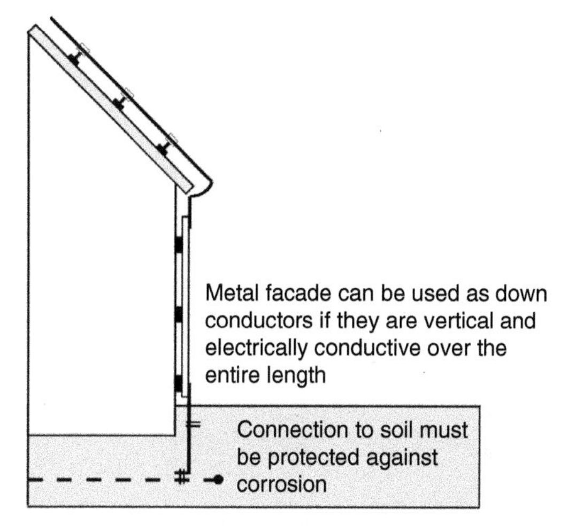

Figure 19.13 Down conductor in accordance with IEC 1024-1, Part 1, Section 5.2.10. Source: Dehn+Söhne [63].

The connection of the down conductor to the grounding system must be as short as possible (Figure 19.11), and the connection points must be protected against corrosion.

19.2.3 Grounding Systems

The grounding system can be in the form of a concrete-footing ground electrode, ring ground electrode, or single ground electrode. The down conductor is connected to this grounding system.

In accordance with IEC 1312-1, the grounding resistance must not be more than 10 Ω. We distinguish here between two types of grounding systems:

1. Type A arrangement
   - Surface ground electrode
   - Buried ground electrode

2. Type B arrangement
   - Ring ground electrode
   - Concrete-footing ground electrode.

1. Type A arrangement: Surface ground electrode
   A surface ground electrode is a ground electrode which is generally installed at a depth of less than 0.5 m and is connected over a length of 5 m to each down conductor (Figure 19.14).
   It can consist of round or flat conductors and be in the form of a ring, Radial, or mesh ground electrode:
   The ground resistance of the surface buried ground electrode is given by:

$$R_E = \frac{\rho_E}{\pi L} \ln \frac{2L}{d} \tag{19.1}$$

   We can calculate the ground electrode resistance with the approximate relationship
   for $L \leq 10$ m

$$R_E \approx \frac{2\rho_E}{L} \tag{19.2}$$

   for $L \geq 10m$

$$R_E \approx \frac{2\rho_E}{L} \tag{19.3}$$

   The symbols have the meanings:

L Length of surface ground electrode

d diameter of surface ground electrode

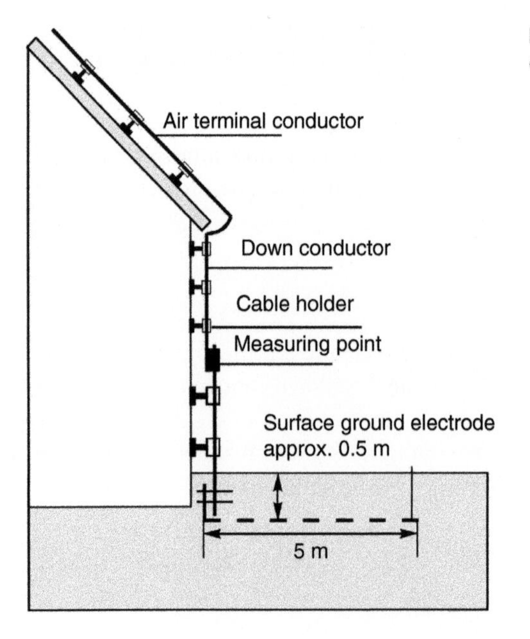

Figure 19.14 Surface ground electrode.

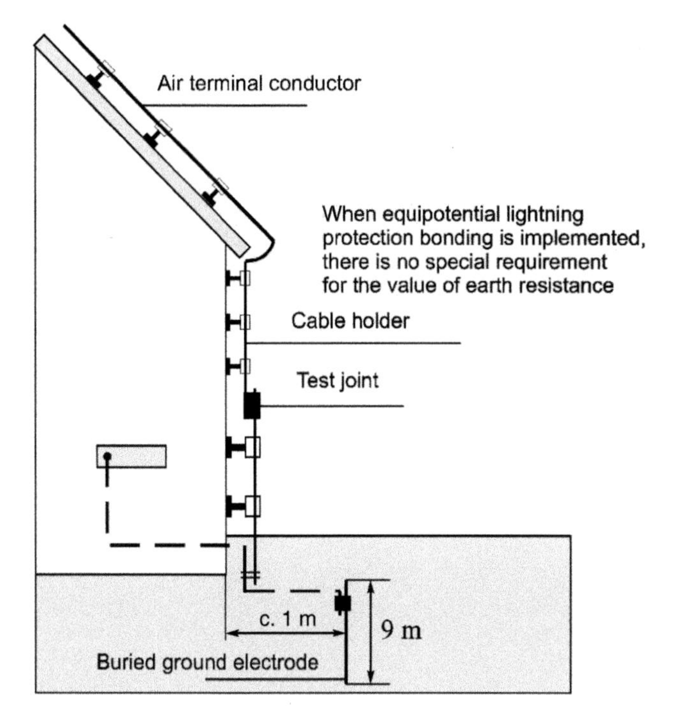

Figure 19.15 Buried ground electrode.

2. Type A arrangement: Buried ground electrode
   It is not possible in all cases to use concrete-footing ground electrodes or ring ground electrodes as lightning protection ground electrodes. In such cases, IEC 1024-1, Part 1 offers the possibility of installing a single ground electrode for each down conductor. As a single ground electrode, either a surface ground electrode with a length of 20 m or a buried ground electrode (Figure 19.15) with a length of 9 m is placed vertically in the ground 1 m away from the foundation of the building. Buried ground electrodes have the advantage that they are placed at further into the ground, where the soil resistivity is less than toward the surface. With vertically placed ground electrodes, frost conditions have no negative effects on the ground resistance. The ground resistance of a buried ground electrode is given by [29]:

$$R_E = \frac{\rho_E}{2\,\pi\,L}\,\ln\frac{4\,L}{d} \tag{19.4}$$

   This can be approximated by the relationship:

$$R_E \approx \frac{\rho_E}{L} \tag{19.5}$$

   The symbols have the meanings:

L Length of buried ground electrode

d Diameter of ground electrode rod

Figure 19.16 shows the ground resistance for buried ground electrodes.

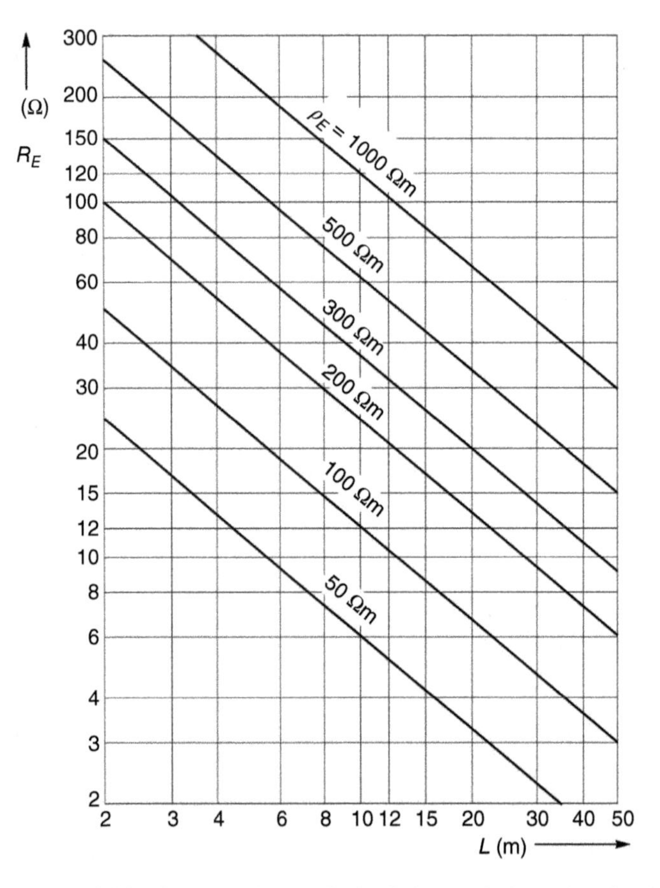

Figure 19.16 Ground resistance for buried ground electrodes. Source: IEC [44].

3. Type B arrangement: Ring ground electrode
   A ring ground electrode is a surface ground electrode at a depth of at least 0.5 m and, as well as possible, as an enclosed ring installed at a distance of 1 m (Figure 19.17) from the exterior foundation of the building.
   The ground resistance of a ring ground electrode is given by [29]:

$$R_E = \frac{\rho_E}{\pi^2 D} \ln \frac{2\pi D}{d} \tag{19.6}$$

   This can be approximated by the relationship:

$$R_E \approx \frac{2 \rho_E}{3 D} \tag{19.7}$$

   Here, the symbols have the meanings:

D	Diameter of ring ground electrode $D = 1.13\sqrt{A}$
A	Area of the enclosed ring ground electrode surface
d	Diameter of grounding cable or half-width of a grounding strip
ρ_E	Soil resistivity in Ωm

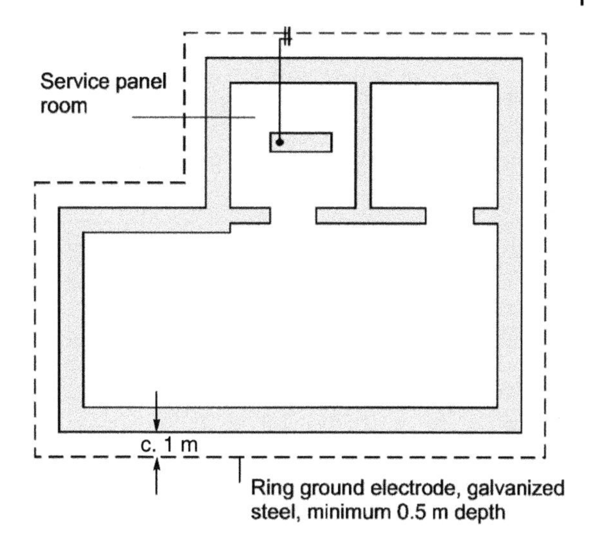

Figure 19.17 Enclosed ring ground electrode.

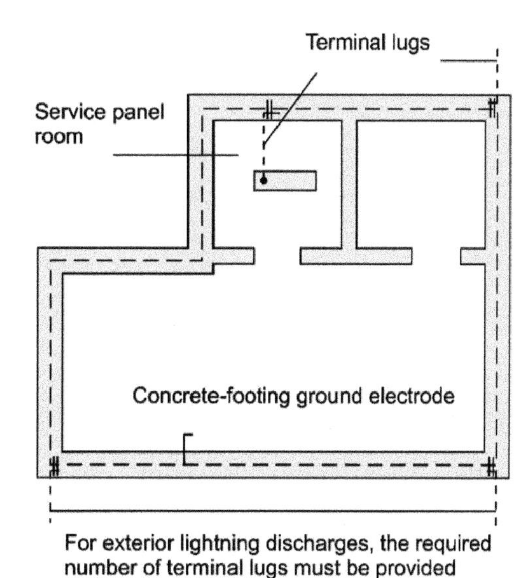

Figure 19.18 Concrete-footing ground electrode.

4. Type B arrangement: Concrete-footing ground electrode
   For optimal functioning, the main equipotential bonding requires an effective and long-term functioning grounding system. The concrete-footing ground electrode is very well suited to this purpose (Figure 19.18).
   Furthermore, it can also be used as a ground electrode for lightning protection, communications systems, and low-voltage systems. It must be in the form of an enclosed ring installed in the foundations of the outer walls of the building. For concrete-footing ground electrodes, steel strips with a cross section of at least 30 mm × 3.5 mm or steel bars of at least 10 mm diameter must be used.
   The ground resistance of a buried ground electrode is given approximately by Ref. [61]:

$$R_E \approx \frac{2\,\rho_E}{\pi\,D} \tag{19.8}$$

$$D = \sqrt{\frac{4\,L\,B}{\pi}} \tag{19.9}$$

The symbols have the meanings:

L	Length of the concrete-footing ground electrode
B	Width of the concrete-footing ground electrode
D	Diameter of the equivalent circuit area

Figure 19.19 shows the ground resistance for steel strip and steel bar ground electrodes.

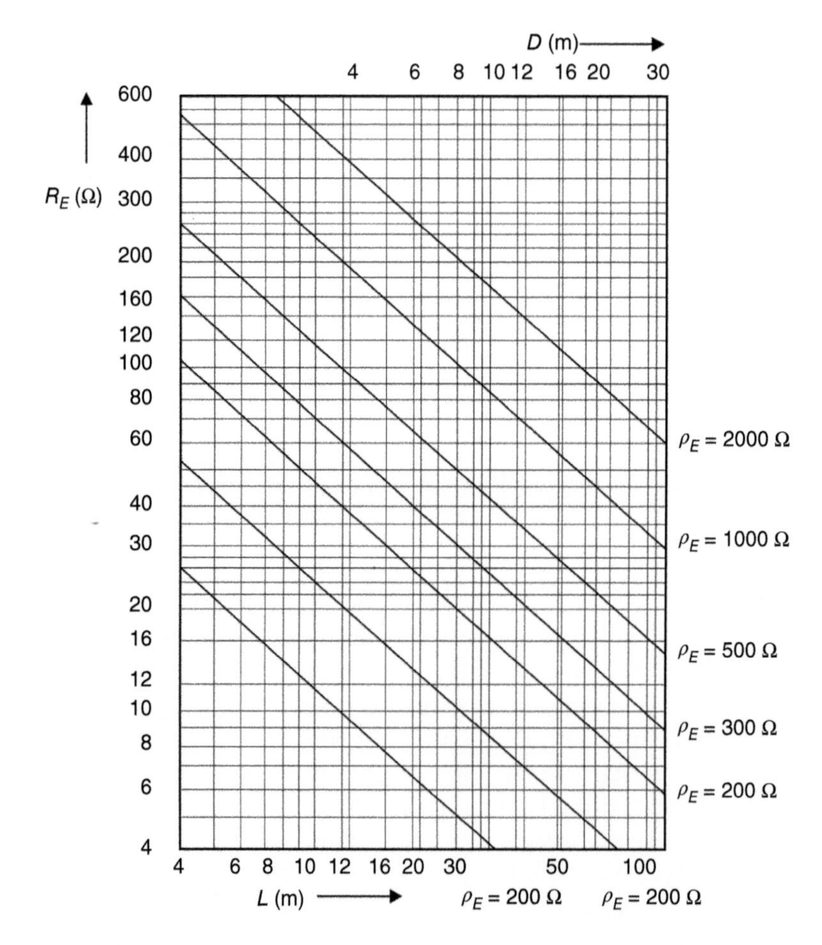

Figure 19.19 Ground resistance for steel strip and steel bar ground electrodes. Source: IEC [44].

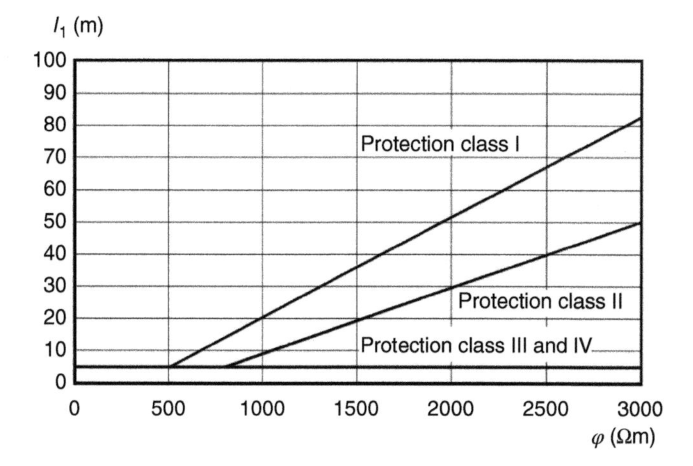

Figure 19.20 Minimum length of a ground electrode.

19.2.3.1 Minimum Length of Ground Electrodes

The minimum length of ground electrodes must not be considered when the measured grounding resistance is less than 10 Ω. For type B (ring or concrete-footing ground electrodes) the average radius r of the area enclosed by the ground electrode may not be less than l_1. The length l_1 can be taken from Figure 19.20.

Here:

$$r \geq l_1 \tag{19.10}$$

$$r = \sqrt{\frac{A}{\pi}} \tag{19.11}$$

When the required value l_1 is greater than the value of r, radial or buried ground electrodes must then be added, for which the lengths l_h (horizontal) and l_v (vertical) are calculated as follows:

$$l_h = l_1 - r \tag{19.12}$$

$$l_v = \frac{l_1 - r}{2} \tag{19.13}$$

The number of additional ground electrodes must not be less than the number of down conductors and must be at least two.

The meanings of the symbols are

r	Average radius
A	Circular area
l_1	Minimum length of a ground electrode
l_h	Length of horizontal ground electrode
l_v	Length of vertical ground electrode

Table 19.4 Soil resistivity for different types of soil.

Type of soil	ρ_E in Ωm (average values)
Marshland	30
Loam, clay, humus	100
Sand (damp)	200
Sand (dry)	1 100
Gravel (damp)	500
Gravel (dry)	3 000
Stony soil	1 000
Granite	2 500
Rock	>10 000
Pure water (13 °C)	56
Pure water (39 °C)	34
Rain water	30–300
Sea water	0.22

The ground resistance of a ground electrode depends on the soil resistivity, the dimensions and the arrangement of the ground electrode and the length of the ground electrode, and less on its cross section [29]. Figures 19.19 and 19.16 show the ground resistance of steel strip, ring and buried ground electrodes as a function of the ground electrode length and the soil resistivity.

The soil resistivity of different soil types is required for the calculation of the ground resistance of a ground electrode. It is usually given in Ωm and is defined as the resistance of a 1 m^3 cube of ground with 1 m edge length, measured between two opposite cube surfaces. Table 19.4 [29, 31] lists frequently measured soil types.

19.2.4 Example 1: Calculation of Grounding Resistances

For a soil resistivity of $\rho = 150$ Ωm, we want to find the ground resistance of a ring and a concrete-footing ground electrode with the dimensions 14 m length and 10 m width and of a buried ground electrode 9 m in length.

$$R_E = \frac{2 \cdot \rho_E}{3 \cdot D} = \frac{2 \cdot 150\ \Omega\text{m}}{3 \cdot 13.3\ \text{m}} = 7.518\ \Omega$$

The diameter of the equivalent ground electrode is

$$D = 1.13\sqrt{A} = 1.13\sqrt{14\ \text{m} \cdot 10\ \text{m}} = 13.3\ \text{m}$$

The ground resistance of the concrete-footing ground electrode can be approximated by

$$R_E = \frac{2\ \rho_E}{\pi\ D} = \frac{2 \cdot 150\ \Omega\text{m}}{\pi \cdot 13.3\ \text{m}} = 7.183\ \Omega$$

The ground resistance of the buried ground electrode is

$$R_E = \frac{\rho_E}{2\,\pi\,L}\,\ln\frac{4\,L}{d} = \frac{150\cdot\Omega\text{m}}{2\,\pi\cdot 9\text{ m}}\,\ln\frac{4\cdot 9\text{ m}}{0.02\text{ m}} = 19.89\ \Omega$$

The ground resistance of the buried ground electrode can be approximated by

$$R_E = \frac{\rho_E}{L} = \frac{150\cdot\Omega\text{m}}{9\text{ m}} = 16.66\ \Omega$$

19.2.5 Example 2: Minimum Lengths of Grounding Electrodes

For a system of protection class II, a soil resistivity of 2000 Ωm was measured. From Figure 19.20, we obtain a minimum length of $l_1 = 30$ m.

From Figure 19.21, the area $A = A_1 + A_2 = 200\ \text{m}^2$. The average radius is then found from:

$$r = \sqrt{\frac{A}{\pi}} = \sqrt{\frac{200\text{ m}^2}{\pi}} = 7.98\text{ m}$$

$$l_v = \frac{l_1 - r}{2} = \frac{30\text{ m} - 7.98\text{ m}}{2} = 11.01\text{ m}$$

This means that an additional buried electrode 11 m in length must be installed.

19.2.6 Exposure Distances in the Wall Area

In accordance with IEC 1024-1, an exposure is too short a distance between a lightning protection system and metal installations or electrical systems for which there is a danger of flashover or breakdown. Exposures of air terminals and down conductors to metal installations of all types must be prevented or eliminated by increasing the distance or by connecting the installations directly to the lightning protection system or through series isolation gaps. The exposure distance is given by

$$d = k_i\,\frac{k_c}{k_m}\,l \qquad \text{mit} \qquad s \geq d \tag{19.14}$$

Here, the symbols have the meanings:

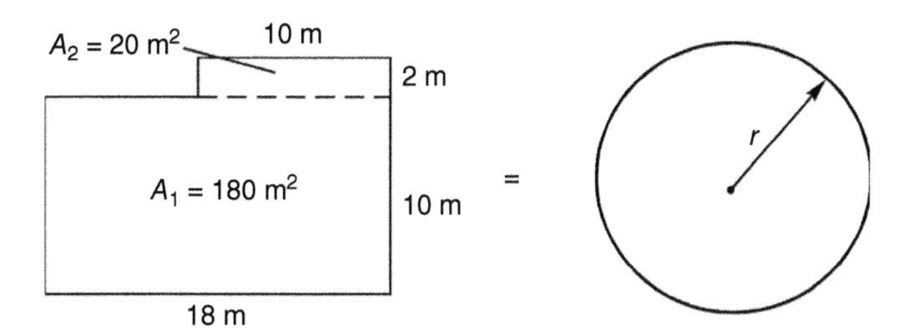

Figure 19.21 Layout of the building.

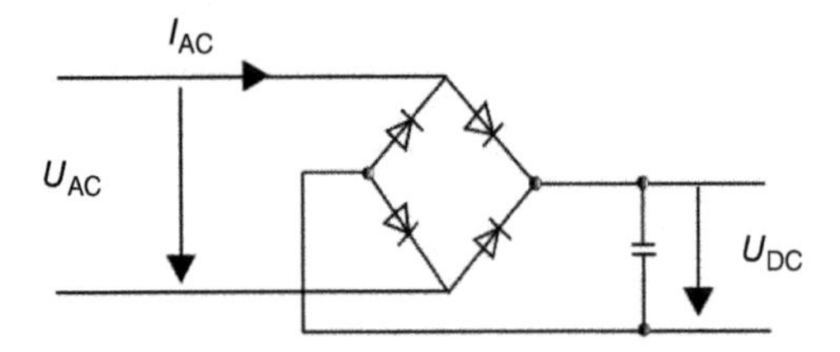

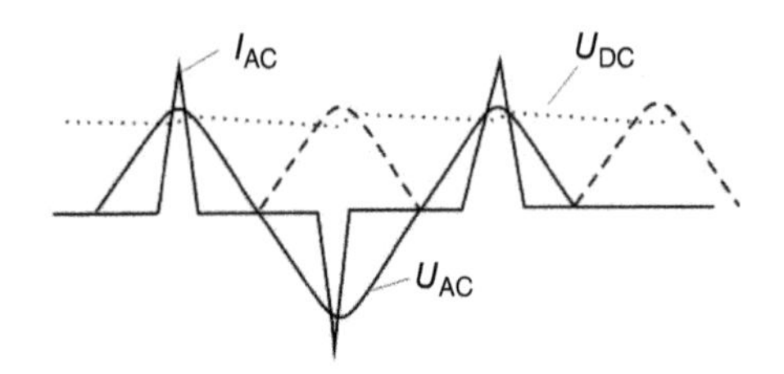

Figure 19.22 Exposure of installations to a lightning protection system with the value k_c.

Table 19.5 Values of the coefficients.

Lightning protection class	k_i	Material	k_m
I	0.1	Air	1
II	0.075	Solid state material	0.5
III and IV	0.05		

s Safety distance in m

d Exposure distance in m

k_c Geometrical arrangement – dependent current distribution coefficient (see Figure 19.22)

k_m Isolating path material-dependent coefficient [32] from Table 19.5

l Length of lightning protection conductor

k_i Lightning protection class-dependent coefficient [32] from Table 19.5

For the determination of the current distribution coefficient k_c there are three possibilities:

- First method
  a) For free-standing lightning conductor poles and conductor cables located between these.
  b) With air terminal conductor on the ridge and down conductors

$$k_c = \frac{c+f}{2\,c+f} \tag{19.15}$$

- Second method
  With a mesh air terminal conductor network on flat roofs, when no ring feeder is provided.

$$k_c = \frac{1}{2\,n} + 0.1 + 0.2\,\sqrt[3]{\frac{c}{h}} \tag{19.16}$$

- Third method
  With a mesh air terminal conductor on flat roofs, when one or more ring feeders are provided.

  The symbols have the meanings:

h	Height or spacing of ring feeder
n	Total number of down conductors
c	Distance from next down conductor
l	Length of air terminal conductor

19.2.7 Grounding of Antenna Systems

In accordance with IEC 1024-1, Part 1, IEC 60 364 and antennas connected to a lightning protection ground electrode, concrete-footing ground electrode, integrated antenna ground electrode, steel constructions, or conductive buried tubular metallic networks (Figure 16.25) must have a minimum cross section of 16 mm^2 Cu (insulated or blank), 25 mm^2 Al (insulated), or 25 mm^2 steel (Figure 19.23).

For room antennas, antennas under a roof or exterior antennas for which the distances depicted in Figure 19.24a are complied with a grounding system is not required. The equipotential bonding must be implemented as in Figure 19.24b.

19.2.8 Examples of Installations

For the technically correct planning of a lightning protection system, the description of the building is of great importance. DIN 48830 gives detailed information

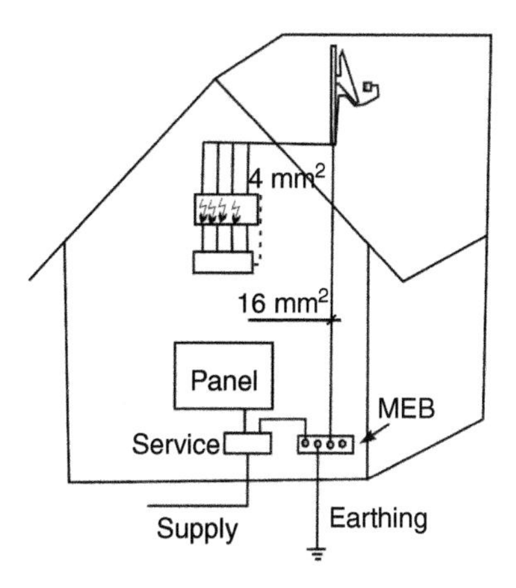

Figure 19.23 Antenna installed on a roof. Source: Dehn+Söhne [63].

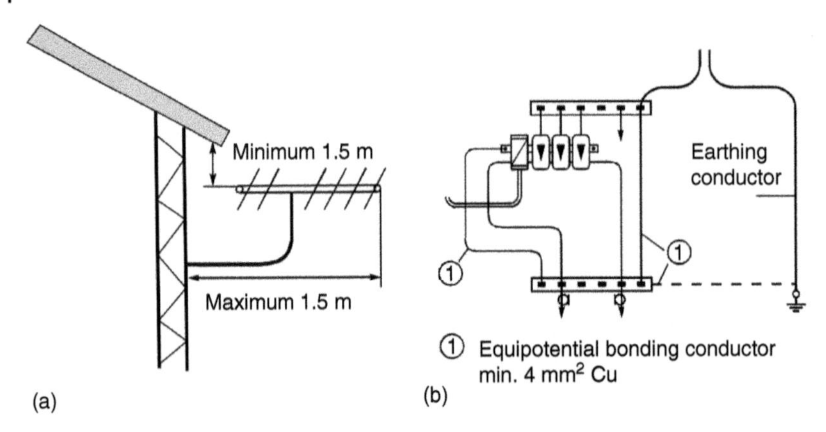

Figure 19.24 (a) Window antenna. (b) Equipotential bonding of antennas. Source: Dehn+Söhne [63].

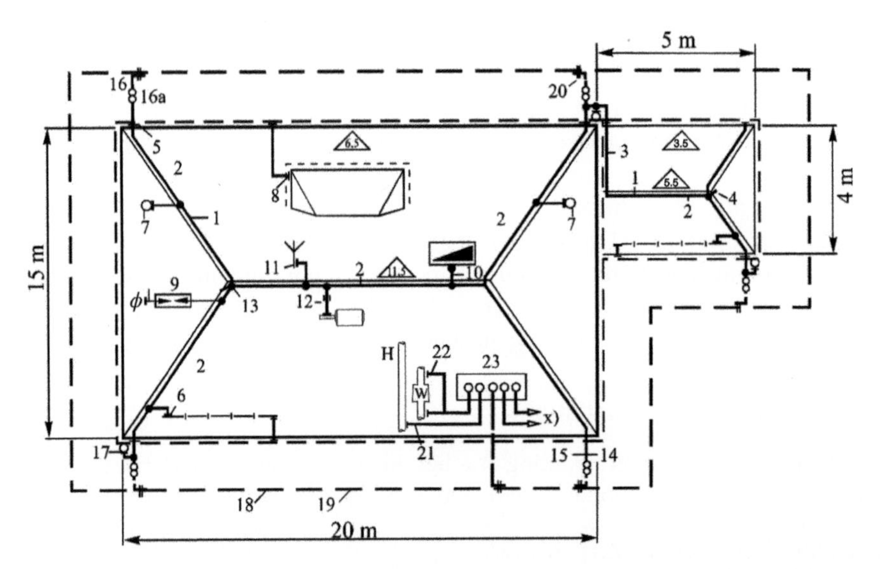

Figure 19.25 Drawing of an exterior lightning protection system. Source: Dehn+Söhne [63].

about the scope of the building description. In the preliminary planning phase, it is necessary to coordinate the lightning protection class to be used with the customer on the basis of the existing specifications. For the installation and realization of the system, the installation plan (Figure 19.25) and the specifications of work and services [33] must be written. Detailed information about planning, installing and testing can be found in the book [34]. The section numbers used are

1	Roof conductors	13	Universal trusses
2.3	Roof conductor holders	14	Steel wire down conductors
4	Lightning conductor peak	15	Cable holders
5	Gutter terminals	16	Ground terminal poles
6	Snow fence terminals	16a	Measuring point
7	Rain conduits	17	Rain conduits
8	KS connectors	18	Grounding blade ditch
9	Series gap	19	Steel wire grounding blade
10	Lightning rods	20.22	Parallel connectors
11.12	Grounding conduits	23	Equipotential bonding busbar

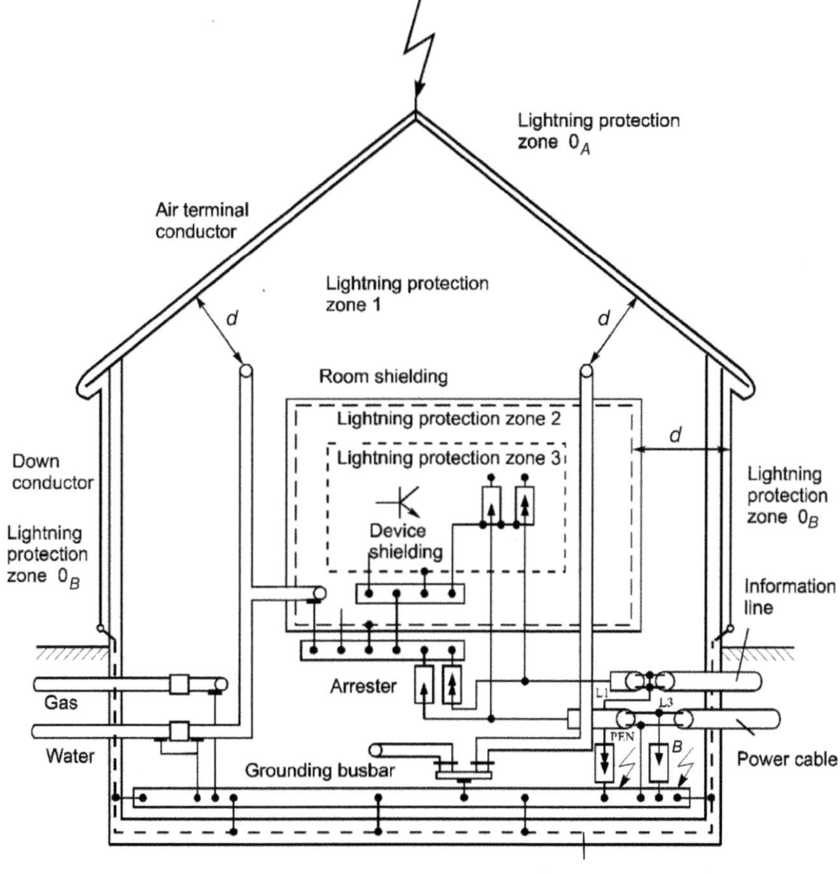

Figure 19.26 Lightning protection zones.

19.3 Interior Lightning Protection

It is necessary to protect electrical installations within buildings against the effects of a lightning discharge current and the resulting electrical and magnetic fields. The main part of the interior lightning protection is the equipotential bonding, to which all metallic tubing, as well as heavy current and information technology systems are connected. For the information technology equipment, the lightning protection zone concept offers the best protection. The principle of this concept is based on room shielding (Figure 19.26).

19.3.1 The EMC Lightning Protection Zone Concept

The buildings are divided into lightning protection zones and matched protective equipment and devices provided from the exterior area up to the most sensitive interfaces (Figure 19.27).

The EMC-oriented lightning protection zone concept has been included in international standards. This concept is especially recommended for buildings with extensive electronic equipment. The principle consists of the step-wise suppression of the electromagnetic fields and their effects which result from defined lightning protection zones.

With the EMC lightning protection concept, we can distinguish between four zones:

1. Exterior lightning protection
2. Building shielding
3. Room shielding
4. Equipment shielding.

Length of outer roof edges	Number of down conductors		
	Symmetrical building	Unsymmetrical building	Ridged roof up to max. 12 m width or length
...20 m	1	1	1
21...49 m	2	2	2
50...69 m	4	3	2
70...89 m	4	4	4
90...109 m	6	5	4
110...129 m	6	6	6
130...149 m	8	7	6

Figure 19.27 Interior lightning protection.

The interfaces between the individual zones must be taken into account in the equipotential bonding, and special down conductors must be installed. Table 19.6 describes the individual lightning protection zones.

The minimum cross sections of the equipotential bonding lines are

Cu 16 mm²
Al 25 mm²
Steel 50 mm².

Table 19.6 Definition of zones.

Zone	Definition
Lightning protection zone 0	Direct effect of lightning possible No shielding against electromagnetic fields
Lightning protection zone 0/E	Zone protected by air terminal against direct effect of lightning discharges No shielding against electromagnetic fields
Lightning protection zone 1	Partial lightning currents cause high-energy transients which can lead to switching operations
Lightning protection zone 2	The electromagnetic field is further attenuated Switching operations and electrostatic discharge processes (ESD) take place
Lightning protection zone 3	The electrostatic field is reduced to a minimum

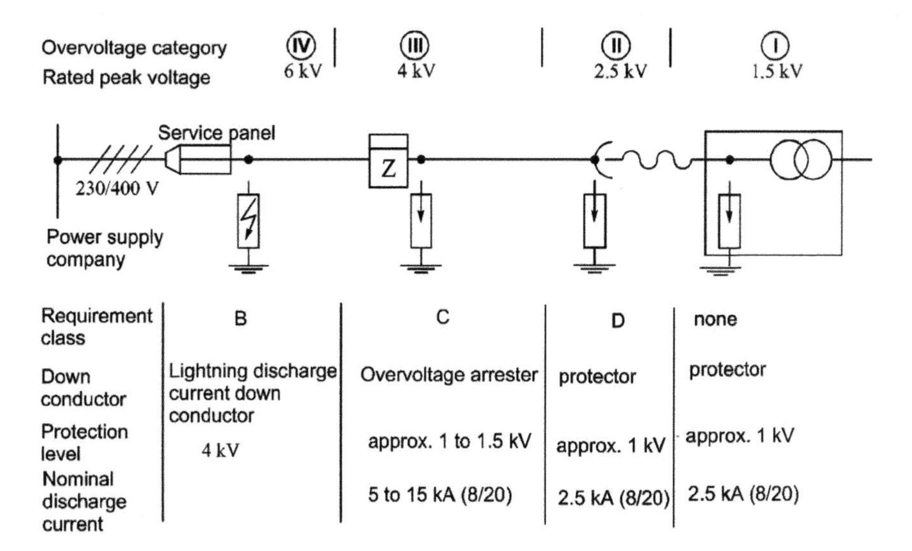

Figure 19.28 Installation locations for overvoltage arresters. Source: Based on Hager [63].

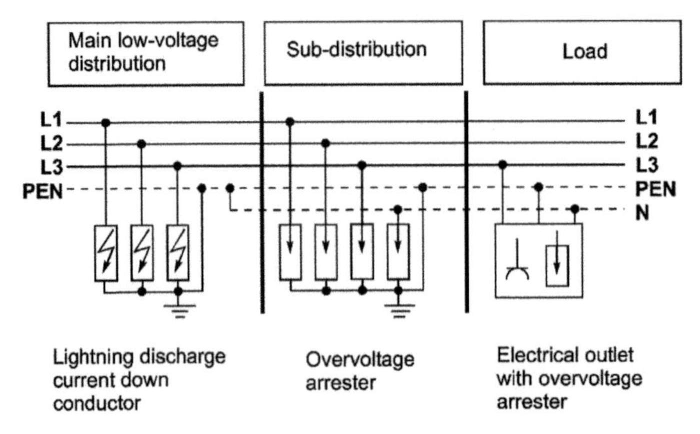

Figure 19.29 TN system with overvoltage arresters. Source: Based on Hager [63].

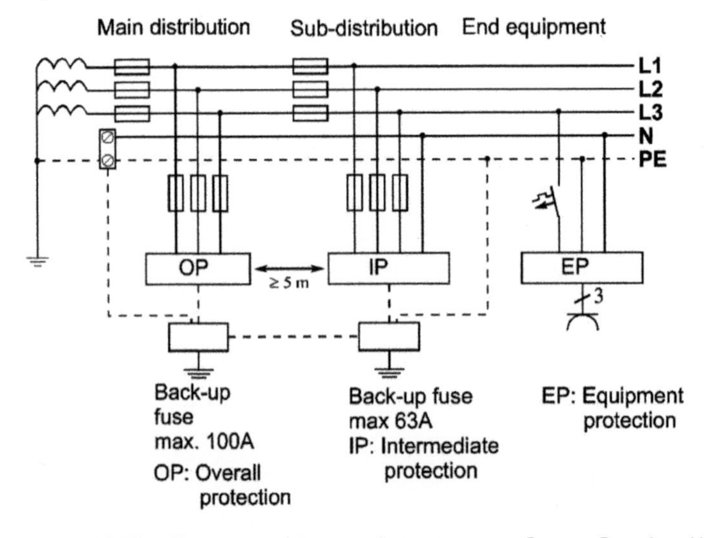

Figure 19.30 TN system with overvoltage arresters. Source: Based on Hager [63].

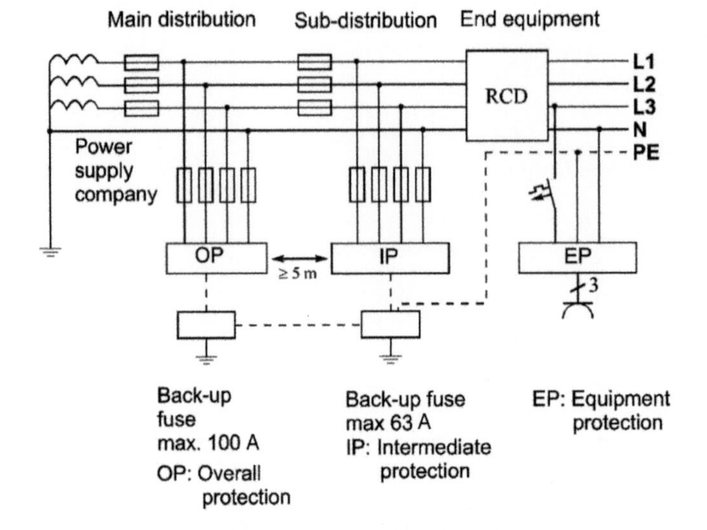

Figure 19.31 TT system with overvoltage arresters. Source: Based on Hager [63].

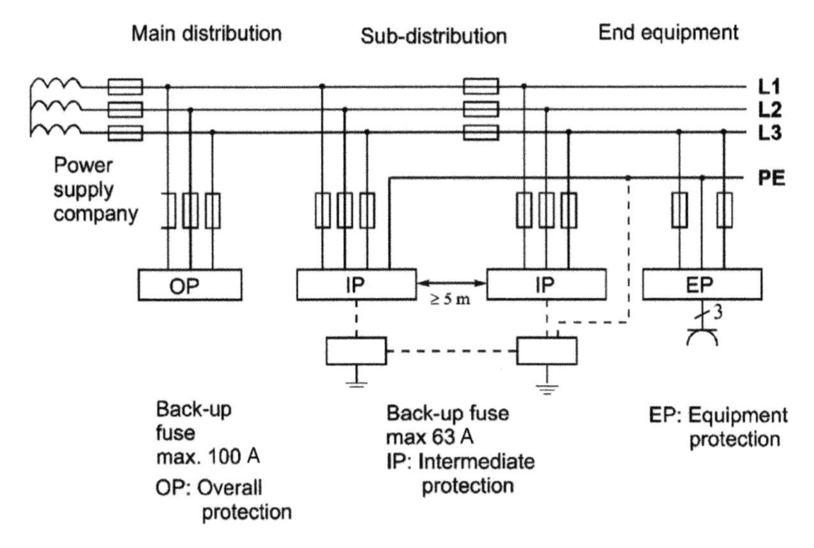

Figure 19.32 IT system with overvoltage arresters. Source: Based on Hager [63].

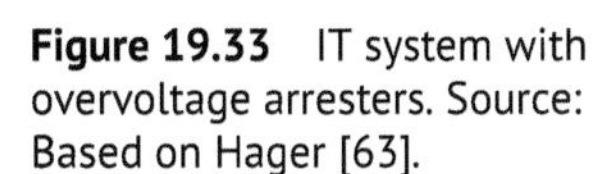

Figure 19.33 IT system with overvoltage arresters. Source: Based on Hager [63].

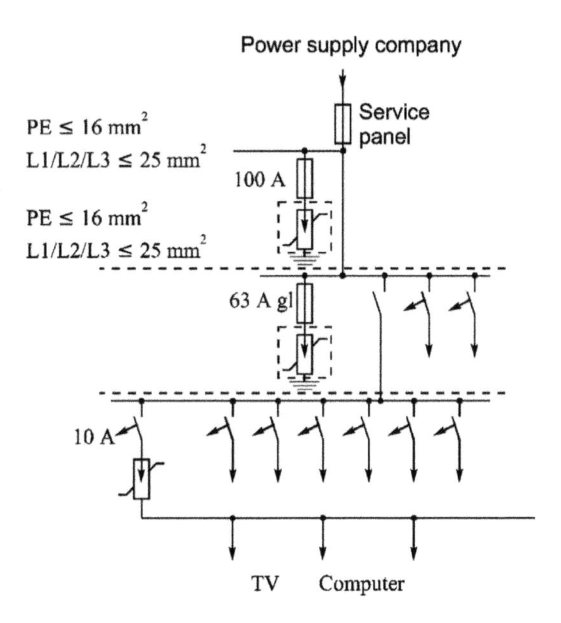

19.3.2 Planning Data for Lightning Protection Systems

For the planning and configuration of lightning protection systems, the new lightning protection standard EN 61024-1 and overvoltage protection standard IEC 1312-1 are available. The decision, which standard to use, lies with the planner and the customer. The specifications of work and services must be written in accordance with the lightning protection systems or the contract procedure for the building industry. This section describes the locations of protective devices in lightning protection systems. According to the overvoltage categories,

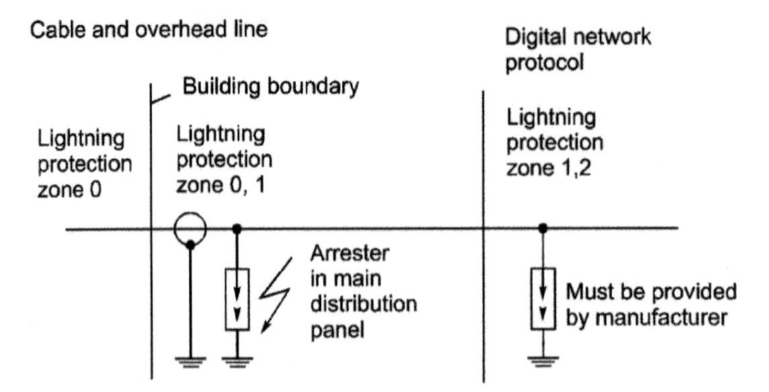

Figure 19.34 Heavy current system, treatment of active lines at the interface. Source: Based on Hager [63].

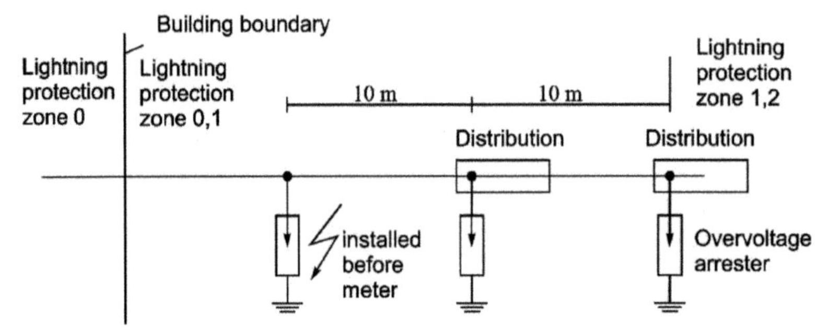

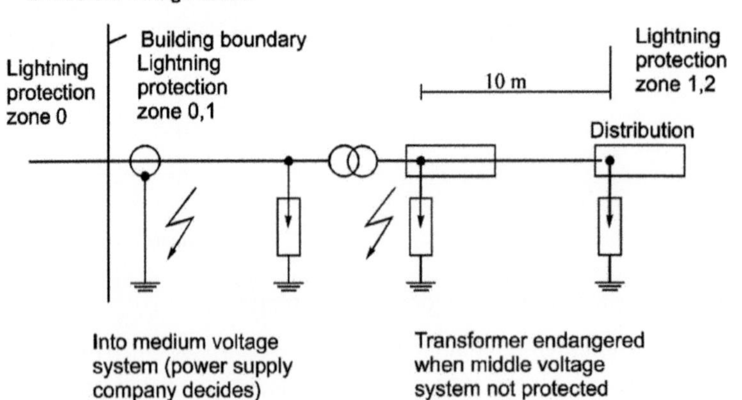

Figure 19.35 Telecommunications system, treatment of active lines at the interface. Source: Based on Hager [63].

down conductors in the main distribution systems are used as overall protection, overvoltage, arresters in the sub-distribution systems as intermediate protection and down conductors in the electrical outlets and the electronic equipment as fine protection. Figures 19.28–19.35 give examples for different low-voltage systems and telecommunications systems.

20

Lighting Systems

20.1 Interior Lighting

20.1.1 Terms and Definitions

- Lamps are technical realizations of artificial light sources primarily intended for lighting purposes, that is, for lighting and illumination. They convert electrical energy into light.
- *Light fixtures*: Light fixtures serve the purpose of influencing the light beam generated by a light source in such a way as to achieve optimal illumination of a system. Light fixtures are therefore electrical operational equipment containing lamps and accessories, which guide the light radiated from the lamps in the required direction.

For the planning and configuration of interior lighting systems, the following definitions, standards, and specifications are especially important [35]:

- *Dry rooms*: Dry rooms are rooms and places in which as a rule neither condensation water nor air saturated with humidity occurs, with no or very little accumulation of nonflammable dust.
- *Damp areas*: Damp areas are rooms and places in which the reliability of the operating equipment can be impaired by humidity, condensation water, or by chemical or similar influences.
- *Wet areas*: Wet areas are rooms and places in which the floors, walls, and equipment are sprayed for the purpose of cleaning.
- *Bath and shower room areas*: Rooms with bathtubs and showers are divided into four areas and are regarded according to type of area and use as dry, damp, or wet rooms.
- *Open air systems*: Protected open air systems are areas protected by roofing against the effects of weather, whereas unprotected systems are exposed to rain.
- *Swimming pools and baths*: Swimming pools and baths are regarded as damp and wet areas.
- *Agricultural operating areas*: Agricultural operating areas are stalls and adjoining rooms, rooms for the Large-scale housing of animals, storage rooms, and supply

Analysis and Design of Electrical Power Systems: A Practical Guide and Commentary on NEC and IEC 60364, First Edition. Ismail Kasikci.

rooms, which are considered at the same time both damp rooms and operating areas subject to fire hazards.

- *Operating areas subject to explosion hazards*: Operating areas subject to explosion hazards are classified as zones according to their condition and the probability of occurrence of a hazardous and potentially explosive atmosphere.
- *Operating areas subject to fire hazards*: In operating areas that are endangered by the presence of dust or fibrous materials, readily flammable materials can accumulate in hazardous levels on the electrical operational equipment. Higher temperatures or arcing on this equipment can lead to the outbreak of a fire.
- *Garages*: Garages with natural ventilation and adjoining rooms for sheltering vehicles are regarded as dry rooms, damp rooms, or operating areas subject to fire hazards.

For the installation of a lighting system, it is necessary to take account of the sections of the Workplace Ordinance and similar standard values for workplaces of relevance for lighting systems. For the installation of a lighting system, IEC 60 364 is of paramount importance for the electrical part.

Of particular interest here are:

IEC 60364, Part 41: Protection against Currents Flowing through the Human Body
IEC 60364, Part 42: Protection against Thermal Influences
IEC 60364, Part 43: Protection of Cables and Lines
IEC 60364, Part 51: Selection and Installation of Electrical Operational Equipment
IEC 60364, Part 559: Light Fixtures and Illumination Systems

20.2 Types of Lighting

20.2.1 Normal Lighting

- The room is uniformly illuminated.
- The arrangement of the workplaces is variable (light incident from the side is recommended).

Examples: Offices and industrial rooms.

20.2.2 Normal Workplace-Oriented Lighting

- The workstations are arranged in fixed zones within the rooms.
- The illumination level is high.

Examples: Production operations, offices.

20.2.3 Localized Lighting

Individual workplaces with greater requirements are illuminated more intensely.

Examples: Test benches and lathes.

20.2.4 Technical Requirements for Lighting

1. *Illumination level*: A sufficient illuminance is required for the illumination level on which the planning and configuration of a system is based. The nominal illuminance refers to an average age and state of pollution. For systems with normal pollution a planning factor of 1.25 is used, for systems with greater pollution a factor of 1.43, and for systems with a high pollution level a factor of 1.67. The decisive plane is 0.85 m above the floor on the horizontal working surface in which the visually oriented work takes place. The ratio of the vertical to the horizontal illuminance is given by $E_v = 1/3 \cdot E_h$. The height of the working plane in sport facilities and sport halls is taken as 1.0 m and in areas of traffic 0.20 m above the floor or ground.
2. *Uniformity of illumination*: All workplaces must have identical illumination. The reflection factors must be complied with. The value $E_{min}/\overline{E} = 1/1.5$ must be adhered to.
3. *Glare restriction*: Lighting systems must not give rise to direct glare or glare resulting from reflection. For the evaluation of direct glare, the quality classes together with the required nominal illuminance are decisive.
4. *Direction of lighting and modeling*: The viewing direction must always be parallel to the light beam axis. A favorable lighting arrangement is when the light is incident from above to the left. Adequate modeling must also be ensured.
5. *Luminous color and color reproduction*: The beam distribution of a lamp is decisive for the luminous color. The luminous color for fluorescent lamps is divided into three groups:
   - *Warm white (ww)*: For relaxation and recuperation, with a high red fraction
   - *Neutral white (nw)*: For trades and industry
   - *Daylight white (dw)*: For certain workrooms and workplaces

   The color reproduction affects the appearance of the illuminated objects. There are six color reproduction property steps. For interior rooms, at least step 3 is required.

20.2.5 Selection and Installation of Operational Equipment

- These must conform with the general rules of engineering practice.
- These must be suited to the intended purpose.
- These must bear a mark of origin.
- These must be marked with the nominal values.
- The effectiveness of the protective measures must be ensured.
- No hazards may result from the use of these electrical systems.

For building installations, the following specifications apply:

Lighting circuits or combined circuits with electrical outlets may be protected only with maximum 16-A-LS breakers. In accordance with the technical conditions for connections, they must have a making and breaking capacity of 6 kA and conform to selectivity class 3. For all other rooms, lighting circuits can be protected with maximum 25 A or less. It must be possible to isolate three-phase circuits by switching so that all ungrounded lines are simultaneously switched.

For selecting light fixtures, the following must be taken into account:

- The permissible normal position
- The behavior of the installation are during fire
- The thermal effect on the environment
- The minimum spacing of spotlights
- The suspension attachments – designed for five times their weight, minimum 10 kg
- The wall boxes for concealed installations
- The through wiring must be implemented with heat-resistant lines.

20.2.6 Lighting Circuits for Special Rooms and Systems

IEC 60 364, Part 700:

1. Bath and shower room areas, Part 701
   Lighting circuits can only be installed in class 2 and 3 areas. Light fixtures in class 2 areas must conform to protection class IP X5 or IP X4. Light fixtures in class 3 areas can belong to protection class IP X0.
2. Swimming halls and swimming facilities, Part 702
   Underwater spotlights accessible from the interior of the swimming pool may be operated only with a safety extra-low voltage.
3. Sauna facilities, Part 703
   Light fixtures of protection class I may be operated only with a safety extra-low voltage or an RCD with 30 mA. The manufacturer's instructions must be followed. The lines and light fixtures must be designed to withstand the high operating temperatures (Table 20.1).
4. Agricultural operating areas, Part 705
   For protection by cutoff only an RCD with 30 mA is permissible. Only Circuit-breakers, and no fuses, can be installed. Light fixtures must belong to protection class IP 54. If the operating condition of the light fixtures cannot be recognized from the operating site, an illuminated signal display is required.
5. Rooms with limited conductivity, Part 706
   Hand lamps may be driven only with a safety extra-low voltage or must have a safety separation from operational equipment. This does not apply for fixed lamps.

Table 20.1 Ambient temperatures of light fixtures.

Type of attachment	Maximum ambient temperature (°C)
Inside, underneath ceiling	140
Outside, on ceiling	75
Outside, direct on wall	60

6. Operating areas subject to fire hazards, Part 720
   In operating areas subject to fire hazards, only lines without a metallic sheath may be used. In case of hazards due to dust or fibrous materials, light fixtures must conform to protection classes IP 4X or IP 5X.
7. Camping vehicles, boats and yachts, Part 721
   For camping vehicles, boats and yachts lighting fixtures of protection class II should generally be used. When different voltages are used, different sockets must be used to clearly identify the different voltages.
8. Overhung constructions, automobiles, and apartments for exhibition, Part 722
   Lighting systems may be operated only with a voltage of 250 V relative to ground. Lamps located in the area of public traffic up to a height of 2 m above the floor must be provided with protection against breakage due to mechanical stressing. Lighting chains with illumination through ribbon cables are permitted for non-supported installation only outside the area accessible by hand. In an outside environment, they must have at least protection against splashing water or the sockets must be located below.
9. Systems in furniture and similar objects of furnishing, Part 724
   When a lamp is installed in the hollow space of a cabinet, and the presence of readily flammable materials near the lamp cannot be prevented, an additional switch must be installed so that the lamp is automatically switched off when the cabinet is closed. Furthermore, the highest permissible power rating of the lamp must be specified.
10. Hoisting gear, Part 726
    Here, the lighting circuit is a special circuit. These special circuits, with safety extra-low voltage, must be connected before an existing disconnect switch. Without safety extra-low voltage, they must be connected through a second disconnect switch. Operation must be possible without a collector wire.
11. Emergency lighting circuits, Part 560
    These circuits for safety purposes must be run separately from the other circuits. For the emergency lighting to remain effective even in case of fire, the line must be fire resistant or correspondingly protected (e.g. F30 line NHXHX).

20.3 Lighting Calculations

In practice, the efficiency method has proven very useful. The calculation will be explained in the following. With this calculational procedure, a uniform distribution of lighting fixtures is assumed. For rooms with furnishings, which have a sustained influence on the lighting conditions, detailed planning with computer programs is necessary. The calculation for the lighting systems requires information about the

- dimensions of the room
- reflection factors of the ceiling, walls, and floor
- type of activity or visual task
- furniture and
- selection and arrangement of lamps and lighting fixtures.

The calculational procedure is as follows:

For each type of light fixture, there exists a utilization factor, in relation to a certain room. This utilization factor depends not only on the technical properties of the light fixtures but also on the room dimensions and the reflection factors of the ceiling, walls, and floor forming the boundaries of the room. First, the room index is determined from the relationship:

$$k = \frac{a\,b}{h\,(a+b)} \tag{20.1}$$

with $h = H - l_p - e$

The symbols have the meanings:

a	Length of room
b	Width of room
h	Height of light spot in m
H	Room height in m
e	Height of evaluation plane above the floor
l_p	Length of pendant or suspension in m

After determining the room index k, the required number of light fixtures for an average illuminance can be determined from:

$$n = \frac{E_n\,A\,\cdot 100}{z\,\Phi\,\eta_B\,M} \tag{20.2}$$

The average horizontal illuminance $\overline{E}$ can be determined from the calculated number of light fixtures from:

$$\overline{E} = \frac{N\,z\,\Phi\,\eta_B\,M\,v}{A \cdot 100} \tag{20.3}$$

Here, the symbols have the meanings:

E_n	Nominal illuminance in lx
A	Floor area of room in m^2
z	Number of lamps per light fixture
v	Reduction factor, taking into account pollution and aging of lamps, light fixtures, and room
η_B	Utilization factor in % according to data sheet, depending on light fixture properties, reflection factors of ceiling, walls and floor and room dimensions, expressed by the room index k
Φ	Luminous of a lamp in lm according to data sheet
$\overline{E}$	Average illuminance in lx
n	Calculated number of lighting fixtures
N	Selected or specified number of lighting fixtures
M	Multiplier for η_B

20.4 Planning of Lighting with Data Blocks

The average illuminance, the required number of light fixtures, and the uniformity of the illuminance can be determined with the use of data blocks. Each data block of a light fixture is divided into six parts, which will now be briefly explained.

20.4.1 System Power

Table 20.2 gives the power consumption for tubular fluorescent lamps with control gear.

20.4.2 Distribution of Luminous Intensity

The light radiation from the light fixtures of interest is shown as a relative luminous intensity distribution curve in polar coordinates (Figure 20.1). The luminous intensity distribution is normalized to a luminous flux of 1000 lm for different reference planes as a function of the angle of radiation.

20.4.3 Luminous Flux Distribution

For the evaluation of the effectiveness of the lighting system in the room, the luminous flux distribution is evaluated. The luminous flux of the lamps consists of the partial luminous fluxes in the lower and the upper half spaces. Light fixtures are identified by a characteristic letter A to E and two characteristic numbers, according to their relative luminous flux distribution. For further classification, in the standard room S has been defined, to which the following light fixture parameters refer.

Table 20.2 Power consumption in W for fluorescent lamps with low-loss conventional (VVG) and electronic (EVG) control gear.

Lamps	VVG	EVG
2 × 11	30	28
2 × 18	42	38
3 × 18	66	57
4 × 18	84	76
3 × 24	90	81
1 × 36	42	36
2 × 36	84	72
1 × 58	66	55
2 × 58	132	110

[52]

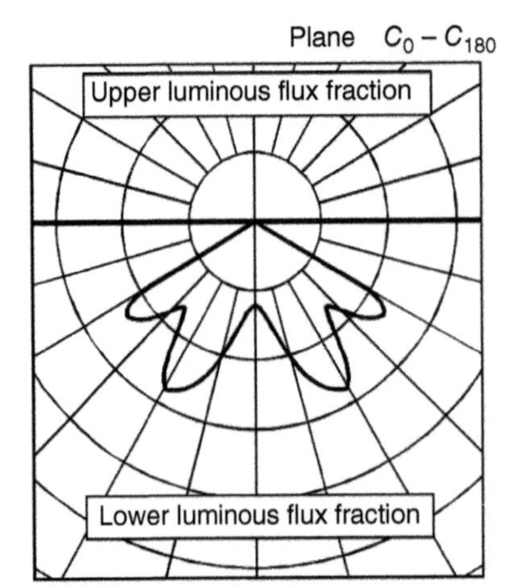

Figure 20.1 Luminous intensity distribution.

$A50$	Identification of the light fixture
$\varphi_u = 1$	Luminance flux fraction from lower half space
$\varphi_{su} = 0.63$	Effective luminance flux fraction in standard room
$\varphi_{so} = 0.01$	Ceiling light current fraction in standard room

20.4.4 Efficiencies

For the planning of lighting systems according to the efficiency method, the utilization factor η_B is decisive. It depends on the technical properties of the light fixtures as well as the dimensions of the room, expressed by the room index k, and the reflection factors of the ceiling, walls, and floor (Table 20.3).

Table 20.3 Utilization factors η_B (%).

Utilization factor η_B in %										
	Ceiling	0.8		0.7		0.5			0.3	0
ρ	Walls	0.5	0.3	0.5	0.3	0.5	0.3		0.3	0
	Floor	0.3	0.1	0.2	0.1	0.3	0.3	0.1	0.1	0
	0.60	37	28	35	28	35	28	28	27	22
	0.80	46	36	44	36	43	37	35	35	29
	1.00	53	43	50	42	50	43	41	41	35
Room index	1.25	61	50	57	49	57	51	49	48	42
k	1.50	67	55	62	54	62	56	62	53	47
	2.00	75	62	69	61	69	64	60	59	54
	3.00	85	70	77	70	78	74	68	67	63
	5.00	93	77	84	76	84	81	75	73	69

[52]

20.4.5 Spacing Between Lighting Elements

The uniformity of the illuminance depends on the luminous intensity distribution of the light fixtures, their spacing (transverse distance x, and longitudinal distance y, measured between the middles of the light fixtures) and the height of the room H (Figure 20.2). The diagram gives the uniformity of the illuminance $\frac{E_{min}}{\overline{E}}$ as a function of the light fixture spacing x for two room heights H. The recommendation is 0.67, i.e. 1 : 1.5.

20.4.6 Number of Fluorescent Lamps in a Room

The number of lamps required to obtain a certain illuminance can be read directly for special room conditions (Table 20.4). The number of lamps determined must be rounded off according to technical and installation requirements. The table does not apply for narrow, elongated spaces, such as floorways.

20.4.7 Illuminance Distribution Curves

Illuminance distribution curves make clear the behavior of the horizontal illuminance on the center line of the utility plane of the room, underneath the light fixture (Figure 20.3).

20.4.8 Maximum Number of Fluorescent Lamps on Switches

The maximum number of fluorescent lamps on switches with positive action contacts, 250 V/10 A can be read from Table 20.5.

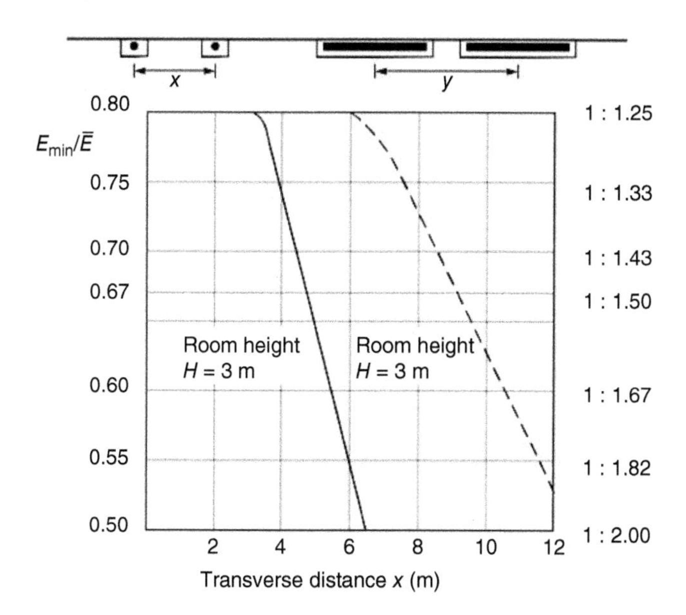

Figure 20.2 Light fixture spacing.

Table 20.4 Number of lamps in a room.

Reflection factors: $\rho = 0.7/0.5/0.2,\ v = 0.8$									
	Lamps 58 W	4100 lm				5200 lm			
	E_n (lx)	300		500		300		300	
	Room height H (m)	3.0	5.0	3.0	5.0	3.0	5.0	3.0	5.0
Floor space A (m^2)	20	3.6	5.7	6.0	9.5	2.8	4.5	4.7	7.5
	30	4.8	7.4	8.0	12	3.8	5.8	6.3	9.7
	40	6.0	8.8	10	15	4.7	7.0	7.8	12
	50	7.2	10	12	17	5.6	8.1	9.4	13
	60	8.3	12	14	19	6.6	9.1	11	15
	80	11	14	18	24	8.3	11	14	19
	100	13	17	21	28	10	13	17	22
	200	23	28	39	47	18	22	31	37

[52]

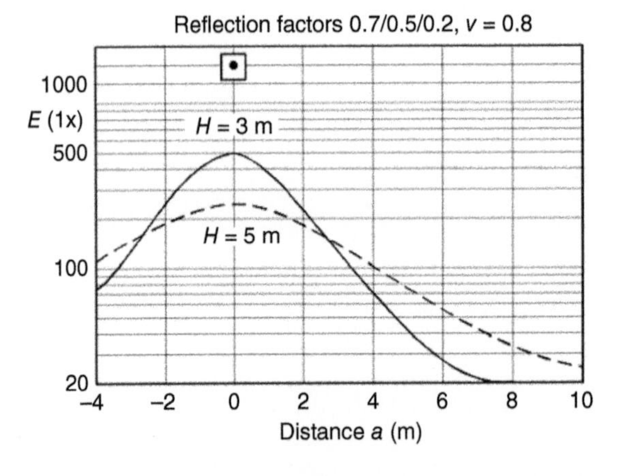

Figure 20.3 Illuminance distribution.

20.4.9 Maximum Number of Discharge Lamps Per Circuit-Breaker

The maximum number of discharge lamps on a circuit-breaker with B or C characteristic can be seen from Table 20.6.

20.4.10 Mark of Origin

Lamp fixtures must satisfy the laws for technical working resources. Light fixtures must bear the testing mark of conformity and the symbol for safety testing, with the following details (Figure 20.4).

Table 20.5 Maximum number of fluorescent lamps.

Tubular fluorescent lamps				
Lamp power (W)	**Conventional control gear with ind.**	**Conventional control gear compensated**	**Conventional control gear with DUO**	**EVG**
18	26	38	50	26
36	22	34	40	26
58	14	22	26	18
		Compact fluorescent lamps		
11	60	76	—	—
18	26	38	50	—
24	26	38	50	—
36	22	34	40	—

[52]

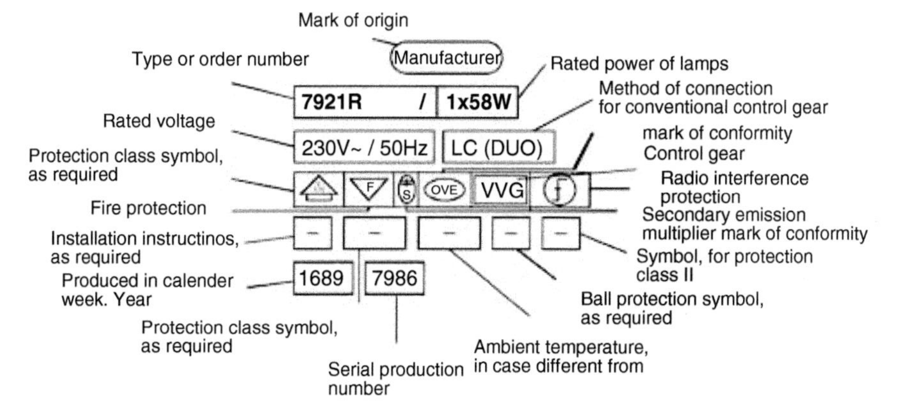

Figure 20.4 Manufacturer's data for light fixtures. Source: Trilux [52].

20.4.11 Standard Values for Planning Lighting Systems

For the rough calculation of the power requirement and the number of lamps required, we can refer to Tables 20.7 and 20.8.

20.4.12 Economic Analysis and Costs of Lighting

The planning and configuring of the lighting system also includes a cost and economic analysis, which will be described briefly in this section [37]. The total yearly costs are made up of the

- Capital costs for the purchase and initial installation and the
- Operating costs for power, replacement of lamps, and system maintenance.

Table 20.6 Number of discharge lamps per circuit-breaker.

Lamp power (W)	Voltage (V)	Condensator capacity (μF)	B/C10 (A)	B/C16 (A)	B/C20 (A)	B/C25 (A)
Fluorescent lamps with conventional control gear						
L18/20	230	4.5	32	51	64	82
L36/40	230	4.5	32	51	64	82
L58/65	230	7	20	33	41	53
High-pressure mercury vapor lamps						
50	230	7	10/19	15/31	18/39	23/49
80	230	8	6/12	9/19	11/24	14/30
125	230	10	4/7	6/12	7/15	9/19
250	230	18	2/4	3/6	3/7	4/9
400	230	25	1/2	2/4	2/5	2/6
700	230	40	-/1	1/2	1/2	1/3
1000	230	60	-/1	-/1	1/2	1/2
Halogen metal vapor lamps						
150	230	20	7/5	11/8	14/10	17/12
250	230	20	7/5	11/8	14/10	17/12
400	230	32	5/3	7/5	9/6	11/8
1000	230	85	1/-	1/1	3/1	3/2
2000	D400	60	1/-	2/1	2/1	3/2
2000	N400	37	—	1/-	1/1	2/1
3500	400	100	—	—	—	—
High-pressure sodium vapor lamps						
50	230	10	16/11	24/17	31/22	38/27
70	230	12	12/8	18/13	23/16	29/20
100	230	12	10/7	16/11	20/14	25/17
150	230	20	5/5	11/8	14/10	17/12
250	230	36	5/3	7/5	9/6	11/8
400	230	45	3/2	4/3	5/4	7/5
1000	230	100	1/-	1/1	2/1	3/2

[52]

Table 20.7 Power requirement for a nominal illuminance $E_n = 500$ lx.

Room size	Floor space (m²)	Power requirement (W/m²)
Average	30	15
Large	150	13

Source: Trilux [52].

Table 20.8 Light flux requirement for a nominal illuminance $E_n = 100$ lx.

Room size	Floor space (m²)	Light flux (lm/m²)
Average	30	225
Large	100	200

Source: Trilux [52].

Capital costs:

$$K_K = n_1 \left[\frac{K_1 \frac{k_1}{100} + K_2 \frac{k_2}{100}}{n_2} \right] \tag{20.4}$$

Power costs:

$$K_E = n_1 \; [t_B \; a \; P] \tag{20.5}$$

Replacement of lamps and system maintenance:

$$K_{LW} = n_1 \left[\frac{t_B}{t_L} \; (K_3 + K_4) + \frac{R}{n_2} \right] \tag{20.6}$$

Total yearly costs:

$$K_G = K_K + K_E + K_{LW} \tag{20.7}$$

The symbols have the meanings:

K_1	Costs of a light fixture
k_1	Capacity costs for K_1 in % for amortization and interest
K_2	Costs for installation material and installation per light fixture
R	Cleaning costs per light fixture per year
n_1	Total number of lamps
n_2	Number of lamps per light fixture
K_3	Price of lamp
K_4	Costs of replacing a lamp

P	Power consumption of a lamp + control gear in kW
a	Costs of electrical power per kWh + basic price
t_L	Useful life of the lamp in h
t_B	Yearly operation period in h

20.5 Procedure for Project Planning

For the planning and configuring of lighting systems, the following recommendations should be considered in the interest of cost effectiveness:

- High luminous efficiency of the light source
  Three-row fluorescent lamps have up to 30% greater luminous efficiency than standard fluorescent lamps.
- Low power loss from the control gear
  Electronic control gear has up to 62% less power loss than conventional control gear.
- High efficiency of the light fixtures
  Specular louvre light fixtures, with operating efficiencies of up to 75%, are preferable to opal recessed light fixtures from the standpoint of using energy efficiently.
- High utilization factors as a result of
  1) Dedicated lighting systems and optimal arrangement
  2) Dedicated room layout
  3) Choice of economic light sources
  4) Computer-optimized planning.

- Floor plans and vertical sections of rooms with windows and doors
- Room state (reflection factors)
- Use of rooms
- Specifications of room temperature (reduced light flux at low temperature)
- Period of operation of the lighting system in hours
- Choice of light fixtures

Planning and configuring the lighting in an industrial hall:

For an industrial system with normal bookbinding work, the lighting planning is to be carried out according to the efficiency method with form sheets. The following data are given:

1. Room dimensions

Room length	$a = 45\,\text{m}$
Room width	$b = 16\,\text{m}$
Room height	$H = 7\,\text{m}$

2. Reflection factors

Ceiling	0.5
Walls	0.3
Floor	0.1

3. Light fixture-type requirement
   Continuous reflector row 1 × 58 W, height of suspension 1 m
4. Requirements for the light source
   Tubular fluorescent lamps 58 W with high luminous efficiency. Light color and step of color reproduction properties must be considered.
5. Definition of problem
   - Planning of the lighting system in accordance with the regulations.
   - Determination of the required number of light fixtures for an average illuminance with the use of the form sheet.
   - Evaluation of the lighting system with regard to uniformity, light color and color reproduction, glare, modeling, and light direction.
   - With what measures can the power consumption of the system be reduced?

Calculational steps for the industrial system:
See Tables 20.9–20.13.

20.6 Exterior Lighting

Exterior lighting must always conform to the requirements and criteria of the Technical Standards Committee for Lighting Technology. The most important standards are listed here.

Table 20.9 Room data.

Length a	45	m
Width b	16	m
Area $A = a\ b$	720	m^2
Height H	7	m
Height of evaluation plane above floor e	0.85	m
Length of pendant or suspension l	1	m
Light spot height $h = H - l - e$	5.15	m
Room index $k = \frac{a\ b}{h(a+b)}$	2.3	
Reflection factors for ceiling/walls/floor	0.5/0.3/0.1	

[52]

Table 20.10 Standard values taken from workspace regulations.

Type of room	Bookbinding work	
Nominal illuminance E_n	300	lx
Light color	ww/nw	
Step of color reproduction properties	2 A	
Quality class of glare restriction	1	
Reduction factor v	0.8	
Uniformity $g_1 = E_{min}/\overline{E}$	1/1.5	

Table 20.11 Data for light fixtures.

Number of light fixture	7921R/58	
Data block number	712	lx
Utilization factor η_B	62	%
Multiplier for η_B M	1	
Number of lamps per light fixture z	1	

[52]

Table 20.12 Data for lamps.

Power per lamp, without control gear	58	W
Power per lamp, with control gear	66	W
Light color	nw	
Step of color reproduction properties	1B	
Light flux per lamp Φ	5200	lm

[52]

Table 20.13 Technical data for lighting systems.

Required number of light fixtures $n = \dfrac{E_n\, A \cdot 100}{z\, \Phi\, \eta_B\, M\, v}$	$\dfrac{300\ \text{lx} \cdot 720\ \text{m}^2 \cdot 100}{5200\ \text{lm} \cdot 62 \cdot 0.8}$	84	
Selected number of light fixtures	N	87	
Average illuminance $\overline{E} = \dfrac{N\, z\, \Phi\, \eta_B\, M\, v}{A \cdot 100}$	$\dfrac{87 \cdot 5200\ \text{lm} \cdot 62 \cdot 0.8}{720\ \text{m}^2 \cdot 100}$	312	lx
Uniformity	$E_{\text{min}}/\overline{E}$	$\geq 1/1.5$	
Spacing of light fixtures		x in m	y in m
		Transverse	Longitudinal
(a) According to data block		9	4.8
(b) Selected/specified		6	1.53

Arrangement of light fixtures: 3 rows of lighting, 29 light fixtures each

Remarks: The power consumption can be reduced with the use of electronic control gear.

The connected load is then $P = N\, z\, p = 87 \cdot 1 \cdot 55\ \text{W} = 4785\ \text{W}$.

The connected load with conventional control gear is: $P = N\, z\, p = 87 \cdot 1 \cdot 66\ \text{W} = 5742\ \text{W}$.

Savings with the use of electronic control gear: 957 W.

- Permanently Installed Traffic Lighting – Illumination of Streets for Road Traffic: General Quality Criteria and Standard Values
- Permanently Installed Traffic Lighting – Illumination of Streets for Road Traffic: Calculation and Measurement
- Standard Values for the Illumination of Pedestrian Crossings
- Illumination of Road Tunnels and Underpasses
- Illumination of Parking Lots and Park Houses
- Stadium Lighting
- Illumination of Sluice Systems
- Good Lighting for Safety on Roads, Pathways, and Public Squares

Exterior lighting systems are calculated with the point lighting method and not with the efficiency method. The luminous intensity and the luminous flux can be taken from the manufacturers' catalogs. The horizontal illuminance is calculated from the equation

$$E = \frac{I\,H}{(l^2 + H^2)\,\sqrt{(l^2 + H^2)}} \tag{20.8}$$

Here, the meanings of the symbols are

E	Illuminance in lx
H	Lighting pole height in m
l	Distance of point of interest from base of light fixture in m
I	Luminous intensity in direction of point of interest in cd

20.7 Low-Voltage Halogen Lamps

General notes on installation:

- The transformer must be designed for the safety extra-low voltage.
- Transformers must be implemented as short-circuit – proof safety isolating transformers.
- Electronic overtemperature protection is built in.
- Protection of the transformer against secondary short-circuits is possible through fine-wire fusing on the primary side.
- When overload protection is present, the lower voltage side is also protected.
- The transformer power must be matched to the actual loading of all connected lamps.
- The lamps may be operated only within the specified limits.
- For dimming mode, it is necessary to observe the manufacturers' special instructions for connecting.
- The dimmer must always be installed on the primary side of the transformer.
- The dimmer must be designed according to the power of all connected lamps, including all connected transformers and any required base load.
- The transformer must be easily accessible.
- The transformer must be installed as closely as possible to the light fixture.
- For safety isolating transformers must be taken into account.

- The voltage drop and the line cross section on the lower voltage side must be taken into account.
- The maximum permissible voltage in the low-voltage installation is restricted in accordance with IEC 357 to 110%.
- Terminals must be relieved from traction, shear, and torsion.
- Junctions must be protected against electric shock.
- At the clamping points permanent low-resistance connections must be made.
- When the cross section of the line is greater than that of the clamping point on the transformer, a line up to 0.5 m long must be led out of the transformer and the required line then connected through a junction box.

Calculation of the maximum line length for a voltage drop:

In 12 V systems, the voltage drop is considerably higher than in 230 V systems with the same power. The high currents and the high-voltage drop lead to heating and power losses in the lines installed. For this reason, low-voltage installations must be carefully planned.

Magnitude of the current:

$$I = \frac{P}{U} \tag{20.9}$$

Voltage drop:

$$\Delta U = \frac{2\,l\,I}{\kappa\,S} \tag{20.10}$$

With these values, we can then calculate the line length for the given lamp power:

$$l = \frac{\Delta U \kappa\,S}{2\,I} \tag{20.11}$$

Calculation of the cross section:

$$S = \frac{2\,l\,P}{\Delta u\,U_n^2 \kappa} \tag{20.12}$$

The symbols have the meanings:

l	Length of line in m
S	Cross section of line in mm^2
I	Current in line in A
ΔU	Voltage drop in V
κ	Conductivity of line; for Cu: 56 m/Ω mm^2
P	Lamp power in W
U_n	Nominal voltage in V
Δu	Voltage drop in %

20.8 Safety and Standby Lighting

20.8.1 Terms and Definitions

- *Back-up lighting*: The backup lighting is a lighting system which is promptly activated in the event of a failure in the power supply to the normal artificial lighting. Here, it is necessary to distinguish between

1) Safety and backup lighting in accordance community facility regulations
2) Standby lighting in accordance with the regulations.
- *Safety lighting*: Safety lighting is required for safety reasons (general safety, accident prevention). It has a protective function.
- *Stand-by lighting*: Stand-by lighting consists of a backup lighting system which takes over the function of the normal artificial lighting in order to enable continued operation for a limited period of time. When no safety lighting is required, a stand-by lighting system can still be installed.
- *Emergency lighting*: Emergency lighting consists of a light fixture with its own or with an emergency source of power, which is used to generate the emergency lighting.

20.8.2 Circuits

- *Floating circuit*: Emergency light fixtures or the emergency-symbol light fixtures connected to emergency lighting power supplies are activated for both normal power and in case of a failure of normal power. The normal power supply is monitored at the main distribution board of the safety power source.
- *Stand-by circuit*: Emergency light fixtures or the emergency-symbol light fixtures connected to emergency lighting power supplies are activated only in the event of a failure of normal power. The power supply for the normal lighting is monitored in the sub-distribution system for this part.

For emergency-symbol or safety lighting systems, in accordance with EN 60598, Part 2.22 and no glow starter or discharge lamps with integrated glow starter may be used; EN 60924 and EN 60925 specify only electronic control gear.

20.8.3 Structural Types for Groups of People

The requirements for safety and backup lighting depend on the use of the room or building (Table 20.14).

20.8.4 Planning and Configuring of Emergency Symbol and Safety Lighting

- *Emergency light fixtures*: Emergency symbols must have white pictorial markings and a green background and be clearly recognizable during the required time in operation. A minimum size is therefore required. The required recognition widths can be determined from formulas or from the manufacturer's information and must be taken into account during planning. In accordance with the accident prevention regulations, the aspect ratio must be 1 : 1 or 1 : 2.
  The calculation of the illuminance requires the:
  - Luminous intensity distribution of the light fixture
  - Luminous flux at the end of the nominal operating life
  - Height of suspension of the light fixture or type of light mounting

Table 20.14 Requirements for safety and backup lighting according to use of the room or building, in accordance with community facility regulations.

Type of structure	E_{min} (lx)	t_u (s)	t_b of SVS (h)	DS for illumination of emergency symbol	DS for SB of RW
Emergency ways in production/office areas	1	15	1	No	No
Exhibition areas >2000 m^2	1	1	3	Yes	Yes
Restaurants with >400 seats	1	1	3	Yes	Yes
Lodging facilities with >60 beds	1	15	3	Yes	No
Garages with >1000 m^2 useable area	1	15	1	Yes	No
Stores with >2000 m^2 sales area	1	1	3	Yes	Yes
Multi-storey buildings >22 m	1	15	3	Yes	No
Schools with floor area >3000 m^2	1	15	3	Yes	No
Places of gathering, movie houses, theater for >100 persons	1	1	3	Yes	No
Meeting rooms for >200 persons	1	1	3	Yes	Yes
Stage scene area	3	1	3	Yes	No
Circus rings, sport race courses	15	1	3	Yes	No
Workplaces with special dangers	min. 15	0.5	min. 1	No	No

DS: Floated circuit, SB: Safety lighting, RW: Emergency way
t_{um}: Switch-over time, t_b: Operating time, SVS: Stand-by voltage source.

[52]

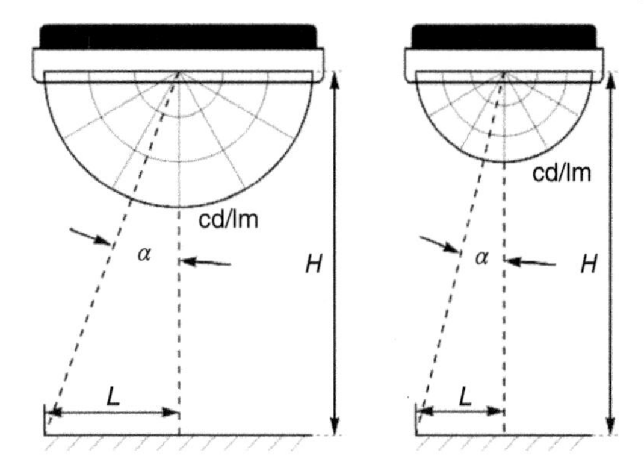

Figure 20.5 Emergency way lighting.

The recognition width of the emergency-symbol light fixture is given by:

$$e = h\,z \tag{20.13}$$

The illuminance (Figure 20.5) can be calculated using the point lighting method, according to the relationship:

$$E = \frac{I\,\cos^3 a}{H^2} \tag{20.14}$$

The spacing between the light fixtures is calculated from the relationship:

$$L = H \tan \alpha \tag{20.15}$$

Here, the meanings of the symbols are

e	Recognition width in m
h	Height of emergency symbol in m
z	Distance factor 200 for internally illuminated emergency symbols,
	100 for externally illuminated emergency symbols
E	Illuminance in lx
H	Height of light fixture – 0.2 in m
a	Light emergence angle in °

$E = 1\text{lx} \times 1.25$	between light fixtures
$E = 0.5\text{lx} \times 1.25$	between light fixture and wall or door

- *Safety lighting*: The safety lighting identifies and illuminates emergency ways and must have the uniformity specified before:

$$g = \frac{E_{\min}}{E_{\max}} \geq \frac{1}{40}$$

at a height of 0.2 m and 1 lx and a nominal operating time of at least one or three hours. The locations of these light fixtures are main entrances and exits, floors, stairs, emergency balconies, emergency tunnels, and workplaces with special dangers. For the planning of emergency-symbol and safety light fixtures, the following points must be taken into account (see Figures 20.6–20.10):

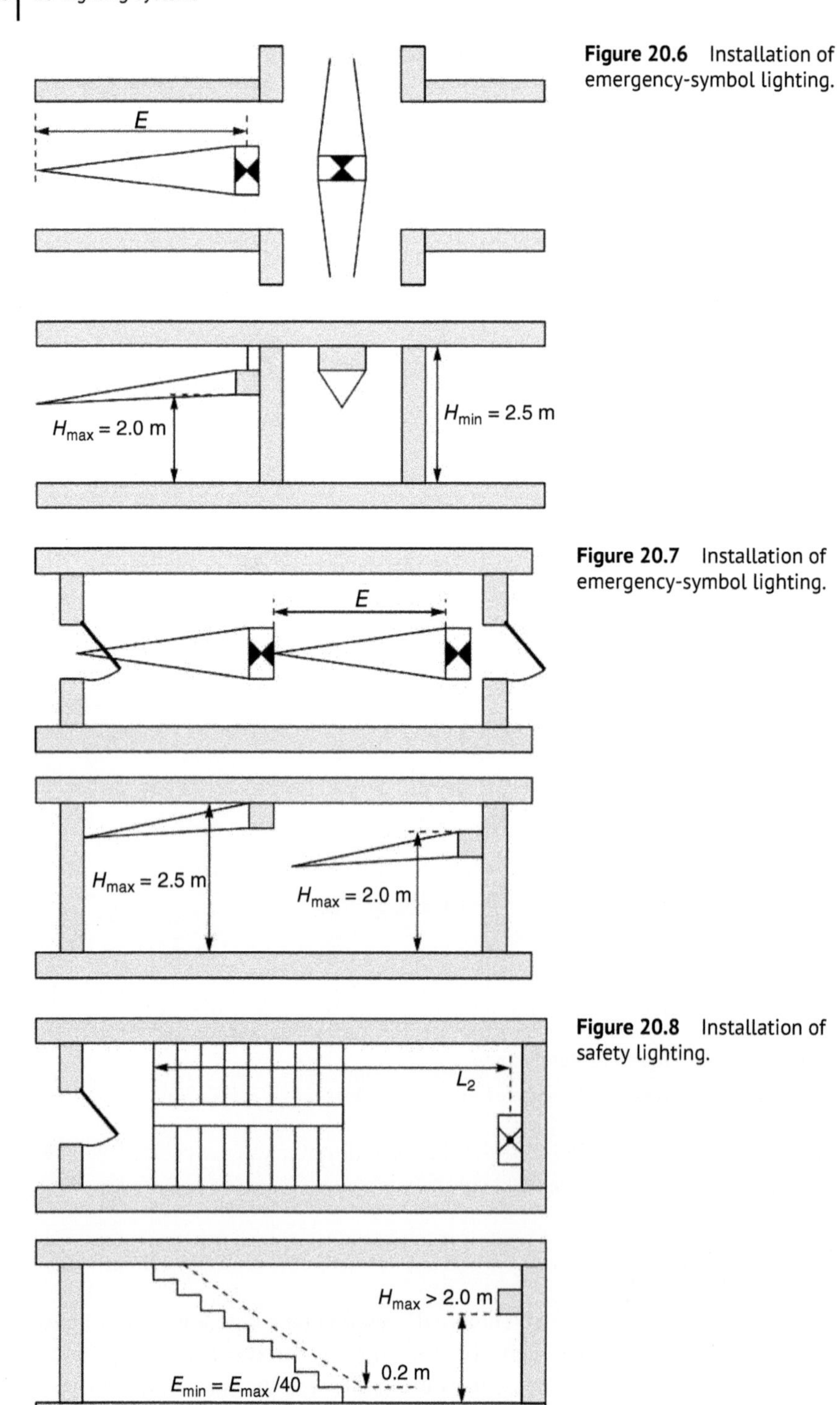

Figure 20.6 Installation of emergency-symbol lighting.

Figure 20.7 Installation of emergency-symbol lighting.

Figure 20.8 Installation of safety lighting.

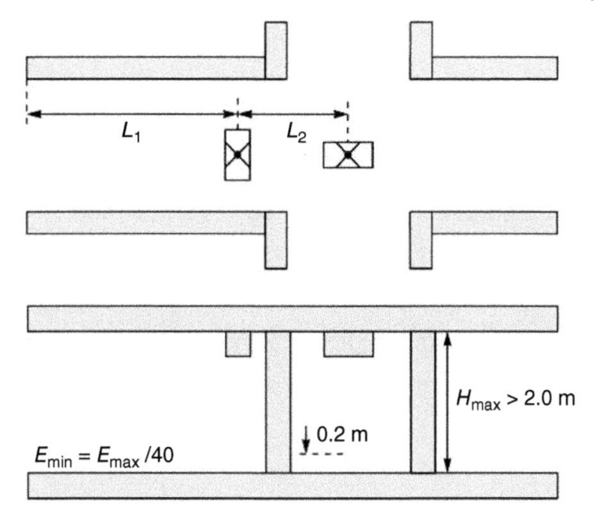

Figure 20.9 Installation of safety lighting.

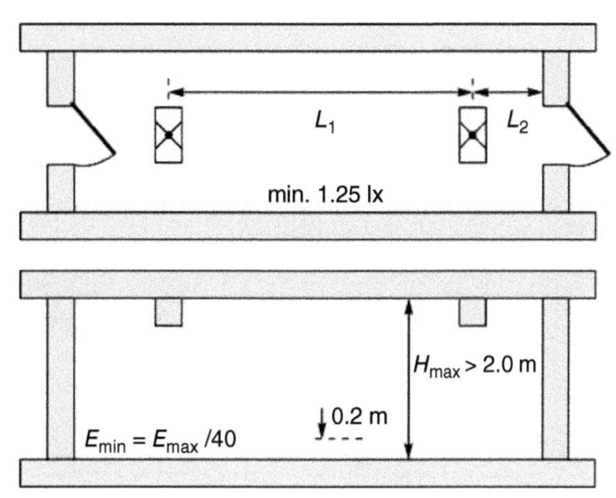

Figure 20.10 Installation of safety lighting.

- Choice of light source
- Covers for light fixtures
- Type of protection
- Protection class
- Light fixture housing (material, design)
- Type of installation (wall, ceiling, pendant, built-in, mounting)
- Version with single battery, central battery system, or groups of batteries

20.8.5 Power Supply

The main distribution system for the safety power supply and the normal power supply must be located in electrical operating areas and isolated from each other so that no arcing can pass between them. Isolation of the walls and ceilings must be carried out with F30 and doors with T30, for a normal fire hazard, and F90, T30 for a greater fire hazard.

20.8.6 Notes on Installation

1. *Electrical operating areas*: Isolation of the walls and ceilings from other rooms must be carried out with F30, for a normal fire hazard, and F90 with an increased fire hazard, and doors with T30 for a greater fire hazard. In this area, there must be no switchgear over 1 kV, no standby generating systems and no operational equipment for other systems.
2. *Battery space*: These rooms must be protected from extreme temperatures and gassing. The width of the gangway must be at least 0.5 m or 1.5 times the cell depth.
   The calculation of the ventilation for the battery space can be made according to the following equation:

$$Q = 0.05\, n\, I\, f_1\, f_2 \tag{20.16}$$

Air intake and exhaust air opening:

$$A = 0.0028\, Q \tag{20.17}$$

Battery charging capacity:

$$P = 4\, U\, I = 4U_{\text{cell}}\, I\, n_{\text{cell}} \tag{20.18}$$

Forced ventilation is not required

with a charging capacity < 3 kW for Pb-acid batteries
with a charging capacity < 2 kW for Ni-Cd batteries.

Here, the meanings of the symbols are:

Q	Volume rate of air flow in m^3/h
n	Number of cell
I	Current in A
f_1	Reduction factor
f_2	Reduction factor
A	Air intake and exhaust air opening in m^2
U	Charging voltage in V
I	Charging current in A

20.8.7 Testing During Operation

- Automatic monitoring of charge in cycles of five minutes
- Automatic function testing (maximum five minutes)
- Automatic operating time testing (minimum 40/120 minutes)
- Automatic monitoring of lines
- Registration and storing of faults (minimum two years), maintenance log book
- Display and printout of test results

20.9 Battery Systems

20.9.1 Central Battery Systems

The most recent version of community facility regulations specifies that in the event of a power failure the standby system must not be supplied from the battery as long as a voltage is still present at the main distribution board of the safety lighting system. Accordingly, the fusing in the main low-voltage distribution system must be designed for the entire safety system distribution. For maintenance purposes, a fuse switch-disconnector is recommended. The central battery system is described in detail as follows (see Figures 20.11–20.15):

1. Components
   - Battery (open, closed, or sealed Pb-acid or Ni-Cd batteries)
   - Charging equipment (recharging to 90% in 10 or 20 hours)
   - Changeover equipment (standby and/or floated circuit)
   - Rectifier for discharge lamps (single/groups/central)
   - Internal displays (central fault, operating mode, ready state, battery discharge warning activated, isolation monitor activated, ventilation monitor activated)
   - External display (central fault, operating mode, ready state)

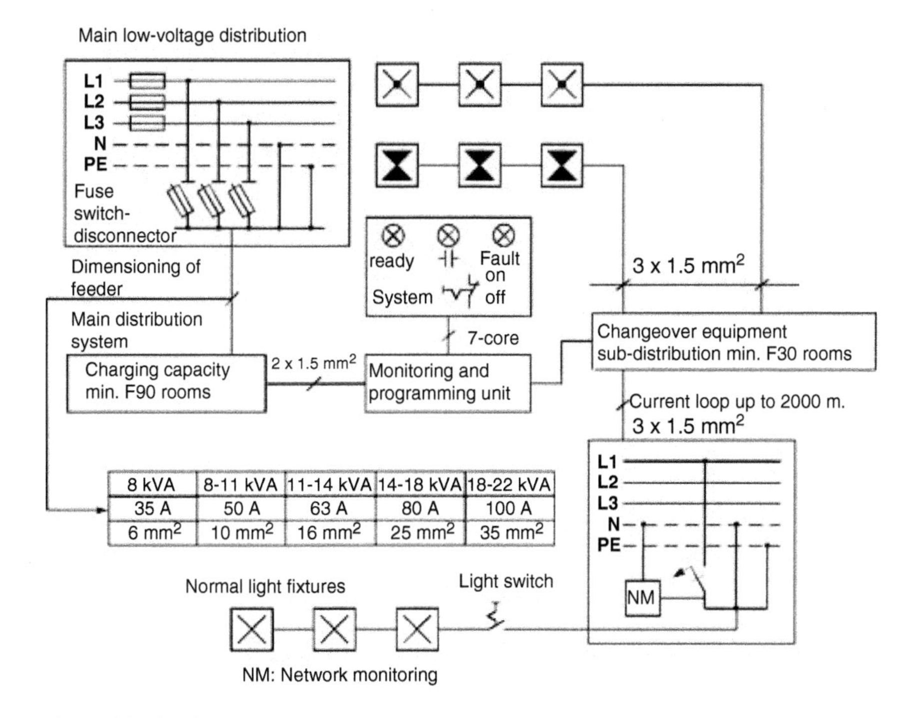

Figure 20.11 Central battery system.

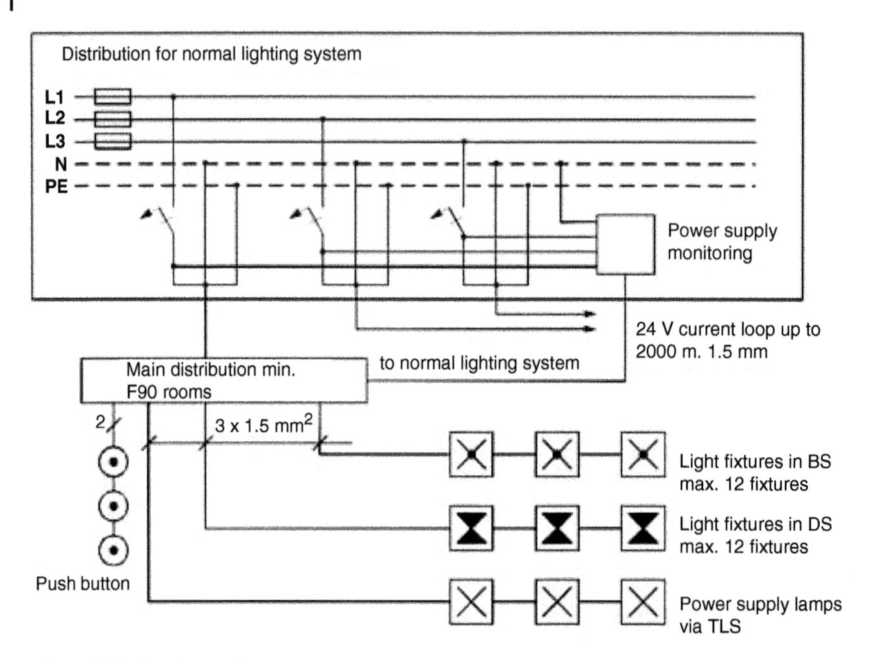

Figure 20.12 Central battery system.

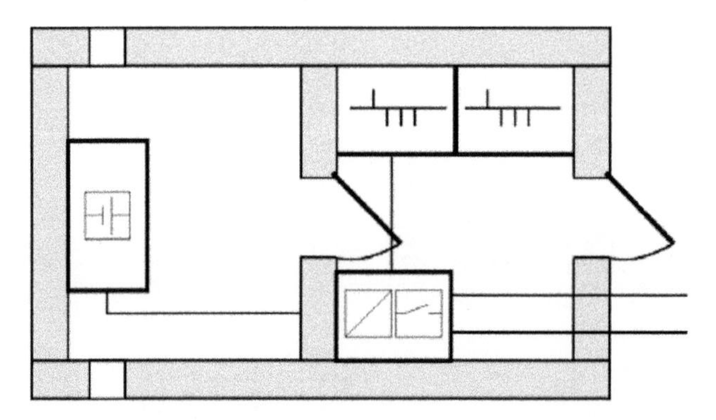

Figure 20.13 Installation of lines.

2. Technical data
   - Power unrestricted
   - Number of light fixtures unrestricted (maximum 12 light fixtures per circuit and maximum 6 A current load per circuit)
3. Monitoring equipment
   - Circuit monitoring
   - Individual monitoring
   - Phase monitoring
4. System properties
   - Phase selection circuit for three-phase connection
   - Battery operation only in the event of a complete power failure

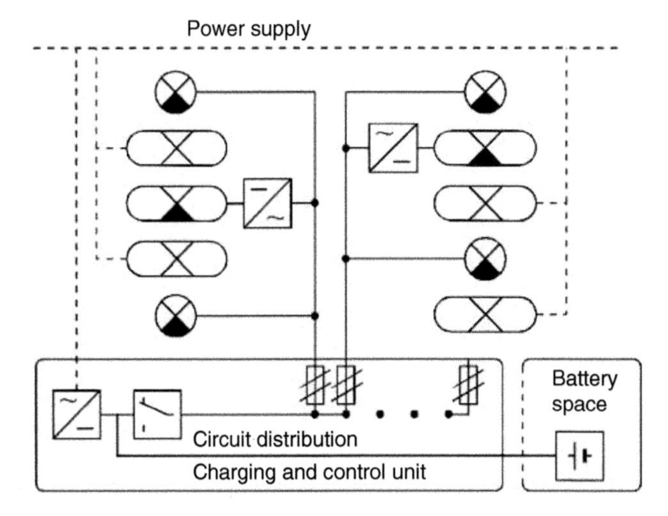

Figure 20.14 Emergency lighting with central battery.

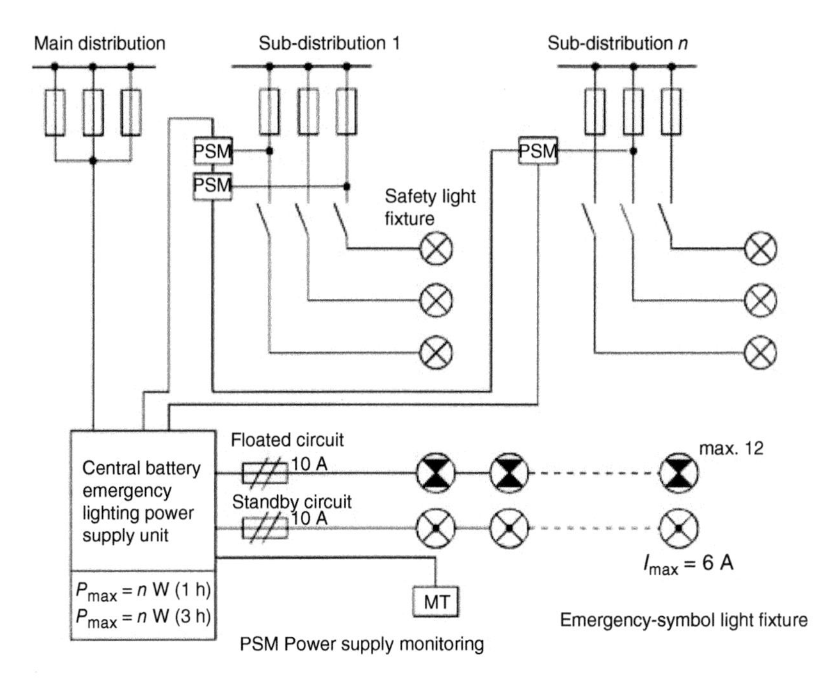

Figure 20.15 Central battery systems.

- Separate monitoring and changeover equipment is required for floated circuit and standby circuits.
- The technical fire protection requirements do not apply for branch circuits of the safety lighting.

The abbreviations have the meanings:

BS	Standby lighting system
DS	Floated circuit
LVMD	Main low-voltage distribution system
UVA	Normal sub-distribution system
N″U	Power supply monitoring
UVS	Safety lighting sub-distribution system
TLS	Staircase light-out query
NL	Normal light fixtures
RL	EXIT
SL	Safety light

For fire protection, it is necessary to take account of the following points (Figures 20.16 and 20.17):

Sections:

2.2.22 Main distribution board for safety power supply is first distribution point in building which is fed directly from standby power source of safety lighting system.

5.2.1.2 Rooms for main distribution board of normal power supply must be separated from rooms with greater fire hazards with at least F90 and from other rooms with at least F30. Doors must be fire retardant.

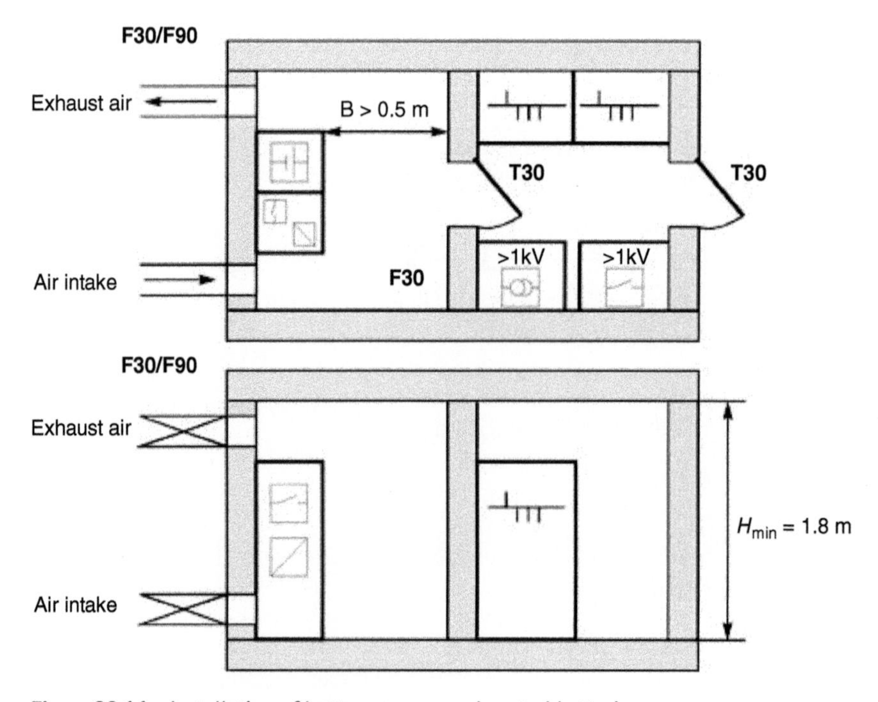

Figure 20.16 Installation of battery groups and central batteries.

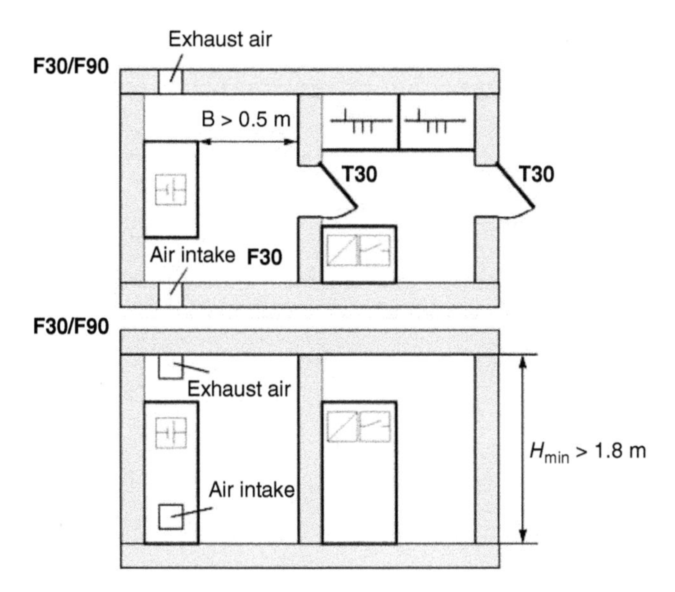

Figure 20.17 Installation of battery groups and central batteries.

5.2.2.1 Sub-distribution systems of the normal power supply must be installed with their own enclosures.

6.6.6 Sub-distribution systems of the safety lighting system must be installed separately from system sections of the normal power supply, with their own enclosures.

6.7.4 For each circuit of the safety power supply, separate cables and separate lines must be used. These cables and lines must be run separately from other line systems. Separate line systems are not required for installation of the branch circuits of the safety lighting system (Figures 20.13 and 20.18).

Lines must be at least F30 up to the first light fixture of the circuit. Thereafter, in the same fire-protected section, F30 is no longer required. Central and sub-stations must be run separately and multicore lines must be run short-circuit and ground fault – proof only for a main circuit and for auxiliary circuits and must not be run through explosion hazard areas (Figure 20.13).

20.9.2 Grouped Battery Systems

Grouped battery systems (Figure 20.19) are described as follows:

1. Components
   - Battery (sealed Pb-acid or Ni-Cd batteries)
   - Charging equipment (recharging to 90% in 10 or 20 hours)
   - Changeover equipment (standby and/or floated circuit)
   - Rectifier for discharge lamps (single/groups/central)
   - Internal displays (central fault, operating mode, ready state, battery discharge warning activated, isolation monitor activated, ventilation monitor activated)

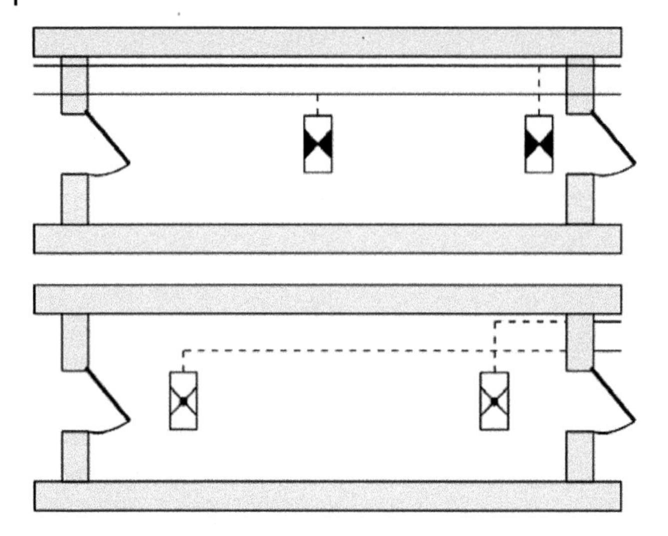

Figure 20.18 Installation of lines.

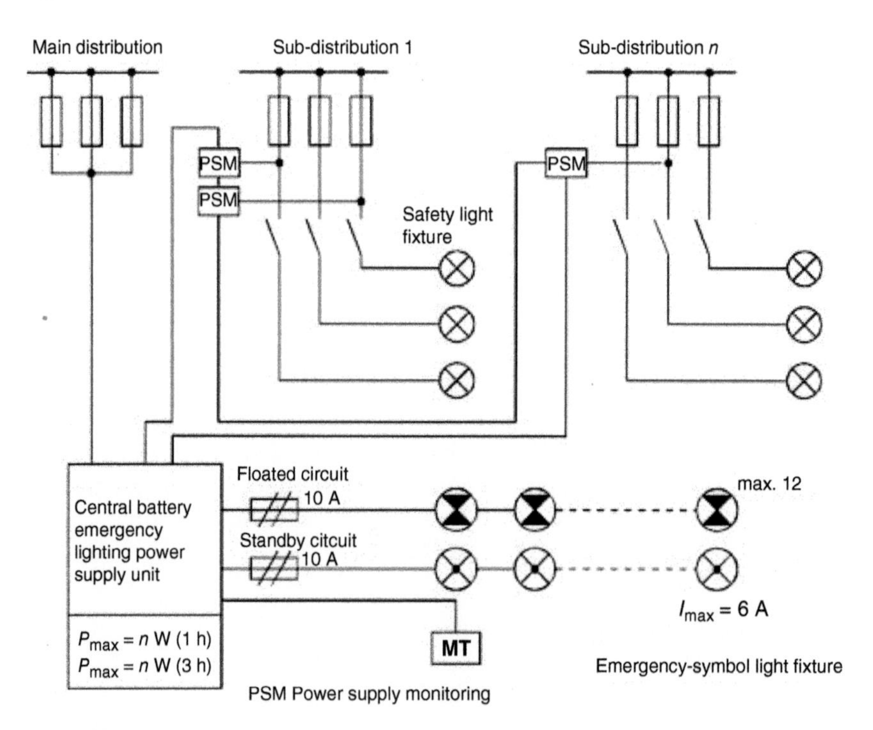

Figure 20.19 Grouped battery systems.

- External display (central fault, operating mode, ready state)
- Technical data

2. Power 900 W for one hour or 300 W for three hours
3. System properties
   - A maximum of 20 safety light fixtures can be connected.

- The current consumption of group rectifiers must not exceed 6 A.
- Separate monitoring and changeover equipment is required for floated circuit and standby circuits.
- The technical fire protection requirements do not apply for branch circuits of the safety lighting.

20.9.3 Single Battery Systems

Single battery systems are rechargeable sealed battery constructions, which have a service life of three years and in accordance with IEC 598-2-22 have a service life of four years. One battery system may not supply more than two safety light fixtures. Single battery light fixtures must be subjected to a function test every seven days and at least once a year to a continuous operation test. The self-monitoring takes place with LEDs. These indicate the status of the light fixture. The display and results must be entered in a testing log book. The testing log book must then be saved for a period of two years. Every seven days, the status of the light fixtures must be documented.

Figure 20.20 illustrates the structure of a safety light fixture.

1. Components
   - Battery (sealed Pb-acid or Ni-Cd batteries)
   - Charging equipment (recharging to 90% in 20 hours)
   - Changeover equipment (standby and/or floated circuit)
   - Rectifier for discharge lamps (single) or electronic control gear
   - Internal displays (battery charge state)
   - External display
   - Monitoring equipment for monitoring light fixtures

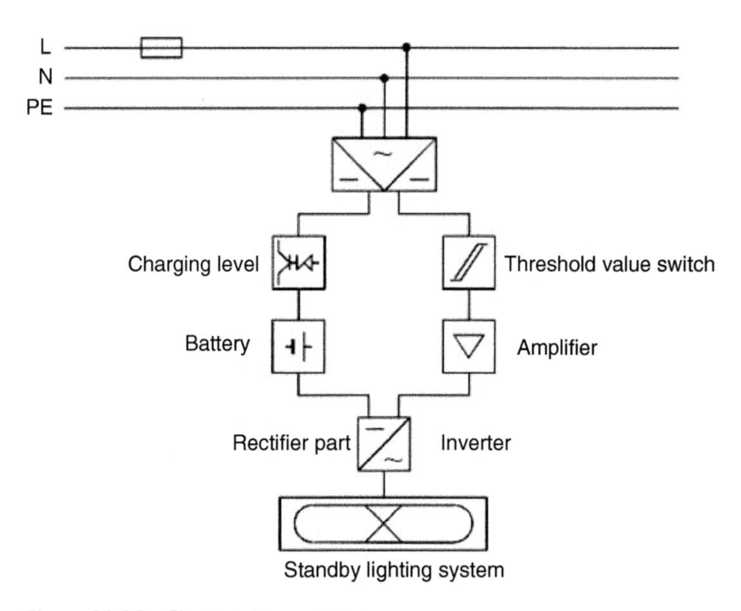

Figure 20.20 Single battery system.

2. Technical data
   - Power unrestricted
   - Maximum two light fixtures connected
3. *Lines*: F30 is not required. No specifications exist for the connection of the power supply units and light fixture connections to a power source.
4. Light fixtures must be distributed over at least two circuits when there is more than one light fixture in the room.

Figure 20.21 depicts a single battery system with floated circuit and backup lighting.

Figure 20.22 shows a single battery system with monitoring system for the connection of one or two light fixtures with fluorescent lamps and electronic control gear and low-voltage lamps with electronic transformer. The electronic control gear must be suitable for DC operations.

Community facility regulations require regular testing for safety lighting systems, such as daily and weekly function tests with the safety lighting system operating fully and without exception a yearly continuous operation test outside of the normal operating time. With the emergency lighting system, single battery emergency light fixtures and single battery emergency light power supply units can be tested through a bus line. Figures 20.23 and 20.24 illustrate the system design of the test equipment.

Single battery systems can be put to good use for on the order of 10–20 units. All single batteries, while they have a high reliability, also have an accumulator which reacts sensitively to high temperatures (i.e. >20 °C). At 30 °C, the accumulator has only half its normal useful life and at 40 °C only one-fourth its normal useful life.

The meanings of the abbreviations are:

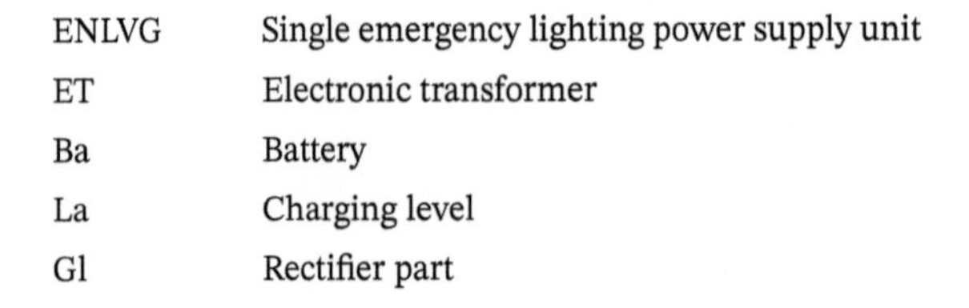

ENLVG	Single emergency lighting power supply unit
ET	Electronic transformer
Ba	Battery
La	Charging level
Gl	Rectifier part

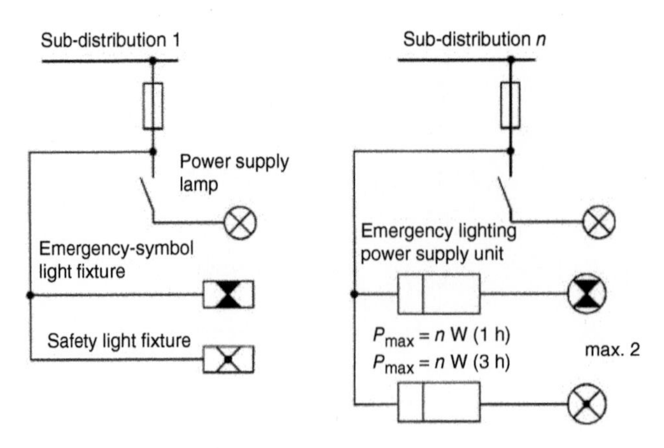

Figure 20.21 Single battery system.

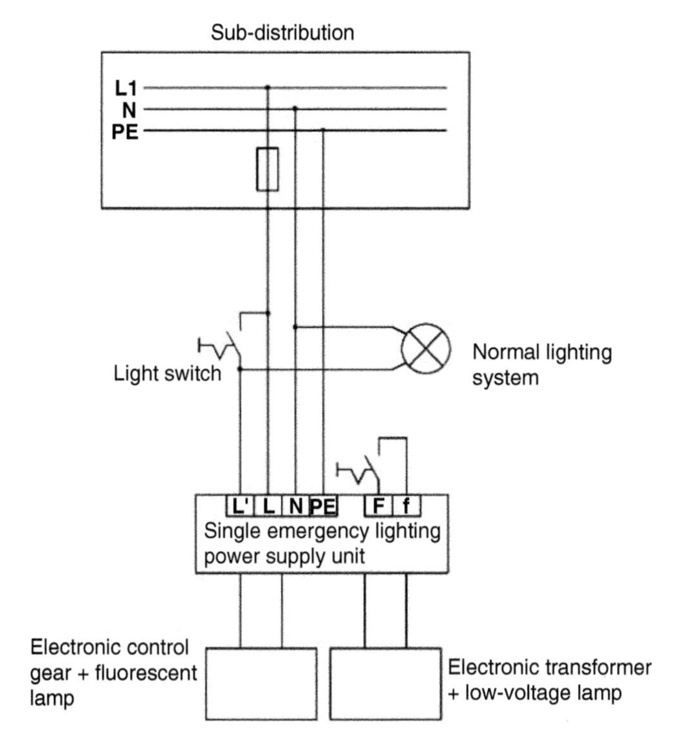

Figure 20.22 Single battery system.

Figure 20.23 Single battery system.

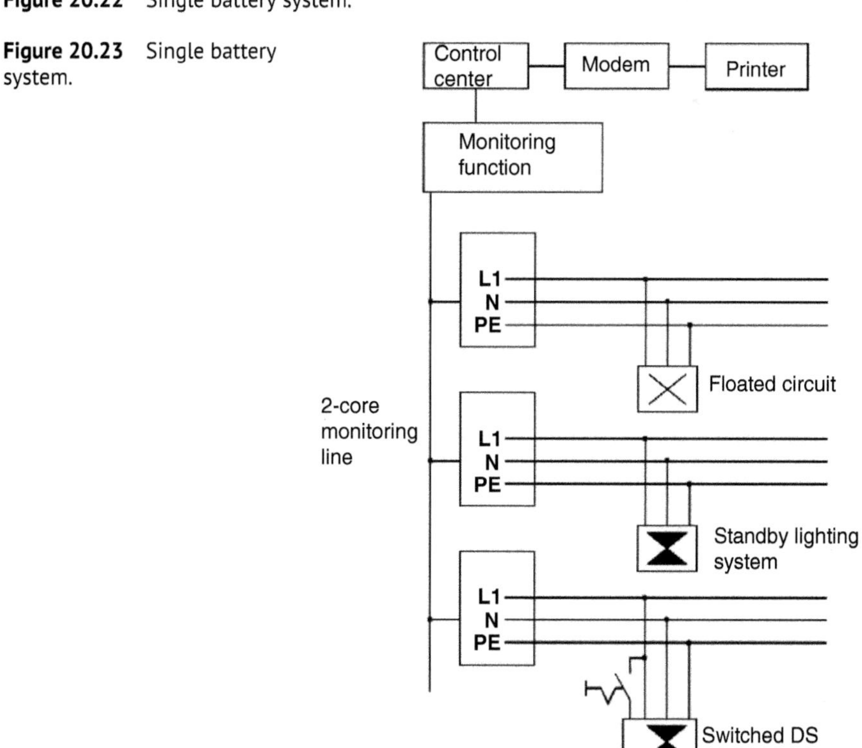

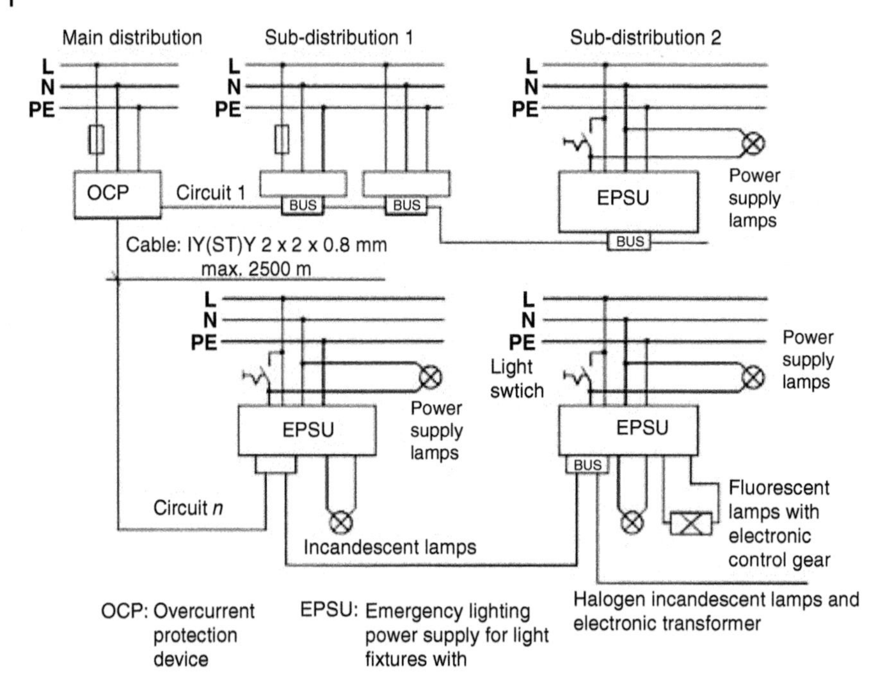

Figure 20.24 Single battery system.

Ve	Amplifier
Schw	Threshold value switch
We	Inverter
SL	Safety lighting system
FC	Floating circuit
SLS	Standby lighting system

20.9.4 Example: Dimensioning of Safety and Standby Lighting

Totally, 100 light fixtures with 55 W and electronic control gear were installed in a lighting system. Calculate:

The total power of the lighting system:

$$P = U\,I = 100 \cdot 55\ \text{W} = 5.5\ \text{kW}$$

The charging capacity (P_L) of the battery:

$$U = 108 \quad \text{cells of 2 V each} = 216\ \text{V}$$
$$I = \frac{5500\ \text{W}}{216\ \text{V}} = 25.46\ \text{A/h}$$
$$P_L = U\,I \cdot 4 = 2\ \text{V} \cdot 2\ \text{A} \cdot 4 = 16\ \text{W} \qquad \text{per cell}$$

16 W per cell for 108 cells gives a total of 1728 W. For power levels ≤ 3 kW, no forced ventilation is required.

The battery space:
We can calculate the ventilation of the battery space as follows:

$$Q = 0.05\ n\ I = 0.05 \cdot 108 \cdot 2\ \mathrm{A} = 10.8\ \mathrm{m^3/h}$$

Air intake and exhaust air opening:

$$A \geq 28\ Q$$

$$A \geq 28 \cdot 10.8\ \mathrm{m^3/h} = 302.4\ \mathrm{m^3/h}$$

When the room is smaller than $28\ \mathrm{m^3}$, the openings can be on the same side.

21

Generators

Synchronous generators are used in power plants. Its structure is similar to the asynchronous machine. The only difference is the number of synchronous turns, no slippage, and it is dependent on the mains frequency and the number of poles. The operating behavior of the generator is briefly explained for sinusoidal currents and voltages both on the fixed network and in isolated operation.

In contrast to the asynchronous machine, there is no relative movement (slip) between the rotor and stator fields, the stationary influence of the exciter field is compensated by the pole wheel voltage U_p expressed. The rotor excited with direct current represents an electromagnet. When the rotor rotates, a voltage is induced in the stator which is connected to stator terminals leading to a current flow. The current flow through the magnetic field built up by the stator current creates a force effect or a torque (generator principle in isolated operation).

Instead, three-phase current from the rigid network is applied to the stator, the rotor is driven, and energy flows into the mains. The rotor north pole runs ahead of the stator south pole. The rotor pulls the stator field behind it (generator). If the rotor is braked (e.g. by a working machine), energy is absorbed from the mains and output on the shaft (motor). The rotor north pole runs the South Pole of the stand. The stator field pulls the fan behind here. You can imagine the frictional connection between stator and rotor as a spring (higher force-longer spring extension-larger pole wheel angle). The possible operating modes can be shown by the simplified equivalent circuit diagram (Figure 21.1). The complete mathematical equivalent circuit diagram of the synchronous generator is very complex and not the subject of this book. Therefore, explanations are only an introduction to the topic.

The pointer diagram of the synchronous machine can be derived from the equivalent circuit diagram. It is based on the consumer counting arrow system. This allows the different load cases (motor, generator, capacitive, inductive reactive power) to be explained on one diagram. The voltage serves as reference arrow and forms together with the load current I_1 the angle φ. The terminal voltage U_1 and pole wheel voltage U_p form the so-called pole wheel angle ϑ (Figure 21.2). This angle depends on the active load.

Analysis and Design of Electrical Power Systems: A Practical Guide and Commentary on NEC and IEC 60364, First Edition. Ismail Kasikci.

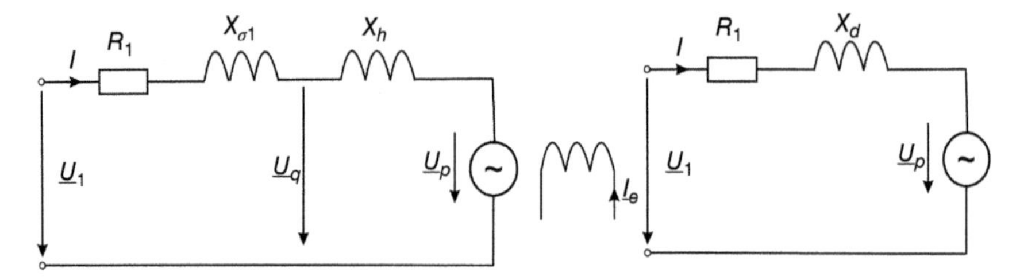

Figure 21.1 Equivalent circuit diagram of a synchronous generator.

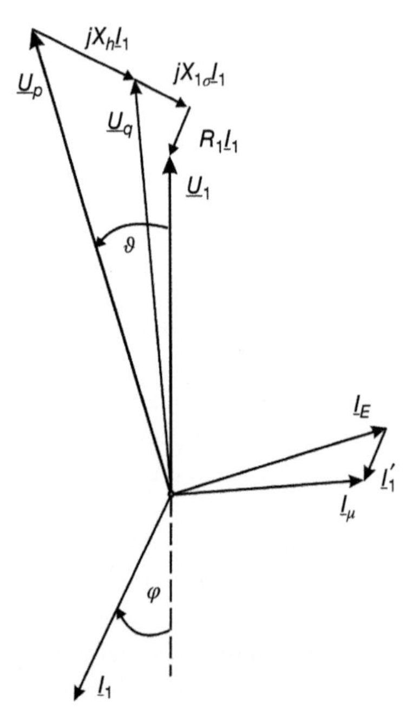

Figure 21.2 Pointer diagram of the synchronous machine.

A three-phase voltage is induced in the conductors of the three-phase windings of the stator, whose frequency f is indicated by ω. For practical purposes, the indication per minute n is important.

$$n = \frac{f}{p} \cdot 60 \tag{21.1}$$

Let us examine the idle synchronous generator. Trigger current I_e mainstream Φ_p this creates the pole voltage u_p.

$$\underline{U}_1 = \underline{U}_p + \underline{I}_1 \cdot (R_1 + j \cdot (X_h + X_{1\sigma})) = \underline{U}_p + \underline{I}_1 \cdot (R_1 + jX_d) \tag{21.2}$$

The main reactance consists of

$$X_d = X_h + X_{1\sigma} \tag{21.3}$$

For practical calculations, the ohmic winding resistance R_1 can be neglected.

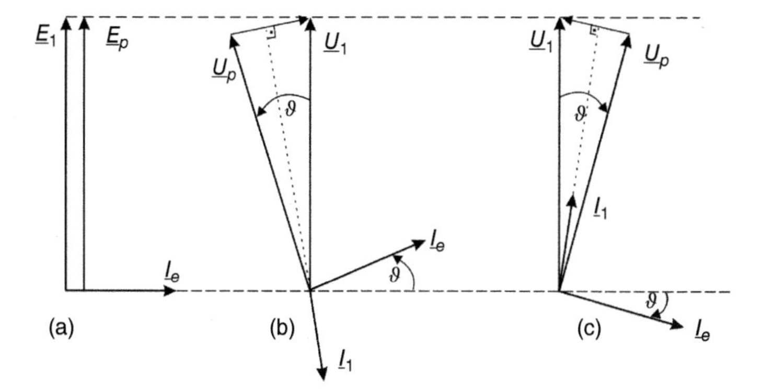

Figure 21.3 Mode of operation of synchronous generators, (a) Idle speed, (b) generator operation, (c) motor operation.

The conditions change if the generator is to work in parallel with other generators in the so-called mains operation. Now terminal voltage and frequency are fixed. However, this is only possible under three requirements:

- Mains and generator must have the same voltage.
- Mains and generator must have the same frequency and direction of rotation.
- The phase difference between mains and generator must be zero.

By running of the generator on the load, the armature flux (anker) φ_a creates $\underline{I}_1$ current due to the stator rotating field. The φ_a stator induces the u_a voltage (armature reaction) in the stator winding. Both flows Φ_p and Φ_a are linked to the same main inductance L_{1h}, since, according to the assumption, the fictitious and actual stator winding have the same winding data. Figure 21.3 shows the simplified equivalent circuit diagrams of the synchronous machine for stator leakage $X_{1s} = 0$ and stator resistance $R_1 = 0$ with current source I_e and voltage source U_p. At the voltage U_1 and the current I_1 entered in the equivalent circuit diagrams after the consumer counting system, the power of the source would be positive, i.e. it takes power from the mains, the machine operates as a motor. However, the pointer diagrams in Figure 21.3 show an operating state in which voltage U_1 and current I_1 are in opposite directions ($\varphi > 90°$). The power of the source is therefore negative, it delivers power to the mains, and the machine works as generator.

Φ_h is the actual main flow present in the air gap (superposition of the two rotating fields). I_μ is a fictitious quantity, just as I_e is a fictitious substitute quantity which actually does not occur anywhere. In the equivalent circuit diagram, it must also be taken into account that the stator current I_1 generates a stray field, represented by $L_1\delta$ or $X_{1\delta} = \omega L_{1\delta}$, and a voltage drop at the ohmic resistance of the winding, represented by R_1.

21.1 Generators in Network Operation

Current–voltage connection can be given from the simplified equivalent scheme.

$$\underline{U}_1 = jX_d \cdot \underline{I}_1 + \underline{U}_p \tag{21.4}$$

The stator current is calculated as follows:

$$\underline{I}_1 = \frac{\underline{U}_1 - \underline{U}_p}{jX_d} \tag{21.5}$$

21.2 Connecting Parallel to the Network

The synchronous machine can only be connected parallel to the network under the following conditions.

1. $n = n_1$ should be.
2. Caution current should be set so that $U_M = U_N$.
3. Phase connection must be L1-L2-L3/UVW.

21.3 Consideration of Power and Torque

A generator working in a fixed network can be given in four tar operating areas. The operating status has been determined according to the phase status of the stator current.

$$I_1 \cdot \cos\varphi_1 = -\frac{U_p}{X_d} \cdot \sin\varphi \tag{21.6}$$

$$P_1 = 3 \cdot U_1 \cdot I_1 \cdot \cos\varphi \tag{21.7}$$

$$P = M \cdot 2\pi n \tag{21.8}$$

$$M = \frac{P}{2\pi n} = 3 \cdot \frac{U_1 \cdot U_p}{2\pi n \cdot X_d} \cdot \sin\vartheta \tag{21.9}$$

Thus, the torque for constant current and voltage as a function depending on the moment of the arc (M_K) can be given (Figure 21.4).

$$M = M_K \cdot \sin\vartheta \tag{21.10}$$

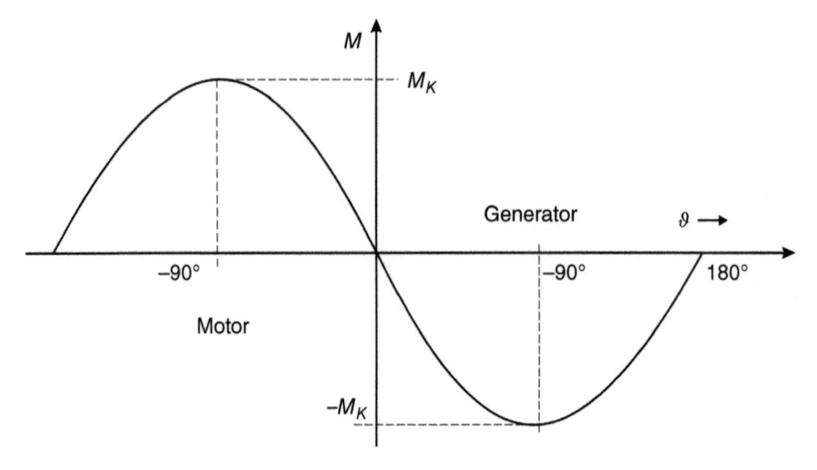

Figure 21.4 Torque as a function of polar wheel angle.

The tipping moment is the highest moment given by the machine $\vartheta = 90°$.

$$M_K = 3 \cdot \frac{U_1 \cdot U_p}{2\pi n_1 \cdot X_d} \tag{21.11}$$

If the turbine torque exceeds the overturning torque M_K, the rotor accelerates and falls out of operation. A stable operation is no longer possible. This means that a stability limit is defined by the tilting moment. If the turbine torque exceeds the overturning torque, the rotor accelerates and falls out of true. A stable operation is no longer possible. This means that a stability limit is defined by the tilting moment. This breaks the relationship between the field of the rotor and the stator field. In nominal operation, the generators achieve a pole wheel angle of up to 30% which is $M_K \approx \cdot M_N$.

Reactive power is determined by the polar spring value and warning system. Over-alerted machine works like a capacitor and gives inductive power. It acts as a reactance at low warning and takes inductive power.

Active power is created by mechanical pressure on the shaft. The superiority of the pole spring relative to the stator field (load angle) causes the grid to power; otherwise, it draws active power from the grid.

21.4 Power Diagram of a Turbo Generator

Figure 21.5 shows the operating states of a turbo generator. Active power can be read in the displayed curves.

- Reactive power can be limited by the excitation current I_e.
- P_{max} is dependent on drive power and cooling.
- Each operating state (P–Q-point) is controlled by the excitation current and drive power.

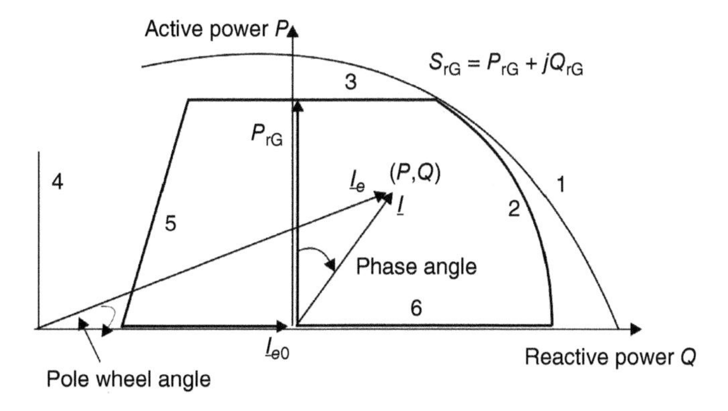

Figure 21.5 Power diagram of a turbo generator, (1) rated power limit value, (2) warning current limit (I_e – limit), (3) pole arc limit, (4) static stable limit, (5) practical static limit (ϑ), and (6) operation of the generator limit.

21.5 Example 1: Polar Wheel Angle Calculation

The values of a generator are given below:

$P_{rG} = 100\,\text{MW}, U_{rG} = 10.5\,\text{kV}, x_d = 0.7\ \Omega, I_e/I_{e0} = 2.3.$

Calculate the polar arc angle.

$$I_k = \frac{I_e}{I_{e0}} \cdot \frac{U_{rG}}{\sqrt{3} \cdot x_d} = 19.9\ \text{kA}$$

$$P_{rG} = \sqrt{3} \cdot U_{rG} \cdot I_k \cdot \sin \vartheta_{rG} \Longrightarrow \sin \vartheta_{rG} = \frac{P_{rG}}{\sqrt{3} \cdot U_{rG} \cdot I_k} = 0.276$$

$$\vartheta_{rG} = 16^\circ$$

21.6 Example 2: Calculation of the Power Diagram

The values of a generator are given below:
$P_{rG} = 100\,\text{MW}, \cos\varphi_{rG} = 0.8, U_{rG} = 10.5\,\text{kV}.$
Calculate the rated power of the generator.

$$S_{rG} = \frac{P_{rG}}{\cos\varphi_{rG}} = \frac{100\ \text{MW}}{0.8} = 125\ \text{MVA}$$

$$P_{rG} = S_{rG} \cdot \cos\varphi_{rG} = 125\ \text{MVA} \cdot 0.8 = 100\ \text{MW}$$

$$I_{rG} = \frac{S_{rG}}{\sqrt{3} \cdot U_{rG}} = \frac{125\ \text{MVA}}{\sqrt{3} \cdot 10.5\ \text{kV}} = 6873\ \text{A}$$

$$S_{rG} = \sqrt{3} \cdot I_{rG} \cdot \cos\varphi_{rG} \cdot (\cos\varphi_{rG} + j\sin\varphi_{rG})$$

$$S_{rG} = 6873\ \text{A} \cdot (0.8 + j0.6)$$

$$S_{rG} = 5498.5\ \text{MW} + j4123.9\ \text{Mvar}$$

22

Transformer

22.1 Introduction

A transformer is an alternating current machine, which is equipped with electromagnetic induction of alternating voltage and current between two or more windings with the same frequency and with generally different values of the voltage and the current. It is therefore a device for the transmission and transport of electrical energy. A distinction is made between main transformers (machine transformers) for transforming the generator voltage to the overhead line high voltage and transformer transformers (distribution transformers) between medium voltage network and consumers (local network). Low power transformers can be found in communications engineering, as transformers in power supplies for the electrical industry and as protective transformers when working with low voltages.

The following relationships are important for the mode of operation of the transformer:

1. *Flooding act*: The law of flow-through describes the relationship between the magnetic field and the electric current generating it. The relationship between induction and current is established via permeability.
   Magnetic induction (flux density):

   $$B = \mu_0 \, H \tag{22.1}$$

   Magnetic field strength:

   $$H = \frac{\Theta}{l} \tag{22.2}$$

   Electrical flow-through:
   In a magnetic field, the line integral over the magnetic field strength H along a self-contained line C is always equal to the total electric current passing through the surface formed by this line.

   $$\oint_C \vec{H} \cdot d\vec{s} = \Theta = I\,N \cong H_1 \cdot l_1 + H_2 \cdot l_2 + \cdot + H_n \cdot l_n \tag{22.3}$$

   In a closed field, the flow rate represents the sum of all currents linked to the induction lines (Figure 22.1).

Analysis and Design of Electrical Power Systems: A Practical Guide and Commentary on NEC and IEC 60364, First Edition. Ismail Kasikci.

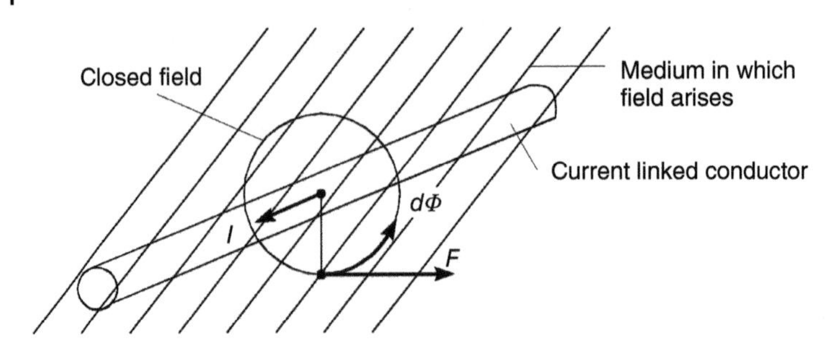

Figure 22.1 Flood law.

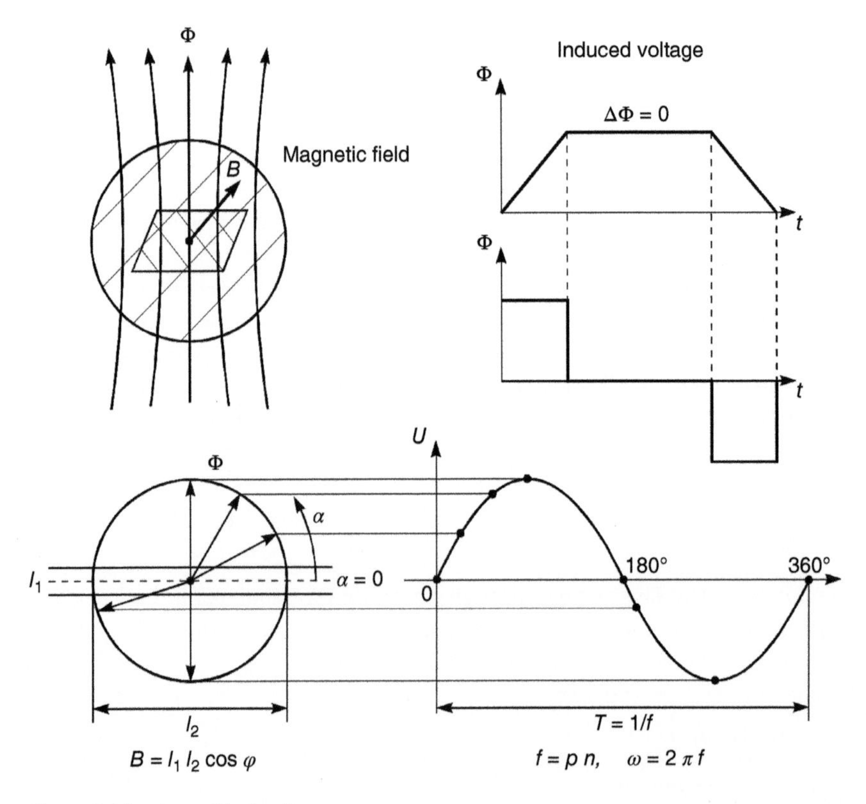

Figure 22.2 Law of induction.

2. *Induction act*: Sinusoidal voltages can be generated by moving windings in magnetic fields or vice versa. If a conductor loop is penetrated by a changing magnetic field, a voltage is induced in this conductor loop (Figure 22.2). The induced voltage is the rate of flux change $\frac{d\Phi}{dt}$ and the number of turns N proportionally:

$$u_0 = -N\,\frac{d\Phi(t)}{dt} = -\frac{d\Psi(t)}{dt} \tag{22.4}$$

The size $N \cdot \Phi(t) = \Psi(t)$ is called flow chaining.

3. *Energy law*: The input power is equal to the output power plus the sum of all losses consisting of active and reactive power losses (Figure 22.3).

$$P_{\mathrm{up}} = P_{\mathrm{down}} + \sum P_{\mathrm{losses}} \tag{22.5}$$

4. *Magnetic scattering*: A part of the river that contributes nothing to the actual purpose, but is interlinked with a winding or even only a part of it, runs outside the iron in the air (Figure 22.4). It causes ohmic voltage drops in the primary winding and heat losses in the iron. The total flow difference causes a magnetic force, which in turn generates flux leakage in the air space between the windings. Potential differences arise within a conductor that result in eddy currents or circular currents.
5. *Transformer principle*: A conductor is moved in the magnetic field or in a fixed coil. In both cases, the magnetic field flow changes, which acts on the conductor (voltage induction).
   If a current changes over time through a coil, a magnetic field is created. A second coil in the vicinity (on a common iron core) is also influenced by this induction (transformer principle, Figure 22.5).

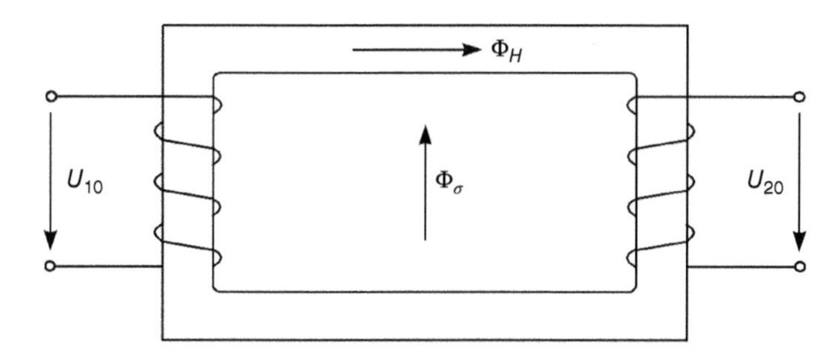

Figure 22.3 Energy law.

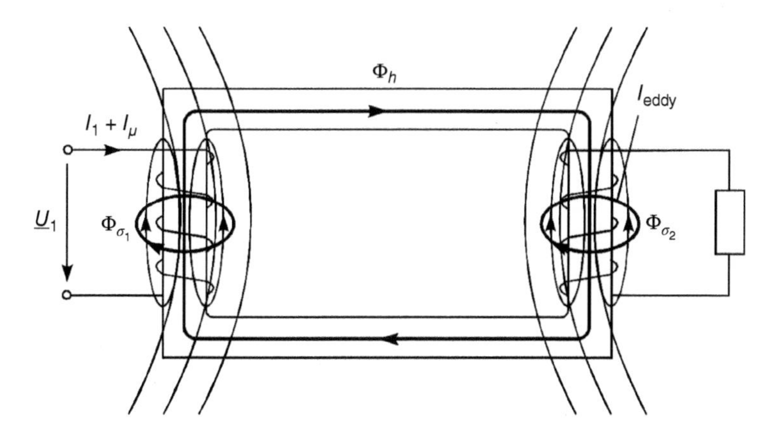

Figure 22.4 Magnetic scattering.

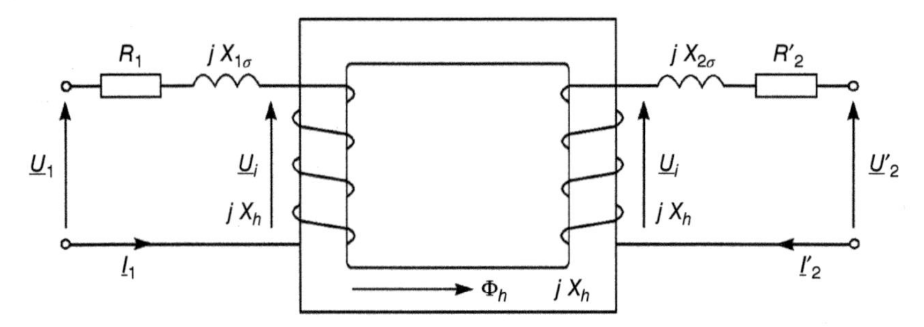

Figure 22.5 Transformer principle.

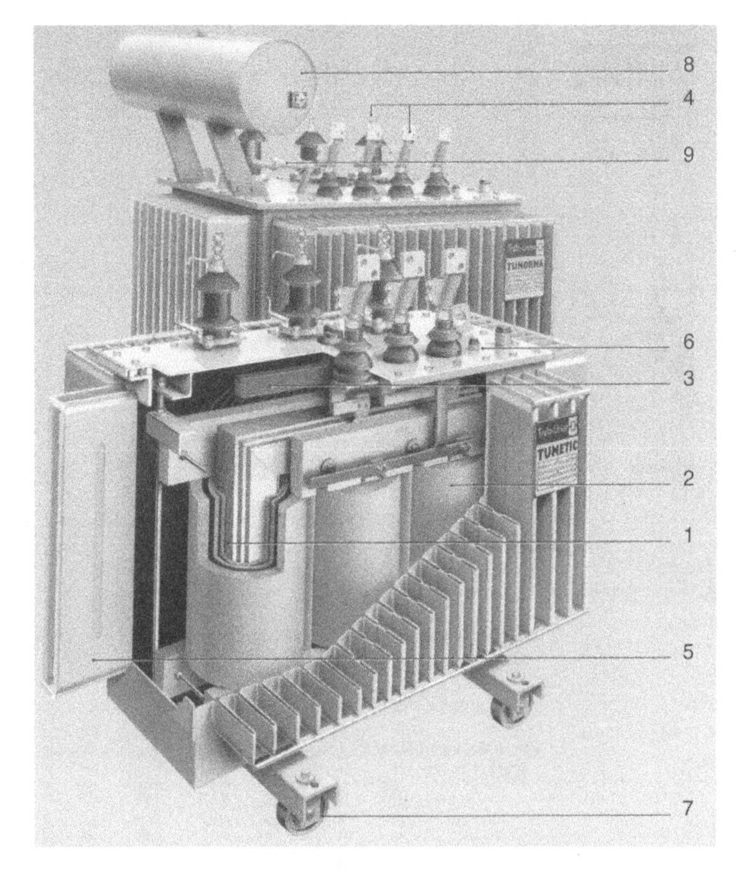

Figure 22.6 Characteristics of an oil transformer. Source: Trafo-Union [67].

It is a deal:

$$u = N_1 \frac{d\Phi}{dt} = N_1 \frac{d}{d} \mathrm{Re}\left[\Phi e^{j\omega t}\right]$$

$$u = \mathrm{Re}\left[j\omega w_1 \; \Phi e^{j\omega t}\right] = \mathrm{Re}\left[\underline{U}\sqrt{2} e^{j\omega t}\right] \tag{22.6}$$

$$\underline{U} = j\frac{\omega}{\sqrt{2}} N_1 \Phi \tag{22.7}$$

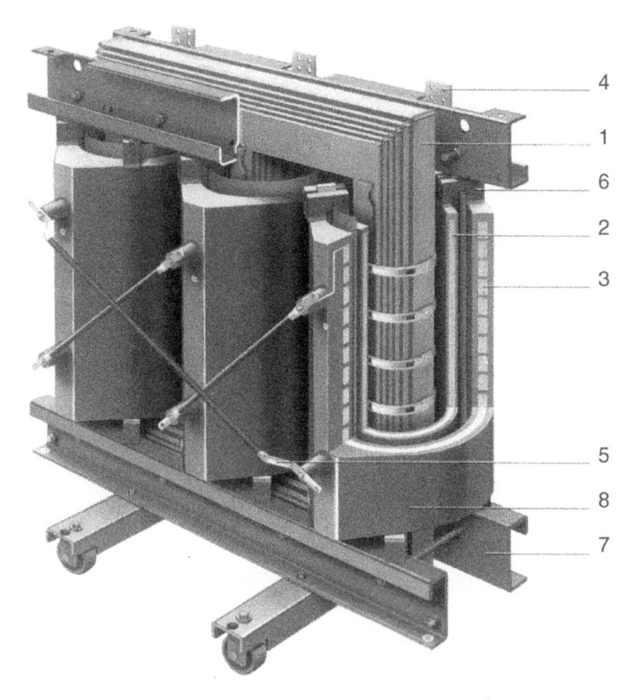

Figure 22.7 Structure and characteristics of a cast resin transformer. Source: Trafo-Union [67].

$$U = |\underline{U}| = \frac{\omega}{\sqrt{2}} N_1 \Phi \tag{22.8}$$

$$U = \frac{2\pi f}{\sqrt{2}} N_1 \Phi = 4.44\, f\, N_1 \Phi \quad [\text{V, Hz, V s}] \tag{22.9}$$

The main components of the distribution transformers (oil and cast resinoutput) from 50 to 2500 kVA are shown in Figures 22.6 and 22.7.

Construction and characteristics of the transformer (according to Figure 22.6): iron core (1), windings (2), diverter (3), feedthroughs (4), boiler (5), boiler cover (6), chassis (7), expansion vessel (8), Buchholz relay (9).

Construction and characteristics of the transformer (according to Figure 22.7): three-leg core (1), US winding (2), OS winding (3), US connections (4), OS connections (5), elastic Spacers (6), press frame and chassis (7), insulation made of epoxy resin/quartz flour mixture (8).

22.2 Core

The core serves the magnetic coupling of the two separate electrical circuits. It is made of cold-rolled, insulated with lacquer, maximum 0.35 mm thick individual sheets (soft material), so that the remagnetization losses remain small.

The use of special additives increases the electrical resistance of the iron in order to keep the eddy current losses to a minimum.

The core must be magnetically as good as possible and electrically as good as possible. bad management. Pressing the core sheets together avoids losses and humming noises. The iron core vibrates (magnetostriction), which is noticeable by humming.

22.3 Winding

It consists of lacquer-insulated copper or aluminum conductors of round or rectangular cross section. The winding is applied to the bobbin with a circular-shaped cross section.

The windings are labeled

1. by energy direction
   - primary winding: the electrical energy, index 1
   - secondary winding: gives the electrical energy off, index 2
2. according to the rated voltage
   - high-voltage winding (OS): winding with highest suspense
   - undervoltage winding (US): winding with lowest voltage.

22.4 Constructions

1. Sheath transformer
   - the iron core surrounds the winding like a sheath
   - has low stray field and low short-circuit current
   - is used as mains, control, and rectifier transformer
   - the coils are spatially separated
2. Core transformer
   - the coil surrounds the legs of the iron core
   - the winding is single layer and has a large circumference
3. Stray field transformer
   - for special gas discharge lamps
   - has large stray magnetic fields

22.5 AC Transformer

22.5.1 Construction

Single phase or AC transformer consists of at least two galvanically isolated windings on an iron core (Figure 22.8). The core is connected with an upper and lower yoke, to form a closed magnetic circuit. The iron core is made of soft iron sheets insulated against each other (small holding of magnetization and eddy current losses).

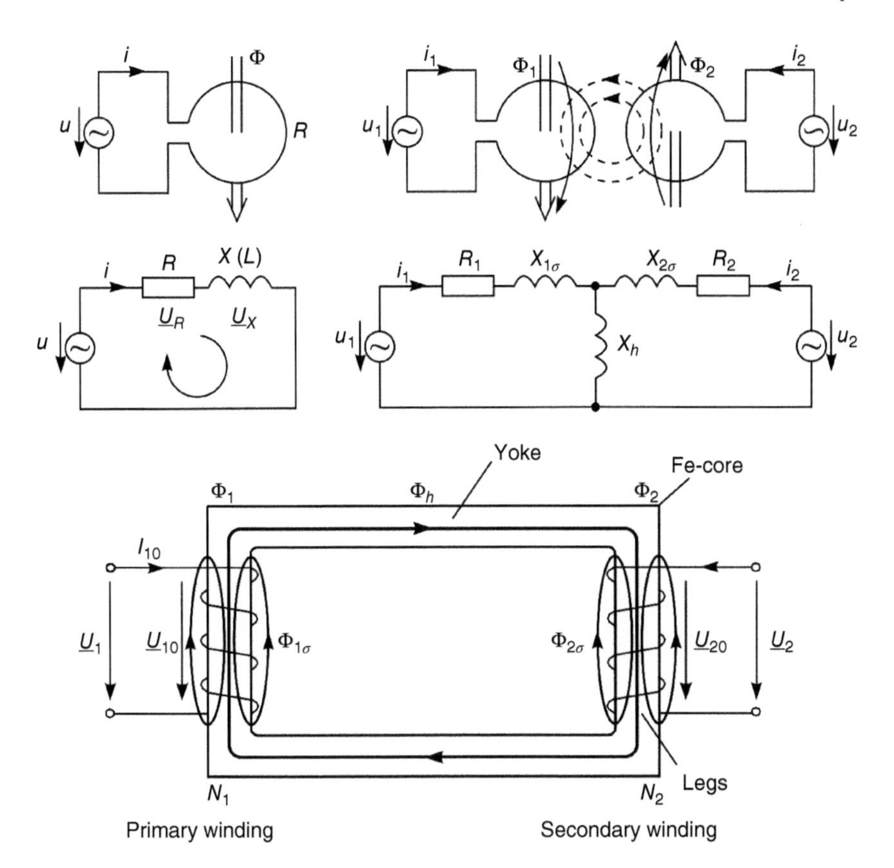

Figure 22.8 AC-transformer.

22.5.2 Mode of Action

The transformer effect is based on the law of induction. An alternating voltage U_1 leaves an alternating current I_1. I_1 causes an alternating field. This alternating magnetic flux induces a voltage in the secondary winding. If a current flow occurs in the secondary winding, e.g. via a resistor, this current generates a magnetic flux that is directed in the opposite direction from the primary coil. The primary coil draws more current from the grid and restores the balance. Thus, the current intensity of the primary coil reacts to the current changes in the secondary coil.

This generates a self-induction in the primary winding and this self-induction in the secondary winding produces an alternating voltage:

$$U_2 = N\frac{d\Phi}{dt} \tag{22.10}$$

The result is then obtained for the primary and secondary induced voltages:

$$U_1 = 4.44 f\, N_1\, \Phi_m \tag{22.11}$$

$$U_2 = 4.44 f\, N_2\, \Phi_m \tag{22.12}$$

22.5.3 Idling Stress

The (U_{20}) is the voltage on the secondary side, when no load is connected. The main transformer equation applies:

$$U_{20} = \Phi\ w\ N \tag{22.13}$$

$$w = 2\ \pi\ f \tag{22.14}$$

22.5.4 Voltage and Current Translation

According to the general law of induction:

Primary coil:

$$U_1 = N_1\ \frac{d\Phi}{dt} \tag{22.15}$$

$$\frac{d\Phi}{dt} = \frac{U_{12}}{N_1} \tag{22.16}$$

Secondary coil:

$$U_2 = N_2\ \frac{d\Phi}{dt} \tag{22.17}$$

$$\frac{d\Phi}{dt} = \frac{U_2}{N_2} \tag{22.18}$$

For a transformer without load (no-load operation), the following behavior occurs in the voltages as well as the number of turns:

$$\frac{U_1}{N_1} = \frac{U_2}{N_2} \tag{22.19}$$

According to DIN VDE 0532, the approximate ratio of voltages is referred to as the transformer ratio:

$$t = \frac{U_1}{U_2} = \frac{N_1}{N_2} \tag{22.20}$$

Except for losses:

$$\frac{U_1}{I_2} = \frac{U_2}{I_1} \tag{22.21}$$

The currents are inversely proportional to the voltages or the number of turns:

$$\frac{U_1}{U_2} = \frac{I_2}{I_1} \tag{22.22}$$

Transformers used for resistance adjustment is called an empedance transmitter:

$$t = \sqrt{\frac{Z_1}{Z_2}} \tag{22.23}$$

22.5.5 Operating Behavior of the Transformer

With the complete equivalent circuit diagram of the single-phase transformer, the following equations can be established in the open-circuit and short-circuit tests (Figure 22.9):

1. *Idle*: At no load, the magnetization power loss (iron losses) measured (Figure 22.10). The no-load current I_0 consists of the magnetizing current I_μ and the active current component I_R (heat losses).

   Idle sizes:
   The no-load current is

   $$I_{10} = \sqrt{I_\mu^2 + I_{\text{Fe}}^2} \tag{22.24}$$

   The following applies to the iron loss current and the ohmic resistance:

   $$I_{\text{Fe}} = \frac{P_{10}}{U_0} \tag{22.25}$$

   The power factor is

   $$\cos\varphi_{10} = \frac{P_{10}}{I_{10} \cdot U_{10}} \tag{22.26}$$

2. *Short-circuit*: The short-circuit voltage u_k is the primary voltage at which a transformer with short-circuited secondary winding already has its own primary current (Figure 22.11). In this test, the short-circuit losses are measured.
   The short-circuit voltage is important for determining
   - of the impedance
   - of the winding power loss
   - of phase shift
   - of the short-circuit current
   - and in the parallel connection of transformers.

   Usually, u_k is given as a related short-circuit voltage as a percentage of the primary voltage. It is a measure of the stress change that occurs under load.

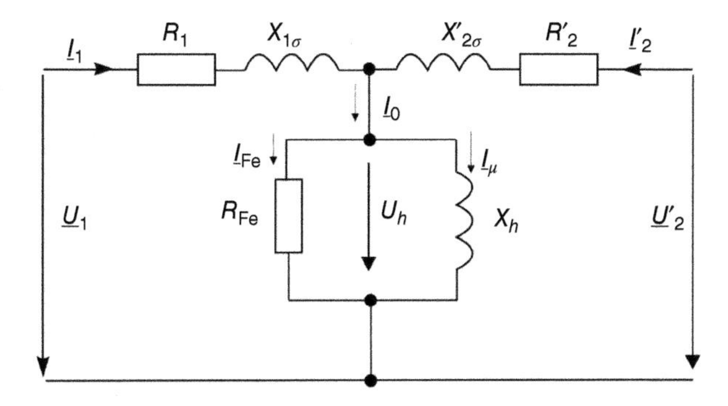

Figure 22.9 Complete equivalent circuit diagram of the transformer.

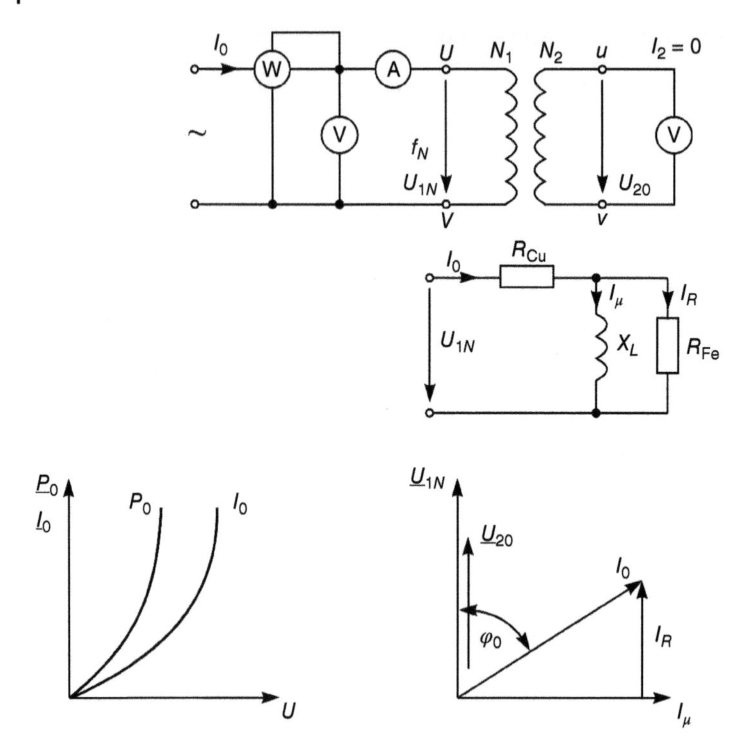

Figure 22.10 Transformer.

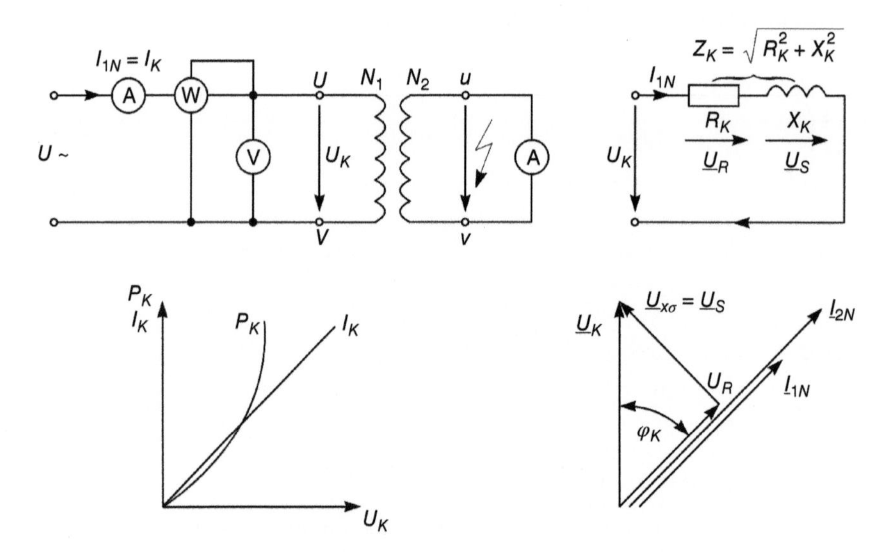

Figure 22.11 Transformer short-circuit.

3. *Short-circuit current*: If a short-circuit occurs on the secondary side of a transformer during operation, the surge short-circuit current i_p can flow first. into the continuous short-circuit current I_k. The amount of i_p depends on the instantaneous voltage value and the magnetic state of the iron core. The worst case, zero crossing of the voltage and saturated iron core, causes the highest surge short-circuit current at the moment the short-circuit occurs. It consists of an alternating current component as a continuous short-circuit current and a direct current component, which results from the collapse of the field and becomes zero after the time $t \approx 5$ seconds. It's a deal:

$$i_p = 2.54\ I_k \tag{22.27}$$

The magnitude of the continuous short-circuit current I_k depends on the short-circuit voltage u_k and the internal resistance Z.

The output voltage depends on:

- of the amount of the charge,
- the magnitude of the relative voltage u_k,
- of phase φ of load current.

Short-circuit variables:

$$\cos \varphi_k = \frac{P_k}{I_k \cdot U_k} \tag{22.28}$$

The total short-circuit impedance is

$$Z_k = \frac{U_k}{I_k} \tag{22.29}$$

The winding resistances and stray resistances amount to:

$$R_1 + R_2' = R_k = Z_k \cdot \cos \varphi_k \tag{22.30}$$

$$X_{1\delta} + X_{2\delta}' = X_k = Z_k \cdot \sin \varphi_k \tag{22.31}$$

An important variable is the short-circuit voltage, which is specified as a percentage, and its value according to DIN 42500 is 4–12%:

$$u_{kr} = \frac{100\% \cdot U_k}{U_n} \tag{22.32}$$

If the percentage of short-circuit voltage is related to the short-circuit current (continuous short-circuit current), the result is

$$I_k = \frac{100\% \cdot I_{\mathrm{rT}}}{u_{kr}} \tag{22.33}$$

Load on the transformer:

At rated load, the load voltage is measured in the output. This terminal voltage depends on the load type and the overload current (Figure 22.12).

Eddy current losses are proportional to the square of the frequency and change quadratically with the magnetic flux density ($P_{\mathrm{vortex}} \sim f^2 \cdot B^2$).

Remagnetization losses are proportional to frequency and change quadratically with the magnetic flux density ($P_{\mathrm{Um}} \sim f \cdot B^2$).

Eddy current losses and remagnetization losses are combined as iron losses.

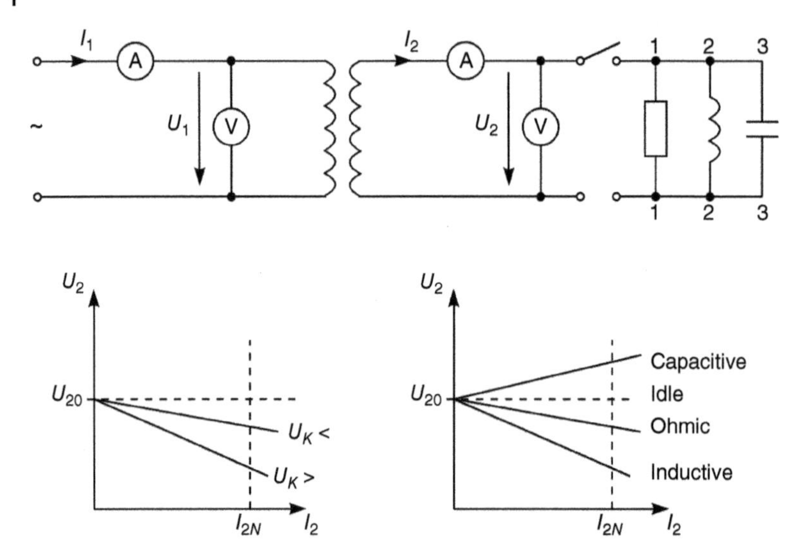

Figure 22.12 Load of the transformer.

22.6 Three-phase Transformer

22.6.1 Construction

The three-phase transformer contains three assembled single-phase transformers, i.e. the voltages are offset by 120° (Figure 22.13). Thus, the sum of the magnetic fluxes in the middle legs is zero at any moment. With the three-phase transformer, u_{kr} is very small, with the LV/HV windings on top of each other on each leg.

22.6.2 Windings

The circuit (Figure 22.14) is the connection of winding strands to a winding. Large letters apply to the high-voltage side (OS) and small letters to the low-voltage side (US).

A distinction is made between three-phase transformers:

1. winding strands in delta connection (D, d)
2. winding strands in star connection (Y, y)
3. winding strands in star/zigzag connection (Y, z)
4. winding strands in open circuit (III, iii).

Note: Zigzag switching requires 15% more material.

22.6.3 Circuit Groups

Vector groups (Figure 22.15) indicate how two windings of a transformer are switched and by which multiple of 30° the pointer of the US follows one of the OS with assigned terminal designation counterclockwise.

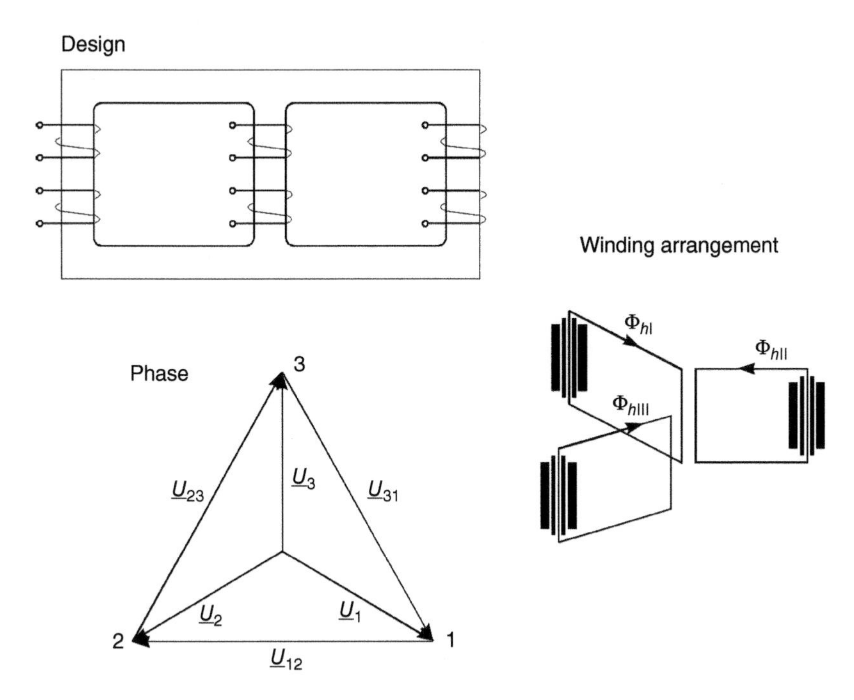

Figure 22.13 Winding arrangement of the three-phase transformer.

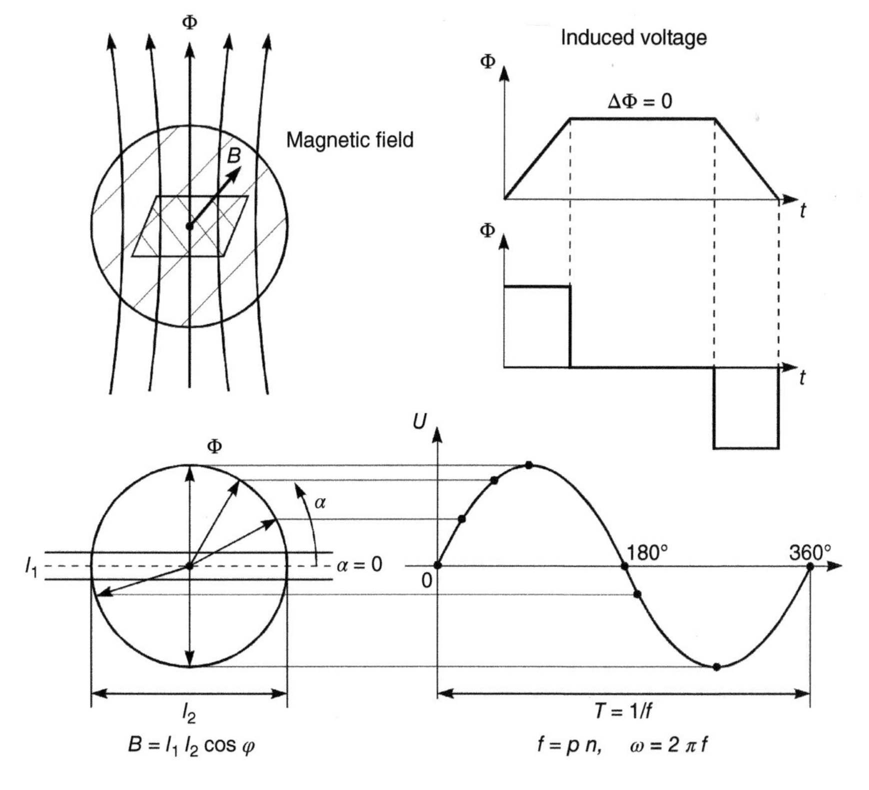

Figure 22.14 Windings circuits.

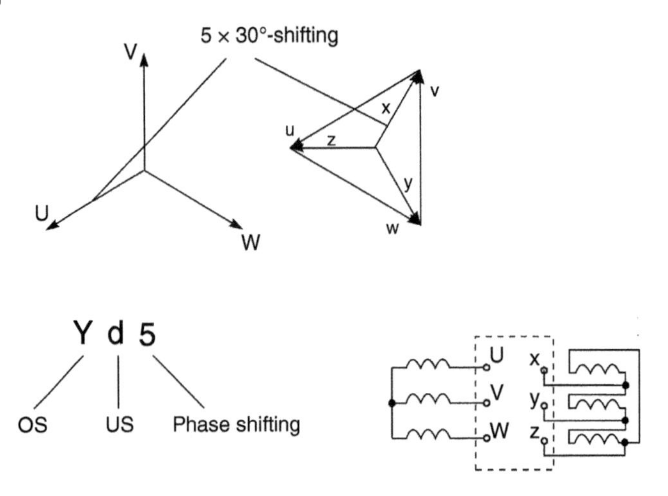

Figure 22.15 Circuit groups.

To be used:

Y, D, z	according to function and manufacture of the transformer
Y	for neutral point and high voltage
D	for decoupling the zero system
z	for low-impedance zero system.

22.6.4 Overview of Vector Groups

The most important switching groups are compiled in DIN VDE 0532. In the power supply, mainly Yy0, Dyn5, Yd5, and Yzn5 are preferred. Each loaded transformer should be in an electrical and magnetic equilibrium (flow compensation). The useful flow of all legs must not be disturbed by additional flow of the load currents. This means that the currents in the output winding must have an equally strong transformer feedback effect on all three strings of the input winding. If the vector group Yyn0 is loaded asymmetrically, there is no flow compensation. The transformer supplies the customer instead of 230 V, e.g. only 150 V. The star point shifts and the neutral conductor have a voltage potential to earth, i.e. Yy circuits are not suitable for single-phase loads. The zigzag circuit provides a remedy. With single-phase loading of the Yzn5 circuit, flow compensation takes place on the legs *U*, *W*, so that no voltage collapse occurs on the secondary side. The operating behavior of both groups is shown in Figure 22.16. With low-voltage networks, the undervoltage side is not designed in a triangle because otherwise the neutral cannot be connected.

Table 22.1 gives an overview of these groups.

22.6.5 Parallel Connection of Transformers

The parallel connection of transformers serves to power increase of an installation (Figure 22.17). To prevent the rated current from being exceeded or the individual transformers from being overloaded, the following conditions must be observed:

Figure 22.16 Symmetrical and asymmetrical load of the transformer.

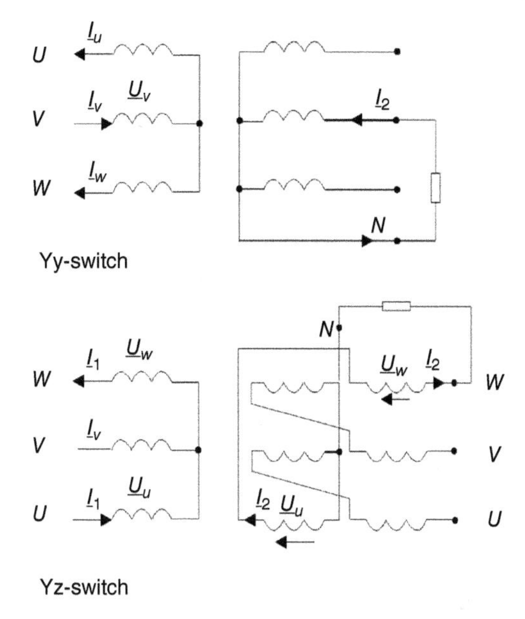

Table 22.1 Overview of vector groups.

Designation	Switch group	Intended use
Triangle star	Dy	For distribution transformers from 315 kVA, star point fully loadable
Star-Star	Yy	For pipelines and distribution transformers low power, star point only loadable up to 10% of I_{rT}
Star-Delta	Yd	For machine transformers and high-power generators
Star zigzag	Yz	For small distribution transformers up to 250 kVA, star point fully loadable

1. same rated frequency and rated voltages (in special cases, this can be deviated from),
2. approximately equal short-circuit voltages (±10% deviation),
3. same number of vector groups,
4. maximum rated power ratio 1 : 3, correct phase connection.

The following equations can be applied when transformers are connected in parallel:

1. With the same short-circuit voltages for the load output of the transformer:

$$S_{L1} = \sum S_{GL} \frac{S_{rT1}}{\sum S_{rT}} \tag{22.34}$$

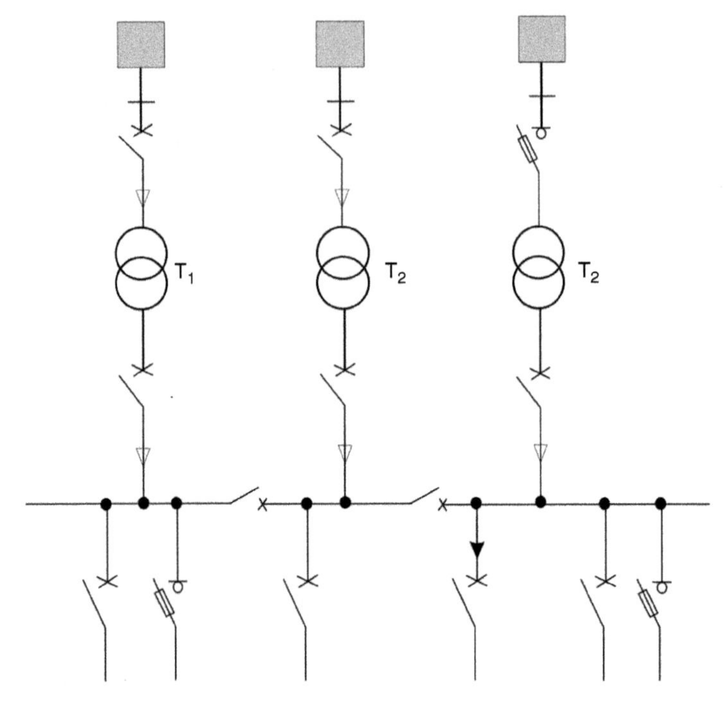

Figure 22.17 Connection of transformers.

2. In case of unequal short-circuit voltages for the load output of the transformer:

$$S_{L1} = S_{rT1} \frac{u_{krm}}{u_{kr1}} \frac{\sum S_{GL}}{\sum S_{rT}} \tag{22.35}$$

For u_{krm}:

$$u_{krm} = \frac{\sum S_{rT}}{\frac{S_{rT1}}{u_{kr1}} + \frac{S_{rT2}}{u_{kr2}} + \cdots} \tag{22.36}$$

It means:

S_{rT1}	Rated power of the first transformer in kW
u_{kr1}	Short-circuit voltage of the first transformer in %
S_{L1}	Load output of the first transformer in kW
S_{rT2}	Rated power of the second transformer in kW
u_{kr2}	Short-circuit voltage of the second transformer in %
S_{L2}	Load output of the second transformer in kW
S_{rT}	Sum of rated outputs in kW
u_{krm}	Average short-circuit voltage in %
t	Transformer transmission ratio
S_{GL}	Total load in kW

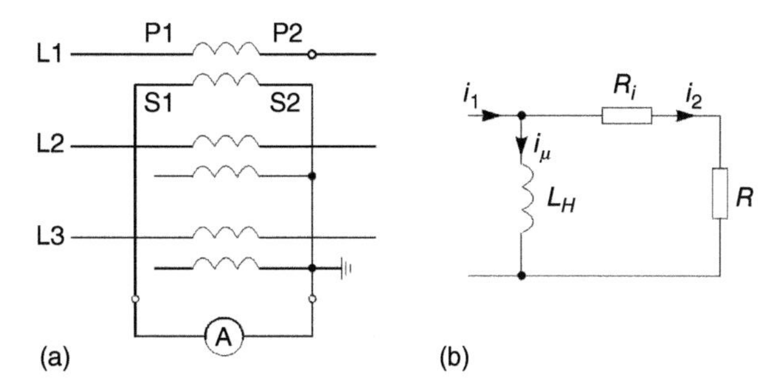

Figure 22.18 Current transformer. (a) Block diagram. (b) Equivalent circuit diagram.

22.7 Transformers for Measuring Purposes

22.7.1 Current Transformers

Current transformers are used for measuring and protection purposes (Figure 22.18). They separate measuring (M) and protective circuit (P) from the primary voltage and protect the devices from overload. Transducers: Class 0.1 M; 1 M, protection transformer: class 5 P; 10 P.

Transformer protection. In the event of overload, they go into saturation and protect the connected measuring instruments.

Measure protection transformers. They transmit high currents without saturation, so that the connected protective relays can reach the short-circuit and excitation currents and switch off the fault according to their tripping values.

The current transformer

- is in line with the network,
- has small power,
- lies in the course of the line on the primary winding,
- is short-circuited on the secondary side via the connected devices, meters, and relays,
- works in short-circuit.

22.7.2 Voltage Transformer

Voltage transformer (Figure 22.19) transforms high voltages to measurable voltages. The rated voltage of the voltage transformer is standardized to 100 V.

- load is called a burden
- works almost idle because the measuring device does not represent a large load; otherwise, there is danger of destruction. Do not short-circuit the
- secondary side
- secondary and primary side
- above 1 kV ground
- standard voltage of your choice : $U_2 = 100\,\text{V}$ or $100/\sqrt{3}\,\text{V}$

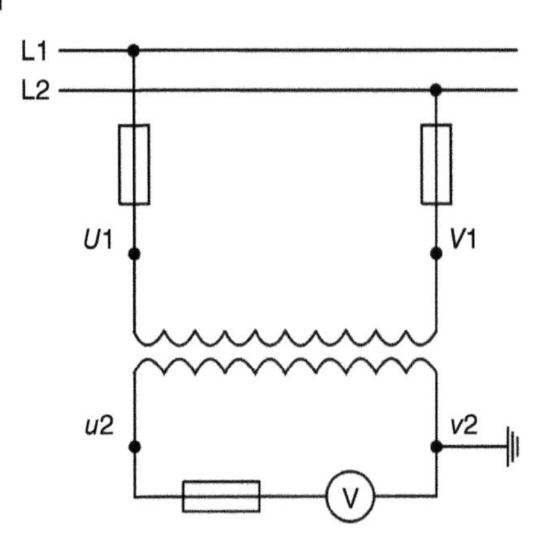

Figure 22.19 Voltage transformer.

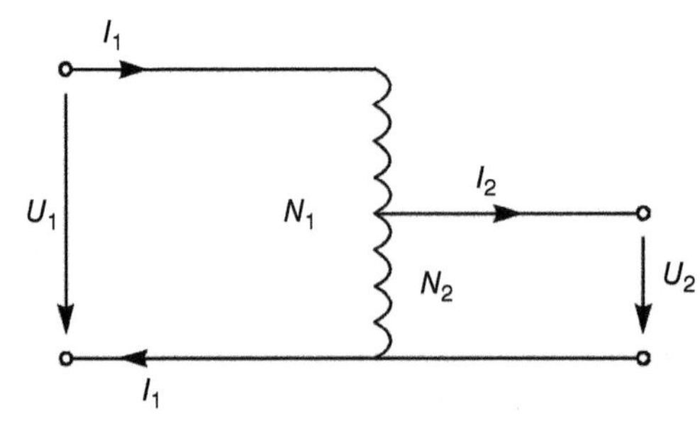

Figure 22.20 Transformer.

Instrum ent transformer: 15 VA Class 0,5 FS 5;
Protection transformer: 15 VA 10P10

Application:

- inductive transducer for HV-level
- capacitive transducer for LV-level

22.7.3 Frequency Transformer

This transformer is an inductive voltage divider and has no galvanic isolation between HV and LV side (Figure 22.20). The through power is transmitted partly conductive, partly inductive.

Application:

- for voltage adjustment of networks
- special transformers for supplying traction motors
- for coupling high-voltage networks
- dry transformers up to 300 kVA
- oil transformers from 300 kVA
- self-cooling (*S*)
- forced cooling
- forced oil circulation.

The size of autotransformers is calculated as follows for known through power:

$$S = P_d \left(1 - \frac{U_{LV}}{U_{HV}}\right) \tag{22.37}$$

22.8 Transformer Efficiency

The magnetic or iron losses consist of the hysteresis and eddy current losses in the iron and dielectric. This losses are independent of the burden. The losses are composed of the current heat losses in the windings. The winding losses increase quadratically with the load. The efficiency of a transformer can be calculated at any load n. It is a deal:

$$\eta = 100\% - \frac{P_0 + n^2\ P_k}{n\ S_{rT}\ \cos\varphi + P_0}\ 100\% \tag{22.38}$$

A transformer has its maximum efficiency at a load to which $P_0 = n^2\ P_k$ applies. This case occurs with a load factor of

$$n = \sqrt{\frac{P_0}{P_k}} \tag{22.39}$$

up. The total losses of a transformer at any load result from:

$$P_v = P_0 + n^2\ P_k \tag{22.40}$$

22.9 Protection of Transformers

Protection against internal faults

1. The Buchholzrelais detects internal damage with gassing or oil flow. There is a message for smaller faults and causes a switch to trip in the event of larger faults.
2. The differential protection compares the input and output currents of the transformer. In the event of a fault (earth, short and short-circuit), it triggers the protective relay. It must be stabilized against the inrush current of the idling transformer.

Protection against overload

1. The thermistor protection signals overload and is used against inadmissible heating of the transformer.
2. The overvoltage protection is achieved by surge arresters.

22.10 Selection of Transformers

The characteristics of the transformer are determined by the requirements of the network. The determined active power is calculated with the power factor $\cos\varphi$ on the rated power S_{rT} to convert. In distribution networks, $u_{kr} = 6\%$ is preferred.

Transformer losses consist of no-load and short-circuit losses. The no-load losses are caused by the constant remagnetization of the iron and are practically constant and load-independent. The short-circuit losses consist of the current heat losses in the windings and losses due to stray fields. They change squarely with the load. Oil transformers and ascarel are preferred; ascare transformers are prohibited. This section discusses the key criteria for selecting distribution transformers in the 502500 kVA power range to supply low-voltage networks.

1. Demand for operational safety
   - Routine tests (losses, u_{kr}, voltage test)
   - Type tests (heating, surge voltage)
   - Special tests (short-circuit resistance, noise)
2. Electrical conditions
   - Short-circuit voltage
   - Vector group
   - Translation
3. Installation conditions
   - Indoor or outdoor installation
   - Special local conditions
   - Environmental protection conditions
   - DIN VDE 0101 or DIN VDE 0100-710, DIN VDE 0108-100
   - Designs: Oil or cast resin dry transformer
4. Operating conditions
   - Load capacity according to DIN 42500 (oil) or DIN 42 523 (dry transformer)
   - Load fluctuations according to DIN VDE 0532 part 10
   - Number of operating hours
   - Efficiency according to DIN 42500 (oil) or DIN 42 523 (dry transformer)
   - Voltage change
   - Parallel operation
5. Characteristics of a transformer with examples
   - Rated power $S_{\mathrm{rT}} = 1000$ kVA
   - Rated voltage $U_{\mathrm{rHV}} = 20$ kV
   - Undervoltage $U_{\mathrm{rLV}} = 0.4$ kV
   - Rated withstand voltage $U_{\mathrm{rB}} = 125$ kV
   - Loss combination according to DIN 42500 T1 B-A′
   - No-load losses $P_0 = 1700$ W
   - Short-circuit losses $P_{\mathrm{k}} = 13\,000$ W
   - Sound power $L_{\mathrm{WA}} = 73$ dB
   - Short-circuit voltage $u_{\mathrm{kr}} = 6\%$
   - Nominal ratio with tapping OS/US = 20 kV ± 2 – 2.5%/0.4 kV
   - Dyn5 vector group
   - Connection technologies, e.g. OS and US-side flange technologies indoor or outdoor installation
   - Compliance with DIN VDE 0532, DIN EN 60076, DIN 42 500, IEC 60076

22.11 Calculation of a Continuous Short-Circuit Current on the NS Side of a Transformer

In low-voltage radiant networks, approximate methods can be used to calculate the short-circuit currents for the transformers. It generally applies with the rated transformer current:

$$I_{rT} \approx k\ S_{rT} \tag{22.41}$$

The factor $k\ (\frac{1}{10^3 V})$ is

at 400 V   1.44

at 525 V   1.1

at 690 V   0.84

Transformer initial short-circuit alternating current (continuous short-circuit current):

$$I''_k \approx I_k = \frac{100\%}{u_{kr}} \cdot I_{rT} \tag{22.42}$$

Figure 22.21 shows the rated data and electrical characteristics of the oil transformers [54, 69].

	Efficiency class 1		Efficiency class 2	
Rated power kVA	Load losses W	Idle losses W	Load losses W	Idle losses W
25	900	70	600	63
50	1100	90	750	81
100	1750	145	1250	130
160	2350	210	1750	189
250	3250	300	2350	270
315	3900	360	2800	324
400	4600	430	3250	387
500	5500	510	3900	459
630	6500	600	4600	540
800	8400	650	6000	585
1000	10500	770	7600	693
1250	11000	950	9500	855
1600	14000	1200	12000	1080
2000	18000	1450	15000	1305
2500	22000	1750	18500	1575
3150	27500	2200	23000	1980

Figure 22.21 Rating data of transformers.

	Efficiency class 1		Efficiency class 2	
Rated power kVA	Load losses W	Idle losses W	Load losses W	Idle losses W
50	1700	200	1500	180
100	2050	280	1800	252
160	2900	400	2600	360
250	3800	520	3400	468
400	5500	750	4500	675
630	7600	1100	7100	990
800	8000	1300	8000	1170
1000	9000	1550	9000	1395
1250	11000	1800	11000	1620
1600	13000	2200	13000	1980
2000	16000	2600	16000	2340
2500	19000	3100	19000	2790
3150	22000	3800	22000	3420

Figure 22.22 Rating data of transformers.

Figure 22.22 shows the rated data and electrical characteristics of the GEAFOL transformers [54, 69].

22.12 Examples of Transformers

22.12.1 Example 1: Calculation of the Continuous Short-Circuit Current

The following data were given for a transformer:

$S_{rT} = 630$ kVA, $u_{kr} = 6\%$, $U_{rT} = 400$ V, $k = 1.45$

Rated transformer current:

$$I_{rT} \approx k\ S_{rT} = 1.45 \frac{1}{1000\ \text{V}} \cdot 630\ \text{kVA} = 913.5\ \text{A}$$

This means that the transformer's initial short-circuit current is alternating:

$$I_k'' \approx I_k = \frac{100\%}{u_{kr}} \cdot I_{rT} = \frac{100\%}{6\%} \cdot 913.5\ \text{A} = 15.225\ \text{kA}$$

22.12.2 Example: Calculation of a Three-phase Transformer

A three-phase transformer is given with the following data.

Apparent power $S_{rT} = 160$ kVA
Voltage levels 20 kV/400 V
Frequency $f = 50$ Hz
Switch group Yyn6
Cross section $A = 130\ \text{mm}^2$

Flux density $B = 1.6\text{T}$

Perform the following calculations:

1. Draw the all-pole circuit!
2. Determine the primary and secondary winding numbers!
3. Determine the primary and secondary current, primary and secondary resistance, and the current densities with $A_{\text{Cu1}} = 1.6\ \text{mm}^2$, and $A_{\text{Cu2}} = 79\ \text{mm}^2$!
4. Determine the remagnetization losses with a loss figure $v_{10} = 0.45\ \text{W/kg}$ and $\rho = 7.9\ \text{kg/dm}^3$. Note that v indicates the losses at 50 Hz, $B = 1\text{T}$ per kg iron mass.
5. Determine the primary and secondary leakage inductance!
6. Specify the T-replacement circuit diagram with the converted parameters!

Question 1:

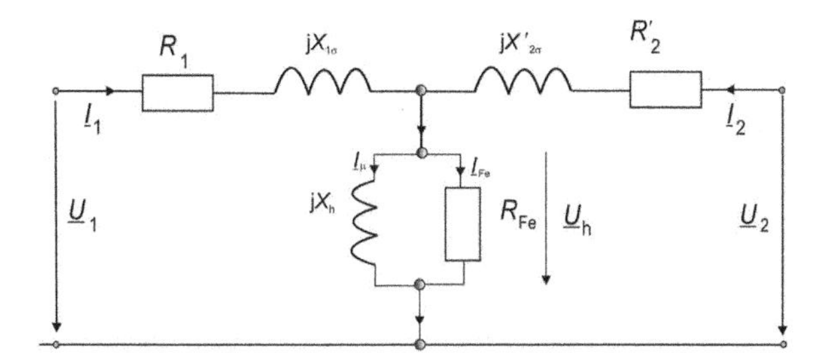

Question 2:

$$t = \frac{U_1}{U_2} = 50$$

Question 3:

$$\Phi = A \cdot B = 0.013\ \text{m}^2 \cdot 1.6D = 20.8\ \text{mWb}$$

Main field voltage: Y-side:

$$U_1 = \frac{U_{1\text{V}}}{\sqrt{3}}$$

$$U_h = U_1 = \omega \cdot N_1 \cdot \Phi \cdot f$$

$$N_1 = \frac{U_1}{\sqrt{2} \cdot 2\pi \cdot f \cdot N_1} == \frac{11\ 550\ \text{V}}{\sqrt{2} \cdot 2\pi \cdot 50\ \text{Hz} \cdot 20.8\ \text{mWb}} = 2500$$

Main field voltage: y-side:

$$U_2 = \frac{U_{2\text{V}}}{\sqrt{3}}$$

$$U_2 = \sqrt{2\pi} \cdot f \cdot N_2 \cdot \Phi \cdot f$$

$$N_2 = \frac{U_2}{\sqrt{2} \cdot 2\pi \cdot f \cdot N_2} = \frac{231\ \text{V}}{\sqrt{2} \cdot 2\pi \cdot 50\ \text{Hz} \cdot 20.8\ \text{mWb}} = 50$$

Question 4:

Apparent power of the transformer:

$$S_{\text{rT}} = \sqrt{3} \cdot U_{1N} \cdot I_{1N} = 3 \cdot U_{2N} \cdot I_{2N}$$

Primary current of the transformer:

$$I_{(1\text{rT}=1N)} = \frac{S_{\text{rT}}}{\sqrt{3} \cdot U_{1N}} = \frac{160\ \text{kVA}}{\sqrt{3} \cdot 20\ \text{kV}} = 4.62\ \text{A}$$

Secondary current of the transformer:

$$I_{(2\text{rT}=2N)} = \frac{S_{\text{rT}}}{\sqrt{3} \cdot U_{2N}} = \frac{160\ \text{kVA}}{\sqrt{3} \cdot 400\ \text{V}} = 231\ \text{A}$$

Current density of the transformer in general:

$$J = \frac{I}{A_{\text{Cu}}}$$

Primary and secondary current density of the transformer:

$$J_1 = \frac{I_{1N}}{A_{\text{Cu1}}} = \frac{4.62\ \text{A}}{1.6\ \text{mm}^2} = 2.89\ \text{A/mm}^2$$

$$J_2 = \frac{I_{2N}}{A_{\text{Cu2}}} = \frac{231\ \text{A}}{79\ \text{mm}^2} = 2.92\ \text{A/mm}^2$$

Calculation of the primary and secondary winding resistance of the transformer:

Average coil length:

$$l_m = d_m \cdot \pi$$

High-voltage side:

$$l_{m,\text{OS}} = \frac{(257 + 203)\text{mm}}{2} \cdot \pi = 723\ \text{mm}$$

Undervoltage side:

$$l_{m,\text{US}} = \frac{(173 + 148)\text{mm}}{2} \cdot \pi = 504\ \text{mm}$$

Resistors:

$$R_1 = \frac{l_{mm,\text{OS} \cdot N_1}}{\kappa_{\text{Cu}} \cdot A_{\text{Cu1}}}$$

$$R_1 = \frac{0.723 \cdot 2500}{57 \cdot 1.6} = 19.82\ \Omega$$

$$R_2 = \frac{0.504 \cdot 50}{57 \cdot 79} = 0.0056\ \Omega$$

Question 5:

Iron volume:

$$V = A_k \cdot (3 \cdot L_K + 2 \cdot L_J) = 0.013 \cdot (3 \cdot 0.48 + 2 \cdot 0.72)\text{m}^3 = 0.0374\ \text{m}^3$$

Iron mass:

$$m = \rho \cdot V = 7900\ \text{kg/m}^3 \cdot 0.034\ \text{m}^3 = 296\ \text{kg}$$

Iron losses:

$$P_{\text{Fe}} = \left(\frac{B_K}{1\text{T}}\right)^2 \cdot v_{10} \cdot m = (\frac{1.6\text{T}}{1\text{T}})^2 \cdot 0.45\ \text{W/kg} \cdot 296\ \text{kg} = 340\ \text{W}$$

Question 6:

The equivalent circuit diagram links the primary variables with the secondary variables.

Iron resistance:

$$R_{\text{Fe}} = \frac{3 \cdot U_h^2}{P_{\text{Fe}}} = \frac{3 \cdot (11\ 547\ \text{V})^2}{340\ \text{W}} = 1176\ \text{k}\Omega$$

Line resistance:

$$R_2' = t^2 \cdot R_2 = 50^2 \cdot 0.0056\ \Omega = 14\ \Omega$$

The secondary quantities must still be converted to the primary quantities using the transmission ratio.

Secondary voltage at idle:

$$U_2' = t \cdot U_2 = 50 \cdot 231\ \text{V} = 11\ 55\ \text{kV}$$

Secondary current at idle:

$$I_2' = \frac{1}{t} \cdot I_2 = 0.02 \cdot 231\ \text{A} = 4.62\ \text{A}$$

Nominal impedance of the transformer:

$$Z_N = \frac{U_{1N}}{I_{1N}} = \frac{U_N}{\sqrt{3} \cdot I_N} = \frac{11\ 550\ \text{V}}{4.6\ \text{A}} = 2510\ \Omega$$

Related string resistors:

$$r_1 = \frac{R_1}{Z_N} = \frac{19.8}{2510} = 0.18\%$$

$$r_2' = \frac{R_2'}{Z_N} = \frac{14}{2510} = 0.6\%$$

$$r_1 + r_2' = 1.4 = 1.4\%$$

Related reactances:

$$x'_{1\delta} = \frac{X'_{1\delta}}{Z_N} = \frac{19.8}{2510} = 2\%$$

$$x'_{2\delta} = \frac{X'_{2\delta}}{Z_N} = \frac{50}{2510} = 2\%$$

$$x_{1\delta} + x'_{2\delta} = 4\%$$

Related iron resistance and main reactance:

$$x_{\mathrm{Fe}} = \frac{X_{\mathrm{Fe}}}{Z_N} = \frac{1176 \cdot 10^3}{2510} = 46\ 853\%$$

$$x_h = \frac{X_h}{Z_N} = \frac{2883 \cdot 10^3}{2510} = 114\ 860.56\%$$

x_{Fe} and x_h can be neglected except during idle. This results in a simplified specify equivalent circuit diagram for rated operation.

23

Asynchronous Motors

Asynchronous motors (ASM) require little maintenance, are simple and sturdy in their construction, and are the most widely used motors. In this chapter, only this type of motor will be dealt with. For other types of motors, the books [65, 66]. ASM have a fixed field winding and are connected to a three-phase power line. The armature (rotor) is arranged according to this. The rotating field revolves spatially with the rated frequency. The rotor moves asynchronously that is not simultaneously with the speed of rotation of the stator field. Instead, the speed of rotation of the rotor trails the speed of rotation of the rotating stator field due to the slip.

23.1 Designs and Types

We must distinguish between squirrel cage (short-circuit) and slipring rotors. For squirrel cage rotors (Figure 23.1), the bars of the cage are in slots in the laminated core and are connected to each other on the front side through short-circuit rings. The winding of the rotor is internally short circuited and not connected to an external network. The transmission of power takes place only through the air-gap field (induction). We can therefore compare an asynchronous machine with a transformer, in which the primary winding (stator winding) is fixed and the secondary winding (rotor winding) rotates. Since by contrast with a transformer, there is an air gap between the stator and the rotor windings, a greater current is required to build up the field in an ASM. This magnetization current is a purely reactive current and must be supplied from the network in both motor and generator operation. The bars are made of copper or aluminum. The stator consists of a housing, laminated core, and winding. The beginnings and ends of the coils are led to the connecting terminal plate. Due to eddy current losses, the laminated cores are isolated and laminated on one side. For slipring rotors, the rotor leads are connected as a normal winding and to each other through two sliprings with resistances. During startup a starting resistor is used which is connected through the sliprings to the rotor circuit. During startup, the individual resistors are switched off stepwise with increasing motor speed. The rotor winding is then short circuited, and the motor functions as a squirrel cage rotor. Due to their more complicated construction and to the starting

Analysis and Design of Electrical Power Systems: A Practical Guide and Commentary on NEC and IEC 60364, First Edition. Ismail Kasikci.

Figure 23.1 Overview of an ASM (stator and rotor). Source: Asea Brown Boveri [56].
1 Motor connection, 2 windings and insulation, 3 heating and ventilation, 4 bearings and lubrication, 5 shaft and rotor, 6 balance and vibration quantity, 7 colors and paint finish, 8 types of construction, 9 rating plates and extra rating plates.

resistors, slipring motors are more expensive and require more maintenance than squirrel cage motors. If direct switch-on is not permissible and the breakaway torque of the squirrel cage motor is too low for the star-delta startup, the use of a slipring motor must be considered.

In industry and trade, ASM are used to drive conveyor systems, textile machines, pumps, and fans. In households, they are used, e.g. as single-phase motors in fan-forced heaters and washing machines.

The equivalent circuit diagram of the ASM is similar to the equivalent circuit diagram of the transformer. The only difference is that the air gap in the ASM causes slippage. At standstill frequencies in the stator and in the rotor are equal, $f_1 = f_2$ and the slip $s = 1$. During operation different frequencies occur, $f_1 \neq f_2$ and $s \neq 1$. Figure 23.2 shows the equivalent circuit of the ASM, where the slip-dependent rotor resistance is shown [65].

23.1.1 Principle of Operation (No-Load)

Connecting the stator winding to the three-phase network results in a rotating magnetic field (rotating field), which induces a voltage in the rotor (armature) given by

$$U_{02} = B\, l\, v \tag{23.1}$$

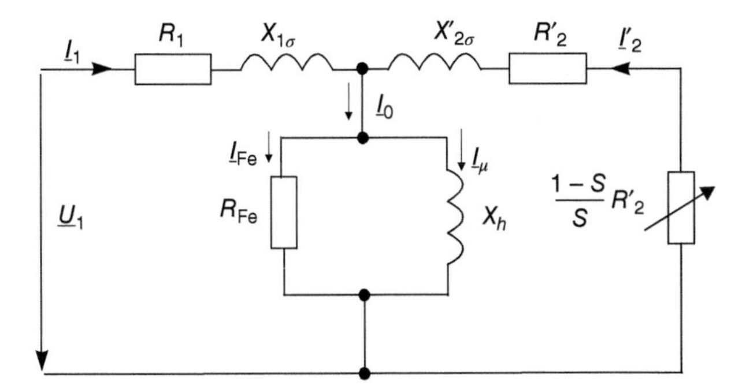

Figure 23.2 Equivalent circuit diagram of the asynchronous machine.

This induction (with the armature circuit closed) causes a current to flow, which in turn results in a magnetic field in the rotor:

$$I_2 = \frac{U_2}{Z_2} \tag{23.2}$$

This current produces a force of deflection F_2 in the direction of the rotating field.

The magnetic poles of the rotor are repelled by the same poles of the stator rotating field and attracted by the opposite poles. The rotor is thus pulled along by the stator.

When $n_2 = n_1$, there is no change in the flux and therefore no torque.

A slip is therefore required for induction:

$$F_2 = B\,l\,I_2 \tag{23.3}$$

This force F_2 produces a torque, given by

$$M = F_2\,r \tag{23.4}$$

23.1.1.1 Motor Behavior

Applying a load to the motor, i.e. braking the shaft, increases the relative speed between the stator rotating field and the rotor. The voltage induced in the rotor then becomes greater, so that the current increases and with it the torque (Figure 23.3).

23.1.1.2 Generator Behavior

Driving the rotor externally reaches a speed of rotation at which all losses are mechanically compensated. The value $\cos\varphi$ is then zero. Purely reactive current then flows in the stator winding. When $\cos\varphi > 90°$, the machine then functions as a generator (Figure 23.3).

23.1.2 Typical Speed–Torque Characteristics

The torque of a motor describes the turning capacity of the rotor. During startup, the torque of the motor first drops, but then increases to a maximum value known as the breakdown torque. This is reached at 85–95% of the full speed of rotation (Figure 23.4).

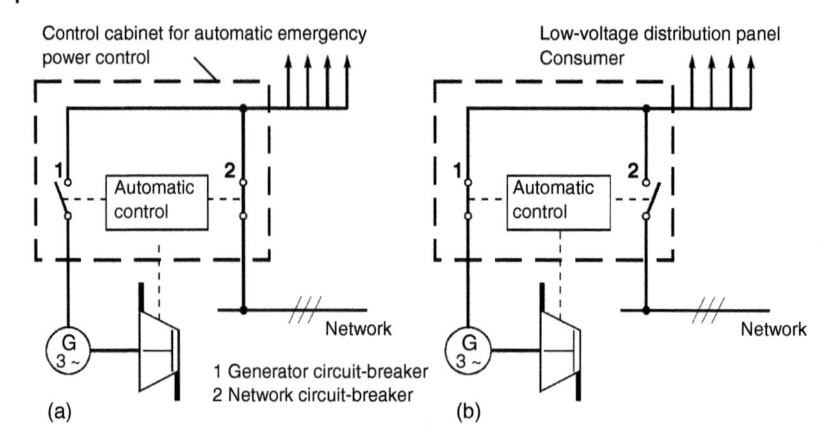

Figure 23.3 Motor and generator operation. (a) Network operation. (b) Emergency operation.

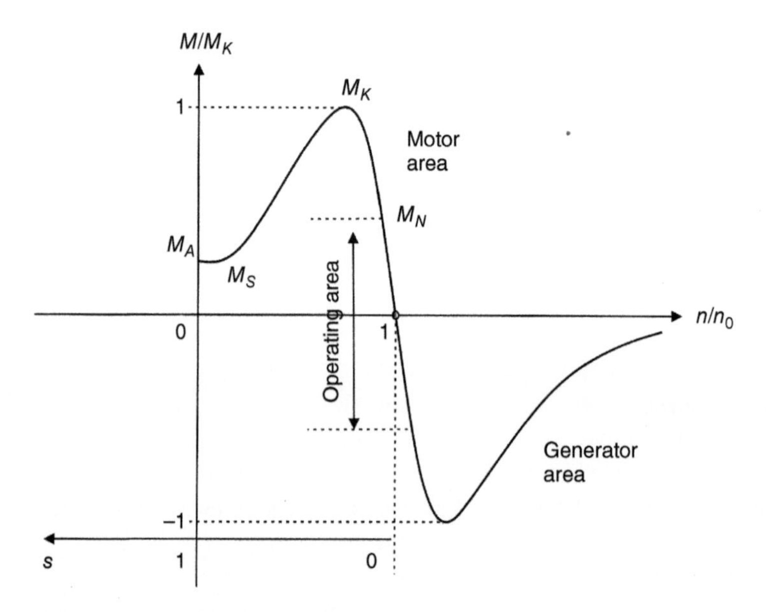

Figure 23.4 Speed–torque characteristics.

The meanings of the symbols are

M_A sl The startup torque (accelerating torque, breakaway torque) is the torque supplied by the motor at rest.

M_S sl The pull-up torque is the smallest torque supplied during startup. In accordance with IEC Publication 34-1, the pull-up torque for motors with a rated power < 100 kW must not be less than 50% of the rated torque and not less than 50% of the breakaway torque. For a rated power > 100 kW, these figures are 30% and 50%, respectively. For single-phase motors and three-phase motors with several speeds, the figure is 30% of the torque.

M_K The breakdown torque is the largest torque which the motor can supply. If this value is exceeded, the motor will stall. It is therefore a measure for the overload capacity of the motor. It must be at least 1.6 M_N and must run 15 seconds long without the motor stalling or the speed of rotation suddenly changing.

M_N The rated torque is the torque supplied by the motor at the rated load and rated speed of rotation.

23.2 Properties Characterizing Asynchronous Motors

23.2.1 Rotor Frequency

Rotor frequency can be given as

$$f_2 = s\, f_1 \tag{23.5}$$

Here:

f_1 Stator frequency
f_2 Rotor frequency
s Slip

23.2.2 Torque

The torque of a motor expresses the turning capacity of the rotor. If the power and speed of rotation are known, the torque can be easily calculated. Along the perimeter of a belt pulley, a certain belt force is present. The product of the force F and the radius of the of the belt pulley r is the torque of the motor M. The power is the work which the motor performs per unit of time, and the work is the product of the force times the distance. The force F thus rotates through n revolutions during one minute and covers a distance (n 2 p r). The rated torque is given by the following relationship:

$$M_N = \frac{9550\, P_n}{n} \tag{23.6}$$

$$M \geq B\, I_2\, \cos\varphi_2 \tag{23.7}$$

$$\cos\varphi = \frac{I_2}{U_2} \tag{23.8}$$

The meanings of the symbols are

M_N Rated torque in Nm
P_n Rated power in W
n Rated speed in min^{-1}

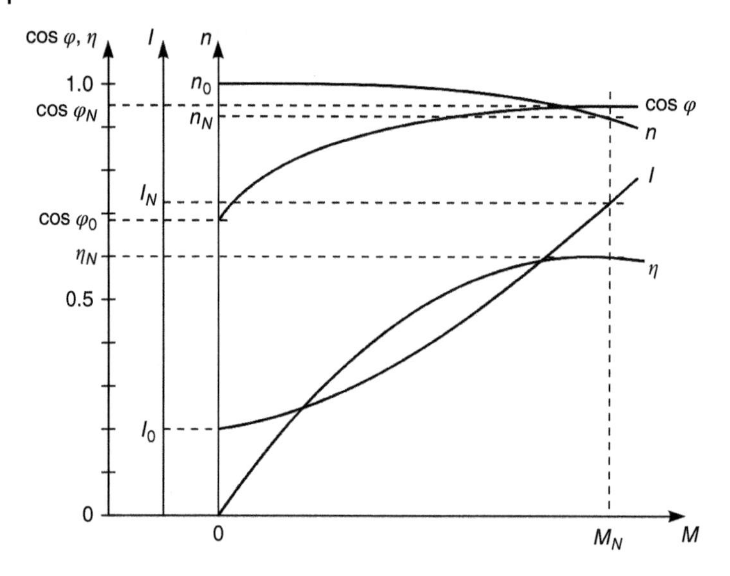

Figure 23.5 Load characteristics.

23.2.3 Slip

The slip is proportional to the load and inversely proportional to the square of the voltage. Figure 23.5 shows the load characteristics. These curves give information about the torque, current draw, and startup of a motor. The speed of rotation changes very little during loading and shows the same behavior as for a DC shunt-wound motor.

$$s = \frac{n_1 - n_2}{n_1} \cdot 100\% \tag{23.9}$$

$$n_2 = n_1\ (1 - s) \tag{23.10}$$

$$n_s = n_1 - n_2 \tag{23.11}$$

23.2.4 Gear System

With a gear system between the motor and the load, the moment of inertia must be recalculated to the motor speed in order to determine the moment of inertia acting on the motor shaft. Here

$$J_L(M) = J_L \left(\frac{n_L}{n_M} \right)^2 \tag{23.12}$$

The load torque characteristic must be recalculated in proportion with the transmission ratio of the gear system and its efficiency. For the load torque:

$$M_L(M) = M_L \cdot \frac{n_L}{n_M} \cdot \frac{1}{\eta} \tag{23.13}$$

When the flywheel moment is known, the mass moment of inertia is calculated from the following relationship:

$$J_L = \frac{\mathrm{GD}^2}{4} \tag{23.14}$$

The meanings of the symbols are

M_L	Load torque in Nm
n_M	Motor speed in min^{-1}
n_L	Load speed in min^{-1}
η	Efficiency of gear system
J_L	Mass moment of inertia of load in kg m^2
$M_L(M)$	Load torque relative to motor shaft in Nm
GD^2	Flywheel moment in kg m^2

23.3 Startup of Asynchronous Motors

The manually operated motor starter and contactors are combined with the overcurrent protection equipment. There are several designs for the combination of contactor and overload protection. As motor protection, motor starters are used for the direct switch-on of squirrel cage motors. Reversing switches are used for both directions. Pole changing switches are provided when there is more than one speed of rotation. The star-delta motor starter is the most common motor starter for limiting the breakaway starting current. The overload protection is in series with the stator winding and must therefore be set to the phase current having the value of the rated motor current divided by $\sqrt{3}$. Short-circuit protection for motors is provided by fuses. Alternatively, motor starting circuit-breakers with thermal and magnetic trip elements provide overload and short-circuit protection.

In accordance with the technical conditions for connection, AC motors up to 1.4 kW and three-phase motors up to $I_a < 60$ A or 8 $I_n < 60$ A can be connected directly.

23.3.1 Direct Switch-On

Direct switch-on requires only a single switch and three terminals. In accordance with the technical conditions for connection of the AC motors up to 1.4 kW and three-phase motors up to 4 kW or with a starting current of 60 V can be connected directly. In general, for the starting torque and the starting current:

$$M_A = (1.5 \cdots 2.8)\ M_N \tag{23.15}$$

$$I_A = (4 \cdots 8)\ I_{\mathrm{rM}} \tag{23.16}$$

The reversing of the direction of rotation of a three-phase asynchronous motor is shown in Figure 23.6. Operationally ready circuit diagrams and data for reversing starters, which are mechanically locked and completely wired, can be taken from the respective manufacturer's information.

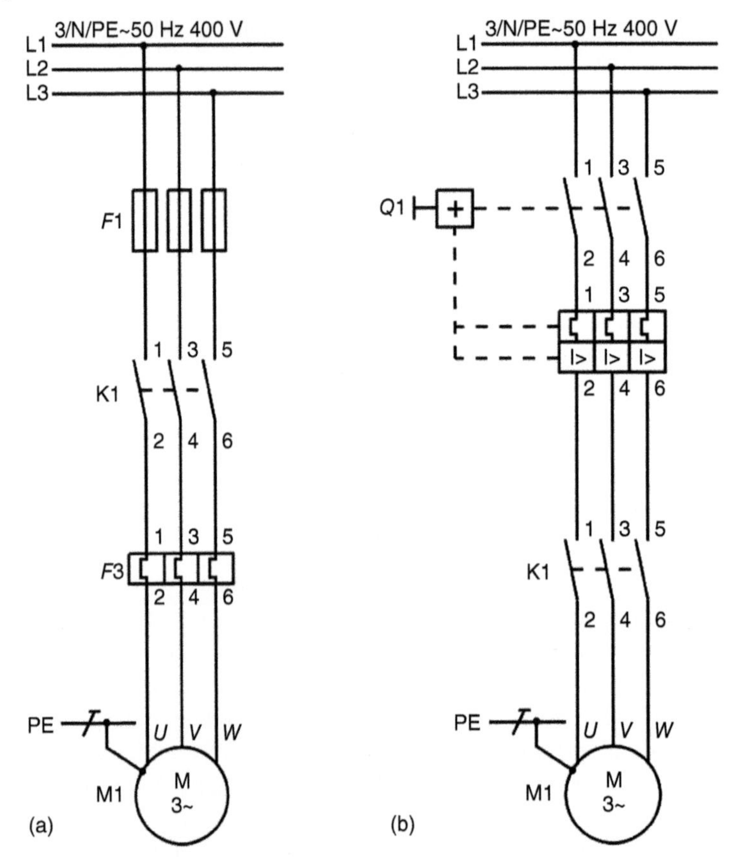

Figure 23.6 Direct switch-on. (a) Fuses with motor protective relay. (b) Fuseless without motor protective relay.

For switching three-phase single-winding ASM, the use of the Dahlander pole-changing switch represents an improvement (Figure 23.7). Pole changing is achieved by switching over and reversing the direction of current in the coil groups. The most important of these circuits are the:

- Delta/star-star for drives with constant torque power ratio: $P_1/P_2 = 1 : 1.4$
- Star-star/delta for drives with constant power power ratio: $P_1/P_2 = 1 : 1$
- Star/star-star for drives with square law torque power ratio: $P_1/P_2 = 1 : 4$ bis $1:8$

23.3.2 Star Delta Startup

Figure 23.8 illustrates typical current and torque Characteristics, and Figure 23.9 depicts the star delta startup.

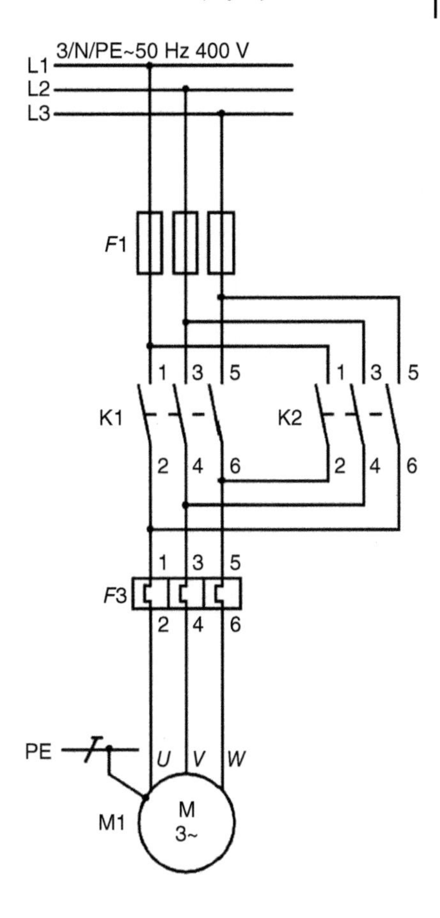

Figure 23.7 Pole changing with a winding, two speeds, one direction of rotation.

The stator winding is first connected in star connection to the network. In this way, the breakaway torque and the breakaway starting current only reach one-third the value for direct switch-on. Following startup, the motor is switched over to a delta connection. This requires a star delta switch and at least six terminals. In general, for the starting torque and the starting current:

$$M_A = (0.5 \cdots 0.9)\, M_N \tag{23.17}$$

$$I_a = (1.8 \cdots 2.5)\, I_{\mathrm{rM}} \tag{23.18}$$

The meanings of the symbols are

M_M	Motor torque
$M_{M\triangle}$	Motor torque for direct switch-on
M_Y	Motor torque for star delta startup
M_L	Load torque (counter-torque)
M_{L0}	Load breakaway torque
M_N	Rated torque
M_S	Breakaway torque

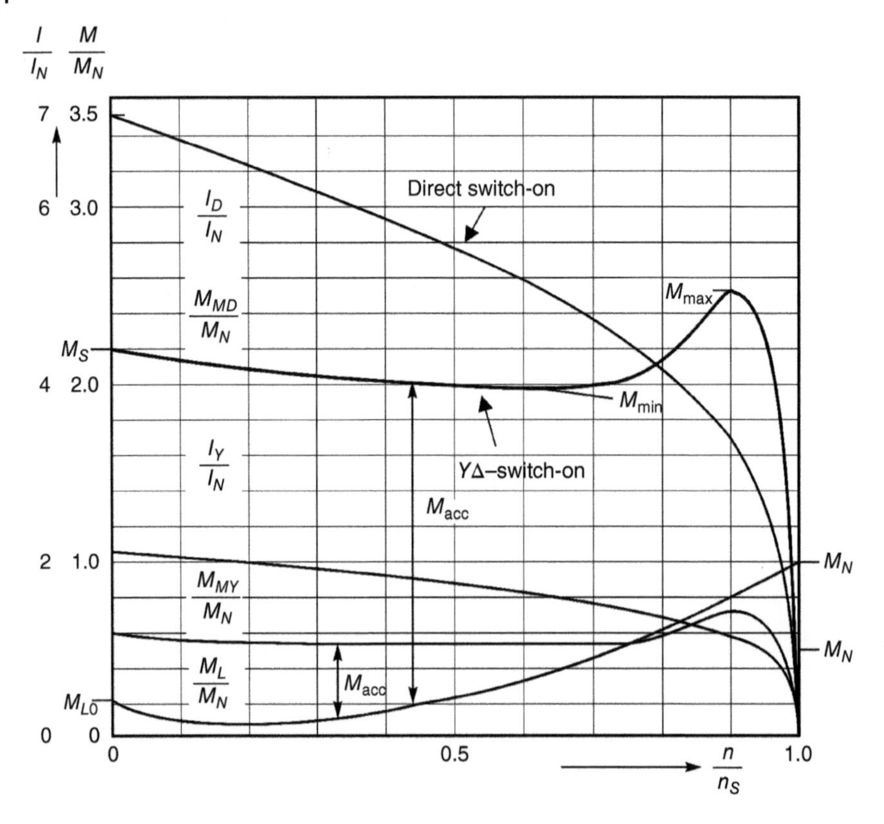

Figure 23.8 Typical current and torque characteristics. Source: Schalten [56].

M_{min}	Pull-up torque
M_{max}	Breakdown torque
M_{acc}	Accelerating torque
I	Current
I_r	Rated current
$I_{\triangle}$	Current for delta connection
I_{Y}	Current for star connection
n	Speed
n_{S}	Synchronous speed

In order to prevent short-circuits while switching from the star phase to the delta phase during startup with star delta motor starters, a star delta time relay with switchover pause must be used. This ensures that the electric arc has died out when the star contactor opens, before the delta contactor closes.

Current, torque, and power conditions for the star delta connection (Figure 23.10) Starting current:

$$I_{A\text{Y}} = \frac{1}{3}\, I_{A\Delta} \tag{23.19}$$

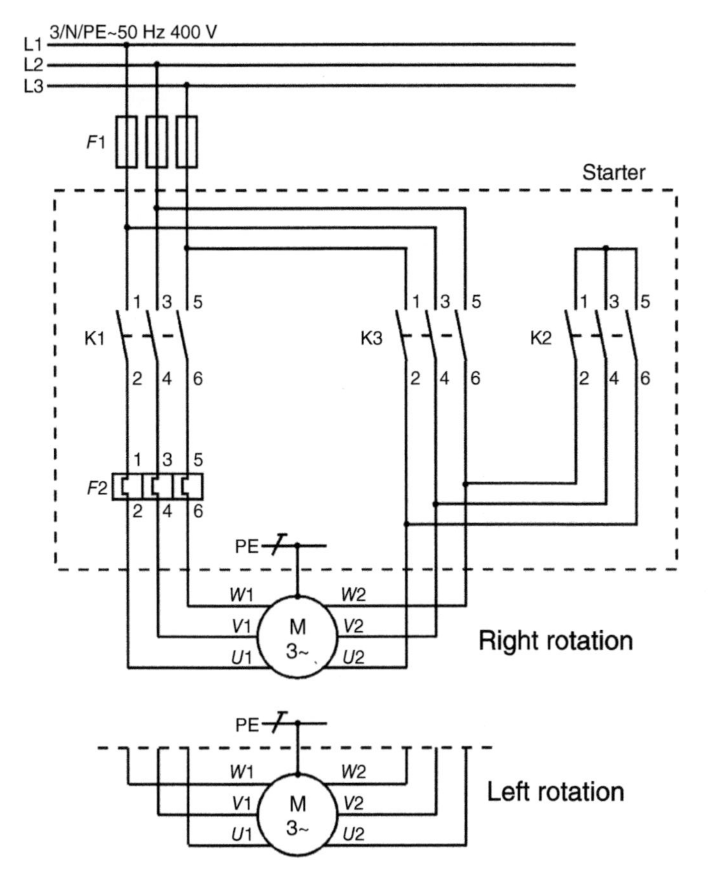

Figure 23.9 Star delta switch with star contactor, delta contactor, and network contactor.

Breakaway torque:

$$M_{AY} = \frac{1}{3}\, M_{A\Delta} \tag{23.20}$$

Power:

$$P_{Y} = \frac{1}{3}\, P_{\Delta} \tag{23.21}$$

Dimensioning the contactors for startup times up to 10 seconds The network contactor K1 must be dimensioned for 58% of the motor power, the star contactor K2 for 33% of the motor power, and the delta contactor K3 for 58% of the motor power (Figure 23.9).

The motor protective relay is in series with the terminal leads to the motor terminals and becomes effective in the star connection (Figure 23.11a). It must be set to 0.58 times the rated current of the motor. The motor protective relay is in the network feeder line (Figure 23.11b). This circuit does not offer complete protection, since its current is changed to 1.7 times the line-to-line current.

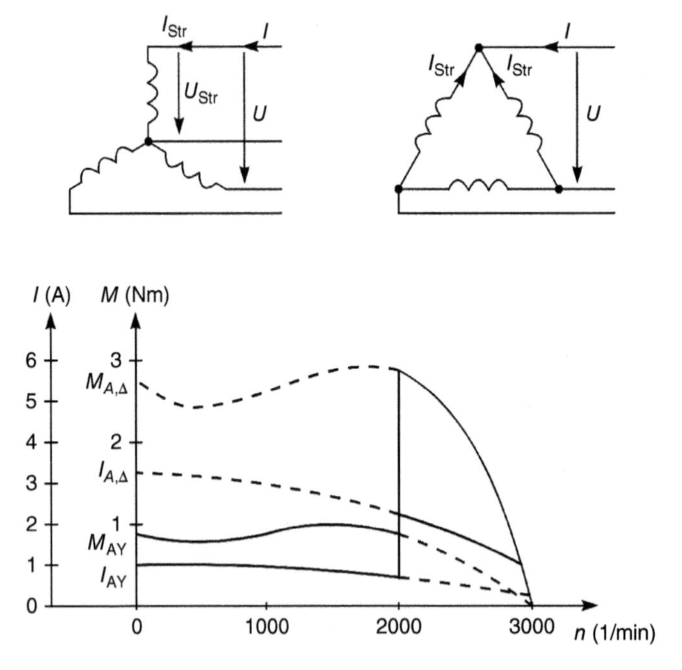

Figure 23.10 Current and torque for star delta startup.

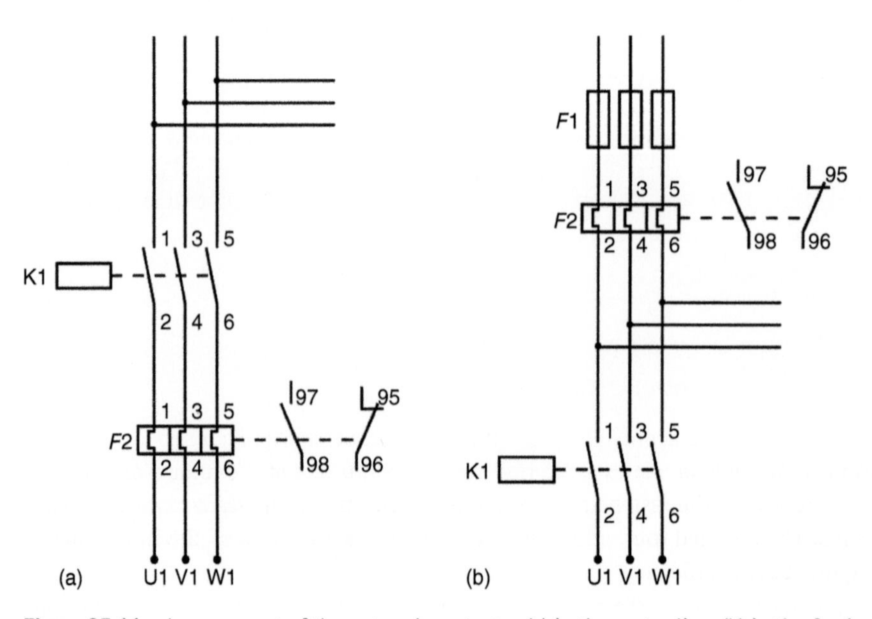

Figure 23.11 Arrangement of the network contactor (a) in the motor line, (b) in the feeder line.

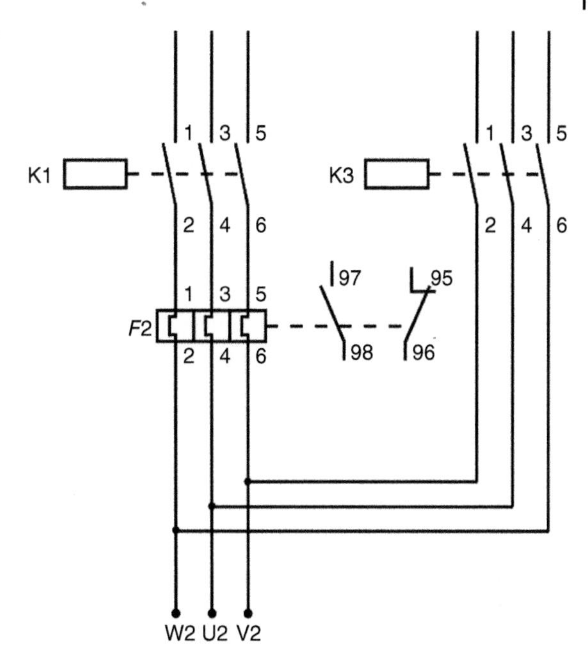

Figure 23.12 Arrangement of the contactors in the delta connection.

The motor protective relay is in the delta star connection. In the star connection, no current flows through the relay. During startup of the motor, no protection is present. This connection is used for high-inertia startup or long-time startup. Figure 23.12 shows the arrangement of the contactors in the delta connection.

23.4 Speed Adjustment

Speed control for AS machines is possible by changing the slip, the number of pole pairs, and the frequency. Here

$$n_1 = \frac{60\,f}{p} \tag{23.22}$$

23.4.1 Speed Control by the Slip

Controlling the slip is possible only with slipring motors. The motor speed depends strongly on the load. An external resistance is connected to the rotor winding. For constant torque, the resistances in the rotor circuit are in the same ratio to the slip speeds. The engaged rotor starter resistance causes a reduction of power and therefore reduces the efficiency.

23.4.2 Speed Control by Frequency

The speed of an AS motor depends on the frequency. In normal three-phase networks, the frequency is 50 Hz and in some countries – principally in

North America – 60 Hz. Increasing the speed is therefore possible only with the use of a frequency converter.

23.4.3 Speed Control by Pole Changing

There are three possible ways to change the number of poles in squirrel cage motors. The stator can be equipped with either

- two or more separate windings
- a three-phase single winding (enabling pole changing) or
- a combination of both types of winding.

For all three cases, speed control is loss-free. Motors with two speeds and separate windings represent poor utilization of the motor capacity, since for each speed only half of the stator winding is used. Motors with two speeds and a three-phase single winding allow better utilization. The Dahlander pole-changing switch is of particular interest here. The stator winding of the motor is composed of six coils. Each phase has two winding sections connected in series. The usual Dahlander circuits are discussed below.

- *Constant torque*: (YY/△)
  The rated torque of the motor is the same for both speeds. The ratio of the rated powers is about 3 : 2. This is obtained by using a winding in double star connection for the higher speed and in delta connection for the lower speed.
- *Variable torque*: (YY/Y)
  The torque varies with the square of the speed. In practice, one speaks of "falling torque" and "square law torque." The ratio of the speed to the rated powers is about 1:5. This is obtained by using a winding in double star connection for the higher speed and in star connection for the lower speed.
- *Pole changing contactors*: Pole changing contactors can be used
  - fuseless without motor protective relays with motor starting circuit-breakers and separate circuit-breakers or
  - with fuses and motor protective relays.

For Y/△ startup (Table 23.1), motors are delta-connected for normal operation. This means that they must always be delta-connected for the operating voltage. When the motor runs with no load or only with a light load, this reduces the power factor of the feeder network. The motor should then be switched to a star connection.

Connection of pole changing motors

Pole changing motors are normally connected as shown in Figure 23.13. Motors in normal design have six terminal clamps in the terminal box as well as a protective conductor terminal. Motors with two separate windings normally have a △/△ connection, but can also be ordered with a Y/Y, Y/△, or △/Y connection.

Motors with a winding in a Dahlander circuit normally have a △/YY connection if they are intended for drives with constant torque. Fan drives have a Y/YY connection. A circuit diagram is shown for each type of motor.

Table 23.1 Comparison of Dahlander circuits.

Winding connection for voltage V	Network voltage V	Direct switch-on	Y/ △ – startup
230	230	△	Possible
400	400	Y	Not possible
400	400	△	Possible
690	690	Y	Not possible
500	500	△	Possible
500	500	Y	Not possible
690	690	△	Possible

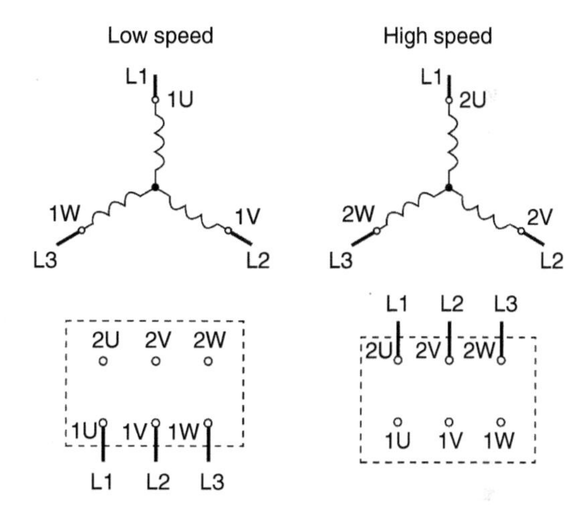

Figure 23.13 Two separate windings Y/Y.

The thermistor protection monitors the temperature of the motor winding (Figure 23.14). The positive temperature coefficient thermistor (PTC) circuit controls an output relay which in turn controls the main contactor. A fault is displayed visually.

23.4.4 Soft Starters

The ASM has a high starting torque and a high starting current. In order to avoid these disadvantages, different startup procedures are used, as already described. However, fine matching to the particular startup is not possible with all of these. Correct setting of the parameters reduces the load on the motor, resulting in less wear, less downtime, and greater operational reliability. With the soft starter to be described briefly in this section, it is possible to set the startup parameters very well today, the main circuit of the soft starter is controlled by a semiconductor and not by mechanical switch contacts. Two thyristors in inverse-parallel connection are employed for each phase, so that the current can be switched at any point in time

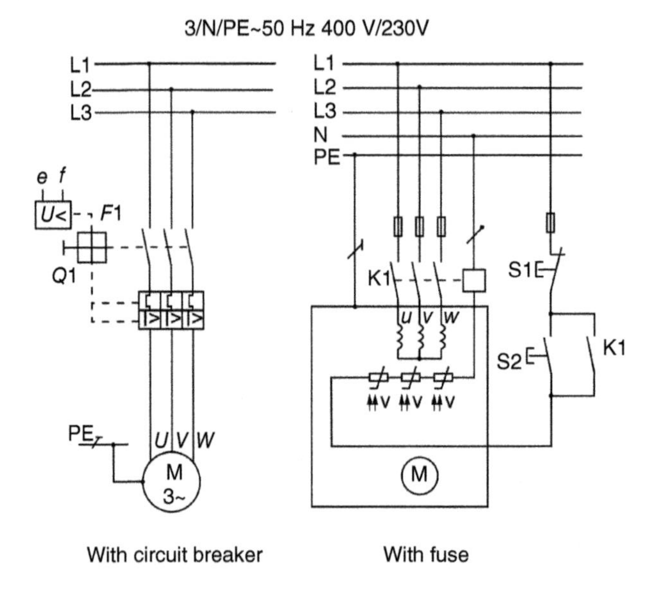

Figure 23.14 Thermistor motor protection. Source: ABB [56].

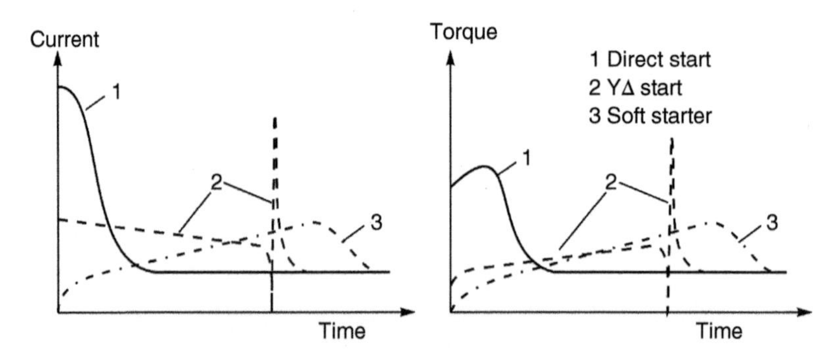

Figure 23.15 Comparison of startup procedures.

during the positive and negative half-waves. An integrated microprocessor controls the conducting period with the help of the firing angle of the thyristor.

The soft starter permits a gentle startup with a reduced starting current. The starting current depends directly on the required stationary breakaway torque and the mass of the load to be accelerated. In many cases, the soft starter minimizes the energy losses because the motor voltage is continuously monitored and automatically matched to the actual requirement. This is especially true when the motor runs with a light load.

Figure 23.15 represents a comparison between different startup procedures.

The choice of the right type of protection is a necessary condition for ensuring that a motor can operate under difficult conditions over a long period of time. The type of protection is given, in accordance with IEC Publication 34-5, by the designation IP, followed by two numbers. Table 23.2 lists important rated motor values for the project planning of ASM. These values can differ slightly from one manufacturer to another.

Table 23.2 Rated motor values for $U_{rM} = 400V$.

		Choice of starters and rated motor values			
P_{rM} (kW)	I_{rM} (V)	Installation-B2 (mm^2)	Current setting (I_e) (V)	Direct startup (I_{rsi}) (V)	Y/△ startup (V)
0.18	0.7	1.5	0.6–1.0	6	—
0.25	0.85	1.5	0.6–1.0	6	2
0.37	1.15	1.5	1.0–1.6	6	4
0.55	1.55	1.5	1.6–2.5	10	4
0.75	2	1.5	1.6–2.5	10	6
1.1	2.9	1.5	2.5–4	16	6
1.5	3.7	1.5	2.5–4	16	6
2.2	5.2	2.5	4–6	20	10
3	6.9	2.5	6–9	25	16
4	9	2.5	6–9	35	20
5.5	12	2.5	9–13	35	25
7.5	16	6	13–18	50	32
11	23	6	18–23	63	40
15	30	10	28–42	80	63
18.5	37	10	28–42	80	63
22	44	10	40–52	100	80
30	59	16	52–65	125	100
37	71	16	60–75	160	125
45	86	25	72–100	200	160
55	104	35	72–100	200	200
75	144	50	102–170	250	200
90	172	70	102–170	315	250
110	209	95	126–210	355	315
132	248	120	180–300	400	400
160	300	150	180–300	500	400
200	372	240	240–400	500	500
250	468	2 × 90	378–630	630	630
315	568	2 × 120	378–630	800	800

Source: Ref. [64]

23.4.5 Example: Calculation of Overload and Starting Conditions

Given the following motor data:

P_{rM} 18.5 kW
U_{rM} 400 V
n 1460 min^{-1}

I_{rM}	35 A
M_A/M_N	2.5
I_a/I_{rM}	6
M_K/M_N	2.9

Calculate all rated motor values for the star delta connection.
In the delta connection:

$$M_A = 2.5\ M_N$$

In the star connection:

$$M_A = \frac{2.5}{3}\ M_N = 0.83\ M_N$$

In the delta connection:

$$I_{a\Delta} = 6\ I_{rM} = 6 \cdot 35\ \text{A} = 210\ \text{A}$$

In the star connection:

$$I_{aY} = \frac{1}{3}\ I_{a\Delta} = \frac{1}{3} \cdot 210\ \text{A} = 70\ \text{A}$$

Nominal torque:

$$M_N = \frac{P_{rM}}{2\ \pi\ n_N} = \frac{18.5 \times 10^3\ \text{Nm/s} \cdot 60\ \text{s/ min}}{2\ \pi \cdot 1460\ \text{min}^{-1}} = 121\ \text{Nm}$$

Tightening torque:

$$M_A = 2.5\ M_N = 2.5 \cdot 121\ \text{Nm} = 302.5\ \text{Nm}$$

Tilt moment:

$$M_K = 2.9\ M_N = 2.9 \cdot 121\ \text{Nm} = 350.9\ \text{Nm}$$

23.4.6 Example: Calculation of Motor Data

A three-phase ASM has the following rated values:
$P_{rM} = 400$ kW, $U_{rM} = 400$ V, 50 Hz, $I_{rM} = 750$ A, $\cos\varphi = 0.86$, $n = 1448$ rpm, 4-pin.

1) Calculate the slip s for the given rated motor speed.
2) Calculate the frequency of the rotor voltage f_L.
3) Calculate the efficiency η!

$$s = \left(1 - \frac{n}{n_s}\right) \cdot 100\% = \left(1 - \frac{1448}{1500}\right) \cdot 100\% = 3.46\%$$

$$f_L = \frac{f \cdot s}{100\%} = \frac{50\ \text{Hz} \cdot 3.46\%}{100\%} = 1.73\ \text{Hz}$$

$$P_1 = \sqrt{3}\ U_{rM}\ I_{rM}\ \cos\varphi$$

$$P_1 = \sqrt{3} \cdot 400\ \text{V} \cdot 750\ \text{A} \cdot 0.86 = 446.87\ \text{kW}$$

$$\eta = \frac{P_2}{P_1} = \frac{400\ \text{kW}}{446.87\ \text{kW}} = 0.896$$

23.4.7 Example: Calculation of the Belt Pulley Diameter and Motor Power

A three-phase motor is supposed to drive a circular saw blade with 550 mm diameter and a cutting speed of 36 m/s. The belt pulley on the saw blade has a diameter of 130 mm. The belt slip is 2.8%.

1) What diameter must the belt pulley on the motor have if its speed of rotation is 995 rpm?
2) How much power must the motor supply if a force of 340 N is required on the motor belt pulley?

Solution to 1:

$$v_2 = \pi\, d\, n_2$$

$$n_2 = \frac{v_2}{\pi\, d} = \frac{36\ \frac{\mathrm{m}}{\mathrm{s}}}{\pi \cdot 0.55\ \mathrm{m}} = 20.84\ \frac{1}{\mathrm{s}}$$

$$n_2 = 20.84\ \frac{1}{\mathrm{s}} \cdot 60 \frac{\mathrm{s}}{\mathrm{min}} = 1250.4\ \frac{1}{\mathrm{min}}$$

with $n_1\, d_1 = n_2\, d_2$

$$d_1 = \frac{n_2\, d_2}{n_1} = \frac{1250.4\ \frac{1}{\mathrm{min}} \cdot 130\ \mathrm{mm}}{955\ \frac{1}{\mathrm{min}}} = 170.21\ \mathrm{mm}$$

Solution to 2:

$$v = \pi\, d_1\, n_1 = \frac{\pi \cdot 170.21\ \mathrm{mm} \cdot 955\ \frac{1}{\mathrm{min}}}{60 \frac{\mathrm{s}}{\mathrm{min}}} = 8.51\ \frac{\mathrm{m}}{\mathrm{s}}$$

$$P = F\, v_1 = 340\ \mathrm{N} \cdot 8.51\ \frac{\mathrm{m}}{\mathrm{s}} = 2894\ \frac{\mathrm{Nm}}{\mathrm{s}} = 2.984\ \mathrm{kW}$$

23.4.8 Example: Dimensioning of a Motor

A pump with a power of 19 kW, $\cos\varphi = 0.8$ is supposed to run five times per hour operating for 60 seconds with a short-term load of 72 seconds (Figure 23.16). The ambient temperature is 50°; installation type C; $I_{an}/I_{\mathrm{rM}} = 7$, loop resistance up to distribution 0.3 Ω; $l = 10$ m

Determine the:

1) Current carrying capacity
2) Backup fuse
3) Voltage drop

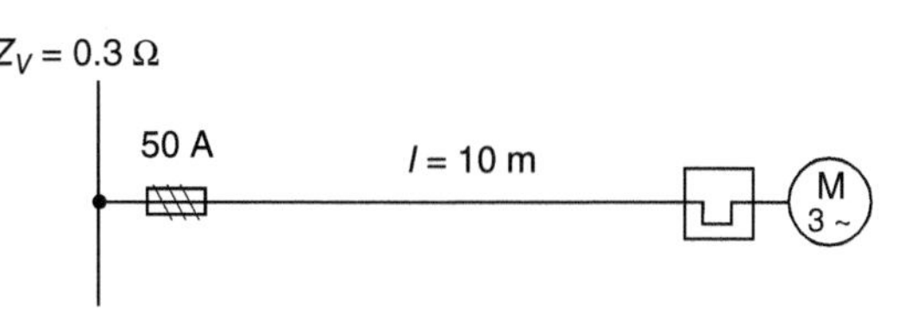

Figure 23.16 Calculation for a motor with operating modes.

4) Short-circuit current
5) Current setting
6) Permissible break time

a) Current carrying capacity

For the load factor n:

$$n = \sqrt{\frac{1}{1 - e^{-t_e/T}}} \quad \text{mit } t_e = 60 \text{ seconds}, T = 72 \text{ seconds}$$

$$n = \sqrt{\frac{1}{1 - e^{t_e/T}}} = \sqrt{\frac{1}{1 - e^{60/72}}} = \sqrt{\frac{1}{1 - e^{\frac{1}{0.833}}}}$$

$$n = \sqrt{\frac{1}{1 - \frac{1}{2,3}}} = \sqrt{\frac{1}{1 - 0.434}} = 1.33$$

The cross section of the insulated lines depends on the minimum time value. For 72 seconds, a line of 4×4 mm^2 is chosen. The new current carrying capacity of the line is then:

$$I'_Z = n\, I_Z$$

$$I_Z = 34 \text{ A}$$

$$I'_Z = 1.33 \cdot I_Z = 1.33 \cdot 34 \text{ A} = 45.2 \text{ A}$$

$$I_B = \frac{P}{\sqrt{3}\, U\, \cos\varphi} = \frac{19 \text{ kW}}{\sqrt{3} \cdot 400 \text{ V} \cdot 0.8} = 34.28 \text{ A}$$

b) Backup fuse:

A 50 V fuse (gL) will provide sufficient short-circuit protection.

c) Voltage drop:

$$\Delta U = \frac{\sqrt{3} \cdot I\, l \cdot 1.12\, \cos\varphi}{\kappa\, S} = \frac{\sqrt{3} \cdot 34.28 \text{ A} \cdot 10 \text{ m} \cdot 1.12 \cdot 0.8}{56 \frac{\text{m}}{\Omega\text{mm}^2} \cdot 4\text{mm}^2} = 2.37 \text{ V} \mathrel{\hat{=}} 0.6\%$$

d) Short-circuit current:

$$R_L = \frac{1.56 \cdot 2\, l}{\kappa\, S} = \frac{1,56 \cdot 2 \cdot 10 \text{ m}}{56 \frac{\text{m}}{\Omega\text{mm}^2} \cdot 4 \text{ mm}^2} = 0.14\ \Omega$$

$$Z_S = 0.3\ \Omega + 0.14\ \Omega = 0.44\ \Omega$$

$$I''_{k1} = \frac{c\, U_n}{\sqrt{3}\, Z_k} = \frac{0.9 \cdot 400 \text{ V}}{\sqrt{3} \cdot 0.44\ \Omega} = 472.37 \text{ A}$$

e) Current setting:

The motor protecting switch is set to the rated current of 34.3 A.

f) Permissible break time:

$$t_{\text{permissible}} = \left(k\, \frac{S}{I''_{k1}}\right)^2 = \left(115 \frac{A\, \sqrt{\text{s}}}{\text{mm}^2} \cdot \frac{4 \text{ mm}^2}{472.37 \text{ A}}\right)^2 = 0.948 \text{ second}$$

24

Questions About Book

24.1 Characteristics of Electrical Cables

1. Explain the concept of active resistance.
2. How are electrical conductivity and electrical resistance related?
3. What quantities determine the resistance of a conductor?
4. Explain the term capacitance.
5. What quantities determine the capacitive reactance?
6. How does the capacitive reactance change with frequency?
7. What is the cause of the inductive reactance?
8. How does the inductive reactance change with frequency?

24.2 Dimensioning of Electric Cables

1. Describe the advantages and disadvantages of the three network types.
2. Draw the equivalent circuit diagram of a line and explain the sizes.
3. Explain the structure of cables and lines.
4. What does NYCWY $4 \times 120\text{mm}^2$ mean?
5. Explain the short circuit resistance of a cable.
6. What is the difference between a line and a cable?
7. What is the function of the ground wire in overhead lines?
8. What criteria determine the selection of lines?
9. Explain the three protection concepts.
10. Draw TN, TT, and IT systems and explain their functions.
11. What does loop impedance mean?
12. What is the maximum permissible touch voltage for humans and animals?
13. Why do you have to ground the star point of the transformer?
14. Explain the earthing methods in LV and HV networks.
15. What are the five safety rules?
16. Explain GIS and air-insulated systems.

Analysis and Design of Electrical Power Systems: A Practical Guide and Commentary on NEC and IEC 60364, First Edition. Ismail Kasikci.

17. What are the components of a medium-voltage system?
18. How can lines and cables be protected against overload and short circuits?
19. Explain the nominal current and release rules.
20. What criteria determine the current carrying capacity of a cable?

24.3 Voltage Drop and Power Loss

1. Explain permissible voltage drop from transformer to the consumer.
2. How is the power dissipation defined?

24.4 Protective Measures and Earthing in the Low-voltage Power Systems

1. Explain the term feed point of an electrical system.
2. Explain the term neutral conductor.
3. Explain the term touch voltage.
4. Explain the term body of an electrical equipment.
5. Explain the term indirect contact with electrical equipment.
6. Explain the term total grounding resistance.
7. Explain the term loop impedance.
8. When is there a short-circuit to the body?
9. When is a short-circuit present?
10. How is the TN-C system constructed?
11. Briefly describe the TT system.
12. Describe the IT system briefly.
13. Which protective devices are permitted in a TN system?
14. Name a protective device that can only protect in the event of an overload.
15. Name a protective device that can only protect in the event of a short-circuit.
16. Where must protective devices be installed to protect against short-circuits?

24.5 Short Circuit Calculation

1. Explain the terms: voltage factor, equivalent voltage source, surge short-circuit current, short-circuit current.
2. Why do you have to calculate the single-pole and three-pole short-circuit current?
3. What does the impedance of a line and impedance of a transformer mean?
4. Describe all types of short-circuit with an example.
5. What role does the short-circuit temperature play in the calculation of KS currents?

24.6 Switchgear

Describe the functions of the following protection and switching devices.

1. Circuit breaker (MCCB)
2. High- voltage fuse (HV)
3. Miniature circuit-breaker (MCB)
4. Relays
5. RCD
6. Earth switch
7. Surge protection device

24.7 Protection Devices

Describe the functions of the following protective devices.

1. Independent time relays
2. Dependent time relays
3. Distance protection
4. Diff-protection
5. Current and voltage transformer
6. What does compensation mean?
7. Why do you have to compensate for electrical installations?
8. Explain the power factor and the displacement factor.
9. What is the difference between the two terms?

24.8 Electric Machines

1. Explain the law of induction and flow law.
2. How does torque arise?
3. Explain the function of a transformer with an equivalent circuit diagram.
4. What does transmission ratio mean?
5. How can you calculate the no-load and short-circuit losses of a transformer?
6. What criteria must be observed when connecting transformers in parallel?
7. What effect does the short-circuit voltage have on the transformer?
8. What are stray field lines?
9. What does vector group mean?

References

1 EU requirements for transformers, Ecodesign Directive from the European Commission Tier 2 - July 1st, 2021.
2 EAFOL-Gießharztransformatoren. Planungshinweise, Transformatorenwerk Kirchheim Energy Management Division Transformers, Hegelstraße 20, 73230 Kirchheim/Teck, Deutschland.
3 Kiank, H. and Fruth, W. (2012). *Planning Guide for Power Distribution Plants: Design, Implementation and Operation of Industrial Networks.*
4 DIN 18015-1 (2020). Elektrische Anlagen in Wohngebäuden - Teil 1: Planungsgrundlagen. ISBN: 978-3-895-78665-5.
5 IEC/TR EN 61000. Electromagnetic compatibility (EMC) - Part 1: General - Section 1: Application and interpretation of fundamental definitions and terms.
6 Siemens. Planning of Electric Power Distribution Technical Principles, Artikel-Nr.: EMMS-T10007-00.
7 DIN 276:2018-12 (2018). Building costs, 2018–12.
8 Kasikci, I. (2016). *Short Circuits in Power Systems.* Wiley-VCH. ISBN 3-527-30482-7.
9 IEC 60479-2 Ed.2.0. Effects of current on human beings and livestock - Part 2: Special aspects.
10 IEC 60364-1:2005. Modified IEC 60364-1:2005/COR1:2009-08 Corrigendum 1 - Low-voltage electrical installations - Part 1: Fundamental principles, assessment of general characteristics, definitions (HD 60364-1:2008).
11 IEC 60364-4-41:2005. Modified IEC 60364-4-41:2005/AMD1:2017-03 Amendment 1: Low-voltage electrical installations - Part 4-41: Protection for safety - Protection against electric shock (HD 60364-4-41:2017 + A11:2017).
12 IEC 60364-4-42:2010. Modified + A1:2014 Low-voltage electrical installations - Part 4: Protection for safety - Chapter 42: Protection against thermal effects (HD 60364-4-42: 2011 + A1:2015).
13 IEC 60364-4-43:2008. Modified Low-voltage electrical installations - Part 4: Protection for safety - Chapter 43: Protection against overcurrent (HD 60364-4-43: 2010).

Analysis and Design of Electrical Power Systems: A Practical Guide and Commentary on NEC and IEC 60364, First Edition. Ismail Kasikci.

14 60364-4-44: 2007 (Clause 442). Modifiied Low-voltage electrical installations - Part 4: Protection for safety - Chapter 44: Protection against overvoltages - Section 442: Protection of low-voltage installations against faults between high-voltage systems and earth (HD 60364-4-442: 2012).

15 IEC 60364-4-44:2007/A1:2015. Modified Low-voltage electrical installations - Part 4-44: Protection for safety - Protection against voltage disturbances and electromagnetic disturbances - Clause 443: Protection against transient overvoltages of atmospheric origin or due to switching (HD 60364-4-443:2016).

16 HD 60364-4-46:2016 + A11:2017 (2021). Low-voltage electrical installations - Part 4-46: Protection for safety - Isolation and switching, 2021-01-22.

17 IEC 60364-5-51:2005. Modified Electrical installations of buildings Part 5-51: Selection and erection of electrical equipment - Common rules (HD 60364-5-51:2009 + A11:2013).

18 60364-5-52:2009. Modified + Corrigendum Feb. 2011 Low-voltage electrical installations Part 5: Selection and erection of electrical equipment Chapter 52: Wiring systems (HD 60364-5-52: 2011).

19 IEC 60364-5-53:2001/A2:2015 (Clause 534). Modified Low-voltage electrical installations - Part 5-53: Selection and erection of electrical equipment - Isolation, switching and control - Clause 534: Devices for protection against transient overvoltages (HD 60364-5-534:2016).

20 IEC 60364-5-54:2011. Low-voltage electrical installations - Part 5-54: Selection and erection of electrical equipment - Earthing arrangements, protective conductors and protective bonding conductors (HD 60364-5-54:2011).

21 IEC 60364-5-55:2001/A2:2008. Clause 551 AMENDMENT 2: Low-voltage electrical installations - Part 5-55: Selection and erection of electrical equipment - Other equipment - Clause 551: Low-voltage generating sets (HD 60364-5-551: 2010 + Cor.: 2010 + A11:2016).

22 IEC 60364-5-56:2009. modified Low-voltage electrical installations - Part 5: Selection and erection of electrical equipment Chapter 56: Safety services Low-voltage electrical installations - Part 5-56: Selection and erection of electrical equipment - Safety services (HD 60364-5-56:2010 + A1:2011).

23 IEC 60364-6:2016/COR1:2017-09. Low-voltage electrical installations - Part 6: Verification (HD 60364-6:2016 + A11:2017) 978-3-540-71849-9.

24 GSALab. SINT Ingegneria Srl Via C. Colombo, 106 - 36061 Bassano del Grappa (VI) - Italy. Web-site: www.sintingegneria.it, Tel.: +39 0424 568457 info@sintingegneria.it.

25 DIN VDE 0276 Teil 1000: 1995-06. Strombelastbarkeit; Allgemeines; Umrechnungsfaktoren.

26 DIN VDE 0100-710: 1994-10. Starkstromanlagen in Krankenhäusern und medizinisch genutzten Räumen außerhalb von Krankenhäusern.

27 DIN VDE 0100-708: 1989-10. Starkstromanlagen und Sicherheitsstromversorgung in baulichen Anlagen mit Menschensammlungen.

28 Kasikci, I. (2018). *Short Circuits in Power Systems, A Practical Guide to IEC 60909-0)*. ISBN-13: 978-3527341368.

29 IEC 60909-0-2016. Short-circuit currents in three-phase a.c. systems - Part 0: Calculation of currents.

30 Oeding, D. and Oswald, B.R. (2018). *Elektrische Kraftwerke und Netze*. Berlin: Springer-Verlag. ISBN 978-3-662-52703-0.

31 IEEE Std. 80-2013. Guide for Safety in AC Substation Grounding.

32 Biegelmeier, G., Kiefer, G., and Krefter, K.-H. (1996). *Schutz in elektrischen Anlagen. Band 2, Erdungen, Berechnung, Ausführung und Messung*. VDE-Schriftenreihe 81. Berlin: VDE-Verlag.

33 Biegelmeier, U.G. and Lee, W.R. (1980). New considerations on the threshold of ventricular fibrillation for AC shocks at 50–60 Hz. *Proceedings of the IEEE* 127: 103–110.

34 Biegelmeier, U.G. and Rotter, K. (1971). Elektrische Widerstände und Ströme in menschlichem Körper. *EM* 89: 104–109.

35 Ferris, L.P., King, B.G., Spence, P.W., and Williams, H. (1936). Effect of electric shock on the heart. *AIEE Transactions on Power Apparatus and Systems* 55: 498–515 and 1263.

36 Dalziel, C.F. (1946). Dangerous electric currents. *AIEE Transactions on Power Apparatus and Systems* 65: 579–585, 1123–1124.

37 IEC TR 60909-1:2002. Short-circuit currents in three-phase A.C. systems - Part 1: Factors for the calculation of short-circuit currents according to IEC 60909-0.

38 IEC TR 60909-2:2008. Short-circuit currents in three-phase A.C. systems - Data of electrical equipment for short-circuit current calculations.

39 IEC 60909-3:2009. Short-circuit currents in three-phase A.C. systems - Part 3: Currents during two separate simultaneous line-to-earth short circuits and partial short-circuit currents flowing through earth.

40 IEC TR 60909-4:2000. Short-circuit currents in three-phase A.C. systems - Part 4: Examples for the calculation of short-circuit currents.

41 EN 60865-1:2012-09 IEC 60865-1:2011. Short circuit currents - Calculation of effects - Part 1: Definitions and calculation methods.

42 DIN VDE 0103:2012-09. Short-circuit currents-calculation of effects, Part 1: Definitions and calculation methods, (IEC 60865-1:2011); German version EN 60865-1:2012.

43 IEC 61936-1:2021. Power installations exceeding 1 kV AC and 1,5 kV DC - Part 1: A.C.

44 DIN EN 50522-2011-11. Earthing of power installations exceeding 1 kV A.C.

45 SIPROTEC. Case Studies for SIPROTEC Protection Relays and Power Quality, 2005, Published by Siemens Aktiengesellschaft Power Transmission and Distribution Energy Automation Division.

46 Hager. Technisches Handbuch, Grundlagen, PR-3-3003-DSP1608 V1810.

47 Schalten, S. (1997). *Verteilen in Niederspannungsnetzen: Handbuch mit Auswahlkriterien und Projektierungshinweisen für Schaltgeräte, Steuerungen und Schaltanlagen*. Siemens. ISBN 3-89578-04-3.

48 Kasikci, I. (2017–2018). Sechsteiliger Fachbeitrag, Zulässige Längen von Kabeln und Leitungen nach Beiblatt 5" der DIN VDE 0100.

49 DIN EN 61439: 2012-02. Low-voltage switchgear and controlgear assemblies, Part 1: General rules.
50 Kasikci, I. and Pantenburg, N. (2018). *VDE Seminar-Projektierung von Niederspannungsanlagen*. Offenbach.
51 Kasikci, I. and Pantenburg, N. (2015). Koordination des Spannungsfalls in Niederspannungsnetzen Teil 1 und 2.
52 Kasikci, I. (2010). *Projektierung von Niederspannungsanlagen*, 3. Auflage. Hüthig-Pflaum-Verlag. ISBN 978-3-8101-0274-4.
53 Beiblatt 5, Zulässige Längen von Kabeln und Leitungen unter Berücksichtigung des Schutzes bei indirektem Berühren, des Schutzes bei Kurzschluss und des Spannungsfalls.
54 Trafo-Union. Operating Manual for Cast-resin Dry Type Distribution Transformer.
55 SIPROTEC. Siemens Schutzgeräte. www.siprotec.de (accessed 13 December 2021).
56 ABB (2014). Motor Guide - Technische Grundlagen der Niederspannungsstandardmotoren, Dritte Ausgabe 2014. ISBN 952-91-0728-5.
57 Günter, S. (2000). *Elektrische Installationstechnik*. Siemens. ISBN 3-89578-061-8.
58 ABB Switchgear Manual, 11th edition, 2006. ISBN 978-589-241112-5.
59 Dominit. PROKON für Kondensatorauslegung.
60 Dominit. *Blindleistungskompensation und Netzqualität*. Condensator Dominit GmbH.
61 IEC 62305. Protection against lightning - Part 1: General principles.
62 IEC 62305. Protection against lightning - Part 2: Risk management.
63 Dehn+Söhne (2017). *Lightning Hand Book*, 4e.
64 Schaltungsbuch, eaton, 2011, Artikelnummer, 165290.
65 Binder, A. (2012). *Elektrische Maschinen und Antriebe, Übungsbuch: Aufgaben mit Lösungsweg*. ISBN 978-3-642-17422-3.
66 Gerling, D. (2015). *Electrical Machines, Mathematical Fundamentals of Machine Topologies*. ISBN 978-3-642-17584-8.
67 Trafo-Union. Siemens Energy Sector Power Engineering Guide Edition 7.0.
68 Trafo-Union. GEAFOL-Gießharztransformatoren, TV1-1995.
69 Trafo-Union. GEAFOL-Planungshinweise.

Index

Analysis and Design of Electrical Power Systems: A Practical Guide and Commentary on NEC and IEC 60364, First Edition. Ismail Kasikci.